Fractography of Glasses and Ceramics V

Fractography of Glasses and Ceramics V

Ceramic Transactions, Volume 199

Proceedings of the Fifth Conference on the Fractography of Glasses and Ceramics, Rochester, New York, July 9–13, 2006

Edited by
James R. Varner
George D. Quinn
Marlene Wightman

The American Ceramic Society

BICENTENNIAL 1807 WILEY 2007 BICENTENNIAL

WILEY-INTERSCIENCE
A John Wiley & Sons, Inc., Publication

Published by John Wiley & Sons, Inc., Hoboken, New Jersey.
Published simultaneously in Canada.

Wiley Bicentennial Logo: Richard J. Pacifico

Library of Congress Cataloging-in-Publication Data is available.

ISBN 978-0-470-09737-3

Contents

SECTION V. FRACTOGRAPHY OF DENTAL AND BIOMATERIALS

SECTION VI. FRACTOGRAPHY OF COMPONENTS

Preface

These proceedings are for the Fifth Conference on the Fractography of Glasses and Ceramics held in Rochester, New York on July 9–13, 2006. This series was started in 1986 by Van Fréchette and Jim Varner of the New York State College of Ceramics at Alfred University, where the previous conferences were held. In 2006 we changed venues from the tranquil Alfred University campus to a more cosmopolitan setting in Rochester.

Times are passing. Some familiar faces and fractographers sadly are no longer able to attend or are no longer with us. We will miss them, but we were pleased that a new generation of rising stars attended the conference. That is a good sign and, to help pass the baton, we arranged for some of the foremost authorities to present invited and keynote presentations. The conference was a marvelous opportunity for the new generation of participants to meet the masters and show them a trick or two.

The organizers especially thank Professor Derek Hull for coming out of retirement and delivering his thought-provoking keynote presentation. Observing, measuring, and interpreting fracture surfaces is a multifaceted process. Professor Hull discussed issues of nomenclature and pattern recognition and reviewed some classical fractographic studies. Another highlight was Dr. Timothy Foecke's presentation on metallurgical aspects of the World Trade Center tower collapses, one of the monumental events of our times. Professor Terry Engelder of Pennsylvania State University reviewed fractographic analyses of the velocity of joint propagation in the crust of the earth, a sometimes controversial topic. Those participants who took the past-conference one-day field tour led by Professor Engelder of saw amazing exposed joint (crack) fracture surfaces in the famous Ithaca Formations of the Appalachian Plateau in the Finger Lakes district of upstate New York. Microscopes were not needed to observe the fracture markings in these formations!

Two sessions focused attention on two hot topics. A special session on Fracture Processes at the Crack Tip brought together some of the leading authorities in the world who shared some controversial findings from atomic force microscopy examination of fracture surfaces. A session on Dental and Biomaterials also reviewed progress in the use of fractography to better understand biomaterial reliability and failure modes.

We are very pleased that there was good international participation from Japan, Brazil, and many countries in Europe. We were fortunate to have Drs. Jan Dusza, Roger Morrell and Robert Danzer, leaders of a European conference series on Fractography of Ceramics and Glasses held in Slovakia, in attendance as contributors to the 2006 Rochester conference. We hope to continue the close cooperation with our European compatriots.

This Fifth Conference on the Fractography of Glasses and Ceramics met our goal of providing a snapshot of the state of fractography of these brittle materials. We thank all of the presenters and all of the participants for making this a lively and successful meeting.

George D. Quinn
James R. Varner
Marlene Wightman

SECTION I

THE FRACTURE PROCESS AT THE CRACK TIP

CHARACTERIZATION OF THE MIRROR REGION WITH ATOMIC FORCE MICROSCOPY

Sheldon M. Wiederhorn, Jose M. López-Cepero* and Jay S. Wallace
National Institute of Standards and Technology
Gaithersburg, MD 20899-8520

Jean-Pierre Guin
CNRS-University of Rennes I
Rennes, France

Theo Fett
Forschungszentrum Karlsruhe and
University of Karlsruhe
Karlsruhe, Germany

ABSTRACT

In this paper we use atomic force microscopy to investigate the roughness of silica glass fracture surfaces in double cantilever beam samples and in the mirror region in a ring-on-ring fracture specimen. We demonstrate a decrease in surface RMS Roughness from about 0.5 nm to 0.4 nm with increasing stress intensity factor ($0.5K_{Ic}$ to about $0.95K_{Ic}$). Roughness occurs as a consequence of elastic interactions of the crack tip with nano scale inhomogeneities in the glass. These results contrast with the usual observation that roughness increases with increasing distance as the crack propagates from the fracture origin into the mist and hackle regions. At low values of K_I, crack surfaces are relatively smooth. It is this relatively smooth area that will be discussed in this paper.

INTRODUCTION

The growth of cracks from small surface flaws is the principal cause of failure in glasses and ceramics; for this reason a substantial effort has been spent characterizing the nature of the surface flaws that initiate failure of glass objects. Such failures are important, not just for structural applications, but for any application in which ceramics or glasses are subjected to stresses as a consequence of the manufacturing process. For example, brittle fracture of ceramic parts in electronic components can occur as a consequence of built in residual stresses, often resulting from the development of thermal mismatch stresses during component manufacture. Once a part has failed, the residual damage responsible for the failure is left behind on the fracture surface. By examining this damage, it is possible to gain insight into the processes and factors that contributed to the failure. The art of examining fracture surfaces to mine information on the causes of failure is known as fractography, for which several excellent texts are available.[1,2,3,4] References [3, 4] are particularly relevant for ceramic materials. When a fracture surface of glass is examined, the most obvious feature surrounding the fracture origin is a flat shiny area called the "mirror" region,[5] which is bordered by an area of growing roughness called the "mist" region, which blends into an even rougher region called the "hackle" region. The transition from the mirror to the mist region is distinct enough that the distance from the fracture

---

* Now at the University of Seville, Department of Materials Science, Seville, Spain

origin to the edge of the mist region, the mirror radius, r_m, can be defined and used to estimate the fracture stress, σ_f. The product of the fracture stress and the square-root of r_m is a constant for each glass, so that σ_f can be determined for failed parts by determining r_m. From such a determination, one can tell if a part has been overloaded, or was unduly damaged before failure.

A closer examination of the boundary between the mirror and mist regions using an electron microscope[6,7] suggests that the border is not sharp at all; roughness builds up gradually with distance from the fracture origin. Near the crack origin, surfaces are smooth and featureless,[7,8] and the crack velocity goes from the slow subcritical values, $< 10^{-10}$ m/s, to values approaching 10^3 m/s.[9] Surrounding the slow growth region is often a semi elliptical boundary, Fig. 1, which marks crack growth velocities less than approximately 0.1 m/s.[9,10] At approximately one-third the distance from the origin to r_m, small scale roughness builds up. The size of the roughness is, however, still too small to be seen using optical microscopy so that the surface looks smooth and featureless. At r_m, the roughness builds up so that features are large enough to be seen optically. Within the semi elliptical mark indicating low velocities, the stress intensity is less than K_{Ic}; for silica glass 0.72 MPa·m$^{1/2}$. Beyond the semi elliptical boundary, the stress intensity factor exceeds K_{Ic}.

A similar scenario was reported for a brittle, glassy, isotropic epoxy resin.[1,8] Within the mirror region, the root-mean-square roughness, R_q, is almost constant; R_q = 40 nm from the fracture origin to a distance of about 400 μm from the fracture origin. From 400 μm to 500 μm (0.8 r_m), R_q increases from about 40 nm to 50 nm. At r_m (600 μm) the roughness increases to between 70 nm and 80 nm. The roughness continues to increase as the crack leaves the mist region and enters the hackle region.[8]

Most studies of the roughness of glass have concentrated on areas close to the mirror-to-mist boundary, within the mist region, or within the hackle region. It is only recently that optically smooth areas, typical of what might be found close to fracture origins have become of interest. Gupta *et al.*[11] measured the roughness of several glasses within the mirror region, and reported the smoothest values they obtained in their study. Fractured surfaces of natural, synthetic or sol-gel silica had a roughness of 0.34 nm, 0.40 nm, or 0.38 nm, respectively. By contrast, fracture surface roughness of natural silica that had been annealed was 0.62 nm. Fracture surfaces for E-glass had a roughness of 0.83 nm. Apparently both the chemical composition and the annealing state of the glass influence the roughness of the glass surface. The surface roughness of inorganic glasses seem to be about two orders of magnitude less than that for the brittle epoxy resin discussed above, probably reflecting differences in microstructure.

Although others have measured the roughness of glass using atomic force, or scanning tunneling microscopy (AFM or STM)[12,13] no-one has studied roughness systematically as a function of position within the slow growth region of the mirror region, or tried to correlate that measurement with roughness measurements made as a function of stress intensity factor or crack velocity on fracture mechanics specimens. The change in roughness within the mirror region is generally thought to be negligible near the fracture origin. For example, AFM data on brittle epoxy resin, described above, suggest that the roughness stays almost constant within the mirror zone.[8] Roughness begins to increase as the crack approaches the transition between the mirror and mist regions of the fracture surface. Similar observations have been made for STM studies on inorganic, non-metallic glasses.[13]

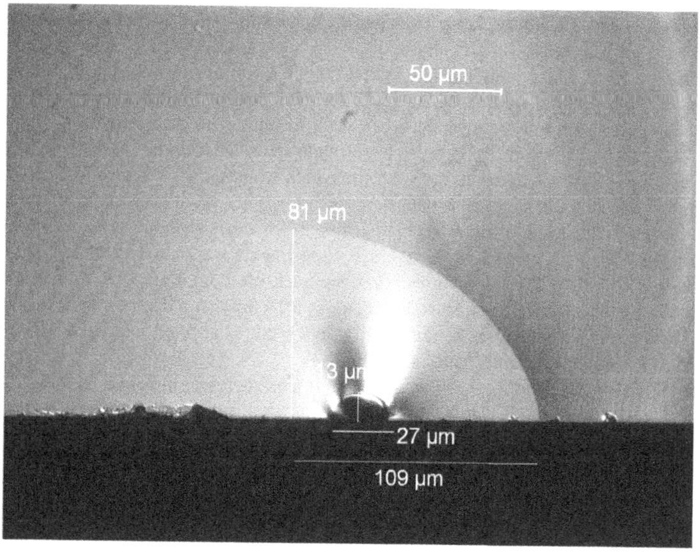

Figure 1: Optical micrograph of the mirror region of the ring-on-ring specimen that survived a 78.4 MPa static stress for 8631 h in water before failure. The dark area 27 μm by 13 μm is the original failure origin. The elliptical shaped surface, 109 μm by 81 μm, separates the region of slow crack growth, within the ellipse, from the region of rapid crack growth, outside of the ellipse.

In this paper we investigate the roughness of the slow crack growth area of the mirror region in silica glass and demonstrate a decrease in surface roughness with increasing stress intensity factor from about $0.5K_{Ic}$ to about K_{Ic}. All measurements are made within the elliptical boundary separating slow crack growth from rapid crack growth. This observation that roughness decreases with increasing crack velocity is in contrast to what has been reported previously for inorganic-nonmetallic glasses. Nevertheless, at some point beyond K_{Ic} the roughness begins to increase in a manner described previously in the literature.

EXPERIMENTAL PROCEDURE

Roughness measurements were made on fracture surfaces in silica glass, C7980 (Corning Glass Works, Corning, NY)*. The RMS roughness is defined as the standard deviation of the height of the surface from the average height of a plane passing through the surface.[14] Roughness measurements were made on two kinds of specimens: a static fatigue ring-on-ring loaded disk that failed in 8,631 h at 78.4 MPa when tested in water, and DCB fracture mechanics specimens that were broken through controlled subcritical crack growth experiments. Roughness measurements were performed using an atomic force microscope (Digital 3100, Veeco

* The use of commercial names is only for purposes of identification and does not imply endorsement by the National Institute of Standards and Technology.

Metrology Group, Santa Barbara, CA). The roughness data from the ring-on-ring sample (Fig. 2) were from 25 measurements made on random 0.5 μm square regions in a 1 μm x 1 μm micrograph. The AFM was operated in the contact mode for both ring-on-ring and DCB samples with tips that had a nominal radius of 20 nm. The images were flattened using zero order line flattening followed by second order plane fitting prior to roughness determination.

Double cantilever beam specimens were made from silica glass slides, (75 x 25 x 1.5) mm, with midline grooves approximately 0.5 mm deep to guide the direction of crack growth; the remaining web thickness within which the crack propagated was about 1 mm. Crack velocities ranged from approximately $1x10^{-10}$ m/s to 10 m/s; applied stress intensity factors ranged from 0.4 $MPa{\cdot}m^{1/2}$ to 0.72 $MPa{\cdot}m^{1/2}$. All experiments were carried out in water at room temperature (≈ 22 °C). Crack velocities, v, were measured on the double cantilever beam specimens using either a traveling telescope (accuracy ±10 μm), or a digital camcorder (accuracy ±10 μm). The camera was capable of taking individual pictures at fixed intervals, which could be varied from ~ 0.04 seconds to many hours.

Initially, a load was applied to the DCB specimens such that the crack velocity was in the range of $1x10^{-11}$ m/s to $1x10^{-10}$ m/s. The crack was then permitted to grow to produce an identifiable region that could be later characterized using atomic force microscopy. Fracture surfaces were produced at higher velocities by increasing the load on the specimen so that the crack velocity was in the range $1x10^{-9}$ m/s to $1x10^{-8}$ m/s and allowing the crack to propagate to failure. The value of K_I increased as the crack grew under constant load and could be calculated for any position on the specimen. The fracture surface roughness could then be determined as function of applied stress intensity factor by selecting the appropriate position on the surface.

AFM height images for square areas 0.5 μm x 0.5 μm, 1 μm x 1 μm, and 2 μm x 2 μm were made at specific areas on the DCB fracture surface, and the corresponding stress intensity factors were calculated. Regardless of the size of the scan-area, the number of scan lines was the same, 512, as were the number of points in each scan line, also 512.

EXPERIMENTAL RESULTS

Roughness of the Ring-on-Ring Specimen Fracture Mirror

An optical image of a portion of the fracture mirror near the origin of the ring-on-ring specimen investigated by AFM is shown in Fig. 1. The entire image is within the mirror region of the fracture surface. The fracture origin was a semi-circular flaw, 13 μm deep and 27 μm wide, and appears near the bottom of the figure. The subcritical crack growth zone extended from the fracture origin to the partial ellipse, 79 μm deep into the glass, seen mainly on the right hand side of the figure. Near the fracture origin within the partial ellipse, all portions of the crack were propagating subcritically in Region I where the rate of crack growth rate is controlled by reaction kinetics.[15] Partial ellipses of this type are commonly observed in glass specimens in which flaws grow stably in a water environment prior to fracture.[16] The left hand portion of the crack was propagating slightly faster than the right hand portion and accelerated rapidly, sweeping around and intersecting with the more slowly moving crack, leaving the elliptical mark shown on the right side of the figure.[16]

Roughness measurements on the subcritical crack growth region within the ellipse in the mirror region are shown in Fig. 2. The dashed line is for data taken on the fracture surface on a trajectory parallel to the tensile surface and the solid line is for data taken on a trajectory perpendicular to the tensile surface of the specimen. The curves in Fig. 2 have slopes of -0.0014 ± 0.0004 for the dashed line and -0.0014 ± 0.0005 for the solid line. At the mid-point of the two sets of data, 50 μm, the separation of the two fitted lines is 0.04 nm, which is slightly less than the instrumental noise, 0.05 nm;[17] the difference between the two curves in Fig. 2 is probably not statistically significant.

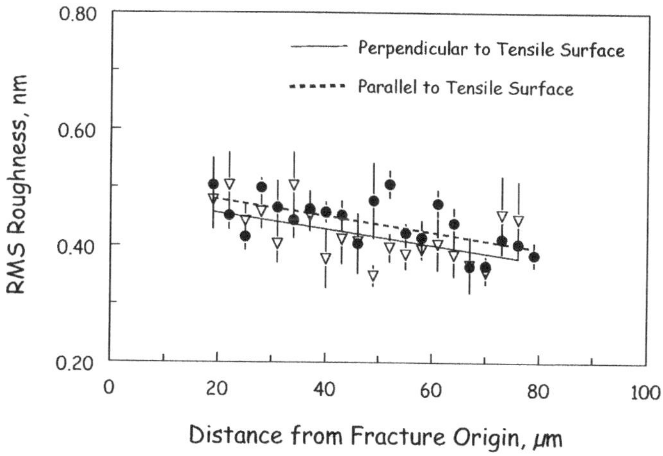

Figure 2: Roughness measurements of mirror region in the ring-on-ring specimen as a function of distance from the fracture origin made on trajectories normal and parallel to the tensile surface.

Roughness of Fracture Mechanics Specimen Surface

Fig. 3 presents the roughness of the DCB fracture surfaces as a function of applied stress intensity factor. As can be seen, the roughness of the fracture surface decreases as the stress intensity factor increases. Because the crack velocity is an exponential function of the applied stress intensity factor, the results in Fig. 3 also mean that the roughness decreases as the logarithm of the crack velocity increases.[17]

The roughness of the fracture surface also depends on the size of the window used for the measurement. The larger the size of the window, the rougher the surface; roughness measured on a 2 μm x 2 μm area is larger than that on a 0.5 μm x 0.5 μm area. This conclusion is supported theoretically by a numerical integration of the fracture surface data as a function of window size,[17] and is expected for fracture surfaces in glasses, which are self-afine in nature.[18]

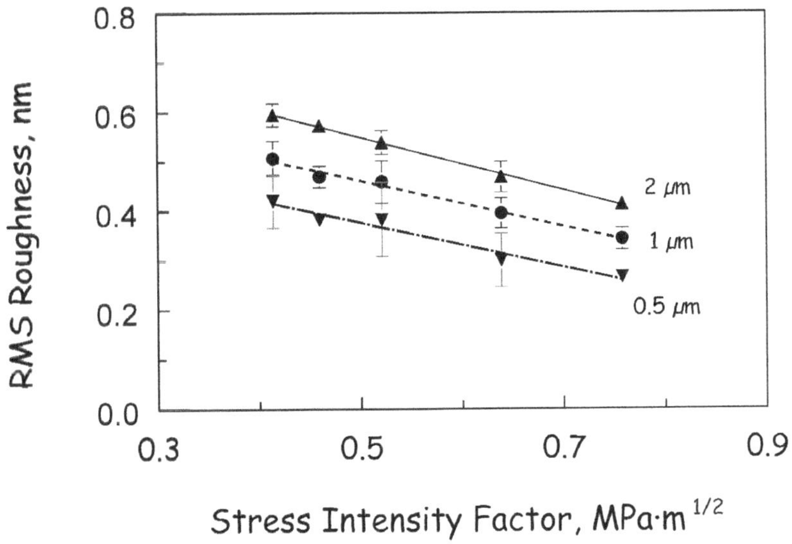

Figure 3: RMS roughness of a fracture surface formed by subcritical crack growth on a DCB specimen tested in water. The number next to each line indicates the length, L_o, of the side of the square area used to measure the roughness.

To compare Fig. 2 with Fig. 3, it is necessary to calculate the stress intensity factor at each point where the roughness was determined on the ring-on-ring sample. Assuming the ratio of the minor to major axis of the crack in Fig. 1 stays constant during growth, $a/c = 0.736$, the stress intensity factor, K_I, was calculated for the data in Fig. 2 using the relation of Newman and Raju.[19] The results are plotted in Fig. 4 along with a straight line representing the roughness data from the double cantilever beam specimen. Both sets of data show a systematic slope of the data, the roughness value of the data decreasing as the applied stress intensity factor increased. The slopes of the two sets of data are slightly different: -0.30 ± 0.07 for the fracture mirror, -0.45 ± 0.05 for the fracture mechanics specimen. The RMS roughness for the surface at the fracture origin of the ring-on-ring specimen is systematically higher than that for the double cantilever specimen.*

* The peak stress intensity factor calculated from the Newman and Raju relation, 0.83 MPa·m$^{1/2}$, is greater than the measured critical stress intensity of this glass, $K_{Ic} = 0.72$ MPa·m$^{1/2}$. Similar inconsistencies in determining critical stress intensity factors by interpreting the partial ellipse as a critical crack size, on the order of 0.1 MPa·m$^{1/2}$ to 0.2 MPa·m$^{1/2}$, have also been found in glass samples by others.[20]

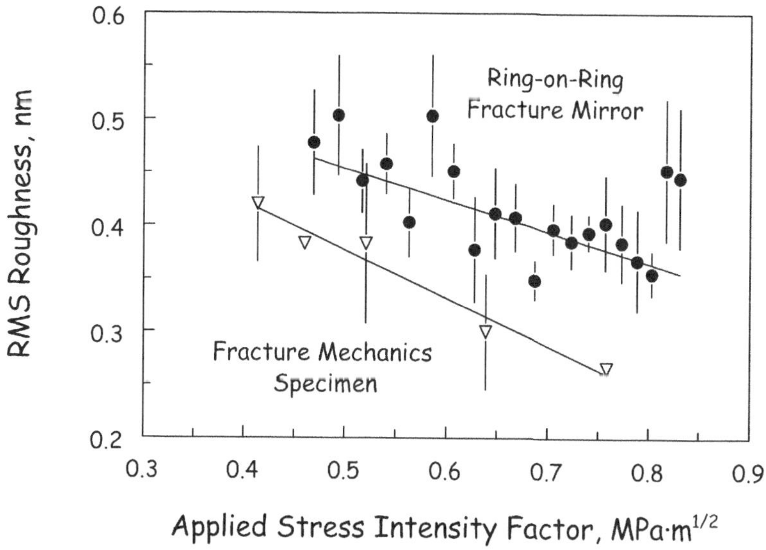

Fig: 4: Comparing the RMS roughness versus applied stress intensity of the mirror region of the ring-on-ring specimen with that of the fracture mechanics specimen data in Fig. 2. All RMS roughness data were taken with a 0.5 μm square window.

DISCUSSION

In his guide to practice, "Fractography of Ceramics and Glasses,"[4] Quinn describes fractography as the "means and methods for characterizing fracture specimens or components." Usually the object of interest is examined optically, first at low magnification and then at high magnification. Damage features are noted and photographs taken to document conclusions on the causes of failure. When higher magnifications are required, scanning electron microscopy is usually used. Atomic force microscopy is a relatively new tool for studying fracture surfaces of materials. It has the advantage of very high in-plane resolution, 1 nm to 5 nm, and an even higher resolution, ≈ 0.05 nm, normal to the plane being investigated. The instrument has a narrow depth of field, <10 μm depending on the instrument; consequently, its use is restricted to relatively flat surfaces. The instrument has been used for determining the roughness of glass surfaces formed either by melting or by fracture.[11,12,13] The importance of corrosion to subcritical crack growth[21,22] and the possibility of cavity formation during crack propagation in brittle materials[23] has also been investigated with the AFM.

The present study is an extension of these earlier works. Here we quantify the initial stages of fracture of a standard ring-on-ring fracture specimen, as a crack grows from a crack origin at slow speeds, i.e., speeds less than 0.1 m/s, and forms the mirror zone around the fracture origin. We compare these data with that from double-cantilever-beam fracture mechanics specimen. For both specimens, the roughness decreased as the applied stress intensity factor increased, Fig. 3 and Fig. 4.

The main difference between the roughness data obtained from the double cantilever specimen and the ring-on-ring fracture origin is the magnitude of the roughness. The roughness obtained on the fracture origin was systematically greater than that from the double cantilever beam specimen at the same value of applied stress intensity factor, Fig. 4. The difference, while small, lies well within the measurement capability of the atomic force microscopes used in this study. Hence, we must conclude that the difference in roughness is real. As noted above, a probable cause for the difference in roughness is a difference in cooling, or annealing when the glass was made. Both sets of samples were made of C7980 glass and presumably had the same composition. The chance is good, however, that the annealing treatment applied to the glass was different, since the size and shape of the samples made from the glass were different.

Differences in annealing treatment can lead to differences in roughness as indicated by Gupta et al.[11] in their study of the roughness of fracture surfaces in silica glass. These authors showed that the roughness of natural silica almost doubled from 0.34 nm to 0.62 nm as a consequence of annealing. Although we do not know the annealing history of the glasses that we studied, such an effect is clearly possible. Consequently, the study of the effect of annealing on the roughness of fracture surfaces is clearly a subject that should be pursued. We are currently quantifying the importance of annealing to roughness by annealing silica glass specimens and then measuring the roughness of fracture mechanics type of specimens.

The decrease in roughness with increase in applied stress intensity factor is consistent with a theoretical analysis presented in a companion paper by the present authors.[24] In this paper, the RMS roughness that develops during fracture was attributed to nanometer size inhomogeneities of the order of 10 nm in dimension within the glass. These inhomogeneities cause residual thermal misfit strains and local modifications of the elastic constants of the glass. Each of these gives rise to stresses that interact with the crack front, causing it to deviate from the mean fracture plane. The analysis[24] leads to the prediction that the roughness, R_q, should be proportional to the inverse of the crack tip stress intensity factor, $1/K_{I,\,tip}$, plus a constant:

$$R_q \propto \frac{a}{K_{I,tip}} + b, \tag{1}$$

where a and b are fit parameters. A fit of equation 1 to the data in reference 24 yielded a straight line relation between R_q and $1/K_{I,\,tip}$, in support of this theory.

In summary, this paper demonstrates that the RMS roughness within the mirror region in silica glass, while very low, does change with position. In contrast to the usual observation that roughness increases with increasing distance from the fracture origin, there is an area within the mirror region where $K_I < K_{Ic}$ that roughness decreases with increasing distance from the crack origin. It is only when cracks approach K_{Ic} that the roughness increases in silica glass leading to the expected transition in surface topography as the crack gets longer.

ACKNOWLEDGEMENT

The authors would like to thank George Quinn for our many fruitful discussions and for his helpful comments on the manuscript.

REFERENCES

[1]D. Hull, *Fractography*, Cambridge University Press, Cambridge, U.K. (1999).

[2]*ASM Handbook, Volume 12, Fractography*, ASM International, Materials Park, OH, (1987).

[3]V.D. Fréchette, *Failure Analysis of Brittle Materials, Advances in Ceramics,* Vol. 28, American Ceramic Society, Westerville, OH (1990).

[4]George Quinn, "Fractography of Ceramics and Glasses," *NIST Recommended Practice Guide, SP* 960-16, to be published, 2007.

[5]J.W. Johnson and D.G. Holloway, "On the Shape and Size of the Fracture Zones on Glass Fracture Surfaces," *Phil. Mag,* 14, 731-43 (1966).

[6]Von E. Gölz, "Übermikroskopische Feinstructuren an Glasbruchfläschen," *Z. Physik* 120, 773-777 (1943).

[7]E.K. Beauchamp, "Mechanisms for Hackle Formation and Crack Branching," pp. 409-445 in *Fractography of Glasses and Ceramics III,* J.R. Varner, V.D. Fréchette and G.D. Quinn, *Ceramic Transactions*, 64 The American Ceramic Society, Westerville, OH (1996).

[8]D. Hull, "Influence of Stress Intensity and Crack Speed on Fracture Surface Topography: Mirror to Mist Transition," *J. Mater. Sci.* 31 1829-1841 (1996).

[9]H.G. Richter, "Fractography of Bioceramics," *Key Eng. Mat.* 223 2157-180 (2002).

[10]H.G.Richter and F. Kerkhof, "Stress Wave Fractography," pp. 75-109 in *Fractography of Glass*, R.C. Bradt and R.E. Tressler, Eds. Plenum Press, New York, (1994).

[11]P. K. Gupta, D. Inniss, C.R. Kurkjian and Q. Zhong, "Nanoscale Roughness of Oxide Glass Surfaces," *J. Non-Crystalline Solids* 262, 200-206 (2000).

[12] J.-F. Poggemann, A. Goβ, G. Heide, E. Rädlein, and G.H. Frischat, "Direct View of the Structure of a Silica Glass Fracture Surface," *J. Non-Crystalline Solids*, 281 221-226 (2001).

[13]D.M. Kulawansa, L.C. Jensen, S.C. Langford, and J. T. Dickinson, "Scanning Tunneling Microscope Observations of the Mirror Region of Silicate Glass Fracture Surfaces," *J. Mater. Res.*, 9 [2] 476-485 (1994).

[14]T.R. Thomas, *Rough Surfaces*, p. 92, Longman, London (1982).

[15]S.M. Wiederhorn, "Influence of Water Vapor on Crack Propagation in Soda-Lime Glass," *J. Am. Ceram. Soc.* 50 [8] 407-14 (1967).

[16]T.A. Michalske, "Fractography of Stress Corrosion Cracking in Glass," pp. 111-142 in *Fractography of Glass,* R.C. Bradt and R.E. Tresseler Eds., Plenum Press, New York (1994).

[17]J.P. Guin and S.M. Wiederhorn, "Surfaces Formed by Subcritical Crack Growth in Silicate Glasses," *Int. J. Fracture,* 140 15-26 (2006).

[18]E. Bouchaud, "Scaling Properties of Cracks," *J. Phys.: Condens. Matter*, 9, 4319–4344, (1997).

[19]J.C. Newman and I.S. Raju, "An Empirical Stress Intensity Factor Equation for the Surface Crack," *Eng. Fract. Mech.* 15 185-192, (1981).

[20]G. Quinn, Personal communication.

[21]Jean-Pierre Guin, Sheldon M. Wiederhorn and Theo Fett, "Crack-Tip Structure in Soda-Lime-Silicate Glass," *J. Am. Ceram. Soc.*, 88 [3] 652-659 (2005).

[22]T. Fett , J.P. Guin and S.M. Wiederhorn, "Interpretation of Effects at the Static Fatigue Limit of Soda-Lime-Silicate Glass," *Eng. Frac. Mech.* 72 [18] 2774-2791, (2005).

[23]Jean-Pierre Guin and Sheldon M. Wiederhorn, "Fracture of Silicate Glasses: Ductile or Brittle?," *Phys. Rev. Let.*, 92 [21] 215502 (2004).

---

[24]Sheldon M. Wiederhorn, Jose M. López-Cepero, Jay S. Wallace, Jean-Pierre Guin and Theo Fett, "Roughness of Glass Surfaces formed by Sub-Critical Crack Growth," submitted to *J. Non-Crystalline Solids*.

INVESTIGATION OF THE SUBCRITICAL CRACK GROWTH PROCESS IN GLASS BY ATOMIC FORCE MICROSCOPY

Jean-Pierre Guin
LARMAUR FRE-CNRS 2717
University of Rennes I
Rennes, France

Sheldon M. Wiederhorn
National Institute of Standards and Technology
Gaithersburg, MD 20899-8500

ABSTRACT

To investigate the possibility of cavity formation in silica glass during subcritical crack growth, the topography of fracture surfaces formed in water at a crack velocity of 8×10^{-11} m/s was mapped using an atomic force microscope. The objective of the study was to determine how well the two halves of a crack in silica glass matched by establishing a three dimensional quantitative comparison of the shape of the two surfaces. This procedure uses a minimization routine to correct first order and quadratic distortions between the two AFM images of the crack surfaces. A three-dimensional image representing topographical differences between the corresponding fracture surfaces followed a Gaussian law centered on zero with a standard deviation, σ, of 0.22 nm. Within the resolution of the technique, 6 nm within the fracture surface and $\approx$ 0.44 nm normal to the fracture surface, no evidence for cavitation was found in silica glass. It is suggested that subcritical crack growth in silica glass occurs in a brittle way by the successive snapping of bonds as the crack advances.

INTRODUCTION

For decades, silicate glasses have been considered to be ideally brittle materials through which cracks propagate by the successive snapping of bonds, without accompanying plastic deformation at the crack tip.[1,2,3] Nevertheless, recent experimental studies of crack motion at very low speeds have questioned this idea, and concluded that crack motion in silicate glasses is accompanied by plastic deformation and cavity formation at the crack tip.

As an example, Guilloteau et al.[4] and Hénaux and Creuzet[5] used atomic force microscopy (AFM) to monitor a moving crack that intersected the tensile surface of a glass specimen that was loaded in four point bending. For both soda lime silicate glass and silica glass, the crack opening displacement exceeded predictions of linear elastic theory by a considerable amount. The excessive crack opening displacement was attributed to a non linear, but reversible process at the crack tip.[5] Silica glass was found to be more consistent with the linear elastic theory behavior than soda-lime silica glass.

More recently, Célarié et al.[6,7] and Pradès et al.[8] investigated the propagation of cracks in an alumino-silicate glass and a commercial grade of silica glass, at speeds as low as 10^{-11} m/s in a controlled environment (room temperature and 42 % relative humidity). These authors used the atomic force microscope (AFM) to image a moving crack that intersected the free surface of a double-cantilever, drilled compression specimen (DCDC).[9] Célarié et al.[6,7] reported the formation of nanoscale cavities ahead of the crack tip. Cavities were reported to be 20 nm long (direction of the moving crack) and 5 nm high (perpendicular to the fracture plane). The cavity dimension normal to the scan plane can not be determined by AFM, because of limitations in the technique. Célarié et al. concluded that crack propagation in glass was primarily a consequence of the nucleation, growth and coalescence of cavities, the same process that is observed in metals, but at a much smaller scale. Repeating these experiments on silica glass, Pradès et al.[8] reported the formation of even larger nanocavities: 150 nm in length and 25 nm in height at coalescence.

Ductile fracture of the type reported by Célarié et al.[6,7] and Prades et al.[8] is expected to leave detectable marks on fracture surfaces as is observed for example in the case of bulk metallic glasses for which cracks also propagate through the nucleation growth and coalescence of cavities;[10,11,12] examination of the fracture surface by a high resolution technique such as atomic force microscopy should be able to reveal the presence of the remnants of cavities. With this in mind, Guin and Wiederhorn[13,14,15] used AFM to search for the remnants of cavities within identical areas on the two fracture surfaces from Double Cantilever Beam specimens.[16] Fracture surfaces were formed by cracks that had propagated in water at speeds ranging from 10^{-11} m/s to 10^{2} m/s; soda lime silica glass and silica glass were used in these studies. Virtual sections passing through the same surface features (nanometer size bumps or depressions) were cut through each of the two fracture surfaces[13,14,15] allowing a two-dimensional, quantitative comparison of the topography of the two surfaces. Within the lateral resolution ($\approx$ 6 nm) of the experiment, fracture surfaces of silica glass were found to match showing a maximum difference between the two surfaces of $\approx$ 0.3 nm.

The results of these studies have been criticized because they represent a 2-dimensional analysis of the surface and only look at a small number of points, 300 to 500, from the fracture surfaces. Such a limited number of points might miss the cavities that are present in the fracture surfaces*. In addition, AFM pictures often contain artifacts due to thermal drift, creep and hysteresis[17,18] that need to be corrected, especially when comparing two scanned surfaces. Hence, the objective of this paper is to present a 3-dimensional analysis of our AFM images for which artifacts have been corrected using a homemade MATLAB routine†. The analysis provides us with a quantitative three-dimensional picture of how well the fracture surfaces match.

EXPERIMENTAL TECHNIQUES

A double cantilever beam technique was used to propagate a crack subcritically at a velocity of 8 x 10^{-11} m/s in a pure silica glass, C7980 (Corning Glass Works, Corning, NY). Details of the technique are given in reference.[13] The fracture surfaces were separated by completing the crack propagation to the ends of the specimen. The two halves were then examined with an optical microscope (Leica Model DMRM, Leica Inc., Deerfield, Illinois) to align the fracture surfaces normal to the optical axis of the microscope and to identify local features that could be used later as targets for AFM investigations.

A Digital III AFM (Veeco Metrology LLC, Santa Barbara, CA) was used to characterize the fracture surfaces after separation. Low magnification AFM images were used as guides for images of higher magnification. By following this mapping procedure on both fracture surfaces,

---

* The authors do not accept this argument, as the sections used for analysis ranged from 250 nm to 500 nm and should have intersected many cavities if they were there.

† The use of commercial designations is for purposes of identification only and does not imply endorsement by the National Institute of Standards and Technology, nor does it imply that the items mentioned are the best ones for the intended application..

identical areas could be imaged and matched.[13,14,15] Before examination, surfaces were cleaned with acetone, then with ethanol, and finally wiped with a tissue made of lint-free, pure cellulose fiber.

The limits of resolution were estimated by a procedure recommended by Bustamante and Keller.[19] The procedure requires two experimental inputs: the probe tip radius and the minimum measurable distance normal to the x-y plane, Δz. The tip radius, 20 nm (60 nm maximum value), was specified by the manufacturer of the tips (DNP tips, Veeco, Santa Barbara, CA), while Δz was estimated by measuring the RMS roughness of a zero nm scan sized window on silica glass. The noise measured on our equipment was found to follow a Gaussian law with an RMS roughness of approximately 0.05 nm. On the assumption that Δz is equal to four times the RMS roughness, $\Delta z = 0.20$ nm. The minimum cavity size detectable, *d*, is given by the equation from Bustamante and Keller:[19] $d = 2\sqrt{2R\Delta z} \cong 6\,\text{nm}$ for a cavity with sharp square sides (10 nm is the maximum value for a 60 nm tip radius). According to this calculation, 150 nm cavities should be easily detected. Areas were scanned with a 2 nm/line resolution, which is below the limit resolution and would easily allow us to see 150 nm sized objects.

The three dimensional analysis of the match of our fracture surfaces was carried out as follows. Each captured image may be represented as a matrix of 512 x 512 elements, $X_{i,j}$, each element having a value equal to the altitude of the corresponding pixel of the surface. After pretreatment of each picture (a plane fit of order one for tilt and offset removal, and a flatten order zero for uneven lines),* the three dimensional analysis was performed using MATLAB software. AFM data are first imported directly into MATLAB as a 512 x 512 matrix using an adapted routine.[20] Then a homemade algorithm[21] based on cross correlation image analysis was applied to our captured images to find their common area. This algorithm also corrects the distortions existing between the two captured surfaces. Such distortions are inherent to scanning probe microscopy; they can be linear distortions (translation and rotation due to in plane and off plane surface misalignment), or non linear distortions (mainly due to drift, creep and hysteresis).[17,18] To correct these distortions, one picture is taken as a reference; the second is considered to be distorted. Once the deformations have been corrected (i.e. the best match has been found after several iterations) the "corrected" image is saved and subtracted from the reference, which gives us a three dimensional error picture representing the mismatch between the two corresponding fracture surfaces.

RESULTS

An example of a micrograph obtained by the procedure described above is shown in figure 1. The crack was propagated in water at about 8 x 10^{-11} m/s. Figure 1a and 1b present the raw images of the upper and lower fracture surfaces as obtained by the AFM. A topographical scale may be found on the right end of the picture as a 256 gray levels scale (black being lower than white). Before treatment, images were plane fitted using an order one filter (offset and tilt of the mean plane were corrected) and flattened with and order zero filter (offset between two consecutive scanned lines were corrected). As illustrated in figure 1 flipping and inverting the two captured pictures (black becoming white and vice versa) makes the visual comparison easier. Figure 2 and figure 3 illustrate the result of the image treatment described above where figure 2a is the picture chosen as reference and figure 2b the corrected one. Figure 2a and 2b show the exact same area on each fracture surface, linear distortions (translation and rotation) between the

* Built in Dimension 3100 software, Version 5.30r2 were used for this purpose.

two pictures as well as quadratic distortions (creep, hysteresis) have been corrected. The size of the final common matching area is 975 nm by 800 nm. The circled area on 2a and 2b highlight a spot where a small particle was adsorbed on one of the two surfaces. It will show up as a darker spot (overlap of matter), on figure 3.

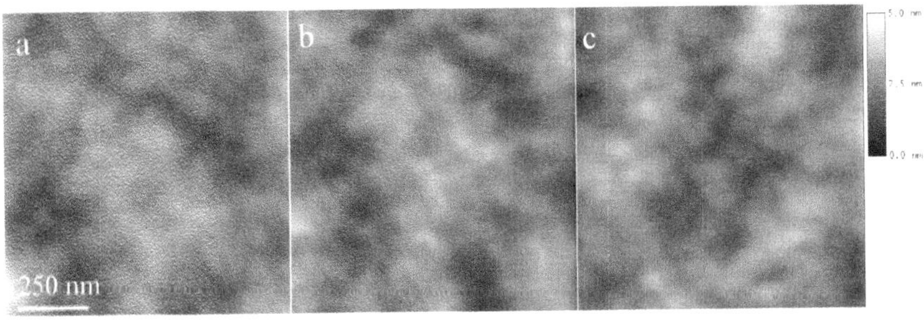

Figure 1: Silica glass, 1 μm by 1 μm square AFM topographical images from opposite fracture surfaces, in water at a crack velocity of 8 x 10^{-11} m/s. The gray scale, upper right hand corner of the figure, indicates the height along the z-axis: a) upper fracture surface taken as reference. b) lower fracture surface that has been flipped over and inverted. c) as-captured lower fracture surface.

Two identical surfaces should match in every detail of their shadowing so that their difference pixel by pixel should result in a uniformly flat picture of 975 nm by 800 nm. Figure 3 is the result of the difference between figure 2a and 2b and is a three dimensional representation of the mismatch between the upper and lower fracture surfaces. The contrast may be interpreted as follows: the brighter the contrast the wider is the gap between the two surfaces; conversely, the darker the contrast the larger the overlap. A zero value means a perfect match (the overall mean plane is equal to zero).

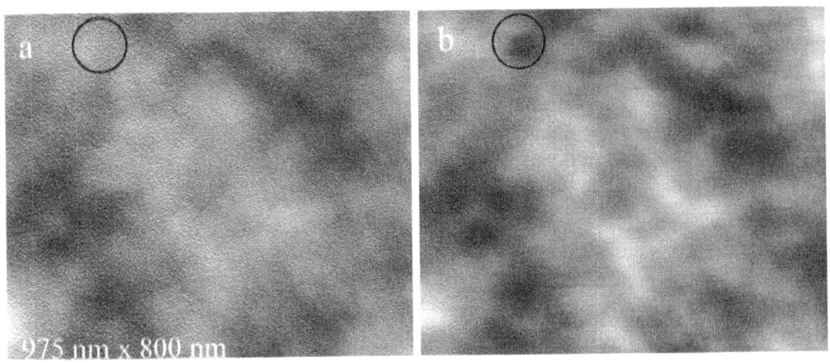

Figure 2: Reference image a) and corrected one b) presenting the best common matching area (975 nm by 800 nm). Black circle highlights area where a particle was adsorbed on one of the surface. The grey scale for this figure is shown in Fig. 1.

To facilitate an interpretation of the results, a histogram of the altitude of the pixels belonging to the mismatch image is shown in figure 3. The pixels range from -1.42 nm to 0.877 nm. The plot of frequency as a function of altitude follows a Gaussian curve centered on 0 with a standard deviation of 0.22 nm, which means that about 95 % of the pixels making up figure lie within two standard deviations of the mean, -0.44 nm to 0.44 nm.

DISCUSSION

From the above results we concluded that within the limits of resolution (2 nm per scanned line) and the conditions used in our study, no indication of cavities could be found on the fracture surfaces of silica glass specimens. These results differ substantially from the work of Célarié et al.[6,7] and Prades et al.,[8] who state that both alumino-silicate glass and silica glass fracture by the nucleation, growth and coalescence of nanocavities. The cavity size reported by Prades et al. for silica glass is approximately 150 nm by 25 nm at coalescence and would have easily been detected in our experiments.

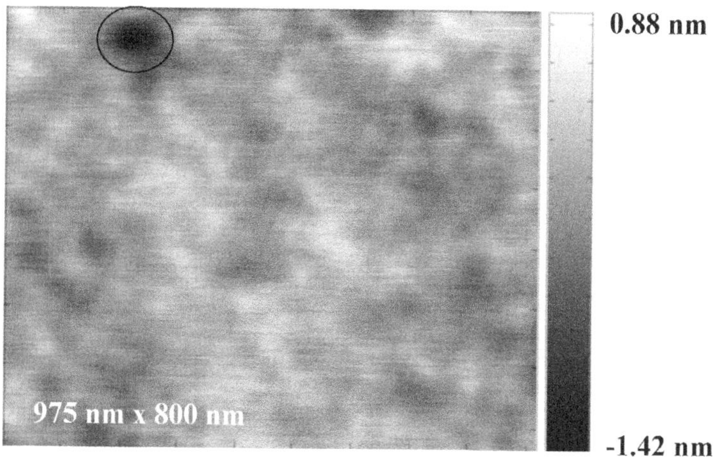

Figure 3: Mismatch picture obtained after subtracting figure 2b from figure 2a. The gray scale, right hand side of the figure, indicates the height along the z-axis. White is equivalent to a gap between matching surfaces, black represents an overlap of matter.

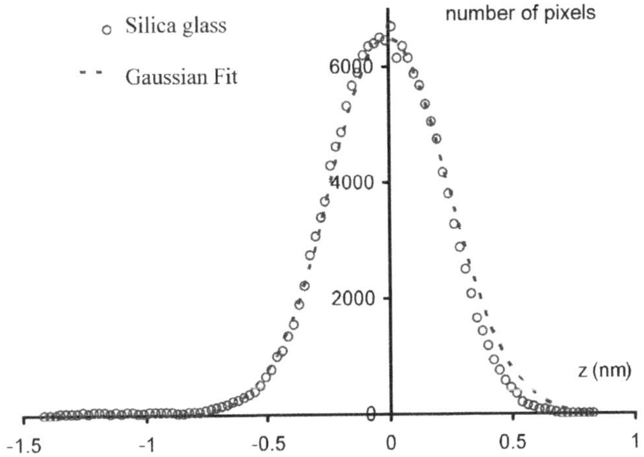

Figure 4: Histogram of the mismatch between the two fracture surfaces (opened circles) showing the repartition of the pixels as a function of there altitude. The histogram follows a Gaussian distribution centered on zero, with a standard deviation of 0.22 nm.

Although we report the results for a single study in this article, the technique has been applied to many different specimens of two glass compositions with the same result: no cavities of the sizes quoted by Célarié et al.,[6,7] or Prades et al.[8] are observed on fracture surfaces of glass.

Mismatch between surfaces: A perfectly brittle fracture would result in two fracture surfaces with an identical topography at the sub-nanometer scale, one being the negative of the other[*]. Therefore, when applying the technique discussed here, the mismatch picture would be uniformly flat. As one can see, the mismatch in figure 3 is not flat, but is close to it. The RMS roughness, $R_q = 0.22$ nm, is just above the level of being detectable[†]. This roughness has several possible origins. The most obvious one, discussed earlier, is the adsorption of nano-particles from the surrounding environment onto the fracture surface. This is usually easy to detect as such particles will show up as dark spots on the mismatch picture (see figure 3). If necessary such areas could be screened out before final analysis.

The sensitivity of the technique used in our study is an important point that has to be considered. How accurately can an atomic force microscope image an area? To answer this question experimentally, two approaches were taken. In one, the trace and retrace images of the same 1 μm by 1 μm square area were compared: one image is captured while the tip is scanning from the left to the right; the second one is captured on the way back. These two images are of the same surface, taken at the same time. In principle the two traces should be identical.

[*] The two surfaces would not be exactly the same as one might expect local atomic rearrangements of the recently broken and dangling bonds.

[†] Since the noise of the AFM is 0.05 nm, a detectible distribution would have to be at least two to three time the background noise of the AFM, 0.10 nm to 0.15 nm.

In the second method, we imaged a 1 μm by 1 μm square area, and then removed the tip from the holder. The tip was then replaced in the holder and again used to scan the same area. We then compared the differences between the two pictures by applying the same image analysis method as described above. The resulting histograms from these analyses were Gaussian shaped with standard deviations of 0.29 nm and 0.27 nm, respectively for the first and second method. These values are close to the standard deviation found for our silica fracture surface, 0.22 nm, which means that our fracture surfaces are similar to identical surfaces that were imaged twice. From this result we conclude that either there are no cavities in the surface, or that they are smaller than the resolution of our technique: normal to the fracture surface about 2 times the standard deviation of the mismatch, say 0.44 nm; within the fracture surface about 6 nm using the method suggested by Bustamante and Keller.[19]

Tip-convolution may account for part of the differences in analyzing opposite surfaces of a crack. Because one surface has the negative topography of the other, and because of the finite geometry of the apex of an AFM tip,[22] holes on one surface do not image in exactly the same way as hills on the other surface. This imaging difference leads to small differences between the two imaged surfaces. Such a convolution phenomenon, for 150 nm by 25 nm size holes would be negligible for a 20 nm tip radius, as is the case for this study.

Cavities at crack tips: Figure 3 shows the three dimensional mismatch existing between the two corresponding fracture surfaces, whereas figure 4 shows the topographic repartition of the pixels making this image. If present, cavities should show up as bright spots on figure 3 indicating the existence of a gap between the two corresponding fracture surfaces. The size of this gap should be of the size and shape of the cavity. Thus, the 150 nm by 25 nm cavity should show up as a white spot with a 150 nm diameter in the plane of view. At the same time, the histogram in figure 4 should show a peak at 25 nm, which is far off of the scale used in figure 4. In fact, there are no distances between the two fracture surfaces larger than 1.5 nm. Both of these observations support the view that no cavities 150 nm by 25 nm are present in the fracture surface of the silica glass specimens.

Another question to be asked is how small does a collection of cavities have to be in order to remain undetected in our experiment? The mismatch of the two fracture surfaces, figure 4, follows a Gaussian curve with a standard deviation, σ, of 0.22 nm. If cavities are present, but too small to be detected they will be lost within the peak of the mismatch curve. The selection of what size a cavity distribution has to be in order to be easily distinguished from the background noise caused by the mismatch distribution will depend on the difference between maxima and the variance of the two distributions. If the cavity-size distribution is assumed to have the same variance as the mismatch distribution, then the minimum detectable size for the cavity distribution can be established by applying a Student-t test to the problem. For 95 % confidence limits, the difference between the maxima of the two distributions would have to be two times the standard deviation of the distributions. Since the mean of the mismatch distribution is zero, the maxima of the cavity size distribution will have a value of 0.44 nm. This distribution should be distinguishable from the mismatch distribution with 95 % confidence. The length of detectable cavities within the fracture plane would have to be greater than 6 nm, which is the lateral detection limit on our AFM. From these considerations we believe that cavities larger than 6 nm by 0.44 nm would be detected if they were present in our experiments.

What about smaller cavities? Molecular dynamic simulations of the dynamic fracture of silica predicts the occurrence of cavities in the nm range[23,24] and suggest that fracture occurs by the nucleation, growth and coalescence of these cavities. This is no surprise as silica is known for its open structure in which nano meter sized inhomogenieties and small voids of about 0.45 nm are present,[25,26] Nevertheless, conditions assumed in molecular dynamic simulations (MD) are hardly comparable to conditions during subcritical crack propagation for two reasons. First, MD simulations in silica are for cracks running at a large fraction of the Raleigh velocity and for strain rates up to 10^{-3} ps^{-1}. These conditions are far from those used in any AFM study of crack growth in silicate glasses.[6-8,13-15] Our crack velocities were approximately 13 to 14 orders of magnitude less than those used in MD simulations.

Furthermore, MD simulations do not consider the action of a corrosive environment on crack propagation, whereas our experiments did, since they take place in the presence of water. Numerous studies in the past 40 years have shown that water has a strong influence on crack propagation in the subcritical crack growth regime.[16,27,28,29,30,31,32,33,34] The kinetics is believed to be controlled by a chemical reaction at the crack tip between the water molecule and stretched Si-O-Si bonds.[35,36,37,38,39] Unless this stress enhanced chemical reaction is built into MD calculation, the observation of cavities in MD studies cannot be deemed relevant to crack growth studies on glass in water or normal air. Therefore, nucleation, growth and coalescence of cavities does not seem to be a relevant mechanism for the subcritical crack growth process, whereas it might be for fast fracture for which chemical reactions are unimportant and mechanical stresses in the vicinity of the crack tip are much higher.

CONCLUSION

The topography of opposing fracture surfaces of a silica glass sample broken by subcritical crack growth in water ($<10^{-10}$ m/s) was imaged by atomic force microscopy. Using a MATLAB routine, first and second order distortions inherent to AFM technology were corrected and the three dimensional map of the mismatch existing between the two parts of the sample was obtained. The main results are:

1. The resolution of the technique we developed was estimated to be 0.22 nm perpendicular to the fracture plane and 6 nm within the fracture plane (depending on the radius of curvature of the probe tip).
2. No traces of the 150 nm by 25 nm cavities reported by Prades et al. were found. Furthermore, within the resolution of our technique no traces of any cavities were found with dimensions greater than 6 nm by 0.44 nm.
3. Although we cannot discount the possibility of cavities less than approximately 6 nm by 0.44 nm, molecular dynamics simulations that discuss cavities of the order of 3 nm to 6 nm are not applicable to subcritical crack growth in glass because the conditions under which these are formed are totally different from those occurring during subcritical crack growth in air or in water. Subcritical crack growth in the presence of water is a chemically driven process enhanced by the presence of a tensile stress at the crack tip. This aspect of crack growth has not yet been built into molecular dynamic simulations.

4. Based on our current experimental results, it is our view that cracks in silicate glasses propagate in a brittle manner by the successive rupture of bonds.

ACKNOWLEDGEMENTS

We acknowledge the help of Dr. Stéphane Roux (Glass Surface and Interface, Unité Mixte CNRS/Saint Gobain, France) who developed the MATLAB routine used in our study for analyzing our AFM data.

REFERENCES

[1]F. W. Preston, "The Mechanical Properties of Glass", *J. Appl. Phys.* 13, 623-63, (1942).

[2]B. R. Lawn, B.J. Hockey and S. M. Wiederhorn, "Atomically Sharp Cracks in Brittle Solids: An Electron Microscopy Study", *J. Mater. Sci.* 15, 1207-1223, (1980).

[3]A. Kelly, W. R. Tyson and A. H. Cotterel, "Ductile and Brittle Crystals," *Phil. Mag.* 15 567-586 (1967).

[4]E. Guilloteau, H. Charrue and F. Creuzet, "The Direct Observation of the Core Region of a Propagating Fracture Crack in Glass," *Europhys. Let.* 34[7], 549-553, (1996).

5. S.Hénaux, F. Creuzet, "Crack Tip Morphology of Slowly Moving Cracks in Glass", *J. Am. Ceram. Soc.* 83, 415-17, (2000).

[6]F. S. Célarié, S. Prades, D. Bonamy, L. Ferrero, E. Bouchaud, C. Guillot, and C. Marlière, "Glass Breaks Like Metals, But at the Nanometer Scale," *Phys. Rev. Let.*, 90 [7], 075504, (2003).

[7]F. S. Célarié, S. Prades, D. Bonamy, A. Dickelé, E. Bouchaud, C. Guillot, and C. Marlière, "Surface Fracture of Glassy Materials as Detected by Real- Time Atomic Force Microscopy (AFM) Experiments," *Appl. Surf. Sci.* 212-213, 92-96, (2003).

[8]S. Prades, D. Bonamy, D. Dalmas, E. Bouchaud, and C. Guillot , "Nano-Ductile Crack Propagation in Glasses Under Stress Corrosion: Spatiotemporal Evolution of Damage in the Vicinity of the Crack Tip," *Int. J. Solids and Structures*, 42 (2), 637-645, (2005).

[9]Janssen, C., "Specimen for Fracture Mechanics Studies on Glass," pp. 10.23-10.30 in *Proceedings of Tenth International Congress on Glass* (Kyoto, Japan July 1974). Ceramic Society of Japan, Tokyo, Japan (1974).

[10]X.K. Xi, D.Q. Zhao, M.X. Pan, W.H. Wang, Y. Wu, and J.J. Lewandowski, "Fracture of Brittle Metallic Glasses: Brittleness or Plasticity," *Phys. Rev. Let.* 94, 125510, (2005).

[11]J.J. Lewandowski, W.H. Wang and A.L. Greer, "Intrinsic Plasticity or Brittleness of Metallic Glasses," *Phil. Mag. Let.*, 85, 77-87, (2005).

[12]G. Wang, Y. T. Wang, Y. H. Liu, M. X. Pan, D. Q. Zhao, and W. H. Wang, "Evolution of Nanoscale Morphology on Fracture Surface of Brittle Metallic Glass," *Appl. Phys. Let.* 89, 121909, (2006).

[13]Guin, J.-P. , and Wiederhorn, S.M., "Crack Growth Threshold in Soda Lime Silicate Glass: Role of Hold-Time," *J. Non-Crystalline Solids*, 316, 12-20, (2003).

[14]J.-P. Guin and S.M.Wiederhorn, "Fracture of Silicate Glasses: Ductile or Brittle?" *Phys. Rev. Let.* 92 [21], 215502 (2003).

[15]J.-P. Guin and S.M.Wiederhorn, "Surfaces Formed by Subcritical Crack Growth in Silicate Glasses", *Int. J. Fract.*, 140, 15-26, (2006).

[16]S. M. Wiederhorn and L. H. Bolz, "Stress Corrosion and Static Fatigue of Glass", *J. Am. Ceram. Soc.* 53 [10], 533-548, (1970).

[17]P. West and N. Starostina, "How To Recognize and Avoid AFM Image Artifacts," *Microscopy Today*, May/June, p 20-26, (2003).

[18]C.Z Cai, X.Y. Chen, Q.Q. Shu and X.L. Zheng "Computer Correction For Distorted STM Images" *Rev. Sci. Instrum.* 63 (12), 5649-5652, (1992).

[19]C. Bustamante and D. Keller, "Scanning Force Microscopy in Biology," *Physics Today*, 48 [12], 32-38 (1995).

[20]Program first developed by Claudio, Rivetti and Mark Young (1995).

[21]Developed by Stéphane Roux (Glass Surface and Interface, Unité Mixte CNRS/Saint Gobain, France).

[22]E.C. Leung, P. Markiewicz, M.C. Goh, "Identification and Visualization of Questionable Regions in Atomic Force Microscope Images", *J. Vac. Sci. Technol.* B 15 (2), 181-185, (1997).

[23]Rountree, C.L., Kalia, R.K., Lidorikis, E., Nakano, A., van Brutzel, L.V. and Vashishta, P. "Atomistic Aspects of Crack Propagation in Brittle Materials: Multimillion Atom Molecular Dynamics Simulations," *Ann. Rev. Mater. Res.* 32, 377-400 (2002).

[24]L. van Brutzel, C.L. Rountree, R.K. Kalia, A. Nakano and P. Vashishta, *Mat. Res. Soc. Symp. Proc.* 703, 117, (2002).

[25]P.K. Gupta, D. Inniss, C.R. Kurkjian and Q. Zhong, "Nanoscale Roughness of Oxide Glass Surfaces," *J. Non-Cryst. Solids*, 262[1-3] 200-206 (2000).

[26]T.P. Swiler, J.H. Simmons, A.C. Wright, "Molecular Dynamic Simulations of Brittle Fracture in Silica Glass and Crystobalite, *J. Non-Cryst. Solids* 182, 68-77, (1995).

[27]S.M. Wiederhorn, "Influence of Water Vapor on Crack Propagation in Soda-Lime Glass", J. Am. Ceram. Soc., 50 (8), 407-414, (1967).

[28]S.M. Wiederhorn, A.G. Evans, and D.E. Roberts, "A Fracture Mechanics Study of the Skylab Windows", pp 829-841 in *Fracture Mechanics of Ceramics, Vol. 2,* Plenum Pub. Corp., N.Y, (1974).

[29]S.M. Wiederhorn, and H. Johnson, "Effect of Electrolyte pH on Crack Propagation in Glass", *J. Am. Ceram. Soc.* 56 (4), 192-197, (1973).

[30]S. Freiman, "Effect of Alcohols on Crack Propagation in Glass", *J. Am. Ceram. Soc.*, 57 (8), 350-353, (1974).

[31]S.M. Wiederhorn, S.W. Freiman, E.R. Fuller, Jr. and C.J. Simmons, "Effects of Water and Other Dielectrics on Crack Growth", *J. Mat. Sci.* 17, 3460-3478, (1982).

[32]G.S. White, S.W. Freiman, S.M. Wiederhorn, and T.D. Coyle, "Effects of Counterions on Crack Growth in Vitreous Silica", *J. Am. Ceram. Soc.*, 70 [12] 891-95, (1987).

[33]S. Takeda, I. Tari, "Effect of Surface Charge on Subcritical Crack Growth in Glass" *Fract Mech Ceram.* 9, 575-587, Plenum Press, N.Y. (1992).

[34]E. Gehrke, Ch. Ullner, M. Hähnert, "Fatigue Limit and Crack Arrest in Alkali-Containing Silicate Glasses" *J. Mater. Sci.* 26, 5445-5455, (1991).

[35]T.A. Michalske, and B.C. Bunker, "Slow Fracture Model Based on Strained Silicate Structures", *J. Appl. Phys*, 56 (10), 2686-2693, (1984).

[36]J.K. West, and L.L. Hench, "Silica Fracture Part I: A Ring Contraction Model", *J. Mater. Sci.* 29, 3601-3606, (1994).

[37]J.K. West, and L.L. Hench, "Silica Fracture Part II: A Ring Opening Model via Hydrolysis", *J. Mater. Sci.* 29, 5808-5816, (1994).

[38]J.K. West, and L.L. Hench, "The Effect of Environment on Silica Fracture: Vacuum, Carbon Monoxide, Water and Nitrogen", *Phil. Mag. A*, 77 [1], 85-113, (1998).

[39]E.D. Taylor, K. Runge and R.J. Bartylett, "Study of the Effect of Hydration on the Tensile Strength of a Silica Nanotube", *Molecular Physics,* 103, 2019-2026, (2005).

FORMATION AND EVOLUTION OF A CONFINED LIQUID CONDENSATE AT THE CRACK TIP IN GLASSES

M. George, M. Ciccotti, V. Ranieri
Laboratoire des Colloïdes, Verres et Nanomatériaux, UMR CNRS 5587
University of Montpellier 2, France

L. Wondraczek
Corning European Technology Center
Corning SAS, Avon, France

A. Dittmar, C. Oelgardt, F. Célarié
Institute of Non-Metallic Materials
Clausthal University of Technology, Germany

C. Marlière
Laboratoire de Tectonophysique, UMR CNRS 5568
University of Montpellier 2, France

ABSTRACT

Crack propagation in oxide glasses at low crack velocities is controlled by stress corrosion. Proper knowledge of the crack tip chemical environment is thus crucial to understand the slow fracture process of these materials. The formation of a liquid condensate in the confined area of the crack tip is theoretically expected and is classically mentioned to explain specific behavior during the crack propagation. Since this condensate is of the nanometer scale and cannot be traced post mortem, it is extremely difficult to observe. This paper reports the experimental evidences of the presence of a liquid condensate at the tip of a crack propagating by stress corrosion in silica glass. The observation has been performed in situ under carefully controlled atmosphere through phase imaging by atomic force microscopy (AFM). The evolution of the condensate has been followed as a function of the relative humidity. The impact of those evidences of the liquid condensate will be discussed and information on its size or evolution kinetic will be reported.

INTRODUCTION

The study of crack propagation at slow velocity in oxide glasses is of great interest on both practical and fundamental bases. Long-term behavior of vitreous material is particularly important considering the large proportion of these materials used in architectural structures and their role in the encapsulation of nuclear wastes. On the fundamental aspect, slow advancing cracks can be accurate probes of the molecular structure of glasses.

Stress corrosion is known to control slow propagation of crack in oxide glasses. The preferential corrosive role of water molecules at the crack tip sustaining tensile stresses is generally acknowledged.[1,2,3,4] Wiederhorn et al. have developed a model that fits well the experimental data and shows how - in the stress corrosion regime - crack velocity is highly dependent on relative humidity,[2,3] and can be limited by transport kinetics of the water molecules towards the crack tip. The physical and chemical environment in which stress corrosion takes place, as well as the mechanisms of transport of the water molecules towards the crack tip are less well-established.[5] Several experimentally observed behaviors,[5,6,7] such as the effect of chemical composition or the disappearance of the transport-limited kinetics phenomenon, are still not accounted by the Wiederhorn model.

Under usual conditions of temperature and pressure, capillary condensation theory predicts the presence of a liquid condensate between the newly created surfaces of the crack which at the vicinity of the crack tip are only distant from a few nanometers[8]. Such a liquid condensate is classically acknowledged to account for the anomalous behavior that can be observed. Its presence will indeed influence largely the chemistry of the corrosion since direct contact between liquid water and glass will replace interactions between isolated gas water molecules and glass made uneasy inside the confined area created at the crack tip in a fragile material.

To our knowledge, observations of the condensate have not yet been reported by other teams. In situ observations that are necessary to reveal such a phenomenon are indeed difficult to perform with the required resolution. We reported previously evidences of the liquid condensate at a crack tip slowly propagating in a silica glass under variable relative humidity.[9] These evidences were obtained by Atomic Force Microscopy (AFM) in situ observations of the crack propagation. In this paper, we will present for the first time the detailed interpretation of these evidences enabled by an extended analysis of the phase contrast.

EXPERIMENTAL

An experimental technique has been developed in the Laboratoire des Colloides, Verres et Nanomateriaux (LCVN) from the University of Montpellier to observe in situ by AFM the propagation of crack in glass at slow velocity under controlled environmental conditions, as described in a previous paper.[10]

Samples of type III silica glass (Suprasil 311, Heraeus, Germany) with a bulk OH^- - content of 200 ppm have been studied using this technique. They were cut into 4x4x40 mm^3 rods with a drilled hole of 1 mm diameter in the centre to fit with doubled cleavage drilled compression (DCDC) geometry. The surface was mechanically and chemically polished with CeO_2 to a RMS roughness of 0.25 nm for an area of 10x10μm^2.

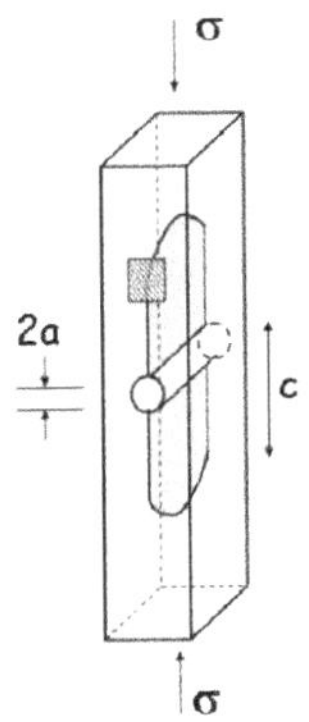

Figure 1 - Geometry of the sample.
The hatched zone corresponds to the zone observed by AFM. The two cracks propagating on both sides of the hole (in grey) are in opening mode (mode I).

DCDC technique allows to create mode I tensile cracks with highly controllable speeds in oxide glasses. For an applied stress σ, the stress intensity factor K_I in our DCDC experiment is given by $K_I = \sigma a^{1/2}/(0.375c/a+2)$, where a is the radius of the hole in the DCDC specimen and c is the length of the crack[11]. The load cell has been interfaced with an AFM (D3100, Veeco) used in tapping mode to observe the intersection between the moving crack plane and the polished surface of the sample, as indicated in figure 1. The neighborhood of the crack tip can thus be imaged for velocities from 10^{-9} m.s^{-1} to $\sim 10^{-12}$ m.s^{-1}. Optical microscopy can be used to record higher crack velocities.

Generation and propagation of the crack were performed at 22.0±0.5 °C in a carefully controlled atmosphere of pure nitrogen and pure water vapor, which allows variations of the relative humidity (RH) between ~ 1 % and 80 % (± 2 %).

RESULTS AND DISCUSSION

Phase Contrast Related to Liquid Layers

To explain how AFM can be used to track the presence of water on a surface, we will firstly review briefly the principle of tapping mode. Tapping mode is currently used to image the topography of surfaces, with lower wearing of both the sample and the tip, but can also be used to obtain information on the local mechanical properties of the surface.[12] The vibrating tip-cantilever system can be modeled as a harmonic oscillator sustaining non-linear interactions with the surface of the sample (van der waals, capillary, electrostatic forces, ...). In tapping mode (also called amplitude modulation mode), this oscillator is driven at a chosen frequency ω, close to the resonance frequency ω_0. The displacement z of the oscillator as a function of time can then be written as:

$$z(\omega,t) = A(\omega)\cos(\omega t+\phi(\omega)) \tag{1}$$

where A is the amplitude of the oscillations and ϕ the phase delay between the measured oscillation and the stimulation of the cantilever.

For a cantilever without interaction with a surface, the phase decrease from 0 to -180° when the frequency goes from 0 to infinity and is equal to -90° at the resonance when $\omega=\omega_0$. The amplitude of the oscillation during imaging (amplitude set point) is determined either automatically or by the operator. The feedback loop of the microscope is then used to maintain this amplitude constant by acting on the vertical position of the cantilever holder. Different signals can be recorded, among which:

- The amplitude variation ΔA (error signal) which should be as low as possible if the parameters of the feedback loop are well defined.
- The vertical displacement of the cantilever, which gives essentially topographical information.
- The phase ϕ of the oscillator, which can be related to dissipative interactions between the tip and the substrate and thus yields useful information on the local mechanical properties of the surface, such as viscosity or adhesion.

The non-linear behavior of the oscillator is difficult to analyze in a quantitative way and focuses the efforts of many researchers.[13,14] One of the main difficulties lays in the existence of multi-stable solutions for the response of the oscillator. These multi-stabilities have been frequently documented to explain the hysteresis commonly observed in amplitude- and phase-distance curves.[13-15] Typically, two stable modes of oscillation can be observed which corresponds to different interactions between the tip and the surface. One of these modes corresponds to an interaction which is mainly attractive. In other terms during the oscillation the tip is not brought closely enough to the surface to feel the repulsive forces. This mode is also called "non-contact" mode (NC). In the second mode, called "intermittent contact" mode (IC), the tip comes closer to the surface and feels both the attractive forces and the repulsive forces. Figure 2 represents two theoretical typical amplitude- and phase- distance curves[13-16] for a driven frequency chosen below the resonant frequency ($\omega<\omega_0$). The solid line curve corresponds to a pure NC mode and the dashed line curve to the transition between a NC and an IC mode. The two modes are easily distinguishable on the phase-distance curves. The NC mode, respectively the IC mode, corresponds to a phase which is below, respectively above, -90°. In both modes, the amplitude and the frequency of oscillations being constant, the power dissipated by the interactions between the tip and the surface is proportional[17] to sin(ϕ). An increase of dissipative interaction between the tip and the surface will thus corresponds in both modes to a phase closer to -90° (which means an increase of the phase in NC mode and a decrease of the phase in IC mode).

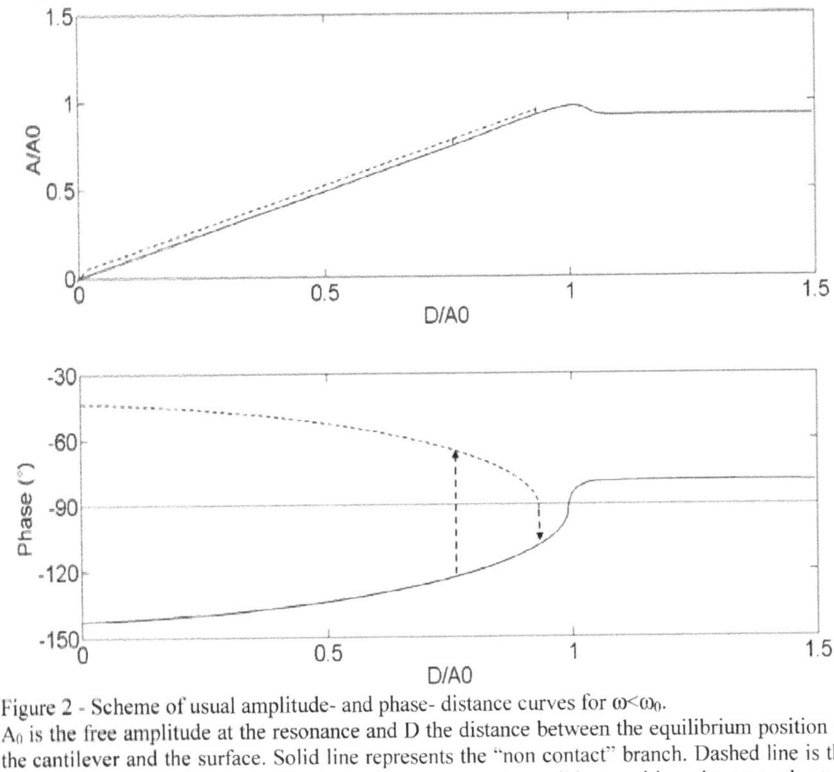

Figure 2 - Scheme of usual amplitude- and phase- distance curves for $\omega<\omega_0$.
A_0 is the free amplitude at the resonance and D the distance between the equilibrium position of the cantilever and the surface. Solid line represents the "non contact" branch. Dashed line is the "intermittent contact" branch. Vertical arrows represent possible transitions between the two branches.

The interactions leading to the predominance of one or the others of these modes are not yet well-established, but the influence of relative humidity has been shown by several experiments.[15,18] Under normal conditions of temperature and pressure, even at very low relative humidity (below 1%), silica glass and oxidized silicon tips are covered by one or a few layers of water molecules[19,20] (water film thickness is typically around 0.5 nm). An increase of the relative humidity leads to the formation of a thicker liquid film[8,19,20] (towards water film thickness of the order of the nanometer). When intermittent contact occurs, the minimal contact distance between the tip and the interface glass sample/water layer is $d_c = 0.165$ nm[19]. For non-contact mode, in our experiments, the minimal distance between the tip and the surface has been found to be around d_c+0.8 nm (around 1 nm). From these values, one can assume that in both modes, the water layers on tip and on glass will connect during the oscillation leading to formation of capillary neck and thus the existence of attractive capillary forces. The strength of the capillary neck is dependent on the volume of water available[9] and so will be the dissipation induced by the formation and destruction of this neck. Another effect of the growth of water layer is that it makes purely attractive NC mode become more and more predominant[15].

Observations of the Condensate at the Crack Tip and Analysis

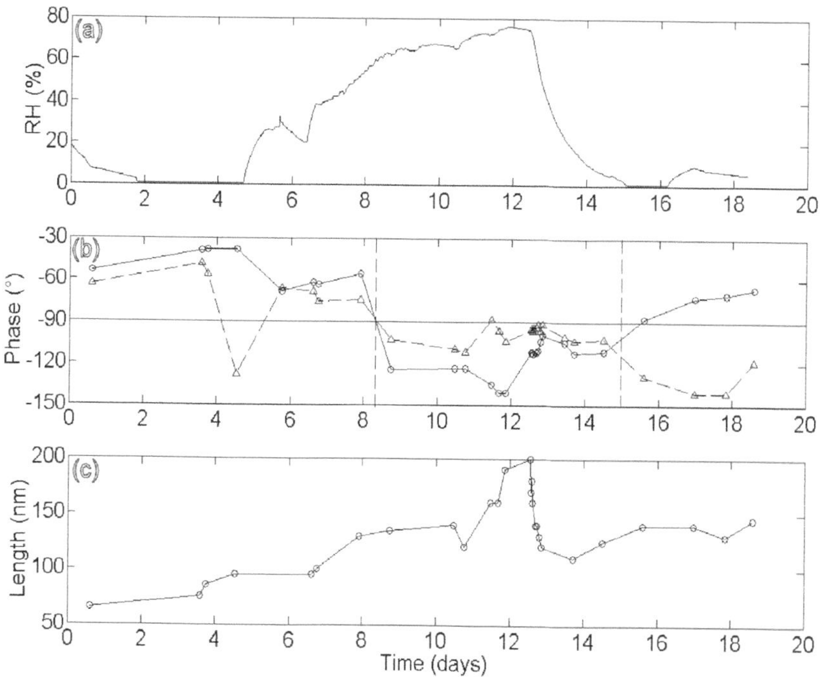

Figure 3 - Curves of relative humidity (a), phase (b) and condensate length (c) vs time. In graph (b), the circles (solid line) correspond to the average phase on the silica glass the triangles (dashed line) correspond to the phase on the condensate.

Tapping mode AFM observations of the crack tip have been performed during an overall cycle of variation of the relative humidity. The relative humidity as a function of time is shown in figure 3a. The velocity of the crack varied as a function of humidity and applied stress between 0.05 and 1 $nm.s^{-1}$.

Figure 4 represents two phase images of the neighborhood of the crack tip at two levels of relative humidity of respectively 1% and 70%. On both images, the crack is vertical, propagating from top to bottom of the picture. The crack tip is characterized by a distinct area showing an important contrast of phase. On the left image, at low humidity, the contrast is negative, whereas on the right image, at high humidity, the contrast is positive. We will show that in both cases this contrast can be associated with an increase of the dissipation, revealing the presence of a liquid condensate at the crack tip.

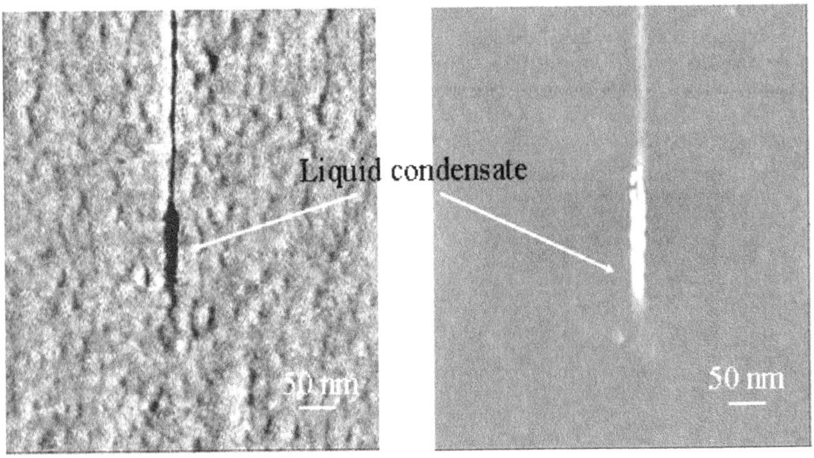

Figure 4 - AFM phase images of the crack tip. (left) RH = 1% (right) RH = 70%. The phase is coded in grey level and increases from black to white.

In figure 3b, the average phase on the silica surface (circles) has been reported as a function of time, in comparison with the maxima phase measured in the area of the crack tip (triangle). During the several weeks of the experiments, the amplitude set point, frequency and amplitude stimulation of the oscillation have varied, but in a range small enough for the variation of the phase to be analyzed qualitatively. The curve presents three areas:

- From day 0 to day 8, both the average phase on the silica surface and the maxima phase at the crack tip are higher than -90° (one point excepted for which the behavior is similar to the points in third area). The AFM works in ICmode. The phase in the area around the crack tip is closer to -90°: this area is more dissipative than the rest of the silicate surface.
- From day 8 to day 15, both the average phase on the silica surface and the maxima phase at the crack tip are lower than -90°. The AFM works in NC mode. The phase in the area around the crack tip is still closer to -90°: this area is still more dissipative than the rest of the silicate surface.

- From day 15 to day 18, the AFM remains in NC mode in the area around the crack tip but works in IC mode on the rest of the silicate surface.

From the explanation given in section 2.1, the variation of phase observed during the cycle of relative humidity variation can be well explained by condensation of water at the crack tip. For low relative humidity, at the beginning of the observations, the amount of water on the surface of the glass is quite low, for the set point amplitude which has been chosen, the AFM operates in IC mode. At the crack tip, the existence of a liquid condensate that has not been dried out by the preliminary drying of the set-up, entails an increase of the capillary dissipative interaction in this area. With the increase of relative humidity, the thickness of water layer on all the surfaces increase. When this thickness is large enough, the AFM switches to NC mode. The area around the crack tip still holds a larger tank of water for capillary forces and we can observe that the dissipation is still larger in this area. The switch between the two modes of operation of the AFM explains the inverted contrast noticed in figure 4. When the atmosphere is dried again, we can see that, after a certain delay, the AFM starts to operate in IC mode again on the silicate glass surface. Yet, it still works in NC mode on the condensate. This leads us to think that there is a delay between drying of the condensate and of the surface.

The persistence of the condensate is confirmed by figure 3c, in which the length of the condensate along the longitudinal direction of the crack has been plotted as a function of time. The length of the condensate as deduced from raw AFM data has been corrected, taking into account propagation speed of the crack and drift of the AFM. Due to the difference of contrast between one mode and the other, the length is difficult to evaluate precisely and the error on the measurement have been evaluated to be of 20 nm. During the experiments, we can yet see that the condensate grows regularly with the increase of the relative humidity from 70 nm to 200 nm. When decreasing the relative humidity, the length begins to decrease but then it stops at a size of around 140 nm even when the RH is below 1%. The size of the condensate has not been observed to start reducing more during the experiment presented in this paper. In the preliminary experiment performed on the same experimental device by Wondraczek et al.[9] what seemed the beginning of the dry out of the condensate has been observed after 5 days in dry atmosphere.

Condensation of water is expected at equilibrium between surfaces as close as the lips of the cracks near the crack tip. The curvature of a condensate meniscus r_K is related to the relative humidity radius by the Kelvin equation[21]:

$$r_k = \frac{\gamma . V}{RTlog(P/P0)} \tag{2}$$

with γ, the surface energy of water; V, the molar volume of liquid water; T, the temperature and P/P_0, the relative humidity.

At ambient temperature, the Kelvin radius will be of the order of the nanometer. Considering the condensate to reach equilibrium (which is relevant since the crack velocity is very low), it can be expected to grow with relative humidity until a distance from the crack tip where the separation between the lips of the crack is around twice the Kelvin radius. The crack in the fragile materials is very sharp: the average angle locally observed between the lips of the crack can be estimated to be 2.5° from measurements of the width of the crack as a function of the distance from the crack tip. With such a geometry, condensate length should increase until around 50 nanometers. Taking into account the lateral overestimation of the length of the condensate due to the bluntness of the AFM tip, this order of magnitude is respected but the experimental condensate is larger than expected. The difference between theoretical length of the condensate and the size of the phase contrast area requires further investigation on the adsorption and capillary behavior of water at the crack tip. It is also useful to remark that the roughness of the crack new surfaces will however play an important role in the size and in the dynamic of evolution of the condensate. Bocquet et al have shown long time dependencies and meta-stabilities of capillary condensate in wetted granulars due to confinement of the system.[22] In a similar way, close-by rough fracture surfaces could certainly account for the difficulty to dry out the condensate.

CONCLUSION

All the results obtained by in situ phase imaging tend to confirm water condensation inside the crack tip, leading to the formation of a confined liquid condensate of which extent evolves with the ambient humidity around the crack. Such a condensation is classically suspected for low velocity crack propagation but had never been experimentally clearly evidenced. The experimental device developed at the LCVN has demonstrated its ability to follow the evolution of this liquid condensate under variable environmental and loading conditions.

This is particularly promising since the existence of a condensate would have a crucial impact on stress corrosion mechanisms. On the purely mechanical point of view, liquid condensation implies the existence of a significant depression inside the condensate. The resulting forces on the crack lips across the whole width of the sample might lead to a non negligible effect of crack closure, directly dependent on the size of the condensate. More importantly, the presence of a confined condensate changes totally the chemistry at the crack tip. Velocity measurements as a function of the applied load have shown different behavior in humid air and in water.[2,5] In low humidity, the presence of a plateau (region where the velocity does not increase with the applied load any more, usually called region II) is related to corrosive agent (water) transport limitation. Condensation of liquid water at the crack tip in the area where gas molecule mobility is usually limited by the narrowness of the crack might explain the gradual disappearance of the plateau when humidity increases.[5] Different effects are expected according to the nature of the cracked glass, since the ion concentrations inside the confined condensate will clearly evolve through ion transfer and lead for instance to pH and surface energy changes or time-dependent behaviors. It would thus be interesting to study phosphate glasses, where anomalous crack velocity measurements have been reported.[6,7] Since the crack tip is the site of enhanced diffusion processes, as shown in soda-silicate glasses by Marliere et al.,[22] we can also wonder what will be the role of the condensate in these processes.

Further experiments will be devoted to understand more precisely the kinetic of the condensate formation under various conditions of humidity and crack velocities. It is believed that experimental data on the liquid condensation at the crack tip will be of important help to further analyze the mechanisms of stress corrosion of oxide materials.

ACKNOWLEDGEMENT

The authors thank Dr Christian Fretigny, Dr Lyderic Bocquet and Dr Michel Ramonda for fruitful discussions.

REFERENCES

[1]R. J. Charles, W. B. Hillig, in *Symposium on Mechanical Strength on Glass and Ways of Improving it.* Union Scientifique Continentale du verre, Charleroi, Belgium p. 511 (1962).

[2]S. M. Wiederhorn, "Influence of Water Vapor on Crack Propagation in Soda-Lime Glass" *J. Am. Ceram. Soc.* 50, 407-414 (1967).

[3]S. M. Wiederhorn, L. H. Bolz, "Stress Corrosion and Static Fatigue of Glass" *J. Am. Ceram. Soc.* 53, 513 518 (1970).

[4]K. M. Davis, M. Tomozawa, "Water Diffusion into Silica Glass: Structural Changes in Silica Glass and their Effect on Water Solubility and Diffusivity" *J. Non-Cryst. Solids* 185, 203-220 (1995).

[5]B. Lawn "Fracture of Brittle Solids",2nd edition, Cambridge University Press, Cambridge (1993), chapters 5 and 6.

[6]T.I. Suratwala, R.A. Steele, G.D. Wilke, J.H. Campbell, K. Takeuchi, "Effects of OH Content, Water Vapor Pressure, and Temperature on Sub-Critical Crack Growth in Phosphate Glass" *J. Non-Cryst Solids* 263&264, 213-227 (2000).

[7]Y-K Lee, M. Tomozawa "Effect of Water Content in Phosphate Glasses on Slow Crack Growth Rate" *J. Non-Cryst Solids* 248, 203-210 (1999).

[8]J. N. Israechvili, "Intermolecular and Surface Forces, 2nd edition, Academic Press, New York,1992

[9]L. Wondraczek, A. Dittmar, C. Oelagrdt, F. Celarie, M. Ciccotti, C. Marliere, "Real-Time Observation of a Non-Equilibrium Liquid Condensate Confined at Tensile Crack Tips in Oxide Glasses" *Journal Of The American Ceramic Society*, 89, 746-749 (2006).

[10]F. Célarié, S. Prades, D. Bonamy, A. Dickelé, E. Bouchaud, C. Guillot, C. Marlière, "Surface Fracture of Glassy Materials as Detected by Real-Time Atomic Force Microscopy (AFM) Experiments," *Appl. Surf. Sci.* 212-213, 92-96 (2003).

[11]M. Y. He, M. R. Turner, A. G. Evans, "Analysis of the Double Cleavage Drilled Compression Specimen for Interface Fracture Energy Measurements Over a Wide Range of Mode Mixities," *Acta Mater.* 43, 3453-3458 (1995).

[12]S. N. Magonov, V. Elings, M.-H. Whangbo, "Phase Imaging and Stiffness in Tapping-Mode Atomic Force Microscopy," *Surf. Sci.* **375**, L385-L391 (1997).

[13]L. Nony, R. Boisgard, J.-P. Aimé, "Nonlinear Dynamical Properties of an Oscillating Tip-Cantilever System", *J. Chem. Phys.* 111,4, 1615-1627 (1999).

[14]R. Garcia, R. Perez, "Dynamic Atomic Force Microscopy Methods", *Surf. Sci. Reports*, 47, 197-301 (2002).

[15]L. Zitzler, S. Herminghaus, F. Mugele, "Capillary Forces in Tapping Mode Atomic Force Microscopy", *Phys. Rev. B*, 66, 155436 (2002).

[16]M. Ciccotti, M. George et al, to be published.

[17]J. P. Cleveland, B. Anczykowski, A. E. Schmid, V. B. Elings, "Energy Dissipation in Tapping-Mode Atomic Force Microscopy", *App. Phys. Letters*, 72, 20, 2613-2615 (1998).

[18]L. Nony, T. Cohen-Bouhacina, J.-P. Aimé, "Dissipation Induced by Attractive Interaction in Dynamic Force Microscopy: Contribution of Adsorbed Water Layers," *Surf. Sci.* 499, 152-160 (2002).

[19]T. Bouhacina, B. Desbat, J.-P. Aimé, "FTIR Spectroscopy and Nanotribological Comparative Studies: Influence of the Adsorbed Water Layers on the Tribological Behaviour," *Tribol. Lett.* 9, 111-117 (2000).

[20]D. Beaglehole, H.K. Christenson, "Vapor Adsorption on Mica and Silicon: Entropy Effects, Layering, and Surface Forces" *J. Phys. Chem.* 96, 3395-3403 (1992).

[21]A. W. Adamson, Physical Chemistry of Surfaces, 3rd edition, Wiley, New York and London (1976).

[22]F. Restagno, L. Bocquet, T. Biben, "Metastability and Nucleation in Capillary Condensation" *Phys. Rev. Lett.* 84, 2433-2436 (2000).

EFFECT OF STRESS GRADIENT AT THE VICINITY OF A CRACK TIP ON IONIC DIFFUSION IN SILICATE GLASSES: AN AFM STUDY

F. Célarié
Institute of Non-Metallic Materials
Clausthal University of Technology
Clausthal-Zellerfeld, Germany

M. Ciccotti, M. George
Laboratoire des Colloïdes
Verres et Nanomatériaux (L.C.V.N.)
University Montpellier 2, CNRS, France

C. Marlière
Géosciences Montpellier (G.M.)
University Montpellier 2, CNRS
34095 Montpellier Cedex 05, France

ABSTRACT

The slow advance of a crack in soda-silicate glasses was studied at nanometer length scales by in-situ atomic force microscopy (AFM) in a well-controlled atmosphere (N_2 and H_2O). An enhanced diffusion of sodium ions in the stress-gradient field at the sub-micrometric vicinity of the crack tip was revealed through several effects: growth of nodules in height images, changes in the AFM tip–sample energy dissipation as detected in phase images. Ex-situ chemical micro-analyses completed the AFM measurements. The nodules patterns revealed a dewetting phenomenon evidenced by "breath figures", i.e. analog to the fogging that occurs when a vapour condenses onto a 'cold' surface [D. Beysens et al., Phys. Rev. Let. 57, 1433 (1986)].

These experimental results were explained by a two-step process: i) a fast migration of sodium ions towards the fracture surfaces as proposed by Langford et al. [J. Mat. Res. 6, 1358 (1991)], ii) a slow backwards diffusion of the cations as evidenced in these AFM experiments (typical time: few minutes). Measurements of the diffusion coefficient of that relaxation process were done at room temperature. Our results strengthen the theoretical concept of a near-surface structural relaxation due to the stress-gradient at the vicinity of the crack tip. Raman and SIMS studies revealed that nodules – for samples studied after exposition to common air - are preferentially covered by an organic overlayer of a carboxylate salt with a long aliphatic chain. The catalytic role of sodium ions in that chemical process is suspected.

PACS INDEXING CODES

62.20.Mk	Fatigue, brittleness, fracture, and cracks
66.30.-h	Diffusion in solids
68.03.Cd	Surface tension and related phenomena
68.37.Ps	Atomic force microscopy (AFM)

INTRODUCTION

The study of mechanisms of rupture in so-called brittle materials is still of great importance both for knowledge of basic mechanisms of bond(s) breaking and for practical reasons such as understanding and possibly correcting the degradation of nuclear waste containers or optical glass fibers or predicting rupture in the upper Earth's crust as well. However, and despite their recent tremendous research work, discussions about the real processes and their characteristic length scales are still intense. In order to try and clarify the debate we aim to study the sub-micrometric vicinity of the tip of a crack running in silicate glass with mobile ions (as sodium cations) and check for possible chemical mass transfer at nanometer range. Soda-lime glasses, despite of their chemical complexity, were good candidates as i) samples with a high homogeneity of composition (a mandatory condition for the double cleavage drilled compression –DCDC- samples we used) can be easily obtained and ii) such local ionic migration was already suspected in complex conditions.[1,2] This paper is based on a study on in-situ AFM and ex-situ complementary experiments in the low crack speed regime. We evidence a slow diffusion process of sodium thanks to the observation of the growth of nodules in the vicinity of the crack tip. The source of the diffusion species are shown to be located at the nanometer vicinity of the crack surfaces. The diffusion zone is evidenced by increased energy dissipation by the AFM tip – sample interaction compared to corresponding "bulk" glass. Measurements of diffusion coefficient were done at room temperature by the study of the advance of the diffusion front by AFM. The latter was visible in height images thanks to a dewetting-like process and in phase images thanks to an increased energy dissipation in the interaction between AFM tip and substrate. The process of nucleation and growth of these nodules was locally investigated and explained by a phenomenon of local condensation of an aqueous phase, the viscosity of which is increased due to an ionic (Na^+) enrichment. It will be pointed out that these observations of the sodium migration in the stress gradient field were possible thanks to an adequation between characteristic times related to both the diffusion process and crack propagation.

EXPERIMENTAL SETUP

Commercial soda-lime glasses[1] were used for this work. The composition of the glass by molar fraction (%) is as follows: SiO_2 : 70.9; Na_2O : 13.2 ; CaO : 10; MgO : 5.4 ; Al_2O_3 : 0.4; K_2O : 0.1. The samples were cut from the same piece of that soda-lime silicate glass. A thermal treatment (530°C) was done before fracture experiments in order to remove residual stresses. Fracture is performed on DCDC (double cleavage drilled compression),[3] parallepipedic (4mmx4mmx40mm) samples designed with a cylindrical hole (radius: 0.5 mm) drilled at the center of two parallel 4mmx40mm surfaces and perpendicularly to them. The hole axis defines the Z-direction. We worked with both the "genuine" surfaces (in contact with air or with tin bath during cooling of the glass melt) without any further polishing or with bulk samples in which the 4×40 mm^2 surfaces were polished. In all cases the measured RMS roughness was less than 0.25 nm for a 10×10 μm^2 scan-size. No change in the presented results was detected versus the nature of the studied surface. These fracture experiments are performed at a constant temperature of 21.0 ± 1°C in a leak-proof chamber under an atmosphere composed of pure nitrogen and water vapour (preceded by a careful out-gassing). The macroscopic relative humidity (RH) can be controlled from 5% to 75% with an accuracy of about 1%. AFM experiments were done in an intermittent contact (tapping) mode. After the AFM experiments, samples were removed from

[1] From Saint-Gobain company.

the glove-box and studied by Secondary Ions Mass Spectroscopy (SIMS) analyses or Raman spectroscopy experiments.

RESULTS

AFM Experiments: Evidence of a Diffusion Process

It must be noted that the phenomena described below were not detectable when relative humidity levels were less than 30%. A typical topographical (height) AFM frame in the neighborhood of the crack tip, for K_I = 0.41 $MPa.m^{1/2}$, v = 1.5 $nm.s^{-1}$ and RH = 45 %, is presented in figure 1.

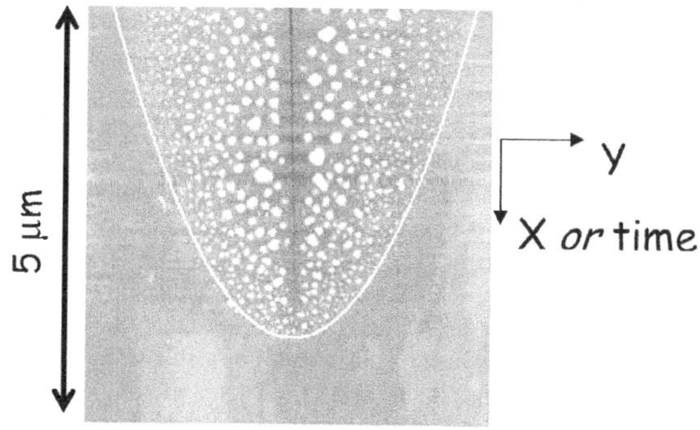

Figure 1: AFM height image in tapping mode for a soda-lime glass at 45% RH. The white line corresponds to the best quadratic fit for delineating the contour between zones with and without nodules. Crack speed: v_c = 1.5 $nm.s^{-1}$. The height signal is coded in grey levels corresponding to a vertical range of 10nm (between black and white levels).

It clearly reveals many nodules typically 200 nm in diameter for the largest (respectively 10nm for the smallest) and 20 nm in height for the highest (resp. 2 nm). It must be emphasized that, in the same conditions of humidity and mean crack speed, this phenomenon was not observed with silica samples[4] and neither with lithium alumino-silicate glasses.[1,5] The X axis (along direction propagation of crack) can be rescaled as a time axis: With chosen values of AFM scan rates the crack speed can indeed be considered as constant during a set of AFM data. The contour delimiting the zones with or without nodules (what will be called "diffusion front") has a parabolic profile characterized by equation $(x - x_0) = B.(y - y_0)^2 = v_c.(t - t_0)$ where v_c is the mean crack speed. This quadratic relation can be interpreted as due to a diffusion process in which the source of diffusing species is located on crack surfaces ($y = 0$ represents the crack lips in the 2D representation provided by AFM techniques). Classical theories of diffusion[4] predict indeed that the local concentration of the diffusion species is a function of variable u with

$$u = (y - y_0)/(2.\sqrt{D_{eff}.(t - t_0)}) \qquad (1)$$

where D_{eff} is an effective diffusion coefficient.

---

[1] SiO_2 mol 70% ; Al_2O_3 mol 20%; Li_2O mol 6%; TiO_2 mol 4%

It must be emphasized that the shape of the parabola of delineation does change with crack speed. However the deduced values of D_{eff} are constant, within experimental error bars, for a given humidity rate whatever the crack speed. More details will be given in.[5]

Careful experiments as detailed in [7] allowed us to estimate the value of the diffusion coefficient at different humidity rates. D_{eff} is equal to 1.0 10^3 nm^2/s for RH=45% and slightly increases with humidity: D_{eff} = 18 10^3 nm^2/s for RH=58%. The comparison with values from literature is delicate as many studies were done by the leaching action of *liquid* water on glass substrates. For instance Lanford et al.[6] measured diffusion coefficient of Na^+ and H_3O^+/H^+ for the hydration of a soda-lime glass the composition of which is similar to that of the glass studied in the present paper. Then the sodium content was profiled in depth by a resonant nuclear reaction method the spatial resolution of which is at least a few micrometers. Typical values for diffusion coefficients were: $D_{Na}{}^+ = 10$ nm^2/s at 90°C. Gehrke's work gave:[7] $D_{Na}{}^+ = 0{,}29$ nm^2/s for soda-lime glass rods of similar composition corroded in distilled water at 23°C. In these studies there was no feedback on pH but its value was supposed constant. More recently, leaching behaviour of sodium from small particles of soda–lime glass in acid solution was studied.[8] Two types of fine glass spheres were investigated: "small", respectively "large", ones with an averaged diameter of 50 (resp. 20) microns. The sodium diffusion coefficients at a temperature of 25°C are $D_{Na}{}^{large} = 0.06$ nm^2/s and $D_{Na}{}^{small} = 0.6$ nm^2/s. These values are very near of those obtained in former studies as in reference 9. It must be noted, too, that the water content of silicate glasses has an influence on sodium diffusion coefficient. For instance, Tomozawa et al.[9] showed that, for $Na_2O\text{-}3SiO_2$ glasses, with increasing water content sodium diffusion coefficient decreases initially to a minimum at 3-4wt % water and then increases. Nevertheless it should be concluded from literature that sodium diffusion inside silicate glasses with no or low content of sodium and without any applied mechanical stress is far below (around three orders of magnitude) values measured in this paper. Please note that these former studies were done by physical or chemical investigations at length scales much greater than those studied in the present paper.

AFM phase images reveal too the presence of similar diffusion front. An example of the profile of the phase signal corresponding to height signal (figure 1) along a line parallel to Y axis is shown in figure 2.

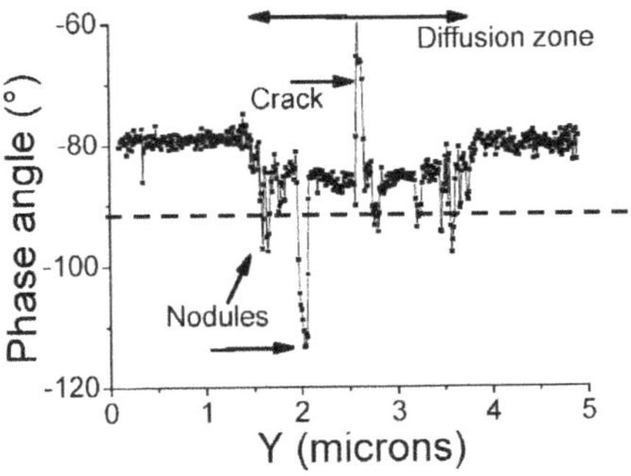

Figure 2: Profiles along the Y direction from the phase image corresponding to height data in fig.1. The dashed line corresponds to a phase angle, ϕ, equal to -90°: when $\phi > -90°$ AFM works in intermittent contact mode; in the opposite case in non-contact mode.

As is well known,[10] the phase signal in AFM tapping mode is a function of changes in the tip–sample energy dissipation. In figure 2 we observe that the phase signal is lower in the diffusion zone (inside the parabolic envelop line as mentioned above) than in the not-disturbed zone and corresponds, when the phase angle is greater than -90°), to an increase in dissipation, the physical origin of which will be explained later. Nodules are related to zones with a phase angle less than -90°: the interaction between the AFM tip and the nodules is therefore purely attractive due to capillary interactions. At the position of the crack line: the value of the phase signal is the one expected for a cantilever "flying" over the crack filled with the standard surrounding gaseous atmosphere.

In order to get more accurate information about the origin and nature of the diffusion species we studied the growth of the nodules by AFM at greater magnification. For that purpose we used a much greater spatial resolution that allows us to work with relatively small times of acquisition (84s between two consecutive images). We observed that in a first step the nucleation of nodules of very small size occurs all over the AFM scanned surface inside the diffusion zone. Then the nodules grow in height and lateral radius before coalescing.. Two nodules (see arrows in figure 3) are first growing independently before coalescing to form one unique bigger nodule the height of which further increases. This behavior can be interpreted by the phenomenon of nucleation and coalescence of droplets (nodules) as in the case of the so-called "breath figures."[11,12] To better understand the origin of these condensation-like patterns we measured by AFM the evolution of the contact angle of nodules on the glassy surface versus time (figure 4).The contact angle is increasing from ~ zero (complete wetting) to 18 ± 3°. This study proves then that nodules grow by a process similar to that involved in breath figures when liquid droplets are condensing on a cold surface. As water plays a major role in local chemistry at the crack tip, the condensation phenomenon evidenced by the observation of nucleation and growth of nodules is likely due to two conjugate effects: i) a local dewetting of the native thin layer continuous water (as evidenced by variation of contact angle) and ii) a condensation from the gaseous atmosphere. This aqueous liquid phase in nodules is likely enriched with sodium ions and consequently has a greater viscosity.

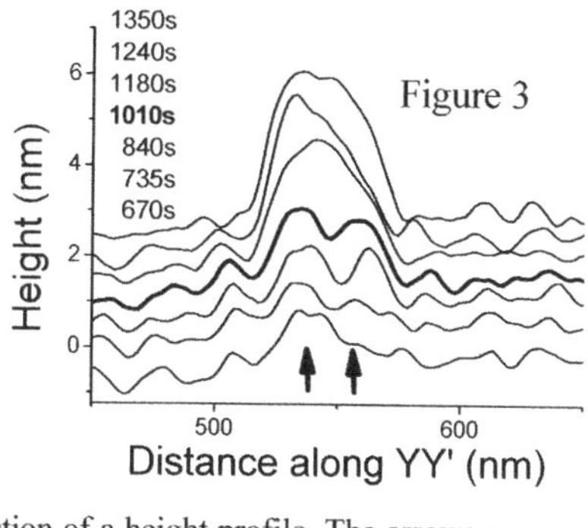

Figure 3: Time evolution of a height profile. The arrows are a guide for the eye to follow on the process of growth and coalescence (bold line) of two nodules. RH=48%, $v_c = 0.21$nm/s.

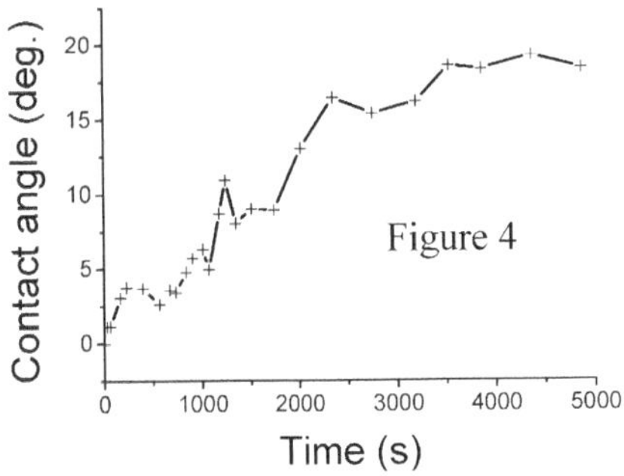

Figure 4: Variation of contact angle of nodules versus time. RH=48%, $v_c = 0.21$nm/s.

Ex-Situ Chemical Micro-Analysis: SIMS Studies

In order to get deeper insight in the chemical nature of nodules we worked as presented now. Mean crack speed was chosen low enough (0.1 nm/s) and humidity high enough (48%) in order to allow an increase of the mean size of nodules with time. These experiments were done at room temperature and in the pure (N_2 +H_2O) atmosphere. Nodules of microscopic size laterally and in thickness were obtained. The sample was then withdrawn from the glove box, removed from the mechanical stage and stayed in standard atmosphere during a maximum of one hour before being put in high vacuum conditions. The (X,Y) surface of the sample was then studied in the vicinity of the crack tip. Firstly SIMS revealed much greater carbon content at the near-surface of nodules as compared with the surface of glass outside of the diffusion zone. Then few tens of nanometers in thickness were removed by ionic sputtering in order to probe the inside of the nodules. We then observed a much stronger signal of sodium at nodules in contrast to the bulk glass. On the opposite, the Si signal was weaker while Ca, Mg and Al ones roughly constant within the accuracy of the experiments.

Ex-Situ Chemical Micro-Analysis: RAMAN Studies

These measurements were done in standard air. The Raman spectra for both bulk glass and nodules for wave numbers between 250cm^{-1} and 3500cm^{-1} are shown in figure 5. They both reveal typical bumps related to the amorphous silicate glass as the diameter of the laser spot is larger than the typical size of nodules and bulk glass is thus slightly lightened. However, narrow peaks are present on the spectrum of nodules. We subtracted the signal related to the glassy part of the spectrum (as deduced from the lower graph in figure 5) from the raw data of nodules). The resulting spectrum is shown in figure 6 (upper curve). In this lower frequency region (σ<1400 cm^{-1}) we observe a strong analogy between the measured spectrum and that of carboxylic salts, such as Na-stearate[13] (or more generally a carboxylate with variable values of aliphatic length, *n*, in (-CH_2-)n) - see figure 6 lower graph. The corresponding peaks in spectrum of nodules are marked by full squares in figure 6. As expected, bands at around 1400 cm^{-1} (star in figure 6), typical of CH_2 scissors and CH_3 asymmetrical deformations, are detected. It must be noted too that the peak at 1610cm^{-1} (triangle in figure 6) can be explained by the absorption of two equivalent C=O bonds in carboxylic salt. We also observed bands at wave numbers around 3000cm^{-1} (see insert figure 5) only present when nodules are lightened by the laser spot.

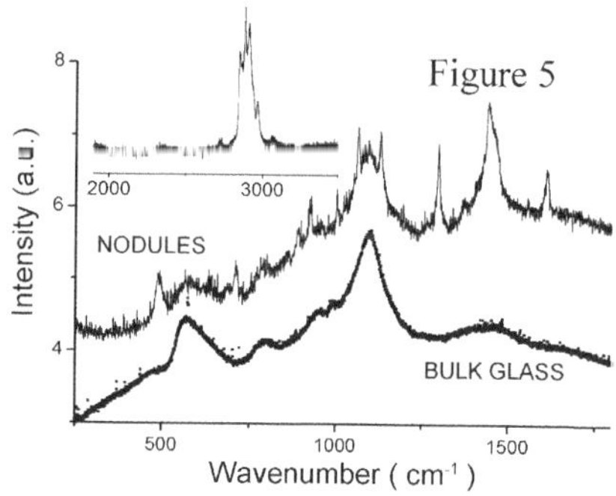

Figure 5: Raman spectra for bulk glass and nodules. The insert corresponds to high wave number spectrum for nodules (no signal was detected for bulk glass).

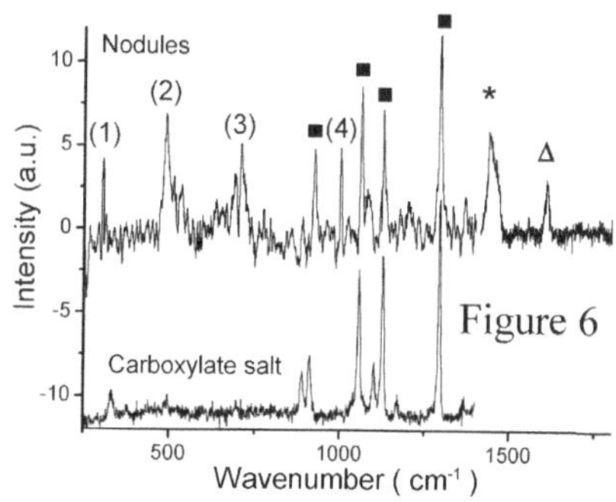

Figure 6: Raman spectrum for nodules (bulk glass contribution was removed). See text for symbols. Lower graph: spectrum for carboxylate salt[15].

These bands are typical of aliphatic C-H stretching vibrations as in $-CH_3$ and $-CH_2$ groups.[14] Thus Raman data reveal the presence of a carboxylic salt only on nodules. From SIMS studies it can be concluded that this compound is located at the near-surface region of the nodules and not in the bulk. The remaining peaks in the Raman spectrum (numbered in figure 6) might be related to the presence of a crystallized inorganic compound in the nodules. A screening of the Raman data bases related to inorganic compounds showed us that the best similarity was obtained with the spectrum of oligoclase (sodium calcium aluminum silicate).[15] It must be emphasized that the usual data bases are pretty poor in mineral compounds based on Na, Si elements. Presence of crystals of cristobalite, devitrite, diopside or tridymite, as usually evidenced in high temperatures experiments on crystallization of soda-lime like glasses[16] is not completely ruled out but fits are of lower quality.

DISCUSSION

The nucleation, growth and coalescence of nodules and the propagation of the diffusion front delimiting the zones with or without nodules were observed at nanometer scale in the vicinity of the crack tip by AFM measurements in tapping mode. Experiments were done in a carefully controlled surrounding atmosphere. Following features were observed:

- These experiments revealed that the diffusion front is related to the process of nucleation of the nodules. These data can be well interpreted by a diffusion process. The source of diffusing species is locating on (in the vicinity) crack lips (in our 2D AFM study; crack surfaces in the case of general 3D conditions). Values of diffusion coefficient were measured at room temperature for different rates of relative humidity.

- By further chemical microanalyses we revealed that nodules are characterized by a higher concentration of sodium element than in bulk glass. We then deduced that Na ions were migrating during the diffusion process.

- The measured values of sodium diffusion coefficient in the vicinity of the crack tip are around three orders of magnitude larger than those deduced from former studies on similar bulk composition of glasses. As stress field is drastically enhanced in the vicinity of crack tip, the role of stress gradient in the origin of this enhanced diffusion process was suspected.

- AFM phase images showed that important changes in the tip–sample energy dissipation occurred. Water was proved to play an important role in that phenomenon. Complementary AFM studies revealed a nanometric process of growth and coalescence of the nodules connected to an increase of the contact angle.

- Raman study showed that an organic compound (constituted by carboxylic salt with a long aliphatic chain) is preferentially formed at the surface of nodule. As Raman studies were done in a standard atmosphere these compounds are likely synthesized or adsorbed on nodules when sample is removed from the glove-box. The presence of an inorganic crystallized compound the growth of which occurs at an undetermined step is suspected too.

The presence of local variation (in the nanometer range) of concentration of sodium ions near by the crack tip is the cause of local variations of surface tension and thus can explain the appearance of local aqueous droplets as it will be shown now. We indeed proved that the advance of the diffusion front was related to a local and slight increase of wetting angle. On the opposite, far away from the crack tip, we did not observe any variation of the roughness of the glass surface: it is covered by a continuous thin layer of adsorbed water (the wetting angle, θ, is equal to zero) the thickness of which is in the nanometer range: from 0.5nm to 1.5nm depending on humidity rate.[17,18] As it is well known the wetting angle is related to interfacial tensions by the Young–Laplace equation $\cos\theta = (\gamma_{SG} - \gamma_{SL}) / \gamma_{GL}$, where γ_{XY} is the interfacial tension between phases X and Y. Subscript L is for liquid, G for gas and S for solid phase. At first it should be mentioned that because of the exchange in sodium and hydronium ions between bulk glass and water overlayer the concentration of H_3O^+ in the liquid phase decreases and then the solution becomes more basic. It is directly connected to the fact that sodium-containing glasses are regarded to be more basic than pure silica glasses.[19] It has been known for a long time that interfacial tension between liquid water and air, γ_{GL}, depends on the nature of the inorganic aqueous salt solution: γ_{GL} increases when the pH of the liquid medium increases.[20] More recently molecular dynamics (MD) calculation[21] showed with more detail that alkali cations are repelled from the liquid/gas interface in case of basic pH and, consequently, cause an increase of interfacial tension, γ_{GL}. As a consequence the concentration in alkali cations near the solid/liquid interface should likely increase or remain constant. Thus γ_{SL} should remain constant or slightly decrease. Experiments[22] and MD calculations[23] showed that there is a small predominance of sodium ions at the interface of Na_2O-$3SiO_2$ glass and air, causing a little increase or no change in interface tension, γ_{SG}. Thus from the Young Laplace equation it can be predicted that the contact angle should increase at the places where chemical heterogeneities (higher concentration of sodium cations) are present. The stability of the continuous wetting film as observed far away from the diffusion zone is due to long-range van der Waals repulsive force.[21] The force can be modified by external disturbances such as the incoming of sodium cations causing local hydrophobic spots that act as nucleation centers for dewetting. As already shown[24] a circular droplets pattern – especially in the case of the thinnest (aqueous) film – may develop by nucleation on more hydrophobic centers, ripen and merge. The nodule pattern is thus predicted to propagate with the diffusion front of sodium ions as observed in our experiments.

A diffusion process is very clearly revealed too by phase images in the AFM intermittent-contact mode. In figure 2 we observed that the phase signal is lower in the diffusion zone (inside the parabolic envelop line as mentioned above) than in the not-disturbed zone It was shown[12] that if the amplitude of the cantilever is held constant, the sine of the phase angle, ϕ, of the driven vibration is then related to changes in the tip–sample energy dissipation (the origin of phase is chosen to be equal to -90° at resonance when the cantilever is far away from the sample). Calculations of dissipated energy are under progress[25] and will not be detailed in this paper. We observe that ϕ is greater than (-90°) almost everywhere on the substrate (except on nodules). The AFM is thus working in intermittent contact mode (*i.e.* the AFM tip undergoes repulsive forces during its vibrating movement at least partially) and a decrease in ϕ corresponds to an increase of dissipated energy during the tip-sample interaction. On the opposite, the nodules are imaged in a non-contact mode (as ϕ <-90°) meaning that the interaction force is attractive all along the movement of the AFM tip as in the case of pure capillary forces. As mentioned in[26, 27] the principal source of energy dissipation during the tip-sample interaction is mainly the adhesion energy hysteresis. The adhesion forces can be, in the present experiments, from two different origins:[28] capillary forces and acid/basic interactions.[21] In the first case the energy dissipation is related to the formation and rupture of a capillary neck between the tip and sample.[29] It is known that the capillary force is proportional to the liquid-vapor interfacial energy, γ_{GL}.[30] As earlier mentioned, γ_{GL} is expected to increase due to an enrichment of the liquid-like phase in cations, we can thus easily explain the decrease in phase angle when the AFM tip is in intermittent contact in the diffusion zone. On the other hand, the higher acid/basic interactions due to the already mentioned difference in acid-basic properties of both the silica AFM tip and sodium enriched zone on the substrate near the crack tip may also contribute to an increase of the dissipation in the diffusion zone. As mentioned by Fowkes et al.,[21] work of adhesion between two silicate glasses (AFM tip and sample) is equal to the sum of two contributions: a van der Waals contribution and an acid-base contribution related to the ability to hydrogen-bonding. That last one is very likely increased due to the more acid character of the silica surface (AFM tip) and the more basic surface in the Na-enriched glassy region. Because of the AFM regime of non-contact over the nodules it can be concluded that nodules are covered by a thicker or/and more viscous liquid-like film than over the rest of diffusion zone.

These comments reinforced our experimental observation of an enhanced migration of sodium ions in the vicinity of the crack tip. The high value of D_{eff} we measured is compatible with the idea of a local enrichment of glassy material in sodium around the crack tip. Indeed a study by Gehrke et al. [9] revealed the influence of the composition of silicate glass with sodium content on the effective diffusion coefficient of sodium. Soda-lime like (17 mol% in Na_2O) and binary glasses (Na_2O and SiO_2) with values of sodium content (12 mol% in Na_2O) near that of the glass studied in the present paper were characterized by $D_{eff} \pm 0.3$ nm^2/s. However, it was shown that an increase of D_{eff} with sodium content in glasses made from 36% Na_2O and 64% SiO_2 are characterized by $D_{eff} = 1.3\ 10^3$ nm^2/s. That value is of the same order of magnitude as the one deduced from our experiments ($D_{eff} = 1$ to 18×10^3 nm^2/s for RH between 45% and 58%). The measured value of D_{eff} by our local AFM measurements is thus fully compatible with the hypothesis of a zone, located in the surroundings of the crack tip, made of a silicate glass with an enhanced concentration of sodium.

Two plausible scenarios for the migration path may be regarded. The first one corresponds to a surface migration of the diffusing species from the crack surfaces or from medium between them. The second scenario is related to a migration through the volume of material from crack surfaces to the outside of the material: in this case a transition between plane strain and plane stress regions makes the observation of that phenomenon possible in these AFM experiments. The first scenario is very unlikely. It must be noted indeed that for usual values of relative humidity in our experiments (RH~45%) the glass surfaces are covered by an aqueous film the equilibrium thickness of which is in the range of few tenths of nanometers, depending on the electrolyte concentration in the water film.[19,20] Thus it could be possible that the migration of alkali species occurs in an aqueous liquid phase. In these conditions mobility of sodium cations in a bulk aqueous liquid phase is well known : $b_{Na} = 3.2\ 10^{11}$mN^{-1}/s.[31] Thanks to the Stokes-Einstein relation, we can calculate at 25°C the diffusion coefficient of Na^+ ions in water, $D_{Na}^{water} = b_{Na}.k_B.T = 1.33\ 10^9$nm^2/s, six orders of magnitude greater than the value we measured. As a consequence the interpretation of our experiments by a surface migration of sodium in the surface aqueous film can be very likely ruled out. Alternatively, we note that the typical value of D_{eff} we measured by our AFM experiments is very near from those obtained in case of diffusion of sodium through a bulk glass with an enriched content in sodium. We thus think that the migration of alkali ions we evidenced around the crack tip occurs through the bulk structure of glass. This effect is likely detected on the free surface of our sample -via a dewetting process- through the transition between plane strain conditions in the bulk to plane stress ones on the investigated surface.

As the observed enhanced migration is localized in the vicinity of the propagating crack the enhanced value of stress around the crack tip is of major importance for the diffusion process. It is indeed known that mechanical stress has a great influence on diffusion processes. McAfee[32] studied the effect of stress upon the diffusion constant of gases (He, H_2 and other gases) in a borosilicate glass at room temperature. It was shown that an enhanced diffusion of helium (and hydrogen but with lower magnitude) is obtained in glass under high tensile stresses. The diffusion of water in silica glass was studied as a function of applied uniaxial stress:[21, 33, 34] the diffusion coefficient increases when a tensile stress is applied in the case of a temperature lower than 250°C. As the ionic volume for Na^+ is ranged between those of H^+ and H_3O^+ ions the occurrence of a similar diffusion process for Na^+ ions in a soda-lime like glass can be predicted. A migration of sodium ions in the applied stress gradient was indeed deduced from electrical measurements[35] on bent soda-lime silicate glass plates studied at temperatures between 75°C and 120°C. More recently, Na peak emissions were detected[36] in vacuum by quadrupole mass spectrometry and surface ionization techniques when soda lime and sodium trisilicate glasses were dynamically fractured. These experiments revealed a fast diffusion (typical times in the range of milliseconds) of Na^+ when soda-silicate glasses were fractured under a high stress gradient in vacuum. These observations are coherent with theoretical results from thermodynamics applied to chemo-elastic solid[37, 38] which allows the calculation of the diffusion coefficient versus the applied stress gradient. The sodium ions are thus predicted to concentrate in the region of maximum tensile stress, i.e. in the nanometer surroundings ahead of the crack tip. As the main contribution to value of stress is mostly due to the $r^{1/2}$ factor (r is the distance from crack tip to the relevant point) the stress gradient should have values of the same order of magnitude in both regimes of slow and fast fracture. Therefore a fast migration of alkali ions from compressive zone to tensile one (at crack lips and near crack tip) should likely occur in our AFM experiments. Thus that fast migration likely causes the presence of a reservoir of sodium ions in the nanometric vicinity of the crack lips which relaxes - in a second step - in a diffusion process with much greater time constants (few minutes) as evidenced in the AFM experiments detailed in this paper.

The observation of an increasing diffusion coefficient with humidity reinforced the conclusions of Agarwal et al.[39] of a faster relaxation in the presence of water vapour. It is also compatible with recent observations of Tian et al.[40] on silica glass. These authors showed indeed that the uptake of water in the structure of the studied silica glass causes a structural relaxation occurring in the near-surface region leading to an initial rapid decrease in the sodium diffusivity followed by a slow increase with the water content. Our results strengthen the theoretical concept of a near-surface structural relaxation.[41] We believe that the stress-gradient present at the vicinity of the crack tip is of major importance to enable the glass network to undergo a structural relaxation causing long-range structural changes, leading to modifications of bonding angles and distances and consequently enhancing mobility of alkali cations. These structural changes may cause a relaxation of stresses at the crack tip, probably coupled to compressive stress due to replacements of Na^+ by H_3O^+ ions[42] leading to the existence of the fatigue threshold.

Raman spectroscopy and SIMS studies revealed that nodules – for samples studied after exposition to common air - are preferentially covered by an organic overlayer of a carboxylate salt with a long aliphatic chain. The presence of a crystallized inorganic compound inside the nodules was evidenced too. The presence of the organic overlayer could result from the preferential adsorption of volatile organic compounds present in the standard air atmosphere on the nodules the higher content in sodium ions could play the role of nucleation center. Another explanation could come from a local conversion of carbon dioxide, present in the surrounding standard atmosphere,[1] to methane or to a longer aliphatic chain : It was indeed shown[43] by using mass spectroscopy that the CO_2 peak disappeared during fracture of a soda-lime and boro-silicate glasses while CH4 peak increased. Reaction $CO_2 + H_2 \rightarrow 2H_2O + CH_4$ was proposed. We thus suspect that the sodium present in nodules could have a catalytic role in that chemical process. Such a behavior could provide a hypothetic complementary explanation to fatigue limit in glasses: thanks to this chemical reaction a chemical compound with a hydrophobic part could be deposited on the fracture surface of the sample. Water molecules might thus not reach the bulk glass and consequently the stress corrosion process at the crack tip might be stopped. In order to check that hypothesis further experiments are needed. The experimental evidence of crystallized mineral compounds in nodules by Raman spectroscopy could provide an alternative explanation to surface crystallization of oxide glasses[18] or observation of nm-scale structures on air-exposed fractured surfaces of soda-lime silica.[44]

---

[1] The occurrence of that chemical process inside the glove-box due to the presence of a residual content in CO_2 may not be ruled out even if very unlikely. Hypothetic migration of residual traces of CO_2 dissolved in bulk glass towards the crack tip could be an alternate explanation too.

CONCLUSION

AFM experiments were done in a gaseous atmosphere composed of a mixture of pure nitrogen/water gases with soda lime silicate glasses. An enhanced diffusion of sodium ions in the stress-gradient field at the sub-micrometric vicinity of the crack tip was evidenced for relative humidity greater than 30%. The study of height and phase images (in AFM tapping mode) clearly evidenced a slow diffusion process thanks to the observation of the growth of nodules in the vicinity of the crack tip and/or variation in phase signal. The source of the diffusion species was shown to be located at the nanometer vicinity of the crack surfaces. That phenomenon related to sodium and hydronium interdiffusion process was detected by AFM through the observation of several effects: a dewetting-like phenomenon evidenced by "breath figures"-like patterns, changes in the tip–sample energy dissipation as observed in phase imaging mode and growth of nodules in height images for lower crack speeds. Ex-situ experiments confirm that sodium ions are the diffusing species. These experimental results have been explained by a two-step process: i) a fast migration of sodium ions towards the fracture surfaces as proposed by Langford et al.[38] (typical time for a dynamic fracture in vacuum: few tens of milliseconds), ii) a slow backwards diffusion of the cations as evidenced in these AFM experiments (typical time: few minutes). Diffusion coefficients of that relaxation process have been measured at room temperature for different relative humidity rates. The corresponding values were found to be much greater than those reported in the literature with glasses of the same composition for similar conditions but were found to be compatible with those obtained for silicate glasses with a higher sodium content.

Raman spectroscopy and SIMS studies revealed that nodules – for samples studied after exposure to common air - are preferentially covered by an organic overlayer of a carboxylate salt with a long aliphatic chain. Such a result could provide a hypothetic explanation to fatigue limit in glasses as this aliphatic chain is hydrophobic and could contribute to stopping the stress corrosion process at the crack tip. Furthermore these studies reveal that surface crystallization of oxide glasses[18] or observation of nm-scale structures on air-exposed fractured surfaces of soda-lime silica[46] stems very likely from the ionic migration due to a residual stress gradient and the related process of interdiffusion with H^+/H_3O^+ ions.

Our results strengthen the theoretical concept of a near-surface structural relaxation.[43] We think that the stress-gradient present at the vicinity of the crack tip is of major importance to enable the glass network to undergo a structural relaxation causing long-range structural changes, leading to modifications of bonding angles and distances and consequently enhancing mobility of alkali cations. Such sodium migration likely causes a relaxation of stresses in the vicinity of the crack tip and a decrease of the local value of K_I factor: this effect could explain, at least partially, the presence of the fatigue limit.

However, much more work is needed in order to better understand the role of the chemical composition of glasses and their physical structure on these stress-gradient-enhanced migration of alkali cations. Another important point would be to better understand the role of sodium migration towards the fracture surface on local chemical and physical properties such as wetting, chemical reaction in the vicinity of the crack tip with standard air.

REFERENCES

[1]B. Nghiem, thesis, University Paris-VI, 1998.

[2]R. Gy, J. Non-Cryst. Solids 316, 1 (2003).
[3]C. Janssen, in: Proceedings of the 10th International Congress on Glass, 1974, p. 10.23.
[4]J. Crank 1975 The Mathematics of Diffusion. Clarendon Press, Oxford/London.
[5]F. Célarié, M. Ciccotti and C. Marlière, to be published (2006).
[6]W. Lanford et al. J. Non-Cryst. Sol. 33, 249 (1979).
[7]E. Gehrke, C. Ullner and M. Hähnert, J. Mater. Sci. 26 5445 (1991).
[8]T. Akai et al., J. Am. Ceram. Soc., 88, 2962 (2005).
[9]M. Tomozawa et al., J. Am. Ceram. Soc. 65, 182 (1982).
[10]J. P. Cleveland et al, Appl. Phys. Let., 72 (1998) 2613.
[11]D. Beysens and C. Knobler, Phys. Rev. Let. 57, 1433 (1986).
[12]J. Viovy, D. Beysens and C. Knobler, Phys. Rev. A 37, 4965 (1988).
[13]Raman Spectra Database of Minerals and Inorganic Materials - the National Institute of Advanced Industrial Science and Technology, Japan.
[14]Clthup N., Daly L., Wiberley S., Introduction to Infrared and Raman Spectroscopy 3rd edition, 1990, Academic Press.
[15]Mineral Spectroscopy Server - California Institute of Technology - Pasadena, California, USA, http://minerals.caltech.edu/FILES/raman.
[16]R. Müller, E.D. Zanotto and V.M. Fokin, J. Non-Cryst. Sol. 274, 208 (2000).
[17]T. Bouhacina, B. Desbat and J.P. Aimé, Tribology Letters 9, 111 (2000).
[18]D. Beaglehole, H.K. Christenson, J. Phys. Chem. 96, 3395 (1992).
[19]F. Fowkes, D. Dwight, D. Cole and T. Huang, J. Non-Cryst. Solids 120, 47 (1990).
[20]J. Randles,. Phys. Chem. Liq., 7, 107 (1977). E. Washburn, Ed. International Critical Tables; McGraw-Hill: New York, 1928; Vol. IV.
[21]M. Mucha et al. , J. Phys. Chem. B, 109, 7617 (2005)
[22]C. Pantano, J. Kelso and M. Suscavage in Advances in Materials Characterization - Vol.15, D. R. Rossington, R. A. Condrate, and R. L. Snyder, Eds., Plenum Press, New York, (1983).
[23]S. Garofalini and S. Levine, J. Am. Ceram. Soc. 68, 376 (1985).
[24]A. Sharma and R. Khanna, J. Chem. Phys., 110, 4929 (1999).
[25]M. Ciccotti *et al.* to be published.
[26]R. Garcia, R. Pérez, Surf. Sci. Reports, 47, 197 (2002).
[27]H. Bodiguel, H. Montes, C. Fretigny, Rev. Sci. Inst. 75, 2529 (2004).
[28]B. Cappella and G. Dietler, Surf. Sci. Reports, 34, 1 (1999).
[29]L. Zitzler, S. Herminghaus, and F. Mugele, Phys. Rev. B 66, 155436 (2002).
[30]J. N. Israelachvili, *Intermolecular and Surface Forces*, 2nd ed. Academic, London, 1991.
[31]S. Pride, Phys. Rev. B 50, 15678 (1994). T. Vora, B. Corry, S.-H. Chung, Biochimica et Biophysica Acta 1668, 106 (2005).
[32]K. McAfee, J. Chem. Phys. 28, 218 (1958).
[33]M. Nogami and M. Tomozawa, J. Am. Ceram. Soc. 67, 151 (1984).
[34]A. Agarwal, M. Tomozawa and W. Lanford, J. Non-Cryst. Solids 167, 139 (1994).
[35]N. Weber and M. Goldstein, J. Chem. Phys. 41, 2898 (1964).
[36]S. Langford, L. Jensen, J. Dickinson and L. Pederson, J. Mat. Res. 6, 1358 (1991).
[37]F. Larché and J. Cahn, Acta Metall. 26, 1579 (1978).

[38]F. Larché and P. Voorhees in Defect and Diffusion Forums. 129-130, 31 (1996), Scitec Publications, Switzerland.
[39]A. Agarwal and M. Tomozawa, J. Non-Cryst. Solids 209, 264 (1997).
[40]L. Tian, R. Dieckmann, J. Non-Cryst. Solids 352 (2006) 679.
[41]M. Tomozawa and M. Hepburn, J. Non-Cryst. Solids. 345-346, 449 (2004).
[42]J.P. Guin and S. Wiederhorn, J. Am. Ceram. Soc. 88, 652 (2005).
[43]R. Baptist and F. Levy, Vacuum 43, 213 (1992).
[44]Y. Watanabe et al., J. Non-Cryst. Solids 177, 9 (1994).

SECTION II

FUNDAMENTAL PHENOMENA

IS THERE LIFE AFTER FRACTOGRAPHY?

Professor Derek Hull
Professor Emeritus Cambridge University and the University of Liverpool
17 Oldfield Gardens
Heswall, Wirral CH60 6TG
United Kingdom

This extended abstract is based on my book "Fractography: Observing, Measuring, and Interpreting Fracture Surface Topography," published by Cambridge University Press, (1999).

One of the difficulties in discussing fractography is the different nomenclature used by scientists and engineers, working in different classes of materials, to describe essentially similar phenomena. Thus, for example, the late Dr. Van Derck Fréchette, one of those responsible for starting this series of conferences on the fractography of ceramics, wrote a number of papers on nomenclature. Many of the expressions he used do not appear in the literature of the fractography of other materials. Nevertheless, there is much to be learnt from a study of the fractography of different materials and it is essential to be familiar with the literature of a subject, which has been actively researched for over 50 years.

Much of the research on fractography was initiated as a result of industrial failures. However, there have been many important and classical experiments, only loosely linked to diagnostic fractography, which provide insights into the subject.

An important aspect of fractography is the interpretation of patterns. These can range from highly ordered patterns, as in crystallography, to chaotic and fractal patterns. The examples in the main paper included (1) the conchoidal and fringe patterns formed on the fracture surface of a medium strength carbon steel shaft as a result of a fatigue failure, (2) the self-similar ordered geometric patterns formed during thermoelastic martensitic transformations, (3) the serrated patterns in slip bands on the surface of polycrystalline copper during fatigue at 20°K indicating the way that cracks are nucleated at temperature below those at which vacancy diffusion occurs, (4) the changes in surface appearance associated with ductile and brittle fracture in notched impact specimens of neutron irradiated mild steel used in pressure vessels, (5) the classical experiments on single crystals of LiF and MgO demonstrating the way cleavage cracks are nucleated at the intersection of slip bands by the interaction of dislocations, (6) the equivalent experiments on body-centred cubic iron crystals with cracks nucleated at the intersection of deformation twins, (7) the arrays of cavities along grain boundaries in a magnesium alloy, used for canning fuel elements, formed during creep deformation by grain boundary sliding and grain boundary diffusion, and (8) the wide range of patterns, including Wallner lines, resulting from the interaction between a growing crack and elastic waves in molybdenum and tungsten single crystals fractured at 20°K.

Research on the fractography of thermosetting and thermoplastic polymeric materials followed the rapid development of plastics for engineering applications. There are two examples in the main paper. First, the discovery that in thermoplastic polymers, such as polystyrene and PMMA, cracks are nucleated in crazes, which form normal to the maximum tensile stress. The words 'crazes' and 'crazing' are used in a different sense to their use in glasses and ceramics although crazes in polymers have a similar macroscopic appearance. In polymers crazing is a well-defined mode of deformation involving extensive drawing of polymer molecules to form load-bearing fibrils normal to the plane of the craze. The fractography of craze-nucleated fracture is dominated by the morphology of crazing. Second, the ability of thermosetting polymers, such as polyester and epoxy resins, that are very brittle in tension, to deform by shear deformation under constrained compressive loading conditions.

Fractography is now used widely for understanding fracture phenomena in composite materials. Thus, for example, fractographic studies have contributed to the achievement of high levels of energy absorption in composite components, made entirely from brittle materials, by creative design of fibre architecture to ensure that progressive fracture occurs on a very fine scale (10μm) throughout the whole structure. Applications include structural components in cars for passenger protection in high-speed collisions and ceramic body armour. In many ways engineers are using biomimetics to achieve effective performance. An example of the application of more conventional fractographic analysis of an engineering failure is the stress corrosion of glass-reinforced plastics used for pipes and pressure vessels. The combined effects of stress and corrosive environment can result in the progressive slow growth of cracks and eventually lead to catastrophic failure after many years in service. Detailed studies of the mechanism of crack growth using fractography have contributed to the development of a model to predict life-times and the influence of material and environmental variables.

Early Iron Age technologists used fracture surface observations as a means of quality control during the production of iron and steel. The first examination of a fracture surface using a light microscope was by Robert Hooke. He was a colleague of Isaac Newton and Christopher Wren. In his beautifully illustrated book *Micrographia*, published in 1665, Hooke illustrates a wide range of objects, from fleas to rocks, which he studied in the microscope he invented. The only fractographic observation was made of an oolitic limestone. In *Micrographia* the fracture surface is illustrated with a splendid engraving and demonstrates the importance of accurate drawings as an essential part of a fractographic investigation. In drawing, one identifies important aspects of a fracture surface.

Equally important is the production and reproduction of high quality photographic images. In my early years I was impressed by the quality of the carbon replica electron microscopy carried out by fractographers such as Peter Forsyth at the RAE Farnborough, England, and Cedric Beachem at NRL, Washington, USA who worked on industrial failures in aircraft and ships respectively. A fractographic image is often the key experimental observation and it is essential that effective ways are found to archive and reproduce these images in this digital age.

In the final part of the paper I refer to a selection of theoretical and experimental studies undertaken in preparation for writing my book on fractography. These include (1) a theoretical explanation of the formation of river lines in glassy solids and rocks, (2) repeat of Sommer's classical experiment on an organic glass, (3) a study of the relation between crack speed, dynamic stress intensity and fracture surface roughness in glassy epoxy resins following the work of Takahashi, (4) modelling the growth of cracks and the development of cracking patterns in cooling magna using the drying of sol-gels, and (5) production of atomically sharp cracks by embedding flakes of mica in the material.

FRACTOGRAPHY AND FRACTAL GEOMETRY: WHAT CAN WE LEARN?

John J. Mecholsky, Jr.
Department of Materials Science & Engineering
University of Florida
Gainesville FL 32611-6400

ABSTRACT

Fracture surface features contain quantitative information about the stress and energy associated with a specific fracture event. The approach presented here is predicated upon the fact that fracture is a fractal process. It is hypothesized that the fundamental unit of fracture at the appropriate length scale is a quantity known as a_0. In turn, a_0 can be related to the fracture energy, γ, and the elastic modulus, E, and through a scaling parameter, the fractal dimensional increment, D*, i.e., $\gamma = \frac{1}{2} a_0 ED^*$. The characteristic markings of mirror, mist and hackle observed on the fracture surfaces of glasses, ceramics and polymers are related to the fractal dimensional increment: $(Y/ Y_j)^{1/2} c/ r_j = D^*$, where c is the crack size, r_j, is either the mirror-mist radius (j = 1), mist-hackle radius (j = 2) or crack branching boundary (j = 3), Y and Y_j are constants related to the initial and propagating crack geometry, respectively. The combination of atomistic modeling, experimental measurements and the application of fracture mechanics and fractal geometry leads to a suggested sequence and organization of the brittle fracture process. By applying fractographic principles combined with fractal analysis and fracture mechanics, several different types of problems can be solved. The combined analyses can be used to determine whether a product has been manufactured properly, to identify toughening mechanisms in composites and to identify the type of loading during fracture. Examples of each application are discussed in terms of fracture surface analysis and microstructural characterization.

INTRODUCTION

Fractography

Quantitative fractography has been used for decades to analyze the failure of industrial and laboratory components.[1-5] The principles of fracture mechanics have been shown to be very powerful tools that complement the observations during fractographic analysis.[6] It is suggested that a non-Euclidean geometry, i.e., fractal geometry, can be applied to fracture surfaces to provide a mathematical tool for describing the irregular surfaces produced during fracture. Using this approach, we discover relationships between the fracture features and the energy required to fracture the sample. The implications of fractal geometry are that there exists a scaling of energy and geometry at all length scales so that the fracture process is fractal in nature.

The observations of the mirror, mist and hackle regions on the fracture surface have been known for many decades (Figure 1). As early as the 1920's, Preston[1] observed that the mirror region contained perturbations from the plane of fracture, and thus was not a "smooth" surface. If we examine the mist and hackle regions of the fracture surface, we notice that if we magnify the mist region, it appears similar to the hackle region (Fig. 2). Thus, the process that produces mist evidently is the same as that which produces hackle. This similarity with different magnification is called scale invariance. I suggest that the perturbations in the mirror region have the same source as mist and hackle. Examination of the fracture surfaces show that they are self-similar, i.e., one region appears statistically the same as another region at the same radial distance from the origin.

Fractal geometry is being used in many fields of science and engineering because it can be applied to describe shapes and processes that are non-linear and seemingly complex.[7] Fractal geometry is a non-Euclidean geometry that exhibits self-similarity (or self-affinity) and scale invariance. Fracture is one of the phenomena that has been modeled using fractal geometry.[8]

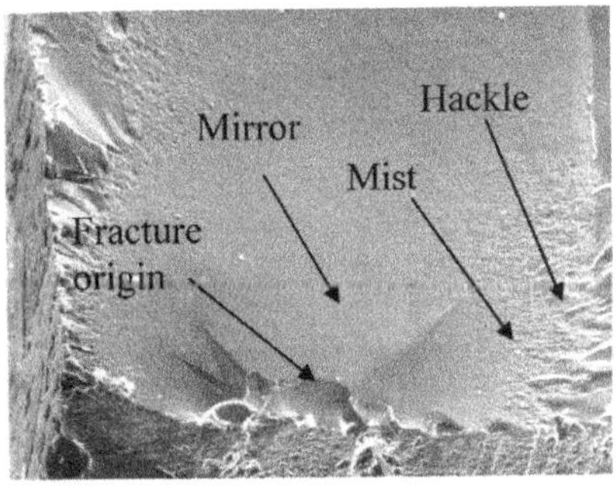

Figure 1 – Scanning Electron Microscope Image of a Typical Fracture Surface in Glass. The characteristic regions are shown in the figure.

Fractal Geometry

Fractal objects are characterized by their fractal dimension, D, which is the dimension in which the proper measurement of a fractal object is made. For example, if we had a plane square, then the only dimension in which to make a meaningful measurement is 2, i.e., the area of the square. The length or volume of a square is meaningless. This same concept is generalized for fractals that have non-integer dimensions. Thus, a plane square with "bumps" out of the plane would have dimension, 2.D*, where D* is the fractional part of the fractal dimension and represents the amount of tortuosity out of the plane. An object with a fractal dimension of 2.1 (D* = 0.1) would be relatively flat and an object with D=2.9 (D* = 0.9) would almost be a volume-filling object. If the same relatively flat "bumpy" plane (with D = 2.1) was measured by a contour line, then the fractal dimension would be D = 1.1, but D* would still be 0.1.

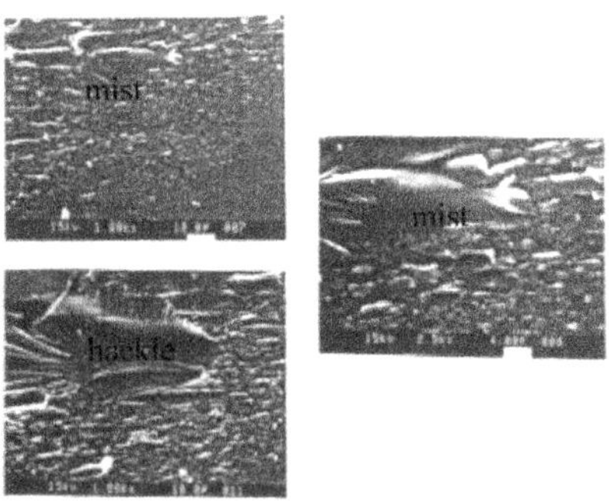

Figure 2 – Glass Fracture Surface in Mist and Hackle Zones. The top left figure is the mist zone. The bottom left is the hackle zone. The figure on the right is the mist zone at a greater magnification. Notice that the mist region magnified appears similar to the hackle region.

The fractal dimension can be measured in a number of ways. One of these has been called the "slit-island technique" in which the length of part, or all, of the contour of an "island" that appears from polishing of an embedded fracture surface is measured. This measurement is made repeatedly using different starting measurement rulers. The highest point of the fracture surface is measured first, and subsequent (lower) areas are measured on subsequent horizontal contours (Fig. 3). Generally, this is a random section on the fracture surface. Thus, an average representation of the roughest portion on the fracture surface is usually obtained. The details of the measurements can be obtained from the literature.[7, 9,10]

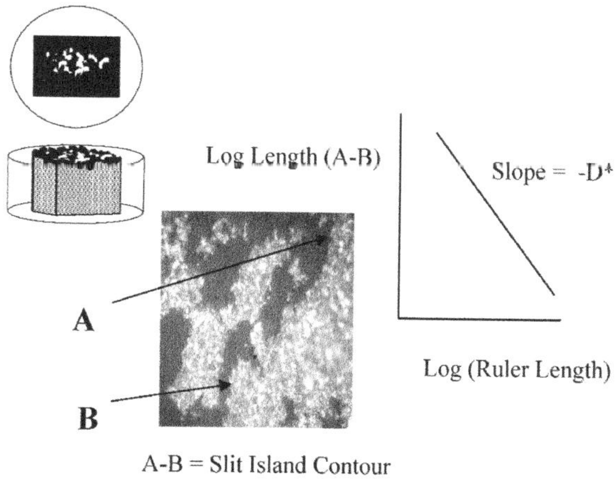

Figure 3 – Schematic of Slit Island Analysis Technique. A replica of the fracture surface is polished until a contour of the fracture surface appears. The length of the perimeter of the contour between points A & B is measured many times using different ruler lengths. The slope of the graph of log length versus log ruler length is –D*.

A proposed sequence of crack propagation as a fractal process has been described in previous work and is summarized here.[11,12] Brittle fracture can be considered a series of quantized, bond reconfiguration events related to the production of free volume at the crack tip. A series of these reconfigurations along the crack front, due to thermal vibrations and lowest energy configurations, grow in magnitude and develop into a mirror-mist-hackle pattern.[12] The pattern that develops is started at the atomic scale with the quantized ring contractions based on minimum energy. These ring contractions occur in discrete steps all along the discontinuous crack front. In some locations, the contractions stabilize and in other locations, two or more rings supplement each other and become larger in a fractal manner. A quantity termed "a_0" represents an indirect measure of the size of the initial crack advance. These enlarged groups, in turn, annihilate and supplement other similar groups either to subside or to percolate into large structures, respectively. The fractal dimensional increment, D*, is a scaling factor for both the fracture energy and the geometry of the fracture surface created. Such structures are suggested to percolate in a self-similar (or self-affine) manner and produce the structure we observe as the fracture surface. At some point, dictated by the type and strength of bonding, more energy is available than necessary for propagation. At this point, the crack branches macroscopically. The angle and nature of the macroscopic branching depends on energy levels, geometric atomic structure, far-field stress pattern, and lowest energy pathways.

Fracture Mechanics and Fractal Geometry

Fundamental relationships can be derived between the fractal dimensional increment, D*, and the fracture toughness of a material in the form of the critical stress intensity factor, K_C:

$$K_C = E a_0^{1/2} D^{*1/2} = Y(\theta)\sigma_f c^{1/2} \qquad (1a)$$

$$\gamma_f = \tfrac{1}{2} E a_0 D^* \qquad (1b)$$

where E is the elastic (Young's) modulus, a_0 is a characteristic parameter having the units of length, $Y(\theta)$ is a geometric constant dependent on geometry of the crack and loading condition, σ_f is the applied stress at fracture, c is the crack size at fracture and γ_f is the fracture energy ($K_c = [2E\gamma_f]^{0.5}$). The relationship between K_C and D^* is based on experimental observations [13-15], and the relationship between K_C and c is based on fracture mechanics and experimental confirmation.[16,17] Note that $D^* \neq 0$[11-13] in the fracture of primary bonds, due to thermal vibrations and the presence of natural defects. The measurement of D^* is an average property of the entire fracture surface and a measure of its tortuosity. The flaw size is a linear measure of the critical flaw area locally around the fracture origin. These regions are all related to the applied (far-field) stress at fracture, σ_f:

$$\sigma_f \, r_j^{1/2} = \text{constant} = K_{Bj} \,/\, Y_j'(\theta) \qquad (2)$$

where $r_j = r_1$, r_2 or r_3 corresponding to the different boundaries, i.e., mirror-mist, mist-hackle and macroscopic crack branching (cf. Fig. 4), respectively, in an analogous equation to Eq. (1a).[2,6,17] K_{Bj} is the crack branching stress intensity and $Y_j'(\theta)$ is a crack border correction factor where θ is the angle from the surface to the interior, i.e., $\theta = 0$ to 90°. Note that K_{Bj} is proportional to K_{IC}, i.e., $K_{Bj} = \lambda K_{IC}$ where λ is 3-5 (for j=1) for most ceramic materials. It can be shown that: [17]

$$[Y'(\theta)^2 / Y(\theta)^2]\, r_j/c = 1/D^* \qquad (3)$$

The relationship in Eq. 3 is valid for all values of j, i.e., j=1,2,3, but the r_j/c will be numerically different as will the $(Y_j'(\theta)^2 / Y(\theta)^2)$ proportionality constants.

The agreement of experimental data with Eq. (3) implies that there is scaling between the crack size and the point of branching. Furthermore, comparison of Eqs. 1-3 shows that the ratio of the energy of crack initiation to that of crack branching is related to r_j / c; there is also scaling between the energy of crack initiation and of crack branching. Thus, analyses considering an energy scaling or a structural (geometric) scaling are equivalent. Moreover, the scaling is linear in energy and geometry.

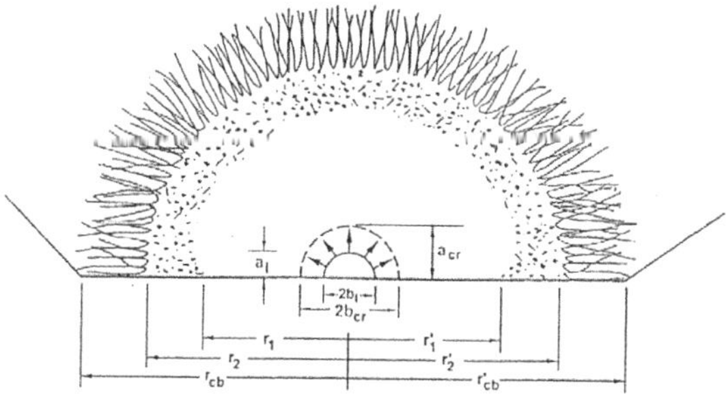

Figure 4 – Schematic of Fracture Surface Features. a_i indicates the initial crack size and a_{cr}, the critical crack size. $r_{cb} = r_3$ in Eq. (2). The primes indicate the possibility of unsymmetric boundaries. (cf. Ref. 6)

Based on the equations derived using fractal geometry, estimates of the theoretical strength of materials can be made[18] without the need for specific potential energy functions describing the bonding of atoms. As mentioned above, fractal geometry provides a framework by which bond rupture can be described as a series of bond reconfigurations. This framework leads to a method by which the theoretical strength, σ_t, of materials can be estimated based on the geometry of the crack tip and the generated fracture surface:[18]

$$\sigma_t = E\, a_0^{\ 1/2} D^{*\ 1/2} / (2\pi a_0)^{1/2} = E\, D^{*\ 1/2} / (2\pi)^{1/2} \tag{4}$$

The estimates shown in Table I are between the commonly quoted values of E/π and E/8.

APPLICATIONS OF FRACTAL GEOMETRY AND QUANTITATIVE FRACTOGRAPHY

The combination of fractography, fracture mechanics and fractal geometry provides a powerful tool for failure forensics, research investigation and production failure analyses. Whether on the production line, in the laboratory or in the field, these tools will be useful in determining whether a failure occurred due to incorrect design, material inferiority or misuse, such as an overload. As examples, I present the analysis of a dental crown, the determination of toughening mechanisms in a glass-ceramic and identification of a wear mechanism in a ceramic bearing material.

Dental Crown

The first example is a dental crown that failed in-situ and was recovered for examination. Several examples of this type of post-mortem failure analysis exist in the literature.[19-21] One example is that presented by Thompson et al.[22] in which several crowns were analyzed and the stress at failure determined from the size of the fracture-initiating crack (Fig. 5). The fractal dimensions were measured for several dental glass ceramics[23] in anticipation of their use in failure forensics. As an example, a crown failed in situ could be analyzed to determine if the failure was due to faulty fabrication or overload. A replica of the failed crown could be made even if only part of the fracture surface is available. The fractal dimension measurement of the crown fracture surface would be compared to the values obtained for a well-fabricated glass ceramic. If the fractal dimension values are the same statistically, then this fact implies that the material was correctly manufactured and that an overload or large crack was responsible for the failure – not fabrication. If on the other hand, the D* values were different, then this would imply an inferior crown material.

Table I – Theoretical Strength, σ_t, Estimates Based on Fractal Geometry.

MATERIAL	K_{IC}(MPa m$^{1/2}$)	γ (Pa m)	E(GPa)	a_0 (nm)	D*	σ_t (GPa)
Silica Glass	0.73	4	70	1	0.11	9
$3BaO$-$5SiO_2$ Glass	0.72	4	64	1.3	0.1	8
Borosilicate Glass	0.71	4	64	1.4	0.09	8
Li_2O - $2SiO_2$ Glass	1.0	7	70	2	0.1	9
$3BaO$-$5SiO_2$ G.C.	1.8	18	90	2	0.22	17
Lithia Disilicate G.C.	3.0	33	137	2	0.24	26
Pyroceram Glass Cer.	2.5	26	120	3	0.13	17
Lithia Borosilicate G.C.	2.7	41	90	5	0.18	15
Zinc Silicate G. C.(#400)	1.6	14	90	6	0.05	8
MgF_2	1.1	4	138	0.2	0.23	26
Alumina (AD 99)	2.9	11	390	0.3	0.21	70
Alumina (AD999)	3.9	19	406	0.3	0.31	89
Alumina (UCC)	2.5	8	393	0.3	0.15	60
Alumina (Lucalox)	4.0	20	393	0.3	0.33	89
CdTe	0.3	1	40	0.3	0.2	7
Zinc Selenide	0.9	6	69	0.4	0.3	15
Alumina (GEND/ND94)	3.9	27	283	0.8	0.23	53
Silicon Nitride	5.1	42	310	0.8	0.33	70
S. C. Calcium fluoride	0.3	0.4	114	0.1	0.07	12
S. C. Spinel	1.2	3	240	0.3	0.09	28
S. C. Silicon (100){110}	1.2	6	130	0.9	0.1	16
S. C. Silicon (110){110}	1.05	3	169	1	0.04	13

In another publication, Thompson et al.[24] showed that the fractal dimension of Lithia-based glass ceramics produced different fractal dimension values depending on the heat treatment and resulting different microstructure. The microstructure that was not microcracked resulted in a toughness value of 3.01 ± 0.12 and a corresponding fractal dimensional increment of 0.24 ± 0.01 whereas a microcracked specimen resulted in a fracture toughness value of 1.31 ± 0.05 with a corresponding fractal dimensional increment of 0.18 ± 0.01. To determine if the fracture stress levels in the lithia-disilicate based crowns were reasonable, a replica of the crown fracture surface would be made and the fractal dimension determined from the replica. If the value of D* was near 0.24, then the material was fabricated as designed. If the value was closer to 0.18, then there was most likely a problem with the fabrication of the dental crown material. By using the knowledge gained from fractal analysis, an additional analytical tool aids in determining the root cause of an unexpected failure. The fractal dimensions of dental materials can be compared to the fracture behavior of other conventional materials (Fig. 6).

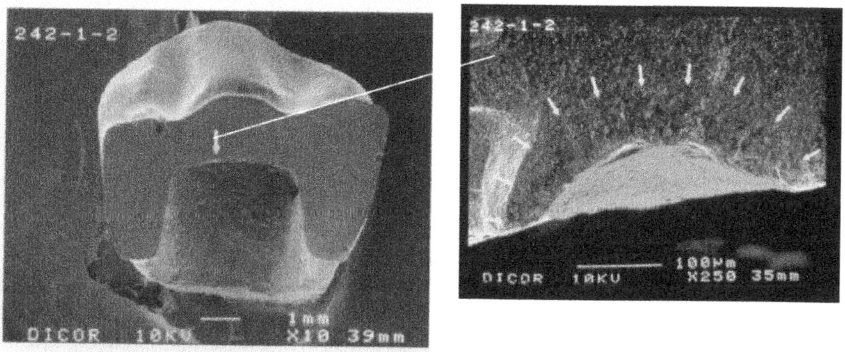

Figure 5. SEM Fracture Surface of a Glass Ceramic Dental Crown. The inset is the fracture origin located in the region shown by the arrow. After reference 22.

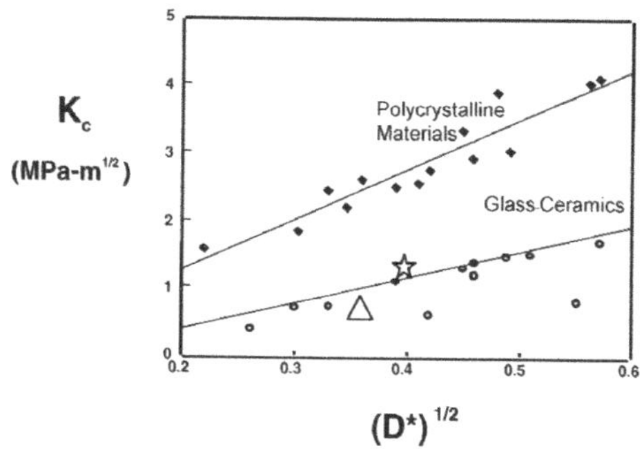

Figure 6 – Fracture toughness as a function of D*$^{1/2}$. The star symbol represents Dicor Glass- Ceramic. The triangle represents dental porcelains. The other materials are listed in references 11, 13 and 15.

Toughening Mechanisms

In research, it is often desirable to have several tools available for analyzing mechanisms of toughening. Several studies[25-30] have examined experimentally the effect of increasing aspect ratios of various particle morphologies on fracture toughness. Faber and Evans[25] concluded that toughness increased monotonically with aspect ratio using a series of hot-pressed silicon nitride compositions. Baik et al.[26] studied mica-based glass-ceramics with high-aspect-ratio crystals and concluded that crack deflection and crack branching caused the observed increase in fracture toughness. Lee et al.[27] demonstrated an increase in toughness of SiC with the addition of 6-wt% Al_2O_3 and 4-wt% Y_2O_3. They concluded that crack deflection and crack bridging from plate-like grains causes the increase in toughness from 4.5 to 8.3 MPa•m$^{1/2}$ with increasing grain size and aspect ratio. In these studies, there were no means available for quantitative characterization of the surface tortuosity or quantitative evaluation of the toughening process.

An example from the literature shows how to apply fractal geometry to aid in this analysis. The objective of a Hill et al.[31] study was to analyze the quantitative relationship between crystal aspect ratio, fracture toughness, and fractal dimension of the fracture surfaces for a 3BaO•5SiO2 glass-ceramic system. As anticipated, the fracture toughness of 3BaO•5 SiO2 glass-ceramics increased with increasing aspect ratio. The fracture toughness increased from 0.7 ± 0.1 MPa m$^{1/2}$ for glass to 2.2 ± 0.6 MPa • m$^{1/2}$ for the glass-ceramic with an aspect ratio (AR) of 8.1. The fractal dimensional increment, *D**, also increased with crystal aspect ratio. The fracture surface morphology changed from intergranular for glass-ceramics with AR values of 1.4 and 3.6, to transgranular for a glass-ceramic with an AR value of 8.1. In addition, evidence of crack bridging was observed in the glass-ceramic with an AR value of 8.1 and not in the other glass-ceramics.[31] The baria-silicate glass-ceramics generally follow the predicted relationship for toughening caused by crack deflection for the lower-aspect-ratio materials but deviated significantly for the material with a mean AR value of 8.1. Based on the above accumulated data, which suggest a difference in fracture toughening behavior between the lower-aspect-ratio materials and that with an aspect ratio of 8.1, the toughness versus fractal dimensional increment (Fig. 7) provides a method for calculating the amount of increased toughening (0.4–0.7 MPa•m$^{1/2}$). This additional toughening increment is attributed to an additional toughening mechanism for the group with an AR value of 8.1, most likely crack bridging. Toughening by the crack-bridging mechanism has been suggested in other studies with materials containing second phases with high aspect ratios (>8).[27]

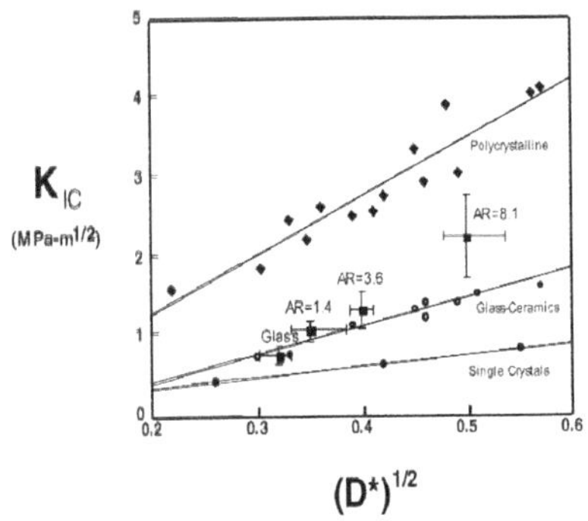

Fig. 7. Fracture toughness as a function of the square root of the fractal dimensional increment for various materials. Baria silicate values[31] are displayed as filled squares. Notice that the value for AR = 8.1 is located well away from the glass-ceramic line. Other data are from Refs. 13-15.

Wear Analysis

Because of the great success of Si_3N_4-Chronidur steel hybrid bearings in the space shuttle Discovery, these types of bearings are being considered for use in the next generation of aircraft engines.[32] The hybrid bearing consists of a ceramic ball and tool-steel-type metal raceways. The space shuttle application was a technical success. Although the hybrid bearing was used successfully in the space shuttle, it was a very specialized application and could not be duplicated for commercial use because of economical reasons. The problem arises from two areas: (1) expense of processing and fabrication of 1.5-inch balls and (2) structural reliability. It is necessary to be able to produce a hybrid bearing at the cost of all steel bearings for commercial applications. In addition, it must be a requirement that these bearings do not fail during use. One of the mechanisms of failure is wear of the ceramic ball in the bearing system.[33] Apparent spallation has been observed in severe tests to determine the expected type of failure, if it ever occurred. In order to determine the type and mechanism of failure that occurs during ball-on-race wear, a ball with apparent spallation damage was examined by making a replica of the wear surface. A fractal dimension measurement was made, and the value was compared to the fractal dimension of the fracture surface of a flexure bar of the same material subject to cyclic loading with a spherical indenter.[34] The fractal dimensional increments were 0.32 and 0.33, respectively (Fig. 8). The fractal dimensions of the two surfaces were statistically the same, indicating that the mechanism of failure for "spallation" in silicon nitride balls is a fracture process and not similar to spallation in metallic balls used in bearing systems.

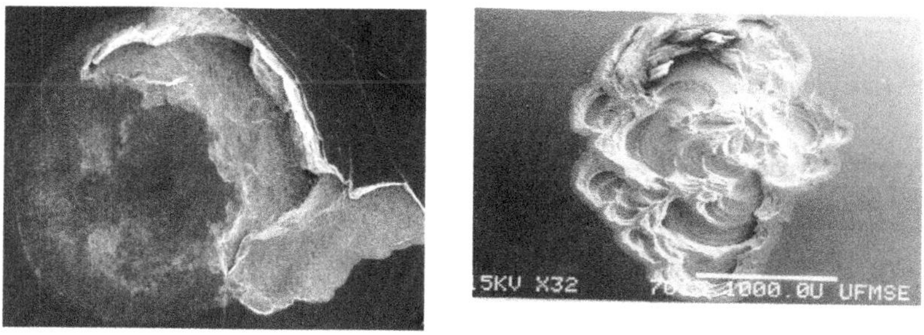

A. B.

Figure 8. Silicon Nitride Wear Surfaces. (A.) Artificially induced in laboratory testing on a flexure bar. [D* = 0.33] (B) Wear surface on silicon nitride ball in bearing test [D* = 0.32].

Crack Branching as a Fractal Process

Fragmentation and high-stress contact events, such as crushing during a grinding process, result in macroscopic crack branching.[35] Crack branching patterns form fractal patterns.[36-39] Their Crack Branching Coefficient (CBC) characterizes these fractal patterns. The CBC is the fractal dimension of the crack-branching pattern. It is to be distinguished from the fractal dimension of the fracture surface (Fig. 9). The fractal dimension obtained from the fracture surface is constant for any loading configuration. In other words, D* is independent of the far-field stress. The CBC is dependent on the far-field stress and changes depending on the loading configuration.

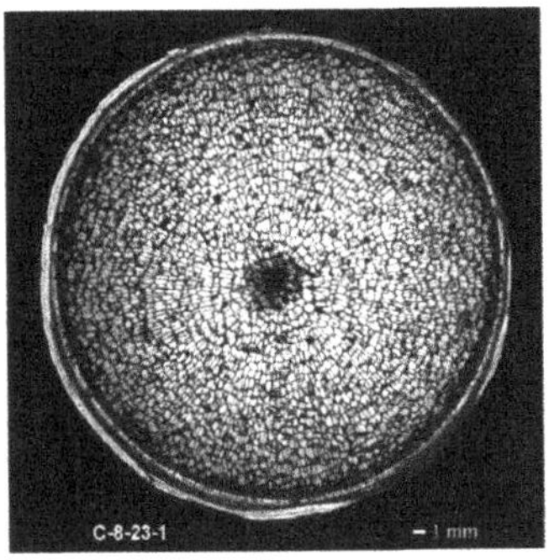

A.

Crack branching pattern

B.

Fracture surface pattern

Figure 9 – Optical micrograph showing ion exchanged glass. (A) Crack branching pattern from which a CBC is obtained. (B) Fracture surface from which D* is obtained.

Mecholsky et al.[38] found a direct relationship between the fractal dimension of the crack branching pattern and the strength of annealed magnesium fluoride disks (Fig. 10). Quinn et al.[39] showed that the fractal pattern of branching in annealed glass disks can be predicted using a computer program developed by them based on fractal geometry. Research relating the macroscopic crack branching patterns in residually stressed glass to fractal dimensions is limited.[36,37] Sakai et al.[36] showed that it is possible to quantify and simulate crack branching patterns in thermally tempered glass using fractal geometry. The situation for ion-exchanged glasses is somewhat more complicated. However, Kooi et al.[37] showed that it is possible to relate the magnitude of the central tension in ion-exchanged glasses to the CBC (Fig.11).

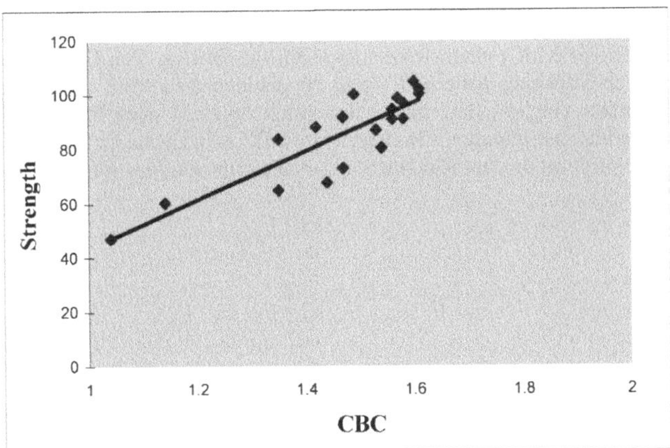

Figure 10 – Strength as a function of the Crack Branching Coefficient (Fractal Dimension). Annealed MgF_2 disks fractured using ring-on-ring loading.[37]

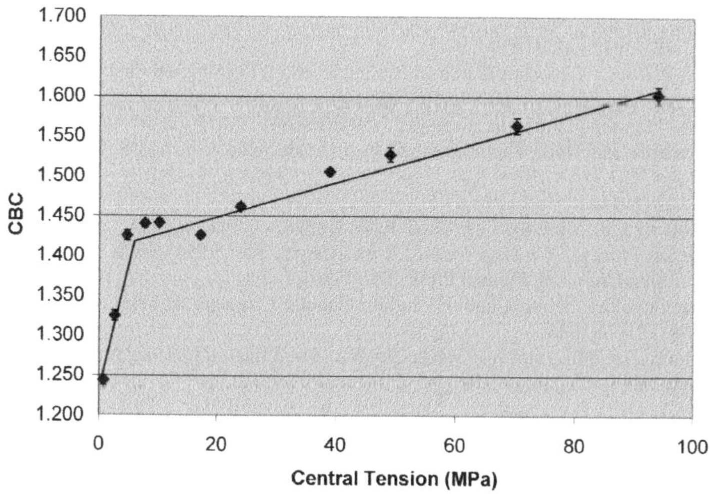

Figure 11 – CBC versus central tension in an ion exchanged glass. Lines were added to aid the eye.

SUMMARY AND CONCLUSIONS

Fracture surfaces exhibit characteristics that can be described using fractal geometry. As a result, the fracture energy required for fracture is related to a characteristic length identified with the structure of the material, a_0, the fractal dimensional increment, D*, identified as a scaling parameter, and the elastic modulus, E: $\gamma = 0.5\ ED^*\ a_0$. The crack-to-mirror size ratio, c/r_1, is equal to the fractal dimensional increment, D*. These two relationships imply that there is a direct relationship between fractography, fracture mechanics and fractal geometry. Thus, fractography, fracture mechanics and fractal geometry can be combined to create a powerful tool for forensic analysis, research investigations and production analyses. Examples were provided that demonstrate how fractal dimension measurements can be used to determine overload versus loading within specifications, to determine the toughening mechanisms in composites, such as glass-ceramics, and to determine modes of failure in brittle materials. Macroscopic crack branching can be described using fractal geometry for annealed as well as residually stressed materials. The Crack Branching Coefficient (CBC) can be used to identify the magnitude of strength and type of loading. The CBC is distinguished from the fractal dimensional increment, D*, which is independent of the loading and magnitude of strength. It is recommended that fractal geometry be incorporated in any forensic investigation of materials failures to insure reliability.

REFERENCES

[1]F. W. Preston, "A Further Note on the Fracture Systems of Glass," J. *Soc.of Glass Tech.* 13, 3-14 (1929).

[2]E. B. Shand, "Breaking Stress of Glass Determined from Dimensions of Fracture Mirrors," *J. Am. Ceram. Soc.*, 42(10) 474-477(1959).

[3]H. Wallner, *Z. Phys.*, "Linienstrukturen an Bruchflächen," 114 [5-6] 368-78(1939).

[4]Ch. De Freminville, *Rev. Met*, 11, 971-1056 (1914).

[5]R.W. Rice "Ceramic Fracture Features, Observations, Mechanisms and Uses," in Fractography of Ceramic and Metal Failures, edited by J. J. Mecholsky, Jr. and S. R. Powell, Jr., *ASTM STP 827*,
Philadelphia PA 5-102 (1984).

[6]S. W. Freiman, J. J. Mecholsky, Jr., and P. F. Becher, "Fractography: A Quantitative Measure of the Fracture Process," *Ceramic Trans.* 17, Am. Ceram. Soc.,(1991) 55-78.

[7]J. C. Russ: *Fractal Surfaces*. Plenum Press, NY (1996).

[8]B.B. Mandelbrot, D.E. Passoja and J. Paullay, "Fractal Character of Fracture Surfaces of Metals," *Nature*, 308, 721-22(1984).

[9]A. Della Bona, T. J. Hill, and J. J. Mecholsky Jr., "The Effect of Contour Angle on Fractal Dimension Measurements for Ceramic Materials," J. *Materials Sci.* 36, 2645-50 (2001).

[10]T. J. Hill, A. Della Bona, and J. J. Mecholsky, Jr., "Establishing a Protocol for Optical Measurements of Fractal Dimensions in Brittle Materials," *J. Materials Science*, 36, 2651-2657(2001).

[11]J. K West, J. J. Mecholsky, Jr, and L.L. Hench, "The Quantum and Fractal Geometry of Brittle Fracture," *J. Non-Crystalline Solids* 260 99-108 (1999).

[12]J. J. Mecholsky, Jr, J. West and L. L. Hench, "The Quantum and Fractal Nature of Brittle Fracture," *Ceramic Transactions* 101, 39-51, The American Ceramic Society (2000).

[13]J. J. Mecholsky, Jr., T. J. Mackin and D. E. Passoja "Self-Similar Crack Propagation in Brittle Materials," *Adv. in Ceramics 22*: Fractography of Ceramics and Glasses, Varner and Fréch ette, eds., The Am. Ceramic Soc (1988).

[14]Y. Fahmy, J. C. Russ and C. C. Koch, "Application of Fractal Geometry Measurements to the Evaluation of Fracture Toughness of Brittle Intermetallics," *J. Mater. Res.*, 6[9] 1856-1861(1991).

[15]J.J. Mecholsky, D.E. Passoja and K.S. Feinberg-Ringel, "Quantitative Analysis of Brittle Fracture Surfaces Using Fractal Geometry," J. *Am. Ceram. Soc* 72[1] (1989).

[16]P.N. Randall, comment on "Plain Strain Crack Toughness Testing of High Strength Metallic Materials" in *ASTM STP 410*, 88-126 (1988).

[17]J. J. Mecholsky, Jr. And S. W. Freiman, "Relationship Between Fractal Geometry and Fractography," J. *Am. Cer. Soc.*74 [12] 3136-38 (1991).

[18]J.J. Mecholsky, Jr., "Estimating Theoretical Strength of Brittle Materials Using Fractal Geometry, *Materials Letters*," 60(20) 2485-2488 (2006).

[19]J. R. Kelly , R. Giordano, R Pober, M. J. Cima, "Fracture Surface Analysis of Dental Ceramics: Clinically Failed Restorations," *Int. J. Prosthodont.* 3(5):430-40 (1990).

[20]B. Taskonak, J. J. Mecholsky, Jr., and K.J. Anusavice, "Fracture Surface Analysis of Clinically Failed Fixed Partial Dentures," *J. Dent. Res.* (3) 277-281 (2006).

[21]S.S.Scherrer , J. B, G.D.Quinn, J. R. Kelly, "Failure Analysis of Ceramic Clinical Cases Using Qualitative Fractography," *Int. J. Prosthodont.* 19(2):185-92 (2006).

[22]J. Y. Thompson, K. J. Anusavice, A. Naman and H. F. Morris, "Fracture Surface Characterization of Clinically Failed All-Ceramic Crowns," *J. Dental Res.*, V.73, 1824-1832 (1994).

[23]J. Y. Thompson, J. C. Cuneo, A. Naman, J. J. Mecholsky, and K. J. Anusavice,

"Relationship Between Fracture Toughness and Fractal Dimension of a Dental Glass-Ceramic," *Trans. Acad. Dent. Mater.*, v. 7 [87], Abstract 8 (1994).

[24]J. Y. Thompson, K.J. Anusavice, B. Balasunramaniam, & J. J. Mecholsky, Jr., "The Effect of Microcracking On The Fracture Toughness and Fractal Dimension of Lithia-Based Glass-Ceramics," *J. Am. Ceram. Soc* [78]11 3045-49 (1995).

[25]K.T. Faber and A.G. Evans, "Crack Deflection Processes--I. Theory," *Acta. Metallurgica* [31] 4, 577-584 (1983).

[26]D.S. Baik, K..S No, and J.S.S. Chun, "Mechanical Properties of Mica Glass-Ceramics," *J. Am. Ceram. Soc.* 78 1217-22(1995).

[27]S. K. Lee, Y. C. Kim, and C. H. Kim, "Microstructural Development and Mechanical Properties of Pressureless-Sintered SiC with Plate-Like Grains Using Al_2O_3-Y_2O_3 Additives," *J. Mater. Sci.*, 29, 5321–26 (1994).

[28]P. Sajgalik, J. Dusza, and M. J. Hoffmann, "Relationship between Microstructure, Toughening Mechanisms, and Fracture Toughness of Reinforced Silicon Nitride Ceramics," *J. Am. Ceram. Soc.*, 78, 10, 2619 –24 (1995).

[29]R.Chaim and V.Talanker, "Microstructure and Mechanical Properties of SiC Platelet/Cordierite Glass-Ceramic Composites," *J. Am. Ceram. Soc.*, 78, 166 –72 (1995).

[30]G. Pezzotti, "On the Actual Contribution of Crack Deflection in Toughening of Platelet-Reinforced Brittle Matrix Composites," *Acta Metall. Mater.*, 41[66] 1825–39(1993).

[31]T. J. Hill, J. J. Mecholsky Jr. and K. J. Anusavice, "Fractal Analysis of Toughening Behavior in 3BaO•5SiO2 Glass Ceramics," *J. Am. Ceram. Soc.*, 83 [3] 545–52 (2000).

[32]I. Hucklenbroich, G. Stein, H. Chin, W. Trojahn, E. Streit, "High Nitrogen Martensitic Steel for Critical Components in Aviation," *Materials Science Forum* v. 318, 161-166(1999).

[33]D. K Shetty and V. K. Pujari, "Load-Bearing Capacity in Quasi-Static Compression and Bearing Toughness of Silicon Nitride Balls", *Tribology Trans.* 47 522-526 (2004).

[34]Z. Chen, J. Cuneo, J. J. Mecholsky, Jr. and S. F. Hu, "Damage Process of Si_3N_4 Bearing Material Under Contact Loading," *Wear* 198, 197-207 (1996).

[35]J. C. Conway and H. P. Kirchner, "Crack Branching as a Mechanism of Crushing During Grinding," *J. Am. Ceram. Soc.*, 69, 603 (1986).

[36]T. Sakai, M. Ramulu, A. Ghosh, and R.C. Bradt, "A Fractal Approach To Crack Branching (Bifurcation) In Glass," pp. 131-146 in *Ceramic Transactions*, 17, Fractography of Glasses and Ceramics II, Edited by V. D. Fréchette and J. R. Varner, The American Ceramic Society, (1991).

[37]J. E. Kooi, S. J. Glass & J. J. Mecholsky, Jr., "Fractal Analysis of Macroscopic Crack Branching in Ion-Exchanged Glass," submitted to J. Mater. Res. (2006).

[38]J. J. Mecholsky, Jr., R. Linhart, B. D. Kwitkin, and R. W. Rice, "On the Fractal Nature of Crack Branching in MgF_2," *J. Mater. Sci.* November, 13, [11], pp. 3153-3159 (1998).

[39]G.W. Quinn, J.B. Quinn, J. J. Mecholsky, Jr. and G.D. Quinn, "Computer Modelling of Crack Propagation using Fractal Geometry," Ceramic Engineering and Science Proceedings. Vol. 26, no. 2, pp. 77-84 (2005).

ACKNOWLEDGEMENTS

I thank Dr. Dann Passoja for introducing fractal geometry and for many years of discussions. I also thank the many former students who have embraced the use of fractal geometry in the study of fracture. Partial financial support was provided by the Office of Naval Research.

QUANTITATIVE THREE-DIMENSIONAL FRACTOGRAPHY IN SAPPHIRE FIBERS

J. M. López-Cepero
J. J. Quispe-Cancapa
A. R. de Arellano-López
J. Martínez-Fernández
Physics of Condensed Matter Dept.
University of Seville
Av. Reina Mercedes s/n, P. O. Box 1065
41080 Seville (SPAIN)

ABSTRACT

The usual methods of choice for fractographic analysis are regular light microscopy, often with interference contrast, and scanning electron microscopy (SEM). However, several other microscopic techniques are also useful. One such technique is confocal microscopy (LSCM); capable of height discrimination, of wide-area imaging and ideally suited to quantitative analysis, LSCM has an excellent synergy with SEM.

In this paper, LSCM will be used to study the fracture surface of single-crystal sapphire fibers under subcritical crack growth (SCG). When SCG is present, several distinct zones may be identified in the fracture surface: the subcritical zone, the mirror zone, and the branching zone. In the branching zone, the fracture follows a restricted set of "easy" crystalline planes on its propagation, resembling a mountain range where the possible orientations of the slopes are restricted. These fibers have already been studied with SEM, but the shortcomings of the electron microscope restrict the possible studies.

In addition to reflection-mode observation, a simple experimental method which allows for fluorescent-mode usage at virtually no economic cost will be introduced. 3D models of the distinct zones of the fracture surface will be built and quantitatively analyzed with profiles and histograms. As compared with SEM, the unique information obtained includes perspective views of the fracture surface, clear SCG-mirror boundary contrast, overall surface orientation statistics, height profiles and easy mirror-zone-limit determination, all with a technique which is fast, non-destructive and which requires almost no sample preparation.

INTRODUCTION

Motivation

Almost every fractographic problem involves analyzing very fine features on the surface of the broken sample. Consequently, the observation of the fracture surface is not done by direct optical inspection, but by using one of several microscopy techniques available to the fractographer.[1] The most widely used is probably conventional optical microscopy, often with the aid of interference contrast equipment; scanning electron microscopy (SEM) comes a close second, especially when the resolution needed goes beyond what optical microscopes may offer. Other techniques include transmission electron microscopy (TEM) of fracture surface replicas, optical profilometry, X-ray tomography, atomic force microscopy (AFM) and the subject of the present paper, confocal microscopy (LSCM, from laser scanning confocal microscopy).[2,3] Each of these techniques has its own strengths and limitations, and it is worth mentioning that there is not much overlap between them. If a given sample is observed by using two different techniques, often the shared abilities among them allow identifying and observing the same regions of the fracture surface, but the unique features of each technique provide different, complementary information.

The present study is focused on the usage of LSCM in a fractographic problem of special interest: tensile fracture of c-axis-aligned sapphire fibers at high temperature. With the aid of a few examples, the technique will be introduced, its special strengths and pitfalls will be identified, and some tips for LSCM use in fracture will be given. Moreover, since extensive SEM characterization of these samples is readily available in previous studies,[4,5,6] comparisons between both techniques will also be made. Thus, the motivation for this paper is twofold: first, demonstrate that LSCM can be a useful technique in fractography and show the kind of information that can be obtained with it; and second, put LSCM in context as a fractographic tool by means of the comparison with SEM, a well-known and widely used technique.

Description of Laser Scanning Confocal Microscopy

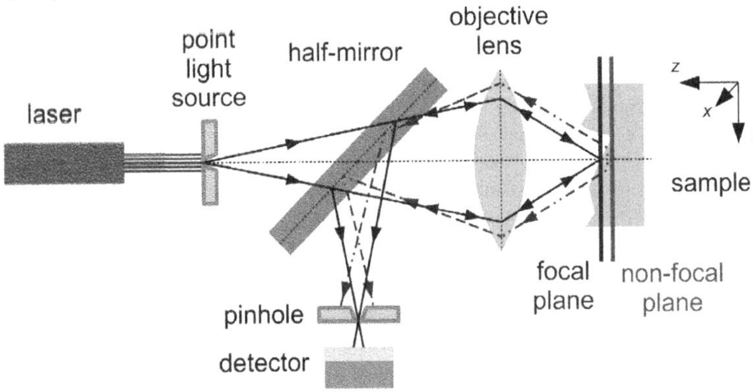

Figure 1: Schematic of the LSCM working principle.

LSCM is an optical microscopy technique with enhanced lateral resolution and depth discrimination capability. A basic idea of its philosophy and working principle will be given here; extensive descriptions can be found in the literature.[3]

LSCM uses a special optical construction, a schematic of which can be found in Figure 1. A laser beam is made into a point light source, which is then focused onto the sample by the objective lens. We will assume that some interaction involving light emission (such as reflection on the surface or fluorescence) occurs as the light coming from the objective lens hits the sample. If the light emission occurs precisely at the point of the sample where the laser beam has been focused, the light rays will follow exactly the same path as the incoming rays. Due to the presence of a half-mirror, a sizable fraction of them will be reflected towards a very narrow diaphragm (the pinhole), positioned so that those light rays will focus at its center. Thus, light will be able to pass through it, reaching a detector and giving a signal. However, if the interaction takes place at a different height in the sample, the returning rays will no longer follow the same path as the incoming ones; once reflected in the half-mirror, they will either not have focused yet or have already focused as they reach the pinhole, and the light will be scattered over a large area of the pinhole. As the central aperture of the pinhole is extremely small, almost all the light (all, in practical terms) will be blocked and there will be no signal on the detector.

The combined effect is that the light will only reach the detector if emitted at the point at which the incoming light rays are focused; the pinhole will discard all light not coming from the focused point. If the laser is then scanned across the sample, only the light from points at a precise height (that of the focused point) will be collected. This defines a plane on the sample (the so-called focal plane). Thus, a confocal image obtained by scanning the laser in the XY directions represents a "slice" of the sample, corresponding to its intersection with the focal plane.

The combined effect of the pinhole and the point light improves the lateral resolution of a confocal microscope. A good rule of thumb is that it is about 80% of the lateral resolution of a classical optical microscope working with light of the same wavelength. Similarly, the Z resolution of the microscope is given as the thickness of the optical slice taken around the focal plane, and is typically about four times the lateral resolution. It should be noted that this Z resolution is not the same as the height resolution when imaging a surface, which, as will be later discussed, is much better.

Height Determination in the Confocal Microscope

Given a single point on the surface —corresponding to a XY point in the acquired images — the recorded luminosity depends on the difference between its height and that of the focal plane; it has a maximum when they are the same, and falls to total darkness when they are very different. The exact analytical expression of the luminosity-vs-Z curve is somewhat complex, but it can be assumed to be symmetrical around the maximum. If a stack of a large number of images with slightly differing Z values is acquired, the height of a given XY point chosen may then be calculated as the luminosity-weighted mean of the Z values. Repeating this procedure for each XY column in the stack yields a height map: a 2D image where the "color" of each point is not associated with luminosity, but with height, much the same as in topographic maps. This height map is used to build the 3D projections of the sample surface, as well as the surface profiles.

Sample Choice

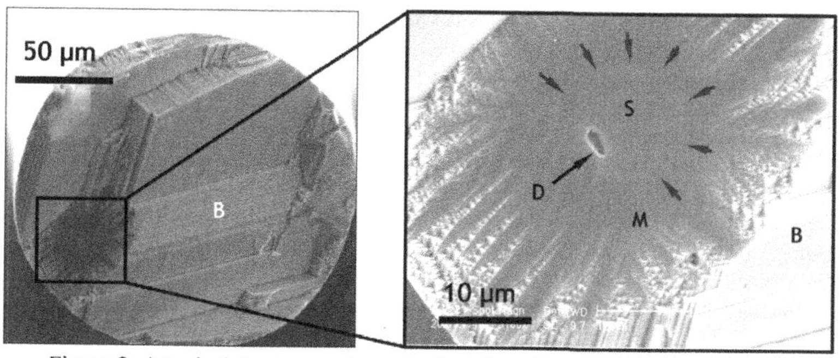

Figure 2: A typical fracture surface of a Saphikon fiber, as imaged with SEM.

Saphikon single-crystal sapphire fibers, grown along the c-axis, were selected for the study. The fibers chosen had several desirable characteristics. First, they exhibited a rich fracture behavior, with evidence of subcritical crack growth, appearance of a mirror plane, and a transition from mirror to branching; the similarity among the observed faceted planes and certain crystallographic planes of sapphire was also noteworthy.[4,5,6] Second, they had previously been studied in depth with SEM, and, thanks to the work of one of the authors, a good number of fibers –already tested and SEM imaged – were readily available.[4] Finally, the fracture of sapphire is, in itself, a very interesting problem which is not completely understood.[7,8,9]

A characteristic SEM fracture surface image of these fibers can be seen in Figure 2. Under high temperature and subcritical tensile stress, the fiber undergoes subcritical crack growth (SCG), as a flat crack (S) grows from the originating defect (D), usually a pore. The transition to catastrophic failure leaves a faint marking on the surface (indicated by arrows), which gives way to a mirror surface (M) and to branching (B).

The SEM observation yields a great deal of information about the fracture surface, but there are still a number of characteristics whose observation could be improved upon. It has been found that the SCG and mirror surfaces are not coplanar, and that branching occurs along well-defined crystallographic planes;[5] however, since the SEM lacks depth resolution, this can not be measured. Furthermore, the size of the identifiable zones (S, M) may be correlated with applied stress at fracture and fracture toughness via simple equations[1]; this process is inconvenient with SEM images because of the lack of an easily quantifiable contrast at the boundaries, especially in the SCG limit, barely noticeable at best and sometimes invisible. The LSCM study will determine whether the confocal microscope can overcome these shortcomings.

EXPERIMENTAL PROCEDURE

Initial Study

The sapphire fibers were tested under tensile stress until fracture at 1200 °C; extensive details about the sample testing methods and results have been published elsewhere.[4,5] After fracture, the fracture surface was identified and gold-coated with sputtering to permit SEM observation. The samples were then observed in a Philips XL-30 scanning electron microscope to obtain the shown SEM images.

Confocal Sample Preparation

The confocal microscope is an optical instrument, so, as long as the sample emits light in the direction of the objective lens when excited with the laser beam, no sample preparation is required. Materials science samples often reflect enough diffuse light for the microscope to be operated in reflection mode. For reflected light studies, gold-coated samples as prepared for the SEM are directly observable without further preparation.

Fluorescent Coating

The fracture surface of the sapphire fibers has a mostly specular behavior. In the surroundings of the fracture origin (defect, subcritical zone and mirror), the slopes involved are near the perpendicular direction to the observer, so a large fraction of the reflected light goes back to the objective lens and through to the detector. Conversely, the branching zone is apparently formed by mirror-like planes at a steep angle to the observer; this means that, once reflected, the light goes away from the objective, making reflection-mode imaging impossible.

A possible way around this problem was using the microscope in fluorescence mode, in which the data acquired matches the distribution of a fluorescent dye in the sample. Thus, a way to stain the sample surface was needed. Desirable characteristics for the process involved ease of coating, good fluorescent yield and uniformity. Usage of commercial fluorescent dyes was to be avoided if possible; since they are commonly used for biological samples, they have special characteristics —useless for this study— which make them expensive and prone to degradation. Finally, the process should ideally be simple enough to apply in any laboratory, with little or no need for protective equipment and manipulation of harmful chemicals.

A process in which a chlorophyll-containing solution was obtained and deposited by evaporation was designed. The chlorophyll is extracted from green leaves, which are cut in small squares; a small number of these leaf pieces is ground in a mortar with a few drops of acetone. Once it has attained a green tint, the acetone is collected in a container and new drops of acetone are poured in. The process continues until the leaf pieces have almost lost their green color, and is then repeated with every remaining set of pieces.

The resulting green liquid, which is a non-saturated solution of chlorophyll and other substances in acetone, should be filtered to remove any small residues and collected in a clean container. The container is then covered and set into a low-temperature heater. Acetone quickly evaporates, leaving the previously dissolved substances on the container walls. Next, a single drop of acetone is poured in the container, rocking it gently so that the drop washes away the deposited substances. In a few seconds, the acetone will reach a deep green color and stop dissolving any more chlorophyll, an indication of solution saturation. The drop is collected in a small vial and the process repeated until the color of the first drop is no longer obtained. The solution is sensitive to light (especially sunlight) and should be stored in a dark place to prevent degradation. Other than that, it has excellent durability, with an observed shelf life of more than 6 months.

The sample surface can then be coated merely by wetting it with a drop of the solution and letting it dry. Since the solution is saturated, the process can be repeated as needed to increase the amount of deposited chlorophyll. To avoid degradation, samples should not be exposed to strong light and should ideally be observed as soon as possible. If needed, the coating can be washed away by cleaning the sample surface thoroughly with acetone.

Confocal Microscope

The confocal microscope used was a Leica TCS-SP2. Experimental settings for reflection and fluorescence mode were similar; a 488 nm Ar laser line was used along with a 20x air objective (0.5 numerical aperture, theoretical lateral resolution about 400 nm). Images were acquired as 1024x1024 pixel slices, with a typical separation between slices of 300 to 500 nm. For fluorescence mode, it was verified that the 488 nm line readily excited chlorophyll fluorescence, which was detected in the red end of the visible spectrum (over 600 nm). As some data deemed of interest (for example, the surface normals in each point) could not be determined with the standard Leica control software, a set of ad-hoc software tools was designed and used.

RESULTS AND DISCUSSION

Reflected Light Mode

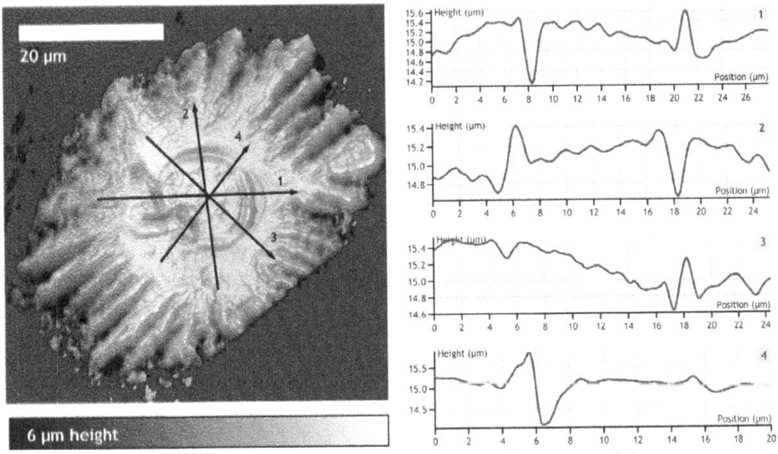

Figure 3: LSCM height map of the D, S and M zones of the fiber in Figure 2 and several associated profiles.

The reflected-light height map of the same fiber imaged in Figure 2 can be seen in Figure 3; Figure 4 shows a 3D projection of the height map. As explained above, the branching zone does not yield signal because of the off-objective reflections. In the visible part, however, many additional details, absent or barely visible in the SEM micrographs, are revealed.

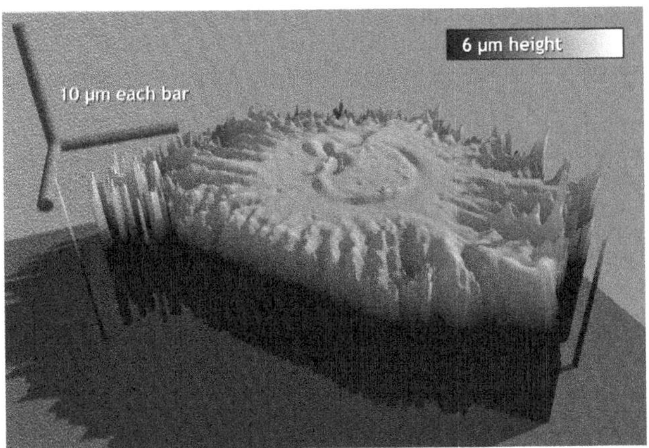

Figure 4: Perspective view of Figure 3.

The most striking feature is a very noticeable relief in the subcritical and mirror zones. The limit between them is much clearer and has an associated height variation, which can be quantified by taking adequate surface profiles (Figure 3). In addition to the characteristic "bump" structure, it can be seen that there is a qualitative difference between the SCG region and the mirror plane, which have different slopes. This has already been reported by using an optical profilometer,[5] but LSCM is able to obtain data for the whole of the surface in a single pass, without the need to choose the desired profile beforehand.

LSCM is especially reliable over extended areas (such as the mirror and SCG planes), but can have trouble at discrete steps. This means that the small bumps in the SCG limit could be an artifact. However, there is evidence to the contrary, since observation of the matching fracture surface yields a mirror valley/hill structure. Besides, if we assume the SCG and mirror zones to be planar and forming an angle to each other, their heights at the SCG boundary will only match in a few points at best. Although the boundary profiles might be somewhat distorted, it can be reasonably stated that the bumps at the SCG limit correspond to a physical feature, most likely a discontinuous, or very steep, transition between one zone and the other.

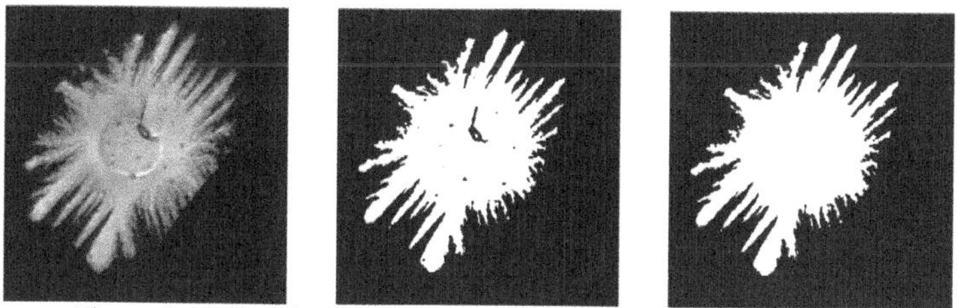

Figure 5: Method for the determination of the mirror zone boundary.

The very marked contrast between mirror and branching in the LSCM images can be used to automatically find their boundary (Figure 5). An LSCM image is converted to grayscale and thresholded; any voids within the main resulting region are filled, yielding a mask of the mirror region. The area comprised by the mask can then be calculated, and its limit can be automatically traced to obtain its perimeter. This procedure represents a tremendous saving of time with regards to hand tracing. For the SCG limit, the enhanced contrast provided by the LSCM image should also prove useful in determining it in a quicker and more reliable way.

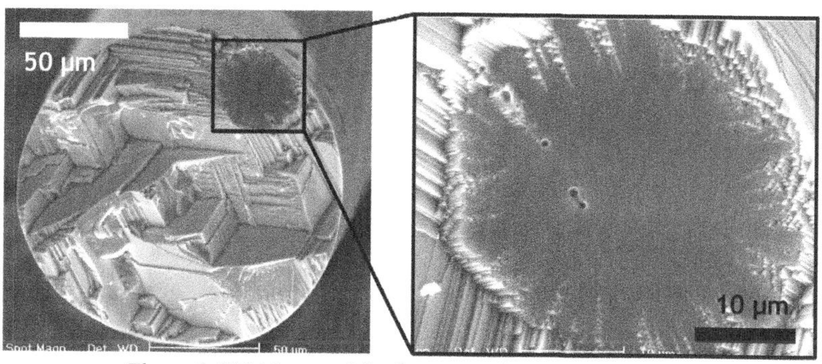

Figure 6: SEM image of the fracture surface of another fiber.
Note that the SCG zone limit is not visible.

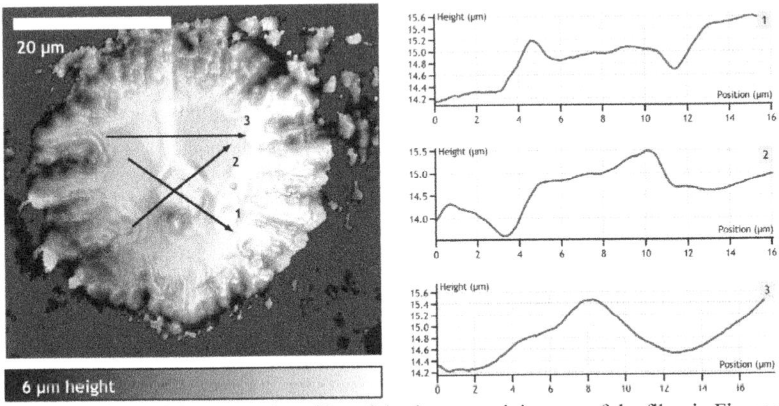

Figure 7: LSCM height map and profiles of the fracture origin zone of the fiber in Figure 6.

Another example can be seen in Figure 6. There, the SEM images show no traces of the SCG limit, which is clearly visible in the LSCM height map (Figure 7). Although the mirror and SCG regions slopes are very close —as could be expected[5]— a transition happens at the SCG limit; in particular, profiles 1 and 2 hint at the existence of a measurable height difference between the two zones. Profile 3 also demonstrates that the mirror plane has significant roughness. It is worth noting that a study based exclusively on SEM micrographs would probably have dismissed the sample as not useful for SCG determination, and these interesting observations would most likely have been overlooked.

Fluorescent Coating

Figure 8: Height map for the whole fracture surface of the fiber in Figures 6 and 7.

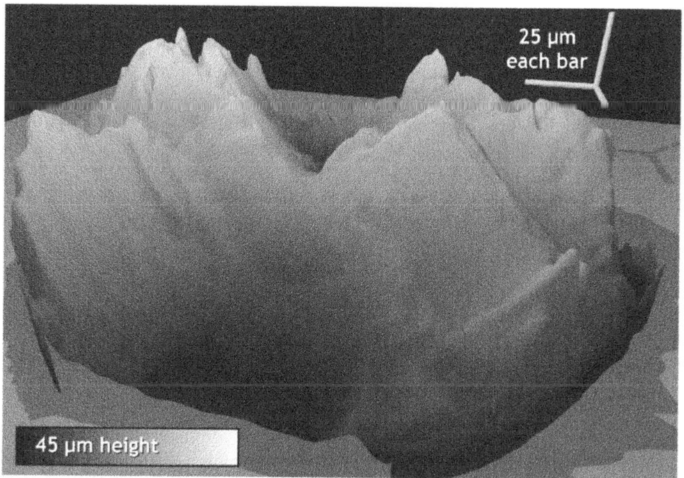

Figure 9: Perspective view of Figure 8.

The sample in Figures 6 and 7 was subsequently coated with chlorophyll for observation of the whole surface; the result can be seen in Figures 8 and 9. The fluorescent coating entails a slight resolution loss, so that the fine detail in the mirror and SCG is lost; however, the branching planes can now be imaged. Since the result is not an image in the classical sense, but a map with numerical values of height, the structure of the surface can be studied in a more reliable way. The height map is especially suitable for data analysis.

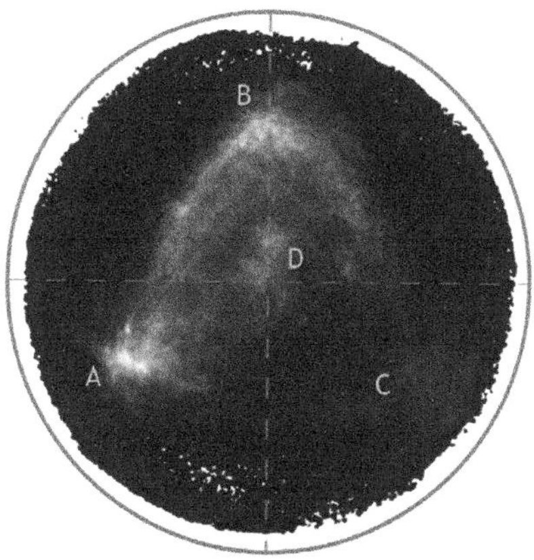

Figure 10: Histogram of the orientations of surface normals of the fiber in Figures 6-9.

For example, if the surface normal in each point is calculated and the orientations plotted in a histogram (Figure 10), four zones are easily discernible: three side maximums, marked as A, B and C, associated to the three predominant branching orientations and a central maximum (D) corresponding to the almost perpendicular subcritical and mirror zones. The C maximum is very faint, since the associated planes are scarce in comparison to the A and B ones. If the histogram is restricted to a few regions within these C planes, however, the maximum can be readily identified.

Existing knowledge of high-temperature fracture implies that the branching occurs in the rhombohedral planes of sapphire.[4] These planes form a family with a threefold symmetry at an angle of 57.6° to the basal plane.[10] Since the raw height and orientation data are available, it can be studied whether the appearance of the A, B and C maximums is consistent with this hypothesis.

Each of the three exact maximum positions was calculated by observation of the histogram data. Then, the orientations corresponding to these three points were obtained. It was found that these orientations are indeed related via a three-fold symmetry to within approximately 1°, and that the average angle to the normal is about 54°. Within the precision of the method —scatter in the histogram has to be taken into account— these results seem to agree very well with the expected theoretical values. Performing such a statistical evaluation over the whole surface with a different technique would probably have been an overwhelming task.

The coating method presented here is not perfect. Deep valleys may appear rounded at the bottom because of the formation of a meniscus, which in turn causes more chlorophyll to be deposited; on the other hand, on sharp peaks the chlorophyll deposition is more difficult because the liquid tends to flow towards the surroundings. However, the obtained results may be considered very positive, since the method permits the imaging of most of the surface and yields consistent results, while being easy, fast, non-destructive and virtually free to try. It is also likely to improve the behavior of most problematic samples; even if a more refined method is needed to get more detailed data, this coating can be applied for a first evaluation before moving on to more expensive alternatives. The identified problems may be addressed by improvements in the surface calculation algorithms and in the coating procedure.

CONCLUSIONS

LSCM has been shown to be a valuable technique for fractographic characterization. It allows fast, robust and non-destructive evaluation of surface height. For problematic samples, a low-cost and simple fluorescent coating method has been introduced. Height determination capabilities, as well as the quantitative nature of the acquisition process, turn LSCM into a very attractive technique, especially when combined with SEM, because of the excellent synergy of their relative strengths and weaknesses.

The high-temperature surface fracture of sapphire fibers was studied with LSCM. Outstanding results include easy SCG limit determination, automated mirror-zone discrimination, calculation of surface height profiles, quantification of the SCG-mirror transition and the characteristic relief in the mirror zones, and statistical evaluation of the fracture surface normals.

ACKNOWLEDGEMENTS

The authors gratefully acknowledge the support of the Spanish Ministerio de Ciencia y Tecnología, project MAT2003-06085. The microscopes used are located at the Microscopy

Service of the University of Seville (Spain), at CITIUS (Center for Research, Technology and Innovation of the University of Seville).

REFERENCES

[1] B. Lawn, "Fracture of Brittle Solids," *Cambridge University Press*, Cambridge, UK, (1993).
[2] J. M. López-Cepero, J. J. Quispe-Cancapa, A. R. de Arellano-López and J. Martínez-Fernández, "Confocal Microscopy for Fractographic Surface Characterization of Ceramics," *Microscopy and Analysis 19* [5], 9-11 (EU) (2005).
[3] A. Diaspro (ed.), "Confocal and Two-Photon Microscopy," *Wiley-Liss*, New York (2002).
[4] J. J. Quispe Cancapa, "Propiedades Mecánicas de Fibras Cerámicas Monocristalinas Fabricadas por Solidificación Direccional," Ph. D. Thesis, University of Seville (2003).
[5] S. A. Newcomb and R. E. Tressler, "Slow Crack Growth in Sapphire Fibers at 800° to 1500°C", *J. Am. Ceram. Soc.*, 76 [10]2505-12 (1993).
[6] S. A. Newcomb and R. E. Tressler, "High-Temperature Fracture Toughness of Sapphire," *J. Am. Ceram. Soc.*, 77 [11] 3030-32 (1994).
[7] S. M. Wiederhorn, "Fracture of Sapphire," *J. Am. Ceram. Soc.*, 52, 485-91 (1969).
[8] S. M. Wiederhorn, B. J. Hockey and D. E. Roberts, "Effect of Temperature on the Fracture of Sapphire," *Phil. Mag.*, 28, 783-96 (1973).
[9] M. Iwasa and R. C. Bradt, "Fracture Toughness of Single-Crystal Alumina," *Advances in Ceramics*, 10, 262-79, (1984).
[10] M. L. Kronberg, "Plastic Deformation of Single Crystals of Sapphire: Basal Slip and Twinning," *Acta Metallurg.*, 5, 507-24 (1953).

LIQUID-INDUCED FRACTURE MARKINGS: AN OVERVIEW

Are Tsirk
Fractography Consultant
17 Argyle Road
Upper Montclair, NJ 07043

ABSTRACT

A brief overview of basic types of liquid-induced fracture markings (LIFMs) encountered in glass is presented, based on numerous observations with several glasses and liquids, as well as various specimen geometries and loadings. To accommodate the variety of markings and the range of fractographic evidence, some modifications and reinterpretations of previously defined LIFMs are made, and some new markings included. Subcavitation hackle is interpreted as bands associated with differential liquid transport, seen as fractographic evidence on inhibition of liquid transport to the fracture front in Region I/II transition as well as in Region II of the curve of fracture velocity V_F vs. the stress intensity factor K_I.

INTRODUCTION

Manifestation of fracture markings due to moisture and liquids depends on the significance of the related environmental effects, recognized for a long time by the phenomena of aging and time-dependent failure. The research by Wiederhorn et al.[1-3] and Schönert et al.,[4] as well as Richter,[5-7] Michalske[8] and Michalske and Bunker[9] has provided a theoretical and quantitative basis for our understanding of the environmental effects (See also Lawn[10]). Before any LIFMs, markings due to moisture were investigated by Fréchette, Quackenbush and Varner[11,12] as well as Richter.[5] The latter are not considered here.

The fundamental basis for LIFMs was established by Michalske and Fréchette,[8,13-19] based on controlled laboratory experiments. In practical situations with irregular geometries, however, a much greater range of LIFMs and patterns can be encountered. Some of these are not recognizable on the basis of the above laboratory experiments, and some require additional considerations for understanding them. Such markings can provide new insights for glass fractography. Since all known LIFMs have been discovered by or under the guidance or influence of Prof. Fréchette, LIFMs have also been termed Fréchette markings.[20]

The objective here is to provide an overview of basic types of LIFMs encountered in glass. It is based on a great number of LIFMs and patterns observed by the author over the past 12 years. Several obsidians, soda-lime and other glasses were used. Most of the fractures were from edge-chipping related to research on the manufacture and use of obsidian tools pertaining to archaeological interests.[21-23] The region near the fracture origin was wetted with liquids including water, human saliva, blood and, in order to drastically vary the liquid properties, various honey-water mixtures at several temperatures. Room temperature was used in all other cases. For better understanding, sonic modulation at 183 Hz was used in some cases. A number of the fracture markings observed were on use-chip scars on obsidian blades of the Aztecs.[22,24] Thus the fractures involved irregular geometries and loadings, variable crack velocities, as well as several liquids and glasses. It was this variability that enabled the observation of a great variety of LIFMs and patterns. The variety of LIFM patterns provides the key for understanding many LIFMs. Though not discussed here, they have been considered for understanding all of the LIFMs. Optical microscopy was used for all observations.

One may pose the question of whether reasonable interpretations on the formation of fracture markings and patterns can be made on the basis of relatively uncontrolled experiments. This is believed possible, by observing the detailed manifestations in various contexts, together with consideration of internal consistency, and still to a large extent with the cited laboratory experiments as a starting point. The readers and time will provide the final judgment.

THE FRACTURE MARKINGS

A brief overview of LIFMs in glass is provided in Table I. An attempt was made to adhere to the definitions in Fréchette.[18] To accommodate the variety of the markings and the broader fractographic evidence, however, modifications have been made and new markings added. "Subcavitation hackle" has been reinterpreted as "linear band features" (See next section). The definition of a scarp has been modified to exclude markings other than moisture-induced or LIFMs (See No.1 in Table 1). Since Fréchette only provided examples of encounter and depletion scarps, general definitions of these are proposed (Nos. 3 and 4 in the table), with deceleration scarps considered as a sub-type of encounter scarps for practical reasons. The term "Sierra marking" is used here for a broad class (No. 7 in the table), requiring neither upstream cavitation nor Taylor instability. The original meaning of a Sierra scarp[18] is retained, but without requiring upstream cavitation.

The appearance of the LIFMs and their "Definition/Mechanism" are noted briefly in the table, along with references to illustrations and previous explanations. All of the markings in the table have been observed or are expected with water in soda-lime glass as well as obsidian, but with some variations.[20] The only marking in the table not observed and identified by the author is the deceleration scarp. The associated mechanisms noted in the table do not, of course, fully explain a LIFM formation. Rather, they are intended to key the reader in to a LIFM designation by providing clues to what is involved.

With regard to designation of LIFMs, two basic points are noted. Firstly, in practice the markings are usually manifested in composite form of two or more LIFM types, and the same marking may be designated as more than one type. For an oval "wet island," for example, the upstream part is a Sierra marking and its downstream part is usually a depletion scarp. The upstream arms of the Sierra marking, associated with an expanding wetted partial fracture front, may be viewed as encounter scarps, which themselves may be escarpment or hackle scarps. Thus, the types of LIFMs in Table I are necessarily overlapping in a sense.

The second point of interest poses a practical dilemma. This relates primarily to a cavitation scarp and markings resembling it. On a specimen broken in bending, for example, the semi-elliptical marking as in Figs. 16 and 17 in Michalske[14] is called a cavitation scarp even though cavitation is associated with only a part of the marking. The author favors such a designation for practical reasons, despite its being imprecise in a strict sense. However, with broader applications in mind, the non-cavitating part could also be called an "abrupt depletion line." In the above case, the complete semi-elliptical marking is indeed a LIFM, even though liquid was responsible for part of its formation indirectly. There are many markings that resemble but are not cavitation scarps. These can be seen with a kink in fracture profile and linear band features just upstream of them. When a depletion scarp is about to form, even a minute stress perturbation may play the major role in producing such a kink as in Fig. 1A. In these cases, the author favors the term "abrupt depletion line," because of the often unavoidable uncertainty regarding the cause of the kink. In such cases, the marking may or may not be a

Table I – Liquid-Induced Fracture Markings (LIFMs)

No.	Marking	Appearance	Definition/Mechanism	Illustration	Reference
1	Scarp	As escarpment scarp, hackle scarp, or cavitation scarp.	A fracture marking formed between wet or moist and dry partial fracture fronts that is caused by liquid (LIFM) or moisture.	—	8, 18, 20
1A	Escarpment Scarp	Usually an escarpment or tilt in section, a line in fracture plane.	A scarp manifested as an escarpment or tilt. Different K_I at the wet and dry partial fracture fronts produce shears in a plane normal to the scarp, at the wet/dry interface. These shears are involved in the manifestation of the marking.	(Figs. 2-27, 2-28, and 2-30)[18]	8, 18, 20
1B	Hackle Scarp	Like a twist hackle, but usually not in fracture direction.	A scarp manifested as a hackle. The same interface shears as for No. 1A, when large enough, are responsible for the hackle.	(Figs. 1a, 3c),[20] (Fig. 4),[22] (Figs. 9B, 9C)[28]	20
2	Cavitation Scarp	Usually a subtle kink in fracture profile, and a smooth arc in fracture plane. Occasionally the kink is absent; termination of linear band features may or may not then suggest the cavitation location. May look like an abrupt depletion line (See text).	A scarp formed due to cavitation at an accelerating fracture front at the Region II/III transition. The cavitation may occur at only a part of the marking, with the rest of it produced indirectly by liquid-induced effects.	(Figs. 16, 17),[14] (Figs. 2-25, 2-26),[18] (Fig. 2a)[20]	8, 14, 18
3	Encounter Scarp	An escarpment scarp, a hackle scarp or a combination.	A scarp formed with wet partial fracture front expanding at the scarp location.	(Fig.2-31, 2-32),[18] (Figs. 13-15)[8]	8, 18
3A	Deceleration Scarp	A special form of an escarpment or a hackle scarp. Rarely identifiable, by Wallner lines or other evidence.	A scarp formed with gradual wetting of the fracture front when it decelerates through Region III/II transition. Wetting retards and enhances fracturing at its upstream and downstream parts, respectively.	(Figs. 24, 25),[8] (Fig. 2-27)[18]	8, 18

No.	Marking	Appearance	Definition/Mechanism	Illustration	Reference
4	Depletion Scarp	An escarpment scarp; extremely rarely a hackle scarp.	A scarp formed with wet partial fracture front retracting at the scarp location.	Fig. 1A, (Fig. 2-33),[18] (Figs. 1b, 2, 3a, 3b, 4)[20]	8, 18, 20
4A	Pointed-looped Scarp Pattern	A scarp resembling an inverted, multi-looped Sierra scarp. The loops may be regular or irregular, smooth or jagged.	A depletion scarp with the appearance described. Suggested mechanism: Initiation due to an instability; persistence related to energy considerations, with limitation of liquid to a fracture front in an overall sense.	(Figs. 2a and b)[20]	—
5	Partition Marking	Looks like an escarpment scarp or a hackle scarp.	Formed with liquid at the partial fracture front on each side. A LIFM, but not a scarp, manifested due to different liquid transport at its sides.	—	—
5A	Partition Hackle	Looks like a twist hackle, but usually does not extend in the fracture direction. Usually the part of a hackle zig-zag or an L-shaped hackle that is not in the fracture direction.	A LIFM manifested as a hackle with different liquid transport at its sides (because lateral liquid transport is prevented before and probably inhibited after the lateral break-through of a hackle). This leads to unequal effective K_I at the partial fracture fronts and to interface shears in a plane normal to the hackle.	Fig. 1b, (Fig. 9B)[28]	20, 28
5B	Partition Escarpment	Observed at some boundaries of linear band features (No. 8). Looks like an especially subtle escarpment scarp. May be laterally diffuse, ill-defined.	A LIFM with the formation mechanism suggested to be akin to that for the starvation (No.8A) and depletion bands (No.8B). With forked fingerlet bands, it is associated with enhanced liquid transport in the bands.	Fig. 3A and 3B	—
6	Zig-zag and L-shaped Hackle Patterns	A hackle with a zig-zag or a L shape. The part not in the fracture direction is a hackle scarp or a partition hackle, with the remaining part of the hackle being a twist hackle.	The turn or turns of the hackle relate to competition of two mechanisms – those for the twist hackle and those for the liquid-induced hackle.	Fig. 1B, (Fig. 9C),[28] (Fig. 1b)[20]	20,28

No.	Marking	Appearance	Definition/Mechanism	Illustration	Reference
7	Sierra Marking	Various forms of escarpment or hackle scarp (or partition escarpment or hackle?*) or their combination. Usually concave in the fracture direction.	Formed when liquid from behind a dry (or starving?*) fracture front catches up with it, to affect parts of the fracture front gradually.	—	18
7A	Sierra Scarp	Manifested as multiple loops, usually contiguous, concave in the fracture direction.	Formed with the liquid behind a dry fracture front, as fingerlets due to Taylor instability, catching up to wet the fracture front gradually. Manifested usually without cavitation upstream.	(Figs. 2-28, 2-30),[18] (Fig. 2)[22]	18
7B	V Marking	Manifested in the form of a V opening in the fracture direction. Each side may be an escarpment or hackle scarp (or partition escarpment or hackle?*) or a combination of these.	A V-scarp is formed with the liquid front, resembling a V opening in the upstream direction, catching up from behind the dry fracture front. Perhaps other V-markings (partition escarpments or partition hackles) could be formed when liquid from behind a starving rather than dry fracture front catches up with it?*	(Figs.2C and 5)[20]	20
7C	Y Pattern	Looks as if a twist hackle splits into a V, with one side of the V being a hackle and the other an escarpment.	Liquid from behind a dry (or starving?*) fracture front catches up with it preferentially at and to one side of the twist hackle upstream of the V.	(Fig. 5)[20]	20
7D	Ψ Pattern	As a Y-pattern but with a tiny twist hackle, appearing to split the V, extending from the twist hackle upstream of the V.	As a Y-pattern, with the mechanism related to the liquid transport at the tiny twist hackle splitting the V unknown.	(Fig. 1d)[20]	20
7E	Slurp Scarp	The hackle offset part of a hackle zig-zag with a miniscule escarpment scarp adjacent to and downstream of the offset.	Liquid from behind a dry fracture front is sucked in to the fracture front preferentially within or at only one side of a twist hackle, causing it to turn as a hackle scarp and to manifest the escarpment scarp.	(Fig. 1c)[20]	20

No.	Marking	Appearance	Definition/Mechanism	Illustration	Reference
8	Linear Band Features	Valleys and ridges appearing as lines or bands in the direction of fracture or liquid transport.	LIFMs having the appearance described. Their onset is suggested to relate to instability in liquid transport leading to "channels" of different liquid transport, which produce shear stresses in a plane normal to the band, causing the valleys and ridges.	Figs. 1A, 2 and 3, (Figs. 2a, 3b, 3c, 7)[20]	20
8A	Starvation Bands	A linear band feature with liquid depletion usually at a scarp or an abrupt depletion line directly downstream of it.	Linear band features with limited liquid supply to the fracture front within the band.	Fig. 1A, 2 and 3, (Figs. 3a, 4c)[20]	—
8B	Depletion Bands	Linear band features with no liquid depletion at a scarp or other fracture marking directly downstream of it.	Linear band features with liquid being completely depleted in the band.	Figs. 2 and 3, (Figs. 3a and 5)[20]	—
8C	Forked Fingerlet Bands	Bands usually resembling escarpment scarps in the form of forked fingerlets, appearing as a "sparse" or "dense" population. Some shoulders of the bands can be partition hackles.	The bands are associated with enhanced liquid transport. Suggested manifestation: due to different liquid supply within and outside a band. At least some of the parts resembling escarpment scarps are suggested to be partition escarpments (No. 5B), the counterparts of partition hackle.	Fig. 3	—
9	Wake Scarp	A scarp downstream of an irregularity such as an inclusion, a twist hackle or wake hackle.	A scarp due to an irregularity affecting liquid transport to the fracture front downstream of it.	(Fig. 9A)[28]	28

* refers to possibility being investigated.

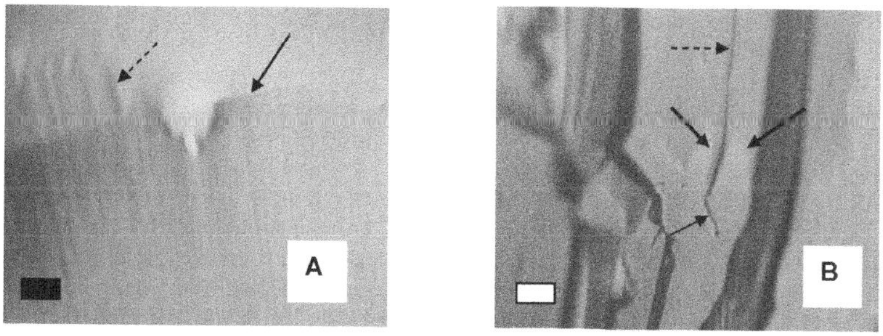

Fig. 1 – (A). Abrupt depletion line (arrow) with adjacent, downstream depletion scarps (dashed arrow). Note its resemblance to a cavitation scarp. The line here is due to a stress perturbation from sonic modulation, and is therefore not a LIFM. A sonic modulation line coincides with the abrupt depletion line and extends to the left of it. Linear band features are seen upstream of both markings. Soda-lime glass fractured with water. Bar = 84 μm. Fracture direction is upward on all photos.

(B). Partition hackle (light arrow) at an L-shaped hackle pattern with a twist hackle (dashed arrow) downstream of it. Liquid was present at each side of the partition hackle, as seen by the depletion scarp (bold arrows). Mexican obsidian (source unknown) fractured with blood. Bar = 25 μm.

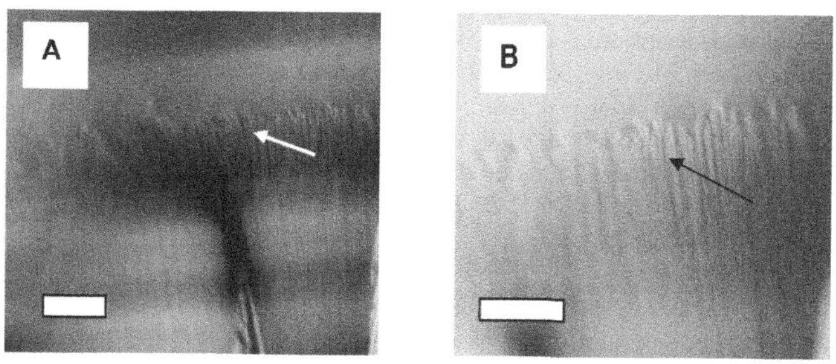

Fig. 2 – Linear band features: Depletion bands (as at arrow) seen among many starvation bands. Same region in (A) (Bar = 140 μm) and (B) (Bar = 88 μm). Mexican obsidian (source unknown) fractured with water, using sonic modulation at 183 Hz. V_F (fracture velocity) is about 3.6 cm/s.

LIFM. The significant point about abrupt depletion lines is that they can easily be mistaken for cavitation scarps (Fig. 1A).

Table I is not intended to include all types of LIFMs, especially their sub-types. In all cases with one or more sub-types indicated, it would be appropriate to add a subcategory "other types." And there are also basic types of LIFMs not considered here.

LINEAR BAND FEATURES: A REINTERPRETATION OF SUBCAVITATION HACKLE

The markings known as subcavitation hackle[18] and transition hackle[8,14] are suggested to be "channels" (usually with diffuse boundaries) corresponding to variable liquid transport to the fracture front, mostly in the fracture direction. Usually manifested as valleys and ridges, they are referred to as "linear band features"[20] or simply "linear bands" here (No. 8 in Table I).

Three kinds of linear band features are distinguished here. Starvation bands (No. 8A in the table) are suggested to be associated with liquid "starvation" at the fracture front, but with no complete depletion of the liquid in the band. In those cases, a complete loss of liquid at the fracture front occurs most often at the scarp directly downstream of them. On a jagged depletion scarp, the scarp is manifested sooner (further upstream) than typical at some locations, as in Figs. 1A and 2 (See also Figs. 3a and 3b in Tsirk[20]), thought to be indicative of liquid "starvation" in the associated bands.

A depletion band (No. 8B in the table) is suggested to be associated with complete liquid depletion at the tiny partial fracture front associated with it. In the case of a cavitation scarp or an abrupt depletion line downstream, the two types of linear band features are usually difficult to distinguish. A depletion band is also associated with liquid "starvation," but to such an extent that the liquid is completely depleted, as in Fig. 2 (See also Fig. 5 in Tsirk[20]). Depletion bands are rather common. However, they would not have been evident with any of the LIFMs considered in Michalske.[8]

Linear band features of a third kind, the forked fingerlet bands (No. 8C in the table) (See Fig. 3), have been observed recently. In an industrial waste glass with many inclusions of variable sizes, such bands are common with water (Fig. 3A and B), occurring usually in lieu of the more usual kinds of linear band features. They are interpreted as bands of enhanced liquid transport, with at least some of their boundaries suggested to be partition escarpments or partition hackle (See No. 5A and 5B in the table). (If one wishes, the sides of starvation bands and parts of depletion bands, discussed above, may also be viewed as partition escarpments that are laterally diffuse, ill-defined.) The forked fingerlet bands may perhaps be parts of a system resembling a river with its tributaries and deltas.

Forked fingerlet bands have occasionally been observed in obsidian with a 1:1 honey-water mixture at room temperature (Fig. 3C and Tsirk:[20] Fig. 5). It is now uncertain whether the upstream ends of the forked fingerlet bands manifested as a V-marking and a Y-pattern (Nos. 7B and 7C in the table, respectively) in the cited figure[20] are in fact scarps (with liquid on one side) or partition markings (with liquid on two sides). V-markings have been observed as some upstream ends of the forked fingerlet bands in the industrial waste glass (with water) in a general region that was wetted. Though previously considered inconceivable, the possibility is presently being studied of whether some Sierra markings (See No.7 in Table I) can also be manifested by liquid gradually reaching a starving rather than a dry fracture front.

Markings that are probably forked fingerlet bands were observed in soda-lime glass with water by Richter in the early 1980's.[25] They are manifested in such a context indeed (Fig. 3D), but under particular conditions unclear at this time.

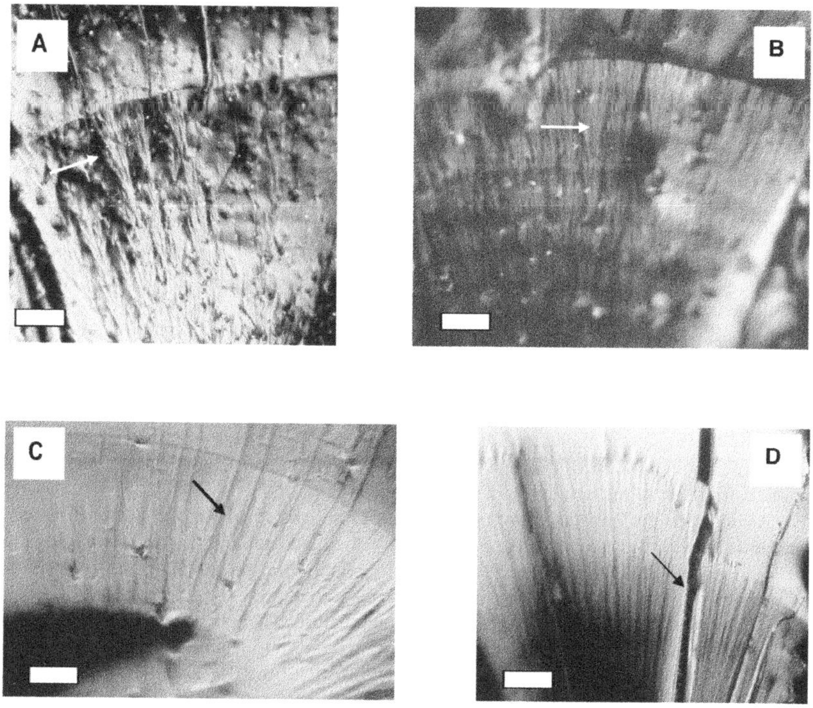

Fig. 3 – Forked fingerlet bands:

(A) and (B). Forked fingerlet bands (as at arrows) seen as a sparse (A) and dense population (B). Lines from sonic modulation at 183 Hz are barely seen in (A) and (B), indicating V_F to be in the range of about 0.8 to 1.2 cm/s. An industrial waste glass with many inclusions, fractured with water. Bar = 140 μm in (A) and 73 μm in (B).

(C). Forked fingerlet bands (as at arrow) in Mexican obsidian (source unknown) fractured with a 1:1 mixture of honey-water at room temperature. Bar = 70 μm.

(D). Forked fingerlet bands (as at arrow) in soda-lime glass fractured with water. In the region to the right of the arrow and the twist hackle, the range of V_F is about 2 to 3 cm/s. Bar = 145 μm.

Forked fingerlet band systems serve as empirical indicators for differential liquid transport with linear band features. Observations suggest that the forked fingerlet bands are manifested in broader contexts than just those usually associated with what have been considered subcavitation or transition hackle.[8,18]

The rest of this section pertains to the first two kinds of linear band features (Nos. 8A and 8B) discussed above, since the forked fingerlet band systems are still being studied.

It is suggested that linear band features are initiated by instability, and that their persistence relates to energy considerations. As always, instability can be triggered by an arbitrarily small perturbation of variable nature. The instability condition, though not known, is expected to relate to the availability of liquid at the fracture front. The V_F for the onset of linear band features is therefore expected to depend on not just that velocity itself, but also on the history of the V_F upstream of the onset location.

Consistent with the interpretation for the linear band features is the observation that they terminate at sustained low V_F, when there is no longer liquid "starvation" at the fracture front needed for their manifestation, and perhaps also by the observation that they sometimes terminate within a looped depletion scarp (as No. 4A in the table) upstream of the scarp. This is indicated by the suggested liquid starvation at the fracture front in an overall sense just prior to the onset of a pointed-looped scarp pattern and by the occasional localized replenishment and lateral redistribution of the liquid within such a looped scarp.

Aside from the general energy considerations suggested above for the persistence of linear band features, the mechanisms responsible for the manifestation of the valleys and ridges are of interest. For the generally accepted normality relation for the maximum tension direction and the fracture surface, the valleys and ridges imply the presence of very localized shear stresses in the plane normal to the direction of a band feature. Two conceivable sources are seen for such shear stresses, both dependent on the differential liquid transport in the "channels." One of these may perhaps relate to the three-dimensional effects of the fluid, including viscous drag, at and near the fracture front. The other relates to the differences in K_I required to advance the minute partial fracture fronts with different amounts of liquid available. Such a mechanism is akin to that discussed in reference to Fig. 8 in Tsirk[20] for wet and dry partial fracture fronts. However, in contrast to the cited discussion, the availability of the fluid along the fracture front would vary gradually here, rather than having an abrupt change as in the wet and dry case. Manifestation of the valleys and ridges undulating in a transverse section (normal to the length of the band feature) and usually not steps or hackles is consistent with the suggested mechanism.

It is not suggested here that linear band features can never be manifested truly as hackles instead of valleys and ridges. For example, Beauchamp[26] has indeed observed linear band features occasionally reverting to hackles and then back to the valleys and ridges of the linear band features. Also, in the special context of the industrial waste glass considered above, liquid-induced hackles (probably partition hackles) have indeed been observed.

The above discussion on linear band features implies that these markings represent fractographic evidence of limitation on liquid transport to a fracture front also in an overall sense. In Michalske's experiments,[8] these markings were observed to start at about the same velocity where the V_F-vs.-K_I curve starts to bend over towards the curve for the dry case in the Region I to Region II transition.[8] The interpretation of linear band features thus points to inhibition of liquid transport playing a role in this transition as well as in Region II. Indirectly, the different nature of liquid transport associated with these features is expected to also have implications on the viscous effects of the liquid on fracture propagation. In reference to the above remarks, incidentally, it is of interest to note that a fracture front may become dry at depletion scarps without cavitation occurring. This is usually so with the edge-chipping cases considered. Limitations on liquid transport are evident in such cases, as are the serious implications on the viscous effects of the liquid.

Also, fracturing with the valleys and ridges is different from that with a planar surface. A greater energy expenditure is expected simply because of the greater surface area, as well as a need for a greater effective K_I in the former case. These factors may influence the V_F-vs.-K_I curve, though to some unknown and perhaps miniscule extent. Michalske[8] does discuss energy expenditure with the formation of subcavitation hackle.

It is suggested that for understanding linear band features, it is relevant to think of what is happening between the traditional wet and dry V_F-vs.-K_I curves.

CONCLUDING REMARKS

Investigation of fractures without strict laboratory controls has its merits. It can provide a greater variety of LIFMs and patterns (the latter not discussed here), as well as fractographic evidence that may otherwise not be evident. Such investigations can be useful in generalizing the necessarily limited experimental results, and they can also serve as a screening process to suggest potentially fruitful avenues of research with controlled experiments.

LIFMs have also been observed in single crystals.[26,27] Hackle scarps are expected but have not yet been observed in crystalline materials.

ACKNOWLEDGEMENTS

I thank Ernie Chrisbocker and Jack Cresson for providing some of the glass material. I gratefully acknowledge the fruitful discussions with Drs. Edwin Beauchamp and the late Prof. M.P. Bieniek. I am grateful to Dr. H.G. Richter for insightful discussions and helpful comments on a draft of this article. My greatest debt is to the late Prof. V.D. Fréchette for enlightening and inspiring discussions over a period of 22 years. I am deeply grateful to the late Don Crabtree for teaching me the production of fractures in knapping, inspiring me to inquire about its scientific aspects, and putting me in contact with Prof. Fréchette even before I got involved with fractography. Finally, I am indebted to the anonymous Aztecs for some of the obsidian fractures, and thank Profs. William Parry and T.H. Charlton for allowing access to them.

REFERENCES

[1]S.W. Wiederhorn, "Influence of Water Vapor on Crack Propagation in Soda-Lime Glass," *J. Am. Ceram. Soc.*, 50, 407-14 (1967).

[2]S.W. Wiederhorn, "Mechanisms of Subcritical Crack Growth in Glass," pp. 549-80 in *Fracture Mechanics of Ceramics*, Vol. 4, eds. R.C. Bradt, D.P.H. Hasselman, and F.F. Lange, Plenum Press, New York (1978).

[3]S.W. Wiederhorn, S.W. Freiman and E.R. Fuller, "Effects of Water and other Dielectrics on Crack Growth," *J. Am. Ceram. Soc.*, 17, 3460-78 (1982).

[4]K. Schönert, H. Umhauer and W. Klemm, "The Influence of Temperature and Environment on the Slow Crack Propagation in Glass," pp. 474-82 in *Fracture 1969*, ed. P.L. Pratt, E.H. Andrews, R.L. Bell, N.E. Frost R.W. Nichols and E Smith, Chapman and Hall, London (1969).

[5]H.G. Richter, "Experimentelle Untersuchungen zur Rissausbreitung in Spiegelglas im Geschwindigkeitsbereich 10^{-3} bis $5x10^{3}$ mm/s," *Wissensch. Ber.*, 9/74, Institut für Festkörpermechanik der Fraunhofergesellschaft, Freiburg (1974).

[6]H.G. Richter, "Zum Einfluß umgebender Medien auf die Rißausbreitung in Glas in einem Zwischenbereich von Bruchgeschwindigkeiten," *Glastech. Ber.*, 56K-1, 402-7 (1983).

[7]H.G. Richter, "Crack Propagation in Glass Under Liquids in an Intermediate Range of Crack Velocities," pp. 219-29 in *Strength of Inorganic Glass*, ed. Charles R. Kurkjian, Plenum Press, New York (1985).

[8]T.A. Michalske, "Dynamic Effects of Liquids on Crack Growth Leading to Catastrophic Failure in Glass," Ph.D. Thesis, Alfred University, Alfred, 1979.

[9]T.A. Michalske and B.C. Bunker, "The Fracturing of Glass," *Sci. Amer.*, 257, 122-9 (1987).

[10]B. Lawn, *Fracture of Brittle Solids*, 2nd ed., Cambridge University Press, Cambridge (1993).

[11]J.R. Varner and V.D. Fréchette, "Fracture Marks Associated with Transition Region Behavior of Slow Cracks in Glass," *J. Appl. Phys.*, 42, 1983 (1971).

[12]C.L. Quackenbush and V.D. Fréchette, "Crack Front Curvature and Slow Glass Fracture," *J. Am. Ceram. Soc.*, 61, 402- 4 (1978).

[13]T.A. Michalske, "Fractography of Slow Fracture in Glass," pp. 121 36 in "Fractography of Ceramic and Metal Failures," eds. J.J Mecholsky Jr. and J.R. Powell Jr., *ASTM Special Technical Publication* 827, American Society for Testing and Materials, Philadelphia (1984).

[14]T.A. Michalske, "Quantitative Fracture Surface Analysis," pp. 652-62 in *Engineered Materials Handbook, Vol. 4: Ceramics and Glasses*, ASM International, Materials Park, OH, (1991).

[15]T.A. Michalske, "Fractography of Stress Corrosion Cracking in Glass," pp. 111-42 in *Fractography of Glass*, eds. R. C. Bradt and R. E. Tressler, Plenum Press, New York (1994).

[16]V.D. Fréchette, "Fracture of Glass in the Presence of H_2O," *Glastechn. Ber.*, 58, 125-9 (1985).

[17]V.D. Fréchette, "Markings Associated with the Presence of H_2O During Cracking," pp. 71-6 in *Fractography of Glasses and Ceramics*, eds. J.R. Varner and V.D. Fréchette, American Ceramic Society, Westerville, OH (1988).

[18]V.D. Fréchette, *Failure Analysis of Brittle Materials*, American Ceramic Society, Westerville (1990).

[19]T.A. Michalske and V.D. Fréchette, "Dynamic Effects in Crack Growth Leading to Catastrophic Failure in Glass," *J. Am. Ceram. Soc.*, 63, 603-9 (1980).

[20]A. Tsirk, "An Exploration of Liquid-Induced Fracture Markings," pp. 87-101 in *Fractography of Glasses and Ceramics IV*, ed. J.R. Varner and G.D. Quinn, American Ceramic Society, Westerville (2001).

[21]A. Tsirk, "On Flintknapping, Presentations at the University of Tartu, Estonia," Unpublished manuscript (1997).

[22]A. Tsirk and W.J. Parry, "Fractographic Evidence for Liquid on Obsidian Tools," *J. Archaeol. Sci.*, 27, 987-91 (2000).

[23]J.C. Whittaker, *Flintknapping: Making and Understanding Stone Tools*, University of Texas Press, Austin (1994).

[24]A. Tsirk, "Fractographic Evidence for Liquid on Aztec Blades from Otumba: An Exploratory Investigation," Report No. 020213 (2002).

[25]H.G. Richter, personal communication, (2006).

[26] E.K. Beauchamp, Personal Communication, (2005).

[27]G.D. Quinn, Personal Communication, (2002).

[28]A. Tsirk, "Hackles Revisited," pp. 447-72 in *Fractography of Glasses and Ceramics III*, eds. J.R. Varner, V.D. Fréchette and G.D. Quinn, American Ceramic Society, Westerville, 1996.

OPTICAL MICROSCOPY OF MULTIPLE MATRIX CRACKING IN A CARBON FIBER-REINFORCED GLASS MATRIX COMPOSITE

Kouichi Yasuda, Koji Hashimoto, Tadashi Shiota and Yohtaro Matsuo
Department of Metallurgy and Ceramics Science
Faculty of Science and Engineering,
Graduate School, Tokyo Institute of Technology
2-12-1-S7-14, Ookayama, Meguro-ku, Tokyo 152-8552, Japan

ABSTRACT

This paper treats the spatial distribution of multiple matrix cracking in fiber-reinforced ceramics matrix composites. In-situ observation was carried out on a carbon fiber-reinforced PYREX glass composite on tensile loading, and the spatial distribution of the matrix cracks was determined by a compare check on its optical micrographs. A stochastic model is developed with an assumption that a new matrix crack is generated only in the effective cell in which the matrix stress remains uniformly high. As compared with the experimental data, the average position of the matrix cracks in the effective cell coincides well with the prediction by the theory; however, the standard deviation is still slightly different from it. The discrepancy is discussed in relation to the underestimation of the effective cell size.

INTRODUCTION

It is commonly recognized that fiber-reinforced ceramic matrix composites have excellent performance in energy absorption during the fracture. For example, their work-of-fracture values are mostly two or three order of magnitude larger than those of monolithic ceramics. Figure 1 shows a schematic illustration of microfracture processes of a composite on tensile loading. After some elastic deformation, first matrix cracking occurs in the composite. If its interfacial bond between fiber and matrix is very strong, the first matrix crack instantaneously cuts through the fibers at the crack plane, and the composite fractures in a brittle manner. If it is not so strong, interfacial debonding occurs at the matrix crack together with yielding elastic elongation of the fibers, and frictional energy is largely dissipated at the debonded interface.[1] With increasing the applied stress, the interfacial debonding gradually extends inside, and at a certain applied stress, the fiber also breaks somewhere in the debonded part. After that, fiber pullout begins to operate accompanying frictional energy dissipation until the fiber is completely pulled out of the matrix socket.[1] Finally, all the fibers are pulled out around the matrix crack, and the composite has fractured in a quasi-ductile manner.

If a lot of matrix cracks are generated one after another in the composite, this phenomenon is called multiple matrix cracking. Obviously, frictional energy dissipation works in parallel at each matrix crack, so that this kind of fracture should result in the maximum performance in energy absorption. However, the related papers on this subject were intended for estimation of either the first matrix cracking stress[2-4] or the ultimate strength of composites,[5-8] and therefore multiple matrix cracking itself has not been fully understood up to now. In this study, a stochastic theory was developed for the spatial distribution of multiple matrix cracking. In-situ observation was conducted during a tensile test for a carbon fiber-reinforced glass matrix composite. Finally, the experimental data was compared to the prediction by the theory.

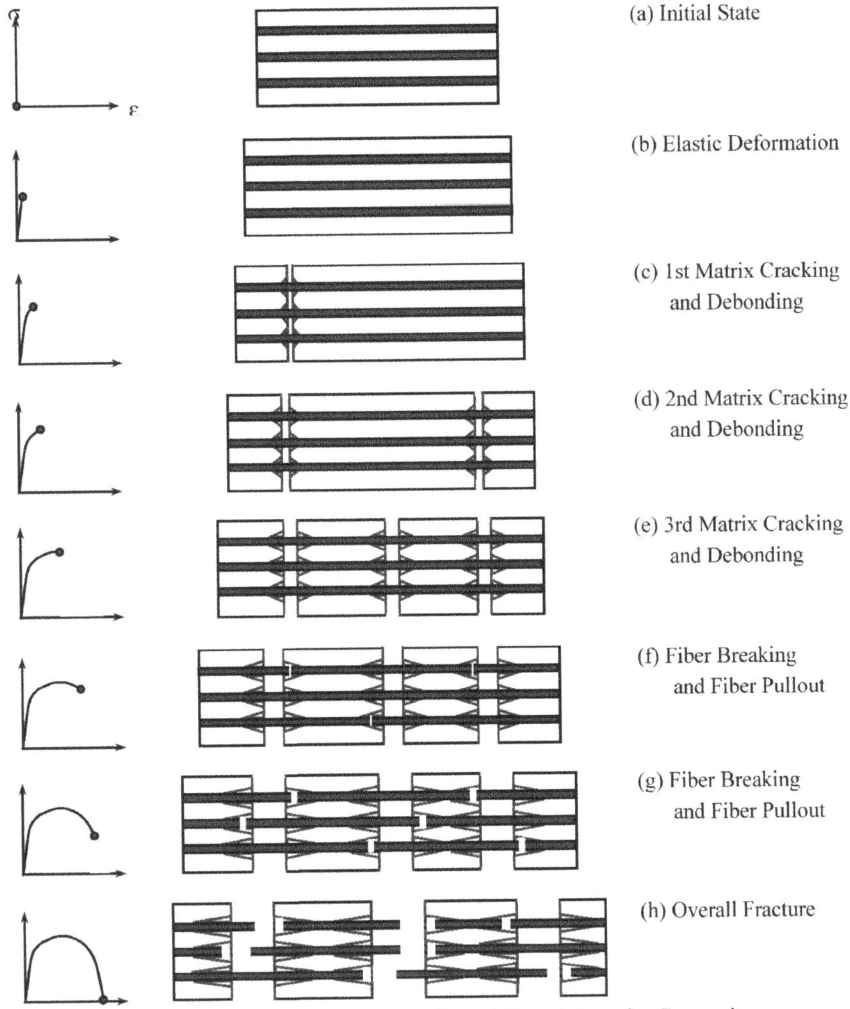

Figure 1 – Microfracture Processes of Fiber-reinforced Ceramics Composites.

THEORY

Spatial Distribution of Multiple Matrix Cracking

Figure 2 shows an illustration of a stochastic model considered in this study. Assume that two matrix cracks have already been in a unidirectional fiber-reinforced composite under a nominal tensile stress $\square_n$. We focus on a region bounded by these matrix cracks with an interval of L, and call it a "cell" in which a new matrix crack will appear upon further loading. The new matrix crack is specified by the x-coordinate along the fiber axis, and the origin is set to be the midpoint of the cell. Assume also that interfacial debonding has extended from the both ends of the cell to the position at a distance of $\square$. Because of degradation in stress transfer at the debonded interface, the tensile stress may be vanishingly small in the matrix around there. As the first order approximation, we assume that the new matrix crack is not generated in an inert region around the both ends of the cell, and also that the inert length ζ is a sum of the debonding length ξ and the stress recovery length η, as shown in Fig.2. Thus, we just have to consider the region from $-(L/2-\zeta)$ to $+(L/2-\zeta)$ in x-coordinate, and define this region as an "effective cell". The inert length ζ shall be quantitatively estimated in the later section.

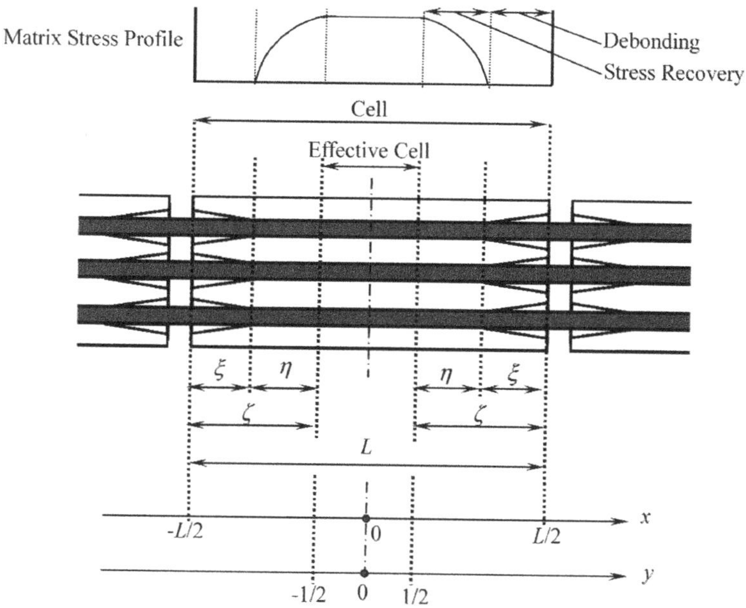

Figure 2 – Stochastic Model of Multiple Matrix Cracking.

If we try to statistically analyze experimental data as it is, we will encounter difficulty in picking up even two or three effective cells of equal size from the data. To average the experimental data collectively, it is necessary to normalize the size of the effective cell. So, we introduce the dimensionless y-coordinate as below,

$$y = \frac{x}{L - 2\zeta} \tag{1}$$

where the domain of y-coordinate is (-0.5, +0.5). From the discussion above, we suppose that the matrix stress is almost uniform in the effective cell, and consequently the probability density function $p(y)$ expressing the spatial distribution of multiple matrix cracking is given by the following equation,

$$p(y) = 1 \tag{2}$$

By using this function, we can easily obtain the expected value $\bar{y}$ and the standard deviation σ_y of the position of multiple matrix cracking in the following equations,

$$\bar{y} = \int_{-\frac{1}{2}}^{\frac{1}{2}} y p(y) dy = \int_{-\frac{1}{2}}^{\frac{1}{2}} y dy = 0 \tag{3}$$

$$\begin{aligned} \sigma_y &= \sqrt{\overline{y^2} - \left(\bar{y}\right)^2} = \sqrt{\int_{-\frac{1}{2}}^{\frac{1}{2}} y^2 p(y) dy - 0} \\ &= \sqrt{\frac{1}{12}} \cong 0.288 \end{aligned} \tag{4}$$

From these equations, it follows that each matrix crack is uniformly distributed over the effective cell; however, their average position should be at the midpoint of the effective cell (viz. the origin in y-coordinate). It also turns out that their standard deviation should be a constant value of 0.288.

Estimation of Inert Length ζ

With reference to the analysis of Marshall and Oliver[9] for fiber push-in test, the similar derivation was made by the present authors for tensile loading on a single fiber in the composite.[1] The resultant equation for debonding length ξ is shown below,

$$\xi = \frac{r_f \sigma_f}{2\tau} - \frac{\sqrt{2E_f \Gamma r_f}}{\tau} \tag{5}$$

where r_f denotes the radius of the fiber, σ_f the applied tensile stress on the fiber, τ the interfacial frictional stress after debonding, E_f the Young's modulus of fiber, Γ the formation energy for interfacial debonding. In a practical calculation, the tensile stress of the fiber (σ_f) should be given by the following relation,

$$\sigma_f = \frac{\sigma_n}{V_f} \tag{6}$$

where σ_n denotes the nominal tensile stress of the composite, V_f the volume fraction of fiber. Note that this equation is not strictly valid at the final stage of fracture of the composite because the intact fiber assumption is broken.

The stress recovery length η is derived here based on the analysis of Budiansky, Hutchinson, and Evans[4] as shown in Fig.3. They showed the matrix stress distribution $\sigma_m(r,z)$ when the tensile stress was applied to a single fiber surrounded by a cylindrical matrix without interfacial debonding as below,

$$\sigma_m(r,z) = \overline{\sigma_m}\left\{1 - \exp\left(-\frac{\rho}{r_f}z\right)\right\} \tag{7}$$

where r and z are cylindrical coordinates, $\overline{\sigma_m}$ is the average matrix stress calculated from the rule of mixture. The dimensionless parameter ρ was defined as,

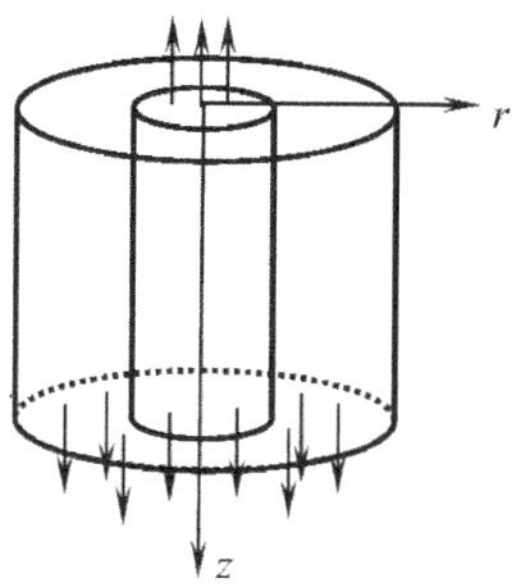

$$\sigma_f = \frac{\sigma_n}{V_f} \quad (z=0)$$
$$\sigma_m = 0$$

$$\sigma_f = \overline{\sigma_f} \quad (z=\infty)$$
$$\sigma_m = \overline{\sigma_m}$$

Figure 3 – Model for Matrix Stress Analysis by Budiansky et al.[4]

$$\rho = \sqrt{\frac{2G_m E_c}{V_m E_m E_f \ln\left(\frac{r}{r_f}\right)}} \tag{8}$$

where G_m denotes the modulus of rigidity of matrix, V_m the volume fraction of matrix, E_c the Young's modulus of the composite. Here, we adopt a triangular array of fibers as shown in Fig.4, and estimate the matrix stress profile passing through the point A (small closed circle) along the fiber axis by superimposing contributions of adjacent three fibers around the point. And we determine the stress recovery length η by the position where the stress level reaches 99.9% of the average matrix stress $\overline{\sigma_m}$. Thus, we obtain the following expression,

$$\eta = r_f \sqrt{\frac{V_m E_m E_f \ln\left(\frac{u}{\sqrt{3}r_f}\right)}{2G_m E_c}} \cdot \ln\left(\frac{1}{1-0.999}\right) \tag{9}$$

where u denotes the side length of the triangular unit cell, and is expressed by the following,

$$u = \sqrt{\frac{2\pi r_f^2}{\sqrt{3}V_f}} \tag{10}$$

Although the derivation of the equation (9) seems to be somewhat rough, it is revealed that the calculated η value by the equation coincides well with the numerical result by 3-dimensional FEM (ABAQUS).

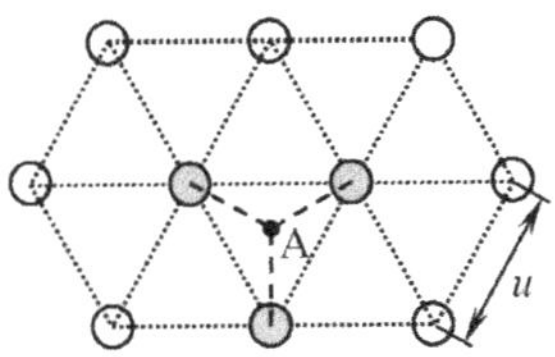

Figure 4 – Triangular Array of Fibers in the Composite Model.

EXPERIMENTAL PROCEDURE

Sample Preparation

Slurry was made by mixing 23 mass% PYREX glass powder (Corning, #7740), 76 mass% distilled water and 1 mass% polyvinyl alcohol in a magnetic stirrer. Carbon fiber (Toray,

Torayca T300B) was immersed in the slurry, and then wound on a spool to make prepreg sheets in which carbon fibers aligned parallel each other. The prepreg sheet was sintered by hot pressing at 750 °C for 90 min under 11MPa in vacuum. The heating rate was 5 °C/min, and the cooling rates were 5 °C/min (750 °C ~ 550 °C) and 3 °C/min (550 °C ~ room temperature). On the way to the room temperature, annealing was performed at 550 °C for 20 min to reduce residual stress in the composite. Thus, the carbon fiber-reinforced PYREX glass composite was obtained.

Rectangular-sectioned specimens were made by cutting out of the composite, and followed by mirror polishing with diamond slurry. The dimensions of the specimens were 4 mm in breadth × 1 mm in thickness × 40 mm in length. As shown in Fig.5, steel tabs were finally attached to the both ends of the specimens with epoxy resin.

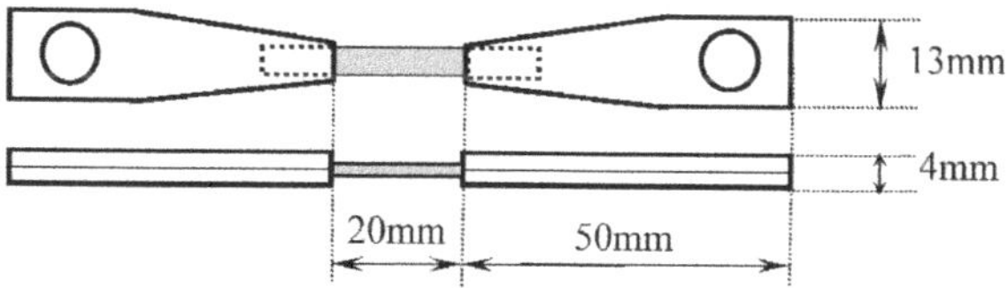

Figure 5 Configuration of Tensile Specimen.

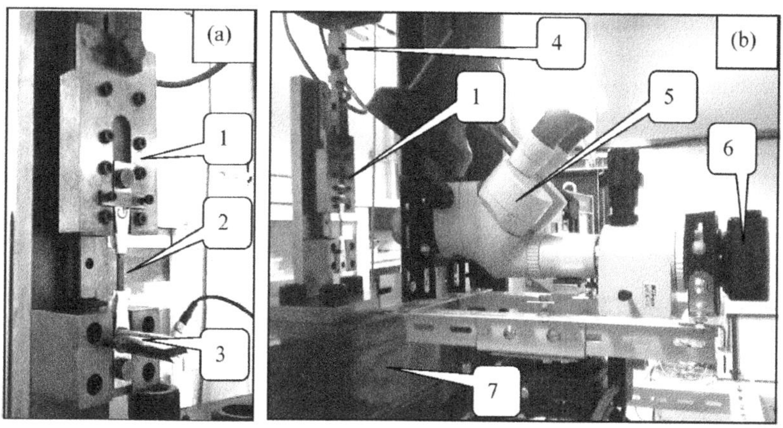

Figure 6 Tensile Loading Mechanism (a) and Setup for In-situ Observation,
1: Tensile Loading Mechanism, 2: Tensile Specimen,
3: AE Sensor, 4: Universal Joint, 5: Optical Microscope,
6: Camera, 7: Crosshead of Material Testing Machine.

In-situ Observation of Multiple Matrix Cracking

Figure 6 shows a tensile loading mechanism and setup for in-situ observation in this study. A linear motion guide (THK co., Ltd, NSR 15 TBA) was mounted in the mechanism designed to eliminate a bending moment caused by misalignment of the loading train. A tensile specimen was pin-jointed to the loading mechanism, and the mechanism was universal-jointed to a materials testing machine (Shimadzu, Autograph DCS-R-10TS). The tensile test was carried out at an ambient temperature under a crosshead speed of 0.005 mm/min. At regular intervals during the test, optical micrographs were taken throughout the gage part with a magnification of 20. Each matrix crack was identified by a compare check on these micrographs. An acoustic emission sensor was also set on the steel tab, and its signal was monitored to check the increasing number of matrix cracks.

RESULTS AND DISCUSSION

Characterization of Specimens

Figure 7 shows the typical microstructure of the composite. White spots are carbon fibers and the other is matrix glass. From an image analysis, the volume fraction of carbon fiber is estimated to be 27 %. The total porosity is 0.6 % measured from the bulk density and true density. 4-point bending test was conducted with the inner span of 10 mm and the outer span of 30 mm under the crosshead speed of 0.1 mm/min. The dimensions of the bending specimen were the same as those of the tensile one. As the result, the Young's modulus is 87GPa, and the bending strength is 320MPa for the composite. According to Marshall and Oliver[9], the fiber push-in test was also performed, and the interfacial frictional stress τ is estimated to be 18MPa and the formation energy of interfacial debonding Γ is estimated to be 0.6 J/m^2. In the push-in test, the penetration depth of an indenter is calibrated from a marking size on a gold-coated carbon fiber at the surface of the specimen.

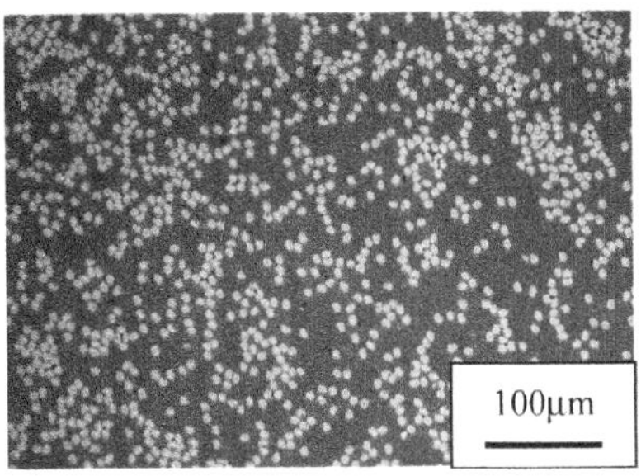

Figure 7 – Microstructure of Carbon Fiber-reinforced PYREX Glass Composite.

In-situ Observation of Multiple Matrix Cracking

Figure 8 shows a tensile specimen just before its overall fracture. It is found that matrix cracks are oriented normal to the fiber axis and almost aligned at regular intervals. Figure 9 shows the change in the average spacing between adjacent matrix cracks. We see that the average spacing gradually decreases with increasing the number of matrix cracks and finally reaches 76μm just before the overall fracture. The minimum spacing observed in this experiment is 47μm, which is nearly equal to 39μm predicted by the ACK theory[7].

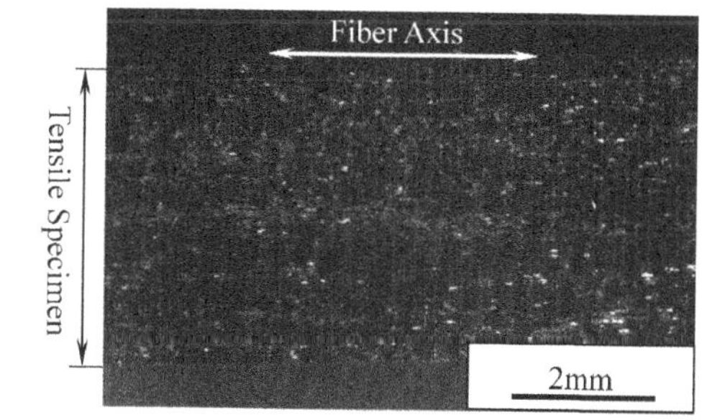

Figure 8 – Multiple Matrix Cracking before Overall Fracture.

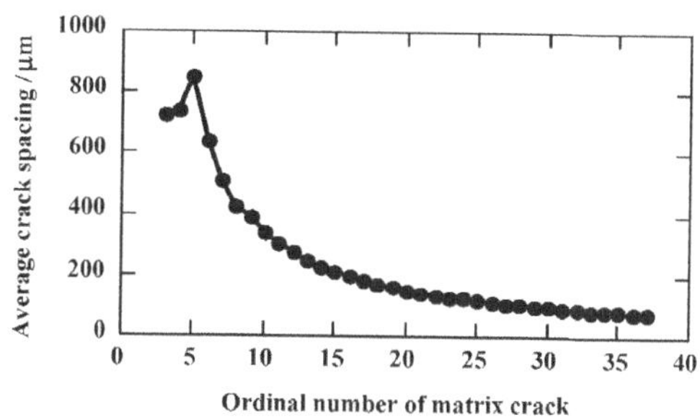

Figure 9 – Change in Average Crack Spacing.

Spatial Distribution of Multiple Matrix Cracking

Figure 10 shows the change in the average position of matrix cracks in the effective cell with increasing the number of the cracks. The open circles denote the average values, and the closed circles the position of each matrix crack in order of appearance. In this plots, we use several material parameters shown in the previous section to calculate the y-coordinate of each matrix crack. From this figure, it turns out that the average position converges to the theoretical value of 0 with increasing the number of matrix cracks although each matrix crack seems to be distributed uniformly in the effective cell.

Figure 11 shows the change in the standard deviation of the position in the effective cell with increasing the number of the cracks. We see that the standard deviation also indicates a convergent change close to a value of 0.255; however, it is still slightly different from the theoretical value of 0.288. The discrepancy must be attributed to the underestimation of the effective cell size in this theory, and also a new matrix crack may be generated at around a relatively high stress region inside the stress recovery length.

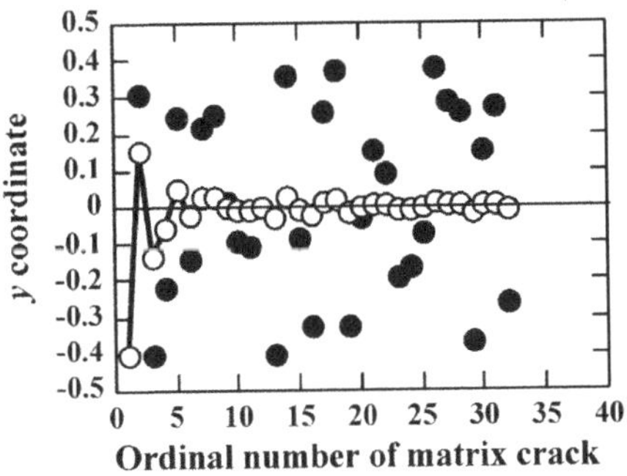

Figure 10 – Positions of Each Matrix Crack (closed circles) and Change in Their Average Positions (open circles).

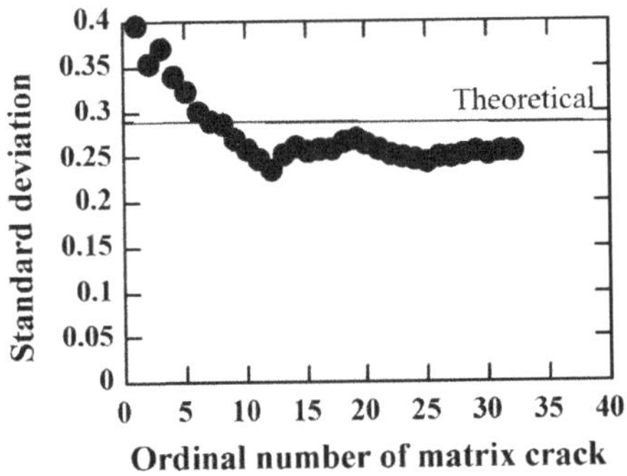

Figure 11 – Change in Standard Deviation of Positions.

CONCLUSION

The in-situ observation of multiple matrix cracking was performed on a carbon fiber-reinforced PYREX glass composite. A stochastic theory for multiple matrix cracking was also proposed with the assumption that matrix cracking did not occur in the inert region around the existing matrix cracks. As compared with the experimental data, the average position of the matrix cracks coincided well with the prediction by the theory; however, the standard deviation was still slightly different from it. The discrepancy was discussed in relation to the underestimation of the effective cell size.

REFERENCES

[1]K.Yasuda, T.Onuma, Y.Matsuo, and T.Shiota, "Evaluation of Dissipation Energies during Debonding and Fiber-Pullout Processes of a C/C Composite," *Proceedings of the 20th International Japan-Korea Seminar on Ceramics*, pp367-370 (2003).

[2]J.Aveston, G.A.Cooper, and A.Kelly, "Single and Multiple Fracture," *Conference Proceedings National Physical Laboratory:* "The Properties of Fiber Composites," pp15-26, I.P.C. Science and Technology Press, Ltd., (1971).

[3]J.Aveston, and A.Kelly, "Tensile First Cracking Strain and Strength of Hybrid Composites and Laminates," *Phil. Trans. R. Soc. Lond.*, A294, 519-534 (1980).

[4]B.Budiansky, J.W.Hutchinson, and A.G.Evans, "Matrix Fracture in Fiber-Reinforced Ceramics," *J. Mech. Phys. Solids*, 34, 167-189 (1986).

[5]H.Cao, and M.D.Thouless, "Tensile Tests of Ceramic-Matrix Composites: Theory and Experiment," *J. Am. Ceram. Soc.*, 73, 2091-2094 (1990).

[6]W.A.Curtin, "Theory of Mechanical Properties of Ceramic-Matrix Composites," *J. Am. Ceram. Soc.*, 74, 2837-2845 (1991).

[7]W.A.Curtin, "Multiple Matrix Cracking in Brittle Matrix Composites," *Act Metall. Mater.*, 41, 1369-1377 (1993).

[8]R.E.Dutton, N.J.Pagano, R.Y.Kim, and T.A.Parthasarathy, "Modeling the Ultimate Tensile Strength of Unidirectional Glass-Matrix Composites," *J. Am. Ceram. Soc.*, 83, 166-174 (2000).

[9]D.B.Marshall, and W.C.Oliver, "Measurement of Interfacial Mechanical Properties in Fiber-Reinforced Ceramic Composites," *J. Am. Ceram. Soc.*, 70, 542-548 (1987).

EVALUATION OF SLOW CRACK GROWTH IN CERAMICS

Roy W. Rice, Consultant
5411 Hopark Dr.
Alexandria, VA 22310-1109

Carl Cm Wu
U.S. Naval Research Lab
Washington, D.C.

ABSTRACT

The susceptibility of a range of ceramics to water driven slow crack growth (SCG) was tested via comparison of flexure strengths in air at 22 °C versus in liquid N_2 at -196 °C. Tested specimens were fractographically examined for fracture origins, fracture mode (i.e. the relative amounts of trans- versus inter- granular fracture) and whether specimen fractures were inconsistent with the assumptions of the test and hence needed to be rejected. Fractography proved to be a valuable addition to the evaluation, e.g. showing frequent intergranular fracture implying SCG in thin grain boundary oxide films in nonoxides with oxide additions and in commercial alumina.

INTRODUCTION

Slow crack growth (SCG) of ceramics, also known by other names such as stress corrosion cracking, is an important factor that occurs in some ceramics with some active agent or corrosive fluid. It consists of crack propagation at stresses below those for fast fracture in the absence of an active fluid, which is water in many cases[1] and is the focus of this work. Note that to be effective, the cracks involved must have a diffusive connection to the external surface for the corrosion fluid to reach crack tips. The essence of SCG is that the atoms of the corrosion fluid chemically interact with the atoms at crack tips where local or global tensile stresses can be highly concentrated to foster local crack growth at stress levels below which bulk fracture of the ceramic does not normally occur. This premature cracking causes limitations in the ability of ceramic components to support as much load as without the occurrence of SCG. Thus knowing whether various ceramics have SCG in the presence of an active agent such as moisture, can be important in using such ceramics. The occurrence of SCG can also depend on some character aspects of the ceramic body, primarily whether the SCG occurs only intregranulary or transgranularly (or possibly in combination).

There is a fair amount of information on SCG in some, mainly common oxide ceramics, but little or no information exists for many, especially nonoxide, ceramics. Thus, a simple method of screening ceramics for SCG was desired. While SCG data is commonly sought via stress intensity versus crack velocity behavior this is more time and material consuming. Also this method may be limited in its utility to sort out intergranular versus transgranular effects. Another approach uses flexure bars tested at different loading rates. SCG susceptibility is indicated by a rate effect upon average strengths. This approach also has the drawback that a substantial number of specimens must be tested to detect the rate effect.

The test method selected in this study was the known technique of tensile testing two sets of flexure bars, one set at a modest (typically room) temperature and the other set at lower temperature where the activity of the corrosion agent (such as water) is zero or close to it. The low temperature test is commonly in liquid nitrogen since it is both effective and practical. The occurrence of SCG in such bi-temperature testing is indicated by strengths at the lower temperature being greater than at room temperature, baring other effects that are generally not common, such as phase changes below room temperature. The extent of SCG is reflected by the magnitude of the strength difference for the two temperatures. Identical test bars are used for both test temperatures. However, there are two limitations to the strength tests alone. First, they can not determine the extent of inter- versus transgranular SCG which can be important as shown later. Second, subsurface failure sources such as pores, or other defects, with no surface connection reflect less or no effects of SCG. Susceptibility to SCG was assessed by the fractographic analysis and a simple comparison of the strengths. An advantage of this combined approach is that only a few specimens are necessary.

It was felt that the above two limitations could be minimized by the application of fractographic examination for fracture origins from flaws without connection to the specimen surface and hence to water. Such strength values should be eliminated since they do not fully reflect effects of SCG on fracture. Also transgranular failure in the area of failure initiation in a specimen means that at least some single crystal SCG can occur. Intergranular fracture initiation means that SCG is less favored in single crystals of the tested material. Intergranular fracture with a fairly continuous grain boundary phase strongly indicates possible SCG due to the intergranular phase.

Previous application of the combined bi-temperature testing and fractographic evaluation has shown utility. Such combined studies of $MgAl_2O_4$(2) showed that single- and polycrystals both exhibited SCG. Such combined evaluations of AlN(3) in contrast indicated that SCG was a result of residual oxide content along grain boundaries and resultant intergranular fracture. Such combined testing of other nonoxides also may be useful. For example, Si_3N_4 made with oxide additives[4,5] showed progressive, often erratic, decreases in crack propagation to zero in fracture toughness type tests as crack size increased, implying that SCG ceased. However, SCG had not ceased over part of this range as shown by delayed failure testing with dead-weight flexure bar tests. The rationale for this difference was that the delayed failure testing used natural, e.g. machining, flaws of which there was a large population of such small to medium flaws such that at least one could cause SCG. On the other hand the growth of single large cracks in fracture toughness type tests could be impeded by patches with little or no oxide phase (from impurities and especially densification aids) pinning sections of larger crack fronts.

EXPERIMENTAL PROCEDURE

A balance of both oxide and nonoxide materials was chosen (Tables I and II) based on suitable size and quality of materials readily available to us. The focus was on samples that had low to zero porosity. The emphasis was on materials for mechanical applications for which there was limited or no SCG indications or data (e.g. ZrO_2 and most nonoxides chosen). In contrast, some materials such as the silicate glass and the commercial (96%) alumina were chosen as calibration via their established SCG. With the nonoxide ceramics, an added factor was whether they were made by hot pressing (or sintering) of powders (with additives) versus chemical vapor deposited (CVD) without additives.

Three-point flexure was selected as the test method. Because of varying shapes and sizes of the starting pieces of material for test specimens, there were some variations in test bar

Table I. Comparison of Oxide Ceramic Strengths at 22 and -196 °C

Material (GS)[1]	Proc.[2]/ Source	Test Dimensions[3]	Strength (σ, MPa)[4]		Δσ[5] (%)
			22 °C	-196 °C	
Glass (0)	Melt	4.9 x 2.6 x 30	114 ± 9 (8)	148 ± 24 (4)	+29
99+% Al_2O_3 (20)	S/Lucalox	4.4 x 2.5 x 30	299 ± 4 (3)	333 ± 20 (5)	+11
CeO_2 (20)	S/Boeing	5.1 x 2.6 x 30	77 ± 4 (2)	86 ± 2 (2)	+12
TiO_2+0.25% Fe_2O_3 (8)	HP/ Ceradyne	5.0 x2.7 x 30	201 ± 25 (3)	302 ± 101 (3)	+50
TiO_2+0.25% Cr_2O_3 (8)	HP/ Ceradyne	5.1 x 2.7 x 30	118 ± 28 (3)	165 ± 23 (5)	+40
TiO_2+0.25% Al_2O_3 (8)	HP/ Ceradyne	5.1 x 2.7 x 30	134 ± 1 (2)	206 ± 29 (4)	+54
TiO_2+0.25% NiO (8)	HP/ Ceradyne	5.0 x 2.5 x 20	170 ± 7 (3)	296 ± 51 (4)	+75
Y_2O_3(+10% ThO_2) (100)	S/GE	4.1 x 2.1 x 20	108 ± 11 (2)	135 ± 18 (4)	+25
ZrO_2 + 5 w/oY_2O_3 (∞)	SkM/Ceres	3.5 x 1.5 x 20	1850 ± 48 (4)	2010 ± 142 (4)	+9
ZrO_2 + 20 w/oY_2O_3 (∞)	SkM/Ceres	4.1 x 2.3 x 20	214 ± 52 (4)	319 ± 76 (5)	+49
ZrO_2 + 10.5 w/oY_2O_3 (20)	HP/WPAFB	5.2 x 2.0 x 20	236 ± 38 (4)	342 ± 20 (4)	+45

1) GS: grain size (approximate, in microns), 0 = glass, ∞ = single crystal.
2) S: sintered, HP hot pressed, SkM skull melted.
3) Specimen width then thickness followed by 3 point flexure span (all in mm).
4) Numbers in () are how many tests are included in the average.
5) Percent change in strength tested in liquid N_2 versus test at room temperature:

$$\Delta\sigma = 100\,(\sigma_{-196} - \sigma_{22}) / \sigma_{22}$$

dimensions as shown in Tables I and II. Test spans of 20 or 30 mm were used at a head travel of ~ 1.3 mm/min. Note that the same test dimensions and loading rates were used for both the room and liquid N_2 temperature tests of a given material. Test bars were diamond machined to the selected dimensions parallel with the long axis of the test bars. A new factor in this study was examination of the fracture surfaces of all tested bars to determine: 1) the mode of failure and 2) whether specimen fracture origins were from typical machining flaws or from defects that were clearly not representative of the material's behavior. Results for observed or probable fracture origins were used in the numerical comparison of strengths. Others were discarded reducing the number of useful specimens from the typical 5 - 10 that were available. Further testing to give more data for those materials with marginal or lower numbers of accepted test results (and to add a few more materials) were planned, but were canceled due to other priorities. However, it is believed that the reduced number of useful test specimens show value of the methodology.

Table II: Comparison of Non-Oxide Ceramic Strengths at 22 and -196 °C

Material (GS)[1]	Proc[2]./Source	Test Dimension[3]	Strength (σ, MPa)[4] 22 <C	-196 <C	$\Delta\sigma$[5] (%)
75%TaC Cermet (5)	Lab Sample	3.4 x 1.6 x20	1367 ± 246 (3)	1333 ± 339 (3)	-2.5
ZrB_2 (40)	HP Ceradyne	3.7 x 1.6 x 20	290 ± 21 (5)	300 ± 48 (2)	+3.4
B_4C (20)	HP Boride Prod.	4.5 x 21 x 20	556 ± 3 (2)	600 ± 44 (2)	+7.9
TiC (?)	?	5.2 x 2.5 x 30	406 (1)	452 ± 4 (2)	+10
TiC (30)	HP Ceradyne	3.4 x 1.6 x 20	393 ± 32 (3)	412 ± 103 (3)	+4.8
ZrC (25)	HP Ceradyne	3.9 x 1.8 x 20	271 ± 4 (3)	257 ± 50 (3)	-5.0
Si_3N_4 (2) (NC 132)	HP Norton	4.7 x 2.4 x 30	990 (1)	960 (2)	- 3
Si_3N_4 (50)	CVD/Synterials	2.7 x 1.1 x 20	347 ± 40 (2)	320 ± 77 (3)	-7.8
MgF_2 (0.5)	HP	5.1 x 2.6 x 30	147 ± 7 (2)	150 (1)	+2.0

Footnotes 1-5 are the same as for Table I.

RESULTS AND DISCUSSION

The strength difference between room temperature and liquid nitrogen is designated $\Delta\sigma$. Before addressing the results, consider that modest changes in $\Delta\sigma$ values occur between liquid nitrogen and room temperatures due to intrinsic changes of the elastic moduli and fracture toughness not related to SCG. The net changes are usually ~ 0.5 % to 1.5 % per 100 °C[(5)] and thus total ~ 1-3%. This is a small contribution and only significant for small $\Delta\sigma$. The second and more complex factor is ascertaining whether two sets of differing values represent statistically different results. The limited number of results for each test series in this paper limit the distinctions that can be made with reasonable probability. However, the application of the evaluation procedure of this paper to stochiometric magnesium aluminate spinel specimens is promising.[2] In that study, comparison of six sets of single crystal specimens (31 total for 2 orientations) gave an average $\Delta\sigma$ value of 33 ± 10 % and 7 sets of polycrystalline samples (32 total for 4 combinations of grain size and porosity levels) averaged 28 ± 8 %. Table I is a summary of data for the oxides tested and Table II is a summary for nonoxides. Both tables have data for some materials with little or no previous testing.

Consider oxides first where silicate glass with $\Delta\sigma$ of ~ 29% represents a typical value for substantial SCG. Typical failure-causing machining flaws and resultant mirrors were found (Fig. 1). On the other hand a $\Delta\sigma$ ~ 11 % for the high purity alumina indicates limited SCG (Fig.2). $\Delta\sigma$ values of ~ 38 & 40 % for commercial 96 % alumina in other tests (3) indicate substantially

Figure 1. Fracture origin of silicate glass in liquid nitrogen (171 MPa). Note the surface flaw origin and traditional mist and hackle surrounding the fracture mirror.

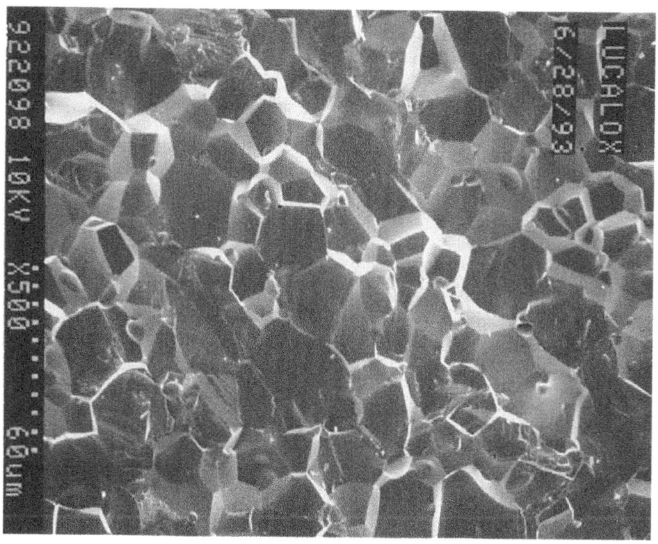

Figure 2. Fracture surface of Lucalox shows predominantly intergranular SCG.

more SCG, possibly enhanced due to the additive/impurity content, much of it at grain boundaries in a material that has mostly intergranular (IG) fracture. CeO_2 with $\Delta\sigma \sim 12$ % indicates limited SCG. Its significant transgranular fracture indicates intrinsic crystalline SCG. Data for TiO_2 with four different additives in Table 1 shows a $\Delta\sigma$ of ~2 times higher than for the silicate glass, and suggests some differentiation of different additive effects. A $\Delta\sigma$ of ~25 % for dense yttria - 10 % thoria (Figure 3) suggests SCG levels near those for silicate glasses. A $\Delta\sigma$ of 49% for fully-stabilized zirconia with 20 w/o yttria single crystals (Figure 4) and a similar value for hot-pressed zirconia with 6 m/o (~ 10.5 w/o) yttria with moderate grain size and mostly transgranular fracture (Figure 5) indicate similar levels of SCG. However, high strength crystals (with 5 w/o yttria partial stabilization) indicate anomalous behavior at the lower temperature, but they have similar fracture markings as the fractures of ZrO_2 crystals with 20% yttria.

Figure 3. Yttrilox with ~ 100% transgranular fracture shows intrinsic SCG in single crystals. The strength in liquid nitrogen was 148 MPa. Note the large grain structure and fracture markings indicative of a flaw size smaller than the grain size.

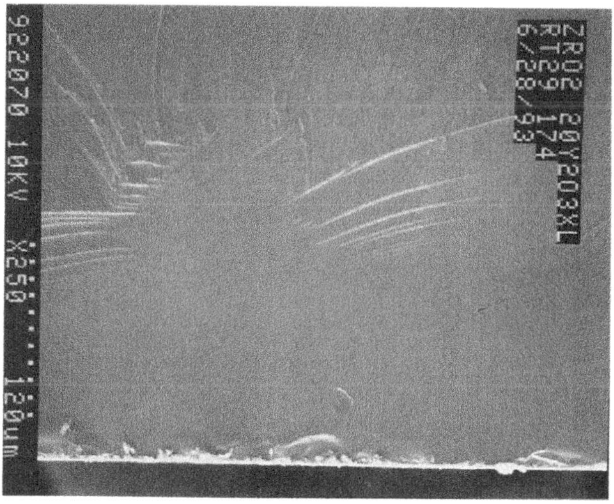

Figure 4. Fracture of fully-stabilized zirconia ZrO_2 + 20 w/o Y_2O_3 single crystal at room temperature (174 MPa). Note the sharp whisker- lance mist and hackle and the machining flaw fracture origin. Note the fracture origin at bottom center.

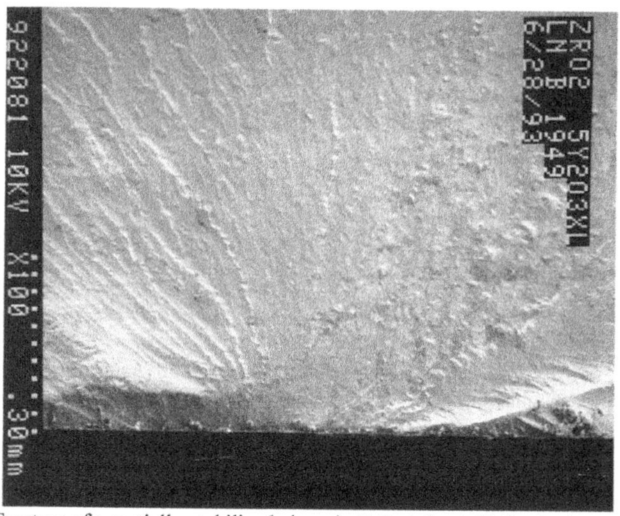

Figure 5. Fracture of a partially-stabilized zirconia ZrO_2 + 5 w/o Y_2O_3 single crystal fractured in liquid nitrogen at 1949 MPa. Note the somewhat dulled whisker-lance mist and hackle.

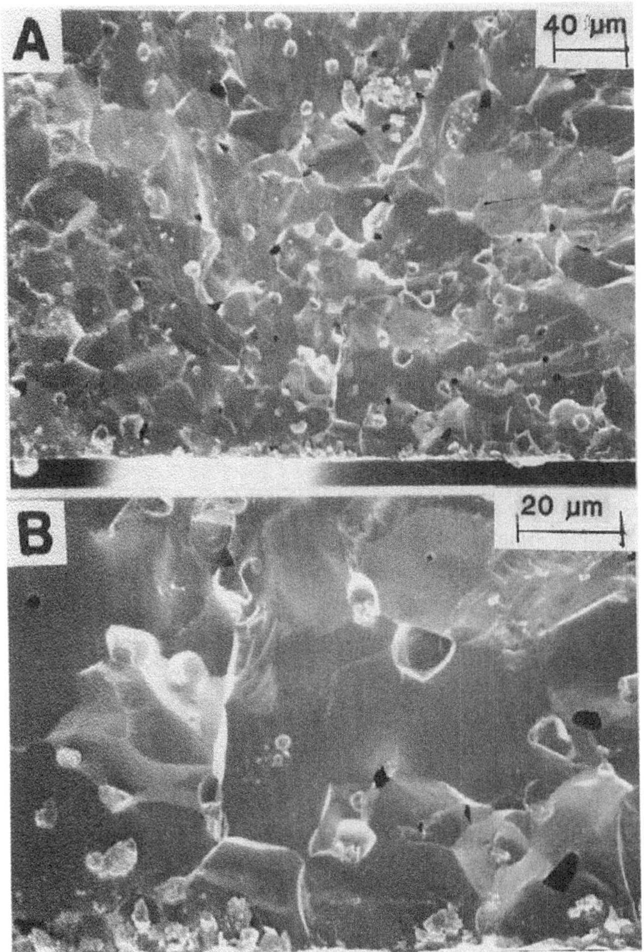

Figure 6. Fracture of ZrB_2 from a somewhat larger grain failing in transgranular fracture, indicating no SCG, at 289 MPa in liquid nitrogen.

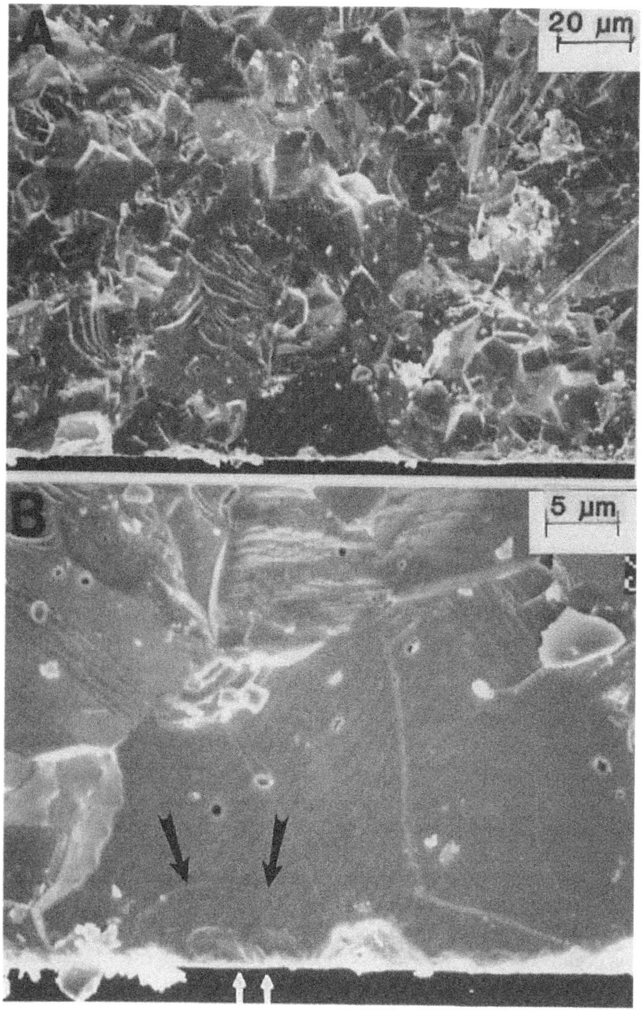

Figure 7. Fracture of TiC from a somewhat larger grain area failing at 289 MPa in liquid nitrogen. Note the predominant transgranular fracture indicating no SCG and the possible impressions of two machining abrasives (white arrows) and resultant machining flaw (black arrows) subsequently smaller than the substantial grain size.

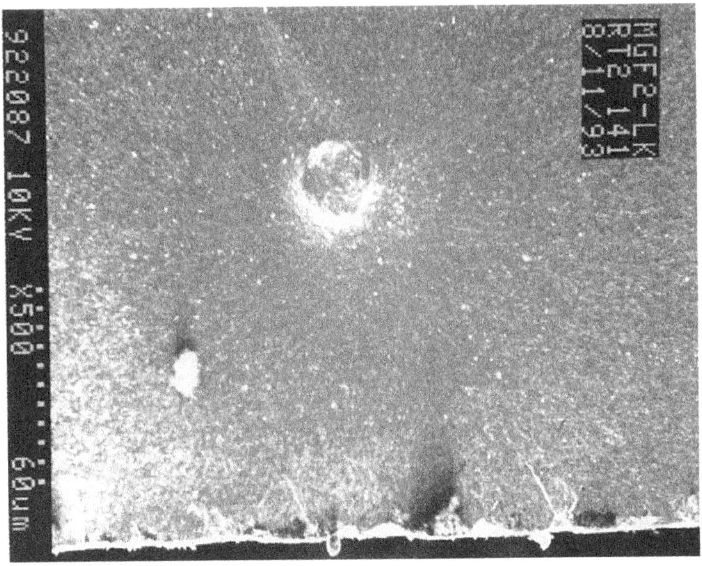

Figure 8. Fracture of hot pressed MgF_2 at room temperature at 141 MPa. Note the isolated internal pore fracture origin which precludes environment driven SCG.

Turning to Table II for nonoxides, they showed at best limited and commonly no evidence of SCG.[6] This is supported for most of these materials by the predominance of transgranular fracture and to some extent larger grain sizes as shown in Figure 6 and 7. One clear exception is the data for MgF_2, which is known to have SCG.[1,7] While it has fine grains and intergranular fracture, which is favorable for SCG, the origins of failure were usually isolated larger pores or pore clusters (Figure 8) that were not connected to the specimen surfaces, thus not providing suitable connection for moisture, hence for SCG.

Though a number of oxides additives have been used as densification aids for Si_3N_4 the aid for the hot-pressed Si_3N_4 tested was based on MgO. This particular silicon nitride is very resistant to SCG because it had only a very small amount of glassy boundary phase.[8] Tests[5,9] for SCG in bulk MgO have been negative (except for some possible SCG due to impurities along grain boundaries), thus supporting the absence of SCG in the hot pressed Si_3N_4 tested.

Results from three other studies support the above results. The first is data for the same test procedure as discussed above, but with dense AlN (hot-pressed or sintered with limited amounts CaO or Y_2O_3 as densification aids) as the test material (3). The $\Delta\sigma$ values ranged from ~7 to 32 % indicating a range of SCG behavior (similar to that of alumina bodies) that generally increased as the residual oxide additive increased. The occurrence of SCG in such AlN was corroborated by both dynamic fatigue and stress intensity versus crack velocity (K-V) tests. Values of n typically averaged ~ 60 with some values nearing 1000. The AlN fractures revealed essentially 100%

intergranular fracture and grain sizes of the order of 5 μm, all indicators that the SCG is due to the amount and type of residual grain boundary phase(s).

The second corroboration of the results of this study are those of SiC. Kotchick and Tressler[10] found no SCG in SiC filaments that are made by chemical vapor deposition (CVD) without any additives. In contrast to the CVD results, both Evans and Lange[11] and McHenry et al.[12] both found that SiC hot pressed with alumina exhibited substantial SCG (e.g. n values of 80 - 100). Evans and Lange further supported residual grain boundary additives as a major source of SCG by demonstrating that SCG does not occur in SiC densified with B additions since they typically do not leave residual phases along grain boundaries. Such behavior of soluble additives is consistent with work of Palm and Greskovitch[13] showing no SCG in silicon nitride bodies sintered with soluble Be based additives. These and additional results are addressed in reviews of fracture mode.[5,6]

SUMMARY AND CONCLUSIONS

A screening test for identifying whether a materials strength is reduced by slow crack growth (SCG) in some environments is by measuring strengths at room and liquid nitrogen temperatures (22 and – 196 °C) has been combined with fractography and microstructural evaluation of the tested specimens. Only a few specimens are necessary. The combination with fractography proved valuable to identify and delete specimens with atypical fracture origins. Though this may (and did) result in discounting significant numbers of specimens, the test typically reflects a real rather than unknown or speculated failure-controlling flaw population.

The materials selected for test proved useful in extending existing data and adding some new materials to the SCG data base, e.g. this paper shows that a number of non oxide ceramics do not exhibit SCG, but may with some oxide boundary phases. Possible effects of additives or concentrating on grain boundaries were indicated as were the possibilities of having variable mixes of grain boundary effects and bulk grain, i.e., single crystal, effects. Much needs to be done to better understand the substantial effects of grain boundary films on SCG. Study of effects of yttria and other rare earth oxide additives for SiC and especially silicon nitride may be of particular use.

ACKNOWLEDGEMENTS

George Quinn, Craig Rice, and Brian Sparenberg are thanked for their assistance in preparing the manuscript.

REFERENCES

[1]S. W. Freiman, "Stress-Corrosion Cracking of Glasses and Ceramics," *Stress Corrosion Cracking- Materials Performance and Evaluation*, R. H. Jones ed., ASM, Materials Park, OH, 337-44, (1992).

[2]R. W. Rice and C. Cm. Wu, "Slow Crack Growth in $MgAl_2O_4$ Single- and Poly-Crystals," *J. Mat. Sci. Let.*, 14, 723 - 27 (1995).

[3]R. W. Rice, C. Cm Wu, K. R. McKinney, S. W. Freiman, G. S. White, L. E. Dolhert, and J. H. Enloe, "Slow Crack Growth in AlN for Electronic Applications," to be published.

[4]K. R. McKinney, B. A. Bender, R. W. Rice, and C. Cm. Wu, "Slow Crack Growth in Si_3N_4 at Room Temperature," *J. Mat. Sci.*, 26 6467 - 72 (1991).

[5]R. W. Rice, *Mechanical Properties of Ceramics and Composites*, Marcel Dekker, New York, (2000).

[6]R. W. Rice, "Ceramic Fracture Mode-Intergranular vs. Transgranular Fracture," pp. 1 – 53 in *Fractography of Glasses and Ceramics III*, Ceramic Transactions, 64, J. R. Varner, V. D. Frechette and G. D. Quinn, eds., Am. Cer. Soc., Westerville, OH, (1996).

[7]J. J. Mecholsky, Jr., "Intergranular Slow Crack Growth in MgF_2," *J. Am. Ceram. Soc.*, 64 (9), 563-66, (1981).

[8]G. D. Quinn, K. Xu, J. A, Salem, and J. J. Swab, "SRM 2100: the World's First Fracture Toughness Reference Material," pp. 499 – 530 in *Fracture Mechanics of Glasses and Ceramics, Vol. 14*, eds. R. C. Bradt, D. Munz, M. Sakai, and K. W. White, Klewer/Plenum, NY, 2005.

[9]W. H. Rhodes, R. M. Cannon Jr., and T. Vasilos, "Stress-Corrosion Cracking in Polycrystalline MgO," pp. 708 -833 in *Fracture Mechanics of Ceramics, II* ,(R. C. Bradt, D. P. H. Hasselman, and F. F. Lange eds.), Plenum Press, New York, (1974).

[10]D. M. Kotchick and R. E. Tressler, "Surface Damage and Environmental Effects on the Strain-Rate Sensitivity of the Strength of Sapphire and Silicon Carbide Filaments," *J. Mat. Sci.*, 10, 608 - 10 (1975).

[11]A. G. Evans and F. F. Lange, "Crack Propagation and Fracture in Silicon Carbide", *J. Mat. Sci.*, 10, 1659 - 64 (1975).

[12]K. D. McHenry, T. Yonushonis, and R. E. Tressler, "Low-Temperature Subcritical Crack Growth in SiC and Si_3N_4", *J. Am. Ceram. Soc.*, 59 (5-6), 262 - 64 (1976).

[13]J. A. Palm and C. D. Greskovitch, "Thermomechanical Properties of Hot-Pressed $Si_{2.9}Be_{0.1}N_{3.8}O_{0.2}$ Ceramic", *Am. Ceram. Soc. Bull.*, 59 (4), 447 - 52 (1980).

PERSPECTIVE ON NEEDS AND OPPORTUNITIES IN FRACTOGRAPHY

Roy W. Rice Consultant
5411 Hopark Dr.
Alexandra, Va 22310-1109

ABSTRACT

This paper outlines needs and opportunities for better application of fractography to silicate glasses, single- and poly- crystalline ceramics, and ultimately to some polymeric, metal, and composite materials. Better identification of mist and hackle boundaries is a basic need, especially for single crystals and larger grain polycrystals. Further study is needed of pores that act as less sharp (i.e., blunter) flaws in glasses and single crystals. The bluntness makes them less serious sources of failure than pores in polycrystals, where grain boundary grooving is seen as a factor.

INTRODUCTION

Fractography has been a valuable tool for development and use of silicate glasses and has much potential for other ceramics as well as application to metals, plastics, and composites. However, as the breadth and diversity of materials have expanded, new challenges to and new possibilities for fractography have arisen. This paper outlines the efforts and the author's vision for advancing fractography of materials.

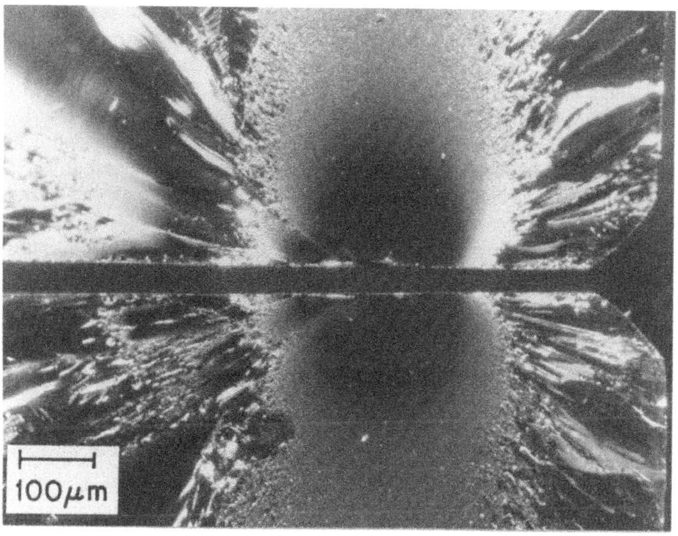

Figure 1. Classic mirror pattern on the mating fracture halves of a silicate glass. Compare with Figures 3 - 7 and 9.

The basic interest in fractography of ceramics and glasses is the formation and character of features referred to as fracture mirrors that form on tensile fracture surfaces. Pertinent background on such mirrors is summarized here. Additional background can be obtained from references 1 and 2. Mirrors, though some times observed by the unaided eye or hand lenses, were typically observed by optical microscopy. They typically consist of a nearly semi- to full circular area of high light reflectance (hence the designation of a 'mirror') that is bounded by a thin band of finer fractures (called mist) graduating to a somewhat coarser band of coarser fractures (called hackle). Failure causing flaw(s) are at the focus of the generally circular mirror and of the approximately radial orientation of the mist and hackle (Fig. 1). The mirror focus on failure causing flaws makes identification of mirrors very useful in finding failure causing flaws as does the fact that the mirror size is typically 10 - 15 times larger than the flaws.

FRACTURE MIRRORS OF POLYCRYSTALLINE CERAMICS

For years optical microscopy gave reasonably consistent fractographic results based on the model that mist was the first formation of features delineating the earlier stages of rapid crack propagation. Mist is followed by hackle and then by macroscopic crack branching provided specimen size and failure stresses allowed complete fracture feature formation. Key to the reliability of such measurements was the belief that they represented observation of a distinct initiation point of mist features on glass fractures. The problem with this interpretation was that higher magnifications by electron microscopes that became available showed that finer mist features not seen by optical microscopy were found well into the mirror region. This led to less confidence in the mist boundary measurements since there was at best a diffuse mirror boundary. This problem was addressed by Hull[3-5] who showed that the boundary could be related to changes in mist feature size. The boundary was approximately coincident with the optically detected mist features. Thus, a boundary based on an increased rate, rather than on initiation, of mist formation was identified. This however left two issues. One was that Hull's observations were based on fracture of a brittle plastic, not of silicate glasses. Second, they clearly did not address polycrystalline and single crystals. Key unknowns of these are respectively issues of inter- and trans- granular fracture and intrinsic anisotropies.[6-8] Mirror size measurements are a major tool in quantitative fractography, and a better understanding of the nature of the mist formation is essential for proper analysis.

Fracture surfaces of polycrystalline ceramics showed features that are clearly similar to fracture mirrors on glasses, especially at finer grain sizes, e.g. by Kirchner and colleagues.[9,10] Good consistency was found for measurements of strength versus the inverse square root of the mirror size of similar alumina bodies by Kirchner and Gruver[9] and this author as shown in Figure 2. Subsequent data of different investigators and materials also showed the same mirror size functional form, but not numerically the same. Other groups obtained different mirror to flaw size (M/c) ratios.[11-13] Similar M/c differences were indicated for different types of glasses.[14] Further data of this author[15] showed strengths of five ceramics of finer grain size also followed similar strength- mirror size relations. The M/c ratios had typical variability, but some M/c ratios were smaller than usual for finer mirror sizes. This was attributed to increasing effects of stresses from anisotropic thermal expansion and elastic moduli.

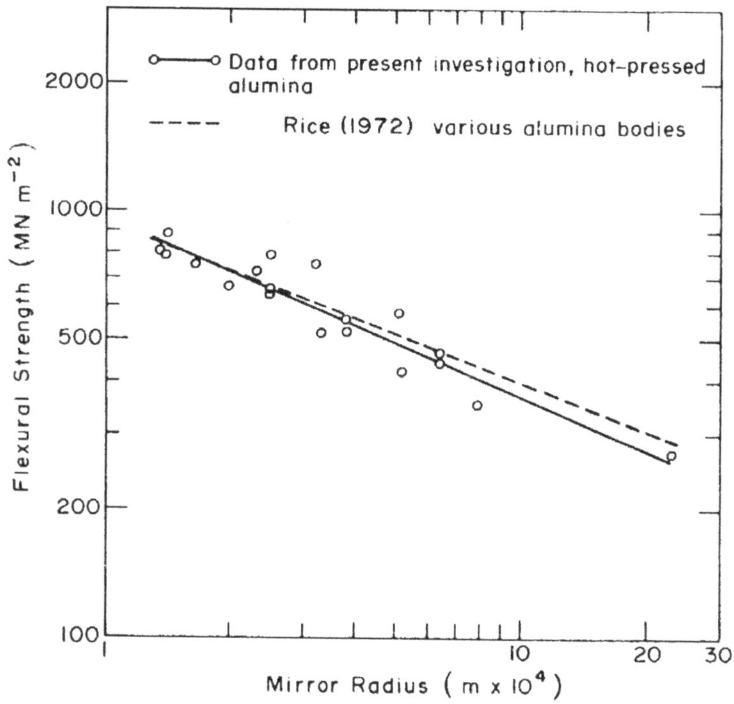

Figure 2. Comparison of uniaxial flexure strength versus mirror radius for fairly dense, finer (mostly 1 -10 micron) grain size polycrystalline aluminas after Kirchner and Gruver[9] and the author's data. The two data sets are in good agreement.

More recent examination of fractures of some ceramics, mostly nonoxides of intermediate grain sizes, showed diffuse patterns of mirror- like features that might be adequate for quantitative fractography.[16,17] However, this may be challenging, since as grain sizes increase, mirror boundaries are harder to observe, especially when flaw sizes are comparable to or less than the grain size. Empirical correlations may be possible, such as based on selective observations that show clearer fracture features may be useful. Figure 3 is an example. Other more fundamental efforts such as development of electronic pattern recognition systems may be feasible and desirable (Fig. 4). Note that the move to single crystalline and polycrystalline evaluation results in reduction or elimination of the bright mirror character, so the term mirror should be used in the historical sense of its use with glasses, or replaced by another term such as mist boundary.

Figure 3. Unusually clear fracture mirror in a large grain Yttria-stabilized zirconia. Mirrors usually are not so clear in such coarse-grained materials. It is not known why the mirror was so clear in this particular specimen, but use of such clearer mirrors can help determine more general mirror patterns.

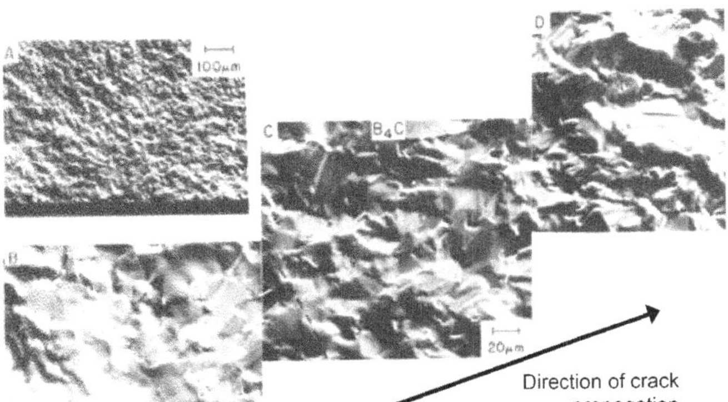

Figure 4. Fracture origin and mirror area in dense Boron Carbide which failed by ~ 100% transgranular fracture. A and B show lower and higher magnification of the fracture origin (bottom center of photo B). B through D are photos in a radial direction showing the origin and a definite increase in roughness with distance from the origin. A mirror boundary is very difficult to quantify. There are also difficulties for mirrors with intergranular fracture.

FRACTURE MIRRORS ON CERAMIC SINGLE CRYSTAL FRACTURES

Study of single crystal fracture patterns offer challenges and rewards in understanding their mirror patterns[1,2,7,8] and other aspects of their behavior such as zigzag crack propagation and crack velocity dependant fracture.[7] The significant anisotropy of fracture due to primary versus secondary cleavage can be a problem, but testing under biaxial stresses can increase secondary cleavage fractures.[2]

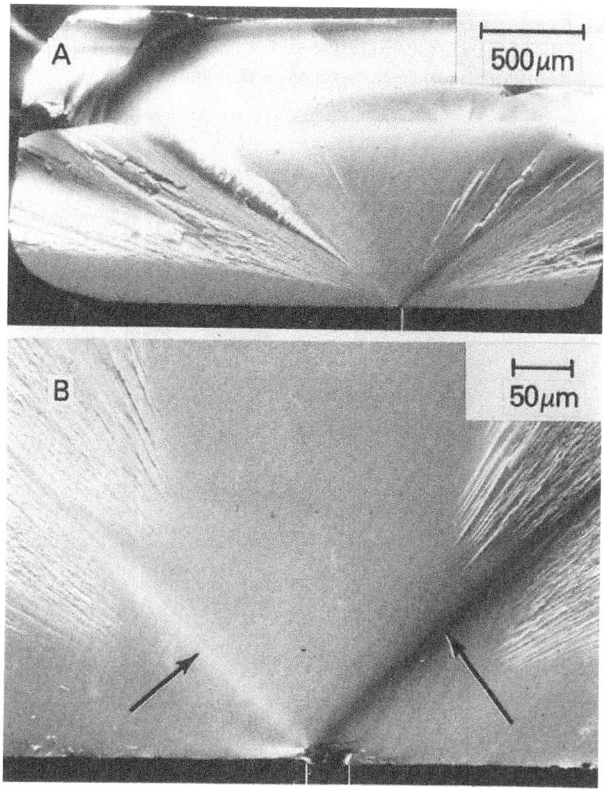

Figure 5. Fracture of a magnesia alumina spinel (cubic) crystal in uniaxial flexure. A and B are low and high magnifications of the fracture with failure nominally on a {100} plane. Note the similarities to glass fractures such as the symmetry of the mirror pattern and the focus of the analogues of mist and hackle back on the origin. Note also the differences such as the fact the mist and hackle analogues are limited to a few symmetric crystallographic sectors and are much thinner and longer. The unusual ridges (arrows) are also centered on the origin. Contrast with Figs. 6 and 9.

Single crystal 'mirrors' exhibit similarities and differences compared to glasses. There are features that clearly focus on the fracture origin, but there are also areas where the fracture features appear to be totally excluded as shown in Figures 5 and 6 for two different cubic crystals. Compare these with Figures 1 and 3. Some testing has also been made with noncubic crystals, e.g., sapphire, which show some increases in mirror complexity over those for cubic crystals.

Key needs for single crystal fractography are for more data and analysis of possible causes or correlating factors for fracture mirror markings. Elastic moduli are clearly candidate factors since they play major roles in mechanical behavior, cover substantial ranges of values for both cubic and noncubic materials, and some correlations are indicated.[10,18]

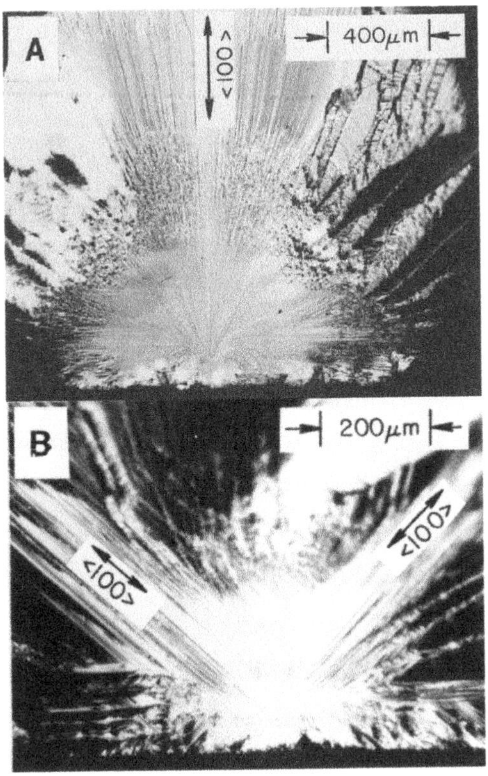

Figure 6. Fracture of MgO single crystals oriented for fracture on a {100} plain with the tensile surface normal to a {100} plane in A, and with the tensile surface normal to a {110} plane in B. Note again the distinctive crystallographic character of the mirror patterns. Compare with Figures 5 and 9.

FAILURE FROM BLUNT FLAWS IN CERAMICS

Ceramic fracture has typically been assumed to be controlled by atomically sharp flaws until temperatures allowing some flaw blunting to occur. However, evaluation of effects of isolated spherical pores on tensile fracture of silicate glass bars showed that failure only occurred from pores if there was a machining flaw associated with the pore.[19,20] Most commonly failure occurred from machining flaws alone whose sizes were substantially smaller than near by pores at typical sharp flaw strength levels. Typical fracture stresses when pores were involved were higher than those expected if a sharp flaw the size of the bubble was the failure source. For glasses, the mirror to flaw size ratio (M/c) for blunt flaws was about half the typical ratio of about 12-14 for sharp flaws. It was postulated that the higher stresses reached in most of these bubbles reflected their smooth flame-polished like finish. The reason for the apparent decrease in the M/c ratio was that the bubbles created stress concentrations. Figure 7 shows a case wherein fracture started from a small surface machining crack rather than a nearby very large pore. Even though it did not initiate fracture, it created a stress concentration which triggered mist and hackle to initiate much earlier than normal as the crack radiated away from the origin.

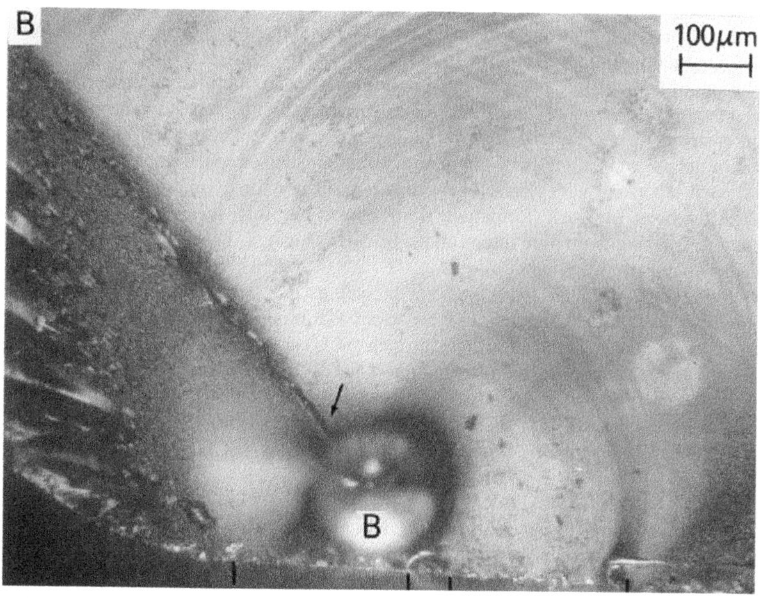

Figure 7. Fracture of a silicate glass specimen from a surface machining flaw (tick marks photo bottom center) and not from a substantially larger spherical near by pore (bubble, B). The pore in the glass was clearly a blunt flaw. Note also the streak of mist and hackle on the fracture surface in the vicinity of the pore attributed to stress concentration around the pore.

The above results for glasses raise the question of effects of single pores on failure of ceramic single- and poly-crystals. Observations of tensile failure of melt drawn sapphire filaments with isolated (pristine-faceted) pores show similar trends for such pores to be less serious flaws than typical sharp flaws such as machining damage.[21-23]

Evaluation of data for polycrystalline ceramics show isolated pores are often fracture origins at stress levels similar to those for failure from machining flaws. The difference between the effects of single pores in glasses and single crystals versus that in polycrystalline ceramics has been attributed grain boundary grooving in the latter.[24,25] Such grooving results in substantially greater stress concentration due to the grove geometry than smooth pores in crystals and glasses, but probably not as much as sharp machining flaws. However, grain boundary grooves have added stress concentration due to elastic anisotropy of grains. The boundary probably offers an easier, i.e., lower stress, path to initiate fracture. Single crystals and glasses lack grain boundaries to both form grooves and offer such favorable paths for crack initiation. There may be a few or many grain boundary grooves around polycrystalline pores. 'As-fired' surfaces may have additional variations in groove types and severities. These groove-flaws inherently increase in size as grain size increases, providing a mechanism to explain decreasing strength as grain sizes increase. They inherently start intergranularly, so much, or all, of the crack formed is intergranular. This hampers fractographic analysis, but as grain size increases, more transgranular cracking should occur. This will facilitate fractography that can test and clarify the suggested failure mechanisms.

Clusters of two or more pores as fracture origins are also often failure sources in glasses, single crystals, and polycrystals. All three forms of material show considerable variability in their failure stress levels. Part of this variability apparently reflects variations of probable parameters of degree and character of the pore distribution such as spacings and intersections of pores with each other and with flaws. There also is probably some variation due to the above discussed differences in crack formation from pores, but this seems reduced, apparently reflecting more cracking from pore intersections or close spacings. This again leaves much more to be done by fractography and other means.

It is also suggested that some phenomena such as R - curve effects may occur due to forming multiple sharp cracks that can in a sense act as a blunter flaw. Such cracks should be further studied in bodies in which slow crack growth (SCG) arises due to residual grain boundary oxide phases. Clearly some further understanding is needed of the effect of thin grain boundary films that lead to intergranular fracture, slow crack growth,[17] and R - curve effects.[26] Such materials may be contrasted with those that have no boundary film from an oxide additive residue and thus have transgranular fracture.

OTHER TESTS AND MATERIALS

Another broad range of materials to consider are composites. One interesting example is Co-bonded WC cermets. A substantial study of the mirror to flaw size ratio as a function of Co content has been published and some findings are shown in Figure 8.[27] This material showed diffuse, but reasonably definable mirrors with similarities to both ceramics and some high strength polycrystalline steel fractures.[1] Single crystal fractures in ceramics and refractory cubic metals have some striking similarities as shown in Figure 9.[2,28]

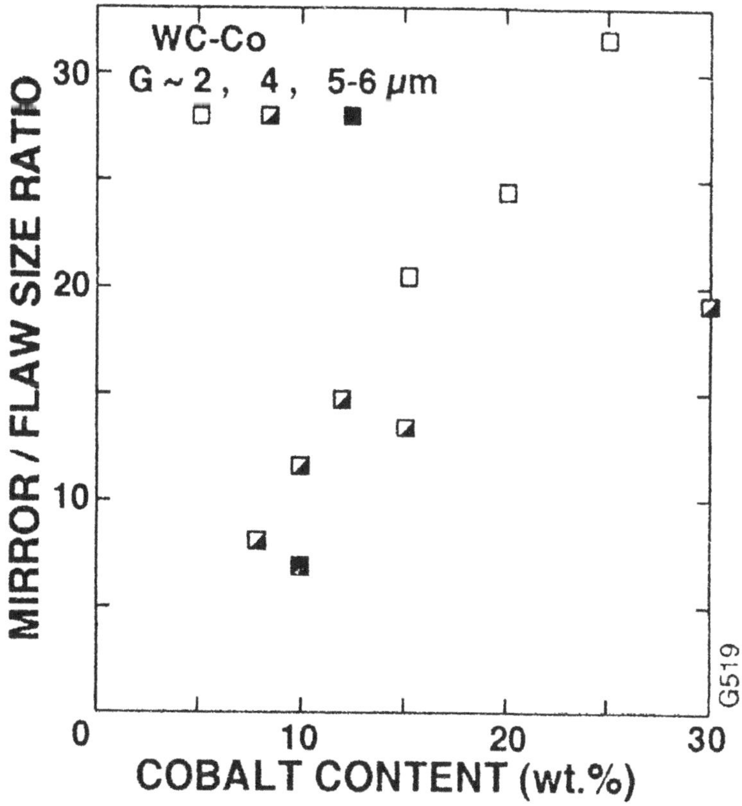

Figure 8. Mirror to flaw size ratio as a function of Co content of Co bonded WC and grain size plotted from data of Luyckx and Sannino.[27] Note a substantial range of M/c values.

This cermet study could also be part of this next and last materials based suggested study, which is the study of the ductile - brittle transition at the crack tip. This is seen as encompassing a variety of tests and materials, of which two candidates might be Si and roofing tar.

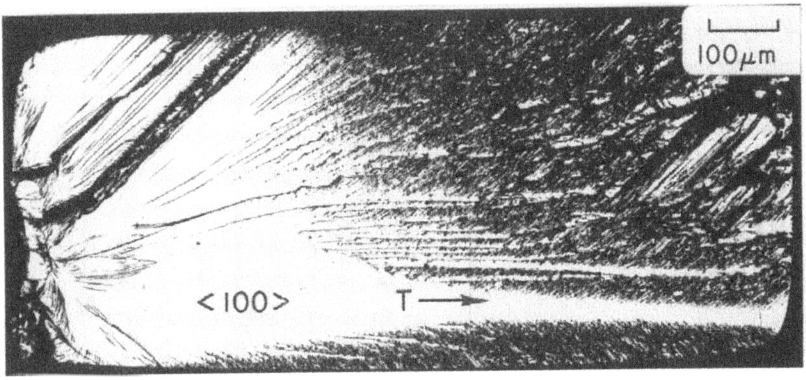

Figure 9. Fracture of a W crystal showing very similar crystallographic features as ceramic crystals, especially MgO as shown in Figure 6. For example, note the tongue (T) of smooth fracture in a <100> direction. After Cordwell and Hull.[28]

Turning to the tools of fractography for new opportunities, in situ microstructural characterization (composition, phases, porosity, grain structure and orientation) fracture origins could be quite helpful. Another tool to aid fractography is ultrasonic stimulation of potential fracture areas. This has proven to be very useful with some glass fractures.[1] Though apparently not very effective for intergranular fracture in polycrystalline ceramics, it may offer much of the benefit in materials with mainly transgranular fracture. Modern electronics would appear to offer the opportunity to sort out the sequence of cracking in many metals, ceramic, polymeric materials and composites. Similarly, more characterization of the fracture process, e.g. via use of acoustic emission, and different stress states should be useful.

SUMMARY AND CONCLUSIONS

Study and understanding fracture mirrors and related features on finer grain polycrystalline ceramic fractures has progressed, but serious problems of recognizing and especially defining fracture mirror boundaries in materials with larger grains must be solved. Fascinating discoveries of single crystal fracture features are important for understanding much more of the fracture process. Very similar fracture features have been observed in some ceramic and refractory metal crystals. The 'bluntness' of isolated pores in ceramic glasses and single crystals explains why such pores are not common fracture origins unless there are machining flaws or other serious stress concentrators associated with the pores. On the other hand, grain boundary grooves in polycrystalline ceramics from firing or other high temperature exposure make flaws out of pores as well as natural grain scale flaws that are of similar 'sharpness' as machining flaws. Substantial benefit is also expected from further study of slow crack growth, especially in ceramics with oxide layer covering grain boundaries. Great benefit is expected from more comparative analysis of fracture behavior of ceramics with that of plastic, metal, and composite materials.

ACKNOWLEDGEMENTS

George Quinn, Craig Rice, and Brian Sparenberg are thanked for their assistance in preparing the manuscript.

REFERENCES

[1]R. W. Rice, "Ceramic Fracture Features, Observations, Mechanisms, and Uses," *Fractography of Ceramic and Metal Failures*, ASTM STP 827, J. J. Mecholsky Jr. and S. R. Powell Jr. eds., 5-103, ASTM (1984).

[2]R. W. Rice, "Perspective on Fractography," *Fractography of Glasses and Ceramics*, Adv. Cer., 22, V. D. Fréchette and J. R. Varner eds., Am. Ceram. Soc., Westerville, OH, 3-56 (1988).

[3]D. Hull, "Influence of Stress Intensity and Crack Speed on Fracture Surface Topography: Mirror to Mist Transition," *J. Mat. Sci.*, 31, 1829-41, (1996).

[4]D. Hull, *Fractography: Observing, Measuring, and Interpreting Fracture Surface Topography*, Cambridge Univ. Press, Cambridge, U. K., (1999).

[5]D. Hull, "Influence of Stress Intensity and Crack Speed on Fracture Surface Topography: Mirror to Mist to Macroscopic Bifurcation," *J. Mat. Sci.*, 31, 4483-92, (1996).

[6]R. W. Rice, "Ceramic Fracture Mode-Intergranular vs. Transgranular Fracture," *Fractography of Glasses and Ceramics III*, Ceramic Transactions, 64, J. R. Varner, V. D. Fréchette, and G. D. Quinn, eds., Am. Cer. Soc., Westerville, OH, 1-53, (1996).

[7]C. Cm. Wu, K. R. McKinney, and R. W. Rice, "Zig-Zag Crack Propagation in $MgAl_2O_4$ Crystals," *J. Mat. Sci. Let.*, 14, 474-77, (1995).

[8]R. W. Rice, A. C. Wu, and K. R. McKinney, "Fracture and Fracture Toughness of Stoichiometric $MgAl_2O_4$ Crystals at Room Temperature," *J. Mat. Sci.*, 31, 1353-60, (1996).

[9]H. P. Kirchner and R. M. Gruver, "Fracture Mirrors in Alumina Ceramics," *Phil. Mag.*, 27(6), 1433-46 (1973).

[10]H. P. Kirchner, R. M. Gruver, and W. A. Sotter, "Fracture Stress-Mirror Size Relations for Polycrystalline in Ceramics," *Phil. Mag.*, 33 (5), 775 - 780 (1976).

[11]G. K. Bansal, "On Fracture Mirror Formation in Glass and Polycrystalline Ceramics," *Phil. Mag.*, 35 (4), 935- 44 (1977).

[12]D. K. Shetty, A. R. Rosenfield, and W. H. Duckworth, "Crack Branching in Ceramic Disks Subjected to Biaxial Flexure," *J. Am. Ceram. Soc.*, 66 (1), C10-12 (1983).

[13]R. N. Katz and K. Cho, "Fracture Toughness Estimation by Fractographic Analysis of Tensile Specimens," *Fractography of Glasses and Ceramics III*, Cer. Trans. 64, J. R. Varner, V. D. Fréchette, and G. D. Quinn, Am. Cer. Soc., Westerville, OH, 91-106, (1996).

[14]J. J. Mecholsky. R. W. Rice, and S. W. Freiman, "Prediction of Fracture Energy and Flaw Size in Glasses from Measurements of Mirror Size," *J. Am. Ceram. Soc.*, 57 (10), 440-43 (1974).

[15]R.W. Rice, "Fractographic Determination of KIC and Effects of Microstructural Stresses in Ceramics" *Fractography of Glasses and Ceramics III*, Ceramic Transactions 17, V. D. Fréchette and J. R. Varner, eds., Am. Cer. Soc., Westerville, OH, 509-545, (1991).

[16]R. W. Rice, "The Difference in Mirror-to-Flaw Size Ratios Between Dense Glasses and Polycrystals," *J. Am. Ceram. Soc.*, 62 (9-10), 533-35 (1979).

[17]R. W. Rice and C. Cm Wu, "Evaluation of Slow Crack Growth in Ceramics," These Proceedings.

[18]S. W. Freiman, J. J. Mecholsky, and P. F. Becher, "Fractography: A Quantitative Measure of the Fracture Process," *Fractography of Glasses and Ceramics II*, Cer. Trans., 17, V. D. Fréchette and J. R. Varner, Am Ceram. Soc., Westerville, OH, 55-78 (1991).

[19]R.W. Rice, "Pores as Fracture Origins in Ceramics," *J. Mat. Sci.*,19, 895-914 (1984).

[20]R.W. Rice, *Porosity of Ceramics*, Marcel Dekker, Inc., New York, New York, (1998).

[21]S. A. Newcomb, "Temperature and Time Dependence of the Strength of C-Axis Sapphire from 800-1500 C," MS thesis, Penn State Univ., May (1992).

[22]S. A. Newcomb and R. E. Tressler, "Slow Crack Growth of Sapphire at 800 to 1500 C," *J. Am. Ceram. Soc.*, 76 (10), 2505-12 (1993).

[23]R. W. Rice, "Corroboration and Extension of Analysis of c-axis Sapphire Filament Fractures from Pores," *J. Mat. Sci. Let.*, 16, 202-05 (1997).

[24]R. L. Coble, "Thermal Grooving in Polycrystalline Ceramics and Initiation of Brittle Fracture," *J. Am. Ceram. Soc.*, 54 (1), 59-60 (1971).

[25]R. W. Rice, *Mechanical Properties of Ceramics and Composites: Grain and Particle Effects*, Marcel Dekker, New York, (1999).

[26]K. R. McKinney, B. A. Bender, R. W. Rice, and C. Cm. Wu, "Slow Crack Growth in Si_3N_4 at Room Temperature", *J. Mat. Sci.*, 26, 6467-72 (1991).

[27]S. B. Luyckx and A. Sannino, "Crack Branching and Fracture Mirrors in Cemented Tungsten Carbide", *J. Mat. Sci.*, 23, 1243-47 (1988).

[28]J. E. Cordwell and D. Hull, "The Brittle Fracture of [100] Axis Tungsten Single Crystals", *Phil. Mag.*, 19 (161), 951- 66 (1969).

SECTION III

FRACTOGRAPHY OF CONTACT DAMAGE IN GLASSES AND CERAMICS

VICKERS INDENTATION FRACTURE IN OPTICAL GLASS COMPOSITIONS

Trevor E. Wilantewicz
Rutgers, The State University of New Jersey
607 Taylor Road
Piscataway, New Jersey 08854

James R. Varner
Alfred University
2 Pine Street
Alfred, New York 14802

ABSTRACT

The Vickers indentation test was used to generate contact damage on polished glass surfaces in a controlled manner. A conventional hardness tester was used to make Vickers indentations at 1 and 5 Kg loads in several different optical and commercial glass compositions. The deformation and fracture behavior were then characterized using optical and scanning electron microscopy. A recording microindenter equipped with optical observation and acoustic emission detection was also used to monitor in-situ the crack initiation behavior of the same glasses up to a 4 Kg maximum load. For the majority of optical glass compositions, a single penny median-radial crack was found to initiate at low loads on the loading half-cycle, followed by the initiation of two additional median-radial cracks at separate instances on either the loading or unloading half-cycle. The final post-mortem geometry of all cracks, regardless of where in the indentation cycle they formed, was essentially half-penny. These glasses appeared to deform by a mainly shear-induced 'flow' of the structure, as indicated by well defined shear-flow lines beneath the indentations. Thus, the usual notion that 'normal' glasses form median-radial cracks only on indenter withdrawal, for relatively small maximum loads, does not hold for most of the optical glasses examined here. Three broad categories of cracking/deformation behavior can be distinguished: 'normal', 'normal/anomalous', and 'anomalous.' Both glass composition (structure), and the ratio G/K (shear modulus/bulk modulus) were found to characterize a glass's relative tendency to undergo shear-flow vs. volume compaction (densification), and hence mostly 'normal' or mostly 'anomalous' behavior, respectively.

INTRODUCTION

Crack formation around Vickers indentations has been studied for soda-lime-silica glass, or closely similar compositions, by many researchers.[1-18] This glass has been considered the archetypical 'normal' glass by some.[8,13,16,17] In sharp indentation testing, the term 'normal' has been used to designate a glass that deforms at more-or-less constant volume, i.e., exhibiting mostly non-densifying behavior, and whose deformation response is mostly shear-flow controlled.[13] This is opposite to 'anomalous' glasses, where densification (compaction) of an open structure is the main deformation response.[13] Vitreous silica is the archetypical 'anomalous' glass.[13] Network modifying ions, in addition to occupying interstitial space, are thought to provide easy paths for shear-flow to occur in 'normal' glasses.[8,9,14,19,20] Although bonds are broken between atoms, they are thought to recombine, however not necessarily in the same structural configurations as in the pristine glass. The densification in vitreous silica, which lacks network modifying ions, is thought to be a result of its open cage-like structure and the ability of the silica tetrahedra to rotate and bend, but not break, and thereby collapse into a more closely-packed structure.[13]

It is apparent that whether a glass behaves 'normal' or 'anomalous' under a sharp indenter is strongly related to its composition. Experiments involving changing glass composition and monitoring the effect on the indentation response have shown this to be the case.[8,13,17,20,21] All of these experiments examined silicate-based glasses, with silica as the network forming constituent in amounts exceeding ~ 57 mol%. The general observation is that as the amount of network modifying ion content is decreased relative to silica, the behavior transitions from 'normal' to 'anomalous.' Although composition was the major determining factor in the transitions observed by Wagner,[21] the maximum load also had some influence. For example, for several glass compositions the formation of ring-cone cracks, which is taken as indicative of 'anomalous' behavior,[13] did not occur as readily as in some of the more 'anomalous' glasses he tested, but rather required a higher load to initiate. In addition, the frequency of ring-cone cracking in these glasses increased as the load increased. In some respects, this behavior isn't that surprising, since even in vitreous silica a finite load is required to initiate ring-cone cracking,[13,17,21] albeit a smaller one. It is apparent that as glass composition changes from pure silica to one containing increasing amounts of modifier ions, the threshold load for ring-cone cracking should increase. Presumably, a composition will be reached where no increase of load will bring about the formation of ring-cone cracks, and such a glass could be considered ideally 'normal.'

The sequence of cracking events about 'normal' and 'anomalous' glasses has also been examined in detail.[7,12,13,17,21] The terminology describing the crack types is worth mentioning. The term median crack has been used to designate a well-developed crack contained on the loading (median) axis of the indenter and that extends well below the damage zone of the contact impression. Such cracks are usually associated with high-load indentations in brittle materials.[10,12] The term well-developed implies the crack extends considerably outside the influence of the damage zone. The initiation of this crack is usually assumed to occur at the base of the damage zone beneath the contact impression.[10,12] Sometimes the term median-vent is used to describe this crack. The shape of this crack usually resembles a full-penny in the initial stages of development.[10,12] On unloading, this crack assumes a half-penny shape, since the elastic compressive stresses which initially constrained the crack from the surface diminish, allowing the crack to 'break-through', leaving behind the crack traces commonly seen on the surface.[10,12] The term radial-crack is used to describe surface localized cracks which do not typically extend below the damage zone of the contact impression.[7] These cracks intersect the surface as soon as they initiate. The term Palmqvist crack is synonymous with radial-crack. Palmqvist cracks have been found in tough cemented carbide materials.[22] When high enough loads are used the radial cracks may continue to grow and coalesce well below the damage zone of the contact impression, forming a half-penny shaped crack system.[22] Sometimes the term median-radial is used to describe situations where both median and radial cracks have formed during the contact cycle, or sometimes just to describe a median crack. Other times this term is used to describe the final half-penny shaped crack system formed by median-crack breakthrough to the surface or by the coalescence of two radial cracks below the surface. The current authors prefer the term median-radial to describe any crack system having a final half-penny geometry, no matter the manner in which this final geometry is achieved, and where the term 'radial' is being used to indicate these cracks extend out radially from the contact impression i.e., radial (Palmqvist) cracks are still considered a separate crack system. Lateral cracks are another crack system which tend to propagate out laterally (parallel) to the surface and typically initiate at the base of the damage zone.[10,17] Shallow lateral cracks may also form, which as their name suggests, initiate much closer to the surface than regular lateral cracks.[17] The shallow lateral cracks may be associated with, or possibly are, the extensions of another set of cracks called secondary radial cracks, as originally coined by Cook and Pharr.[17] Secondary radial cracks are different than radial cracks in that they tend to initiate along the sides of the contact impression rather than the corners, and are slightly inclined to the loading axis.[17] Finally, ring and cone cracks are frequently found around sharp contacts in 'anomalous' glasses. A surface crack usually initiates from a flaw near the outside of the contact region and propagates unstably around the contact site forming a ring, whereby further increase in load causes the ring crack to propagate downward in a stable manner.[13] At a critical load the ring crack flares unstably into a truncated cone geometry, which then propagates stably with increasing load.[13] The system is commonly referred to as the ring-cone crack system. Other crack types may exist which do not fit into any of the above classifications.

The nomenclature used in the descriptions of cracking behavior that follows is the same used by the authors who are cited. Arora et al.[13] found that 'anomalous' (densifying) glasses, including vitreous silica and borosilicate glass, formed ring-cone cracks at low loads on the loading half-cycle when contacted with a Vickers diamond. All of their work was performed in a dry nitrogen environment. First a ring crack was observed to initiate unstably on the surface, then with increasing load this crack grew stably downward, then flared into the full-blown ring-cone crack system at a critical load. Median-radial and lateral cracks were occasionally found to initiate on loading as well, although Arora et al. described this aspect of the behavior as 'variable.' On unloading the median-radial and ring-cone crack geometry was found to change very little, however it was observed that a part of the cone crack would break-away, and leave a circular surface crater. Arora et al.[13] also pointed out that in rare instances when ring-cone cracking was suppressed, a well-developed set of median-radial cracks would initiate on unloading. Cook and Pharr[17] also investigated, in-situ, the crack initiation behavior of vitreous silica with a Vickers diamond, but in laboratory air environment. They found cone cracks to initiate early on loading, at ~ 1.8 N, followed by radial crack initiation at ~ 9.2 N, which was then followed by secondary radial cracks at ~ 23 N. Cook and Pharr,[17] however, did not observe lateral cracks on loading like Arora et al.[13] did. On unloading, Cook and Pharr[17] observed the cone crack to close, followed by closure of the radial cracks, which was then followed by initiation of lateral cracks, which was then followed by initiation of another set of secondary radial cracks. Cook and Pharr[17] also examined an 'anomalous' borosilicate glass. Radial and secondary radial cracks were observed to initiate on loading at loads of ~ 16 N and ~ 21.6 N, respectively. The threshold load for cone cracking was found to be ~ 5.6 N, which was about three times greater than the threshold load in vitreous silica. Apparently, the network modifier ions present in the borosilicate glass delay the build-up of the surface elastic tensile stresses responsible for ring-cone cracking. If it is assumed that a critical amount of densification needs to be achieved before the surface elastic tensile stresses can build to the critical level required for ring-cone cracking, then it is clear the borosilicate glass does not densify as readily compared to vitreous silica, and hence can be considered less 'anomalous.'

There is somewhat more agreement regarding the cracking sequence observed in 'normal' glasses contacted with a Vickers diamond. Cook and Pharr[17] found no cracking on loading for a soda-lime-silica glass indented up to 40 N. On unloading, however, radial cracks initiated at ~ 26 N along with smaller secondary radial cracks. The cracks they observed did not extend below the depth of the plastic zone of the contact impression, and were thus the surface localized radial (Palmqvist) cracks, according to the authors. The authors used a depth-of-focus technique to determine this. This was followed by lateral crack formation near complete unloading of the indenter. Arora et al.[13] found that median-radial cracks, having a penny-like geometry, initiated on loading, although they did not state what load this occurred at. However, the maximum load used was 40 N in their experiments. Lateral cracks were observed to initiate near complete unloading of the indenter. Lawn et al.,[7] in a study on soda-lime-silica glass, in a variety of test environments, found that radial cracks initiated during the unloading cycle, or just after complete unloading, for maximum loads up to 6.8 N. Judging from data by Lawn and Swain,[12] who also monitored the formation of cracks in-situ in soda-lime-silica glass in laboratory atmosphere, the threshold load for median-radial cracking on loading was ~ 50 N, and at a load of ~ 500 N these median-radial cracks were observed to break through to the specimen surface. When lower maximum loads are used such break-through behavior usually occurs on unloading.[3,4,7,10,17,21] Marshall and Lawn[10] observed similar behavior as Lawn and Swain,[12] except the threshold for median-radial cracking on loading was ~ 20 N, and that breakthrough to the surface occurred at loads in excess of 150 N. Wagner[21] found that for maximum loads up to 13.8 N (1.4 Kg) radial cracks initiated on unloading for both the 'normal' and 'anomalous' glasses he tested. Similar to Cook and Pharr,[17] Wagner[21] used a depth-of-focus technique to determine that the cracks were surface localized. In addition, for vitreous silica Wagner[21] found that radial cracking occurred on unloading for 89% of the tests, and in 11% of the tests these cracks form during loading. Ring-cone cracks were found to initiate early on the loading cycle.

In summary, past work suggests that for Vickers indentation median-radial cracks initiate on unloading for silicate-based glasses containing at least 57 mol% silica and for relatively small maximum loads (< 20 N). When high loads are used, median-radial cracks can initiate on loading. For 'anomalous' glasses the crack initiation behavior is more variable, and it appears that the ring-cone crack system and its formation on the loading cycle is the only unique cracking characteristic of 'anomalous' glasses, since median-radial, lateral, and radial cracks have been observed in 'anomalous' glasses as well. The sub-surface damage zones show drastic differences between the two glass types. A network of intersecting shear-flow lines have been observed to comprise the damage zones beneath indentations in 'normal' e.g., soda-lime-silica glasses,[8,9,13,14] whereas in vitreous silica shear-flow lines are not observed in the damage zone,[8,9,13,14] but instead density contours brought about by pressure-induced changes in refractive index are observable.[14]

The aim of the current work was to investigate in-situ the crack initiation behavior of glasses containing substantially less silica, or none at all, compared to those examined in the past, and which to the authors knowledge have not been examined before.

EXPERIMENTAL PROCEDURES

A recording microindenter was built so that crack initiation could be investigated in-situ. A schematic of the system is shown in Figure 1. The whole assembly rested on a sturdy table, and four rubber pads were positioned between the steel base plate of the indenter and the table in order to dampen any vibrations. The entire system was computer controlled. The soft-PZT piezoelectric actuator[†] pushes the Vickers diamond into the glass sample while the load and displacement of the diamond are simultaneously monitored as a function of time by the strain-gauge load cell[‡], and two capacitance displacement gauges,[*] respectively.

[†]Model PZL-60, Burleigh Instruments, Inc., Fishers, New York, USA

[‡]Lebow Model 3189-50, Eaton Corp. Lebow Products, Troy, Michigan, USA

*Model HPT-40, Capacitec Corporation, Ayer, Massachusetts, USA

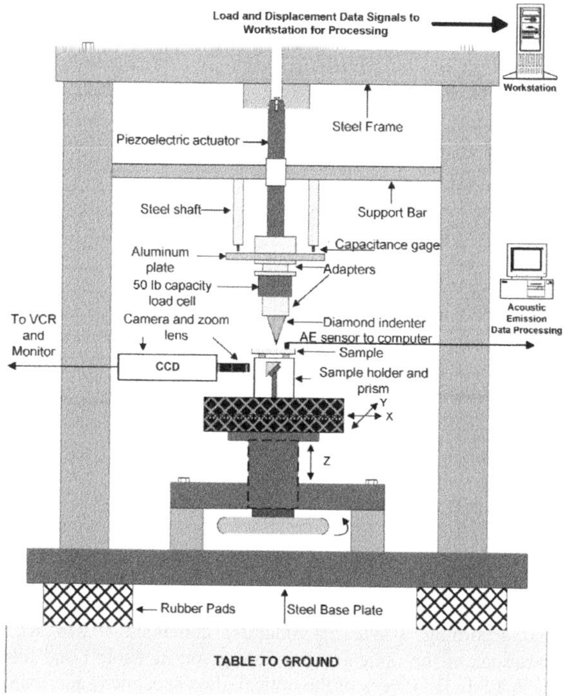

Fig. 1. Schematic diagram of the recording microindentation system used in the current work.

The crack initiation process was examined in real-time by means of an optical glass prism situated below the transparent glass samples. The prism re-directed the light from the fiber-optic lamps used to illuminate the test piece, into the CCD camera. The CCD camera was attached to a video monitor which allowed real-time observation of the indentation process. A VCR allowed the entire process to be taped for later analysis. A PAC MISTRAS-2001 acoustic emission system‡ was used to detect any acoustic emissions resulting from the crack initiation tests, in-situ. The acoustic emission system aided in the detection of cracking.

Sample Preparation

Nine optical glass compositions were tested, as were a commercial float glass, Corning 7740 Pyrex™ borosilicate glass, and a vitreous silica†. Six of the optical glasses contained silica and are designated by the prefix 'S', while three compositions contained no silica and are designated by the prefix 'NS.' The Suprasil 312 vitreous silica specimen was flame fused from synthetic

‡Physical Acoustics Corporation, Princeton, New Jersey, USA
†Suprasil 312—Heraeus Quarzglas GmbH, Hanau, Germany

Table I. Optical Glass Compositions (Mol %).

Components	Glass								
	S1	S2	S3	S4	S5	S6	NS1	NS2	NS3
SiO_2	81	67	44	13	8	6			
ZrO_2					6	6	6		
Na_2O	6	7							
K_2O	5							23	
CaO					27	10			6
ZnO		10				7	5		
BaO			27						
PbO	5			58					
MgO									
SrO									
B_2O_3		11	25	22	51	60	60		
Al_2O_3	5	6	6	6				7	
La_2O_3					18	13	18	6	
Y_2O_3							7		
P_2O_5								64	7
Nb_2O_5							5		
KHF_2	1								
LiF									9
CaF_2									8
MgF_2									9
BaF_2									12
SrF_2									26
AlF_3									23
Total (Molar %)	103	101	102	99	100	102	101	100	100

Table II. Float and 7740 Pyrex™ Borosilicate Compositions (Mol %).

	Float Glass (mol %)	Pyrex™ Borosilicate Corning Code 7740 (mol %)
SiO_2	70.3	81
Al_2O_3	0.4	2
B_2O_3		13
CaO	9.3	
MgO	6.1	
Na_2O	13.7	4
K_2O	0.1	
RO_2	2.41	20.25
R_2O_3	0.01	3.75
RO	0.53	
R_2O	0.47	1.00

silica particles, and according to the manufacturer contained ~ 216 wt. ppm OH. The approximate compositions of the optical glasses are given in Table I and that of the float and borosilicate glasses in Table II. Pieces of the optical glass specimens approximately 30 x 25 x 4 mm were ground and polished to achieve a flat, near-optical quality surface with few scratches and pits. The float and Pyrex samples were cut from glass sheets. The Pyrex sample was polished on one side, while no preparation was necessary for the float glass. The vitreous silica specimen was in the form of a large disk ~ 30 mm diameter by ~ 10 mm thick. Grinding was conducted in distilled water using SiC paper. First, 320-grit, followed by 400-grit, then 600-grit SiC paper was used. Polishing was conducted using non-petroleum based commercial diamond spray solutions, in sizes of 15 μm, 6 μm, 3 μm, and 1 μm. A final polishing step was performed with a commercial cerium oxide compound combined with distilled water. The only exception to the above procedure was for specimens S3 and NS1, where a polishing oil was used for the grinding steps, and for the polishing steps diamond pastes of 15 μm, 6 μm, 3 μm, and 1 μm were used along with the oil. The oil was thought to represent an inert medium which would reduce water interaction with these glasses. These specimens were used only for the in-situ crack initiation experiments. Additional specimens of glass S3 and NS1 were prepared via the water-route, and were used for the rest of the tests to be described. For glass NS2 an additional specimen was prepared via the oil-route described in order to see the effect, if any, of polishing medium on crack initiation behavior. As it turns out, the results showed no effect of polishing medium on crack initiation behavior, and there were no visible signs of chemical attack on the surfaces of any of the glasses after polishing via the water-route. All specimens were cleaned after polishing in reagent grade alcohol in an ultrasound bath and then annealed to remove any residual stresses. The alcohol helps remove dirt and water from the surfaces. Only the vitreous silica specimen was not annealed. All specimens were then stored in a drying oven kept at ~ 100 °C prior to testing. All tests for the float glass specimens were performed on the atmosphere side.

Crack Initiation Tests

Tests were performed on the polished sides of samples. A maximum load of 4 Kg was used for specimens S1, S3, S5, S6, NS1, float, and vitreous silica. A maximum load of 3.5 Kg was used for samples S2, S4, NS2, and NS3, and a 1 Kg maximum load for the Pyrex™ borosilicate glass. For all tests the indenter displacement rate was 0.2 μm/s for both the loading and unloading cycles. Specimens were tested in laboratory air. The temperature and relative humidity between testing specimens was 23 ± 2 °C and 28 ± 11%, respectively. An air conditioner was used to help keep the humidity of the laboratory air low. Ten indentation tests were performed per sample. Post-test microscopy followed the crack initiation tests. Additional crack initiation tests using a 1 Kg maximum load and 0.2 μm/s displacement rate were used to check the reproducibility of certain aspects of the crack initiation behavior for all samples.

Cross-Section Morphology

Thin specimens, approximately 1 mm thick, were ground and polished. Indentations using a conventional dead-weight hardness tester were made at loads of 1 Kg and 5 Kg, with a 5 s dwell under load. The indentations were made in a straight line such that the median-radial crack traces were more-or-less parallel to one another. The specimens were then flexed until they fractured in half. This occurred via the median-radial cracks propagating and cleaving through the centers of the indentations, along one of the indentation diagonals. The resulting indentation cross-sections were examined under an optical and a scanning electron microscope.

Mechanical Properties

The Vickers hardness was measured using a conventional dead-weight hardness tester with a load of 200 g, and a dwell time of 15 seconds. The vitreous silica specimen was not tested at this load because cracking caused unreadable indentations. Instead, the recording microindenter was used to make 100 g load indentations for hardness determination. Ten tests were made per specimen. The elastic moduli of the glasses was measured using the pulse-echo technique, which measured the time of flight for both shear and longitudinal sound waves to propagate through the specimens. Glass density was measured using the Archimedes method, with kerosene as the immersion liquid. Indentation toughness (K_C) was measured by making indentations at loads ranging from 0.5 Kg to 2 Kg, depending on the glass, and measuring the length of the major crack traces on the surface, from the center of each indentation, immediately after indentation. Measurement of the cracks commenced within seconds after making an indentation, and it usually took on the order of one minute to measure all four crack traces on the surface for each indentation. Toughness values for vitreous silica and 7740 Pyrex™ borosilicate were not calculated since these glasses densify significantly, which invalidates the use of the technique. Equation 1, after Anstis et al.,[23] was used to calculate the toughness.

$$K_C = 0.016\left(\frac{E}{H}\right)^{1/2}\left(\frac{F}{c^{3/2}}\right) \tag{1}$$

K_C = indentation toughness ($N/m^{3/2}$)
E = Young's Modulus (N/m^2)
H = hardness (N/m^2)
F = indentation load (N)
c = crack length from center of indentation to crack tip (m)

The Vickers hardness (H_V) was used for H in Eq. 1.

RESULTS

Mechanical Properties

Table III shows the results of the mechanical properties testing. Poisson's ratio (ν) is shown along with the ratio of shear modulus to bulk modulus (G/K). The representative values of fracture toughness for vitreous silica and Pyrex™ borosilicate glass shown in Table III are from Wiederhorn[25].

Table III. Mechanical Properties Data (all errors ± one std. dev.)

Glass	ρ (g/cm^3)	H_V (GPa)	E (GPa)	ν	K_C ($MPa\sqrt{m}$)	G/K
S1	2.50	4.71 ± 0.11	63.3 ± 1.3	0.156 ± 0.022	0.78 ± 0.08	0.884
S2	2.46	5.36 ± 0.09	70.1 ± 0.5	0.182 ± 0.013	1.01 ± 0.11	0.814
S3	3.55	6.23 ± 0.12	88.1 ± 0.6	0.267 ± 0.006	0.51 ± 0.02	0.545
S4	6.16	3.04 ± 0.07	48.6 ± 0.8	0.268 ± 0.011	0.26 ± 0.01	0.543
S5	3.47	7.61 ± 0.11	112.8 ± 1.8	0.264 ± 0.010	0.65 ± 0.02	0.570
S6	3.74	7.81 ± 0.14	114.2 ± 2.8	0.269 ± 0.016	0.69 ± 0.04	0.544
NS1	4.29	8.02 ± 0.26	123.0 ± 2.0	0.288 ± 0.019	0.69 ± 0.03	0.490
NS2	2.57	3.63 ± 0.05	49.7 ± 0.6	0.249 ± 0.013	0.44 ± 0.02	0.601
NS3	3.66	4.31 ± 0.13	78.9 ± 1.2	0.285 ± 0.009	0.35 ± 0.02	0.490
float	2.52	5.28 ± 0.05	73.6 ± 1.4	0.221 ± 0.019	0.72 ± 0.02	0.687
Pyrex™*	2.23	5.35 ± 0.11	62.7	0.20	0.798‡	0.750
Suprasil 312	2.23	6.78 ± 0.40	74.6 ± 0.4	0.17 ± 0.1	0.758‡	0.844

*-E and Poisson's ratio from Ref. 24

‡-from Ref. 25

Crack Initiation

The crack initiation results are summarized in Table IV for both the 4 Kg and 1 Kg maximum load tests. M-R stands for median-radial cracks. The median-radial crack initiation loads listed are for the first median-radial cracks observed to initiate.

The following descriptions are for the 3.5 Kg and 4 Kg maximum load tests. Glasses S1, S3, S4, S5, S6, NS1, NS2, and NS3 formed median-radial cracks readily on the loading half cycle i.e., at low loads. Usually what appeared to be two median-radial cracks initiated simultaneously, and at opposite indentation corners, followed by the initiation of two other median-radial cracks at separate instances. For vitreous silica, it was believed that some median-radial cracks initiated at loads near 1 Kg. The numerous ring-cone cracks which began initiating early on the loading cycle (< 300 g) obscured the view of the median-radial crack initiations, and thus made it difficult to observe the exact load at which these cracks initiated. Additional median-radial cracks initiated on unloading, usually accompanying the partial detachment/closure of the dominant cone crack. For glass S1, large ring-cone cracks began initiating on loading at around 2 Kg. For glass S2, large ring cracks initiated in 60% of the tests beginning at loads greater than 1 Kg, but a major cone-crack only formed in one of these tests. Median-radial cracks initiated on unloading in this glass. For the 7740 Pyrex™ borosilicate glass ring-cone cracks began initiating at low loads on the loading cycle (< 300 g), with median-radial cracks initiating on unloading. It is believed that median-radial cracks would have initiated on loading if a higher maximum load was used (only a 1 Kg max. load was used in the current work), similar to what Cook and Pharr[17] observed for an 'anomalous' borosilicate glass. In addition, for glasses S3, S4, S5, S6, NS1, NS2, and NS3, lateral cracks initiated on the loading cycle, usually after the first median-radial cracks initiated. For glasses S4 and NS3 the lateral crack growth on loading was substantial. For the float glass, median-radial cracks initiated on unloading in 80% of the tests, however in 20% of the tests initiation of these cracks took place on loading, near the maximum load. In addition, post-test optical microscopy revealed the presence of small partial ring cracks around the corners of indentation sites in float glass and glass NS2, although for the latter they were not as numerous or as pronounced. For all glasses lateral cracks initiated on the unloading cycle. In addition, shallow lateral cracks were observed to initiate, usually after the deeper lateral cracks but sometimes before, in all of the optical glasses and vitreous silica, with the exception of glass S2 and the Pyrex™ borosilicate glass. These cracks frequently chipped away portions of the surface, particularly in vitreous silica, glass S1, S3, NS1, NS2, and NS3.

The following descriptions pertain to the 1 Kg maximum load tests. The crack initiation tests carried out at 1 Kg maximum load reproduced some of the behavior seen in the 4 Kg tests. Most notably, the initiation of median-radial cracks on the loading cycle and lateral cracks on the unloading cycle was reproduced, however there were some differences. For glasses S1, S3, S5, and S6 not all of the median-radial cracks initiated on the loading cycle for all of the tests. That is, some median-radial crack initiation took place on unloading in these glasses. In addition, lateral crack development was not as frequent or as well developed on loading as in the 4 Kg tests for those glasses where lateral cracks initiated, with the exception of glasses S4 and NS3, where the frequency and intensity of lateral cracking was still high. In addition, surface chipping from shallow lateral cracks was not as frequent compared to the 4 Kg load tests. For the float glass, all median-radial crack initiation took place on unloading. For glasses S1 and S2, no ring-cone cracking took place on loading, since the threshold load for this was considerably higher than 1 Kg. In addition, for glass S2 only 30% of the tests initiated median-radial and lateral cracks on unloading. For the rest of the tests no cracks formed. For the vitreous silica specimen well-developed ring-cone cracks still developed on loading, and median-radial cracks were believed to initiate near the maximum load as well. On unloading additional median-radial cracks initiated accompanying the partial closure/detachment of the cone crack.

The crack initiation behavior suggested that glasses S1 and S2 were 'anomalous' as were the vitreous silica and borosilicate glasses. Although the remainder of the optical glasses median-radial cracked readily on loading, a characteristic usually thought associated with 'anomalous' glasses,[13,17] there was no evidence of ring-cone cracking, and as the sub-surface cross-sections will show, the damage zones beneath these glasses showed extensive shear-flow line formation, a characteristic of 'normal' i.e., volume conserving, glasses.

Table IV. Crack Initiation Behavior Summary. (all errors ± one std. dev.)

	LOADING CYCLE					UNLOADING CYCLE		
Glass	Max. Load (Kg)	M-R Cracking	M-R Initiation Load (g)	Lateral Cracking	Ring-Cone Cracking	M-R Cracking	Lateral Cracking	Shallow Lateral Cracking
S1	1	yes	317 ± 200	no	no	yes	yes	no
	4	yes	436 ± 268	no	yes	no	yes	yes
S2	1	no	/	no	no	yes*	yes*	no
	3.5	no	/	no	yes	yes	yes	no
S3	1	yes	817 ± 130	yes*	no	yes	yes	yes
	4	yes	574 ± 216	yes	no	no	yes	yes
S4	1	yes	40 ± 13	yes	no	no	yes	yes
	3.5	yes	43 ± 18	yes	no	no	yes	yes
S5	1	yes	311 ± 116	yes*	no	yes	yes	yes
	4	yes	285 ± 80	yes	no	no	yes	yes
S6	1	yes	482 ± 163	no	no	yes	yes	yes
	4	yes	480 ± 176	yes	no	no	yes	yes
NS1	1	yes	257 ± 80	yes*	no	no	yes	yes
	4	yes	similar to 1 Kg	yes	no	no	yes	yes
NS2	1	yes	133 ± 33	yes*	no	no	yes	yes
	1 (oil polished)	yes	89 ± 35	yes*	no	yes	yes	yes
	3.5	yes	132 ± 29	yes	no	no	yes	yes
NS3	1	yes	instantaneous	yes	no	no	yes	yes
	3.5	yes	instantaneous	yes	no	no	yes	yes
float	1	no	/	no	no	yes	yes	no
	4	yes*	3531 ± 231	no	no	yes	yes	no
Pyrex™	1	no	/	no	yes	yes	yes	no
Suprasil 312	1	yes	~ 1 Kg	no	yes	yes	yes	yes
	4	yes	~ 1 Kg	no	yes	yes	yes	yes

*-indicates infrequent occurrence

Indentation Cross-Sections

Figure 2 is a cross-section of a 1 Kg indentation site in glass S1. Median-radial and lateral cracks are apparent, and the damage zone is devoid of shear-flow lines. An SEM image of another damage zone is shown in Figure 3, which more clearly shows the lack of shear-flow lines, a feature consistent with glass S1's 'anomalous' crack initiation behavior. A cross-section of a 5 Kg indentation site in glass S1 is shown in Figure 4, where a sub-surface cone crack is apparent, as are median-radial and lateral cracks. The median-radial cracks appear to have been restricted in their development by the presence of the cone crack.

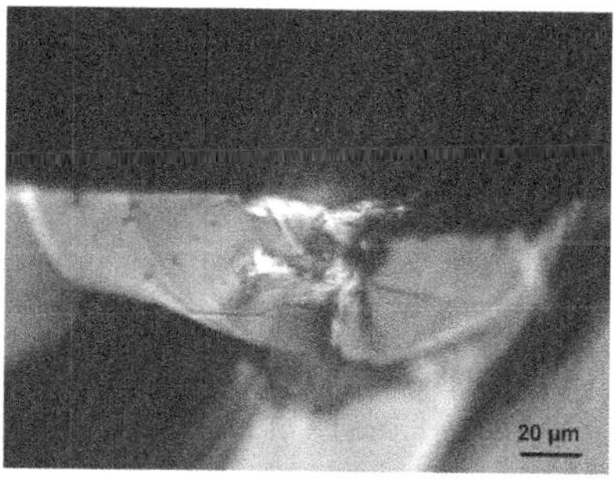

Fig. 2. Cross-section of a 1 Kg indentation Site in glass S1. Refl. light, BF.

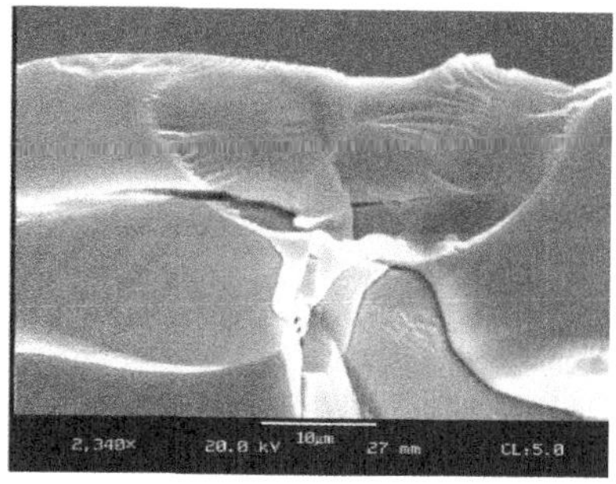

Fig. 3. SEM image of damage zone beneath 1 Kg indentation site in glass S1. No shear-flow lines are apparent. Scale bar is 10 µm long.

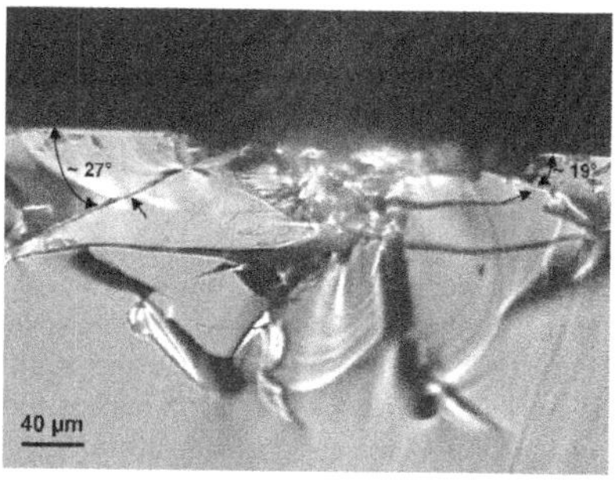

Fig. 4. Cross-section of a 5 Kg indentation site in glass S1. Refl. light, BF.

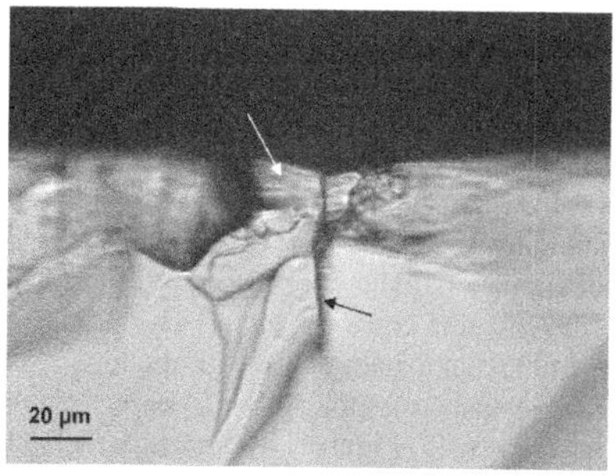

Fig. 5. Cross-section of a 1 Kg indentation site in glass S2. White arrow indicates damage zone, and black arrow a median-radial crack. Trans. light, BF.

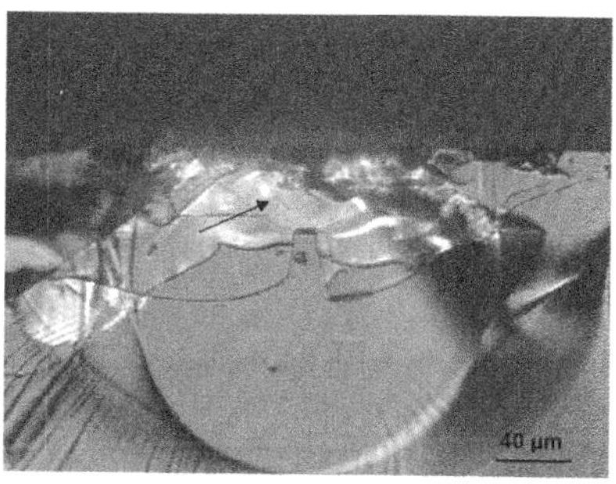

Fig. 6. Cross-section of a 5 Kg indentation Site in glass S2. Black arrow in center indicates damage zone. Refl. light, BF.

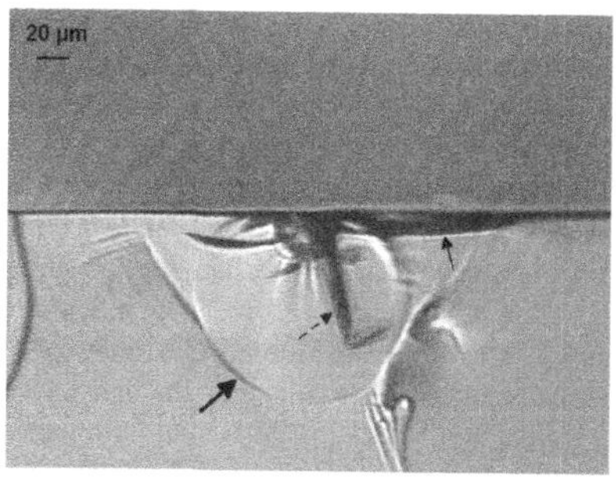

Fig. 7. Cross-section of a 1 Kg indentation site in glass S3. See text for arrow descriptions. Trans. light, BF.

Figures 5 and 6 are cross-sections of indentation sites in glass S2 from 1 and 5 Kg indentation sites, respectively. There is a lack of shear-flow lines in the damage zones in both indentations, however the damage zone for the 5 Kg indentation site contains many microcracks. A well-developed cone crack is apparent in Figure 6, as is a penny-shaped median-radial crack arrest line. The lack of shear-flow lines in the damage zone, and the appearance of cone-cracking in this glass, are consistent with 'anomalous' behavior.

Figure 7 is a cross-section of a 1 Kg indentation site in glass S3. A full-penny median-radial arrest line is apparent (large arrow), as is a median-radial vent crack extending down from the base of the damage zone (small broken arrow), lateral cracks, and a shallow lateral crack (small arrow). The full-penny crack eventually broke through to the surface, resulting in a nearly half-penny final crack shape. Other cross-sections revealed the presence of two median-radial cracks in the plane of fracture. Figure 8 shows a SEM cross-section image of a 1 Kg indentation site in glass S3. The large black arrow indicates a shallow lateral crack, while the two smaller arrows indicate deep lateral cracks. In between the two types of lateral cracks additional lateral cracks are present. The damage zone is comprised of numerous microcracks, but shear-flow lines are also apparent. A higher magnification image, illustrating in more detail the damage zone, is shown in Figure 9. The small black arrows designate fine shear-flow lines.

Figure 10 is a cross-section of a 1 Kg indentation site in glass S4. A well-defined damage zone containing many fine shear-flow lines is apparent. The top center portion of this zone contains a heavily-damaged core region of material. The arrest lines indicate that two in-plane median-radial cracks had initiated and which were bisected by another median-radial crack extending from the base of the damage zone (median-radial vent). The two in-plane median-radial cracks, taken together and before break through to the specimen surface, formed a nearly penny-shaped crack system. The actual shape of each of the individual cracks appears to resemble a half-penny, but with the long axis perpendicular to the surface. The final shape of the crack system resembles a half-penny. Shallow lateral cracks have removed the top part of the damage zone. Figure 11 is a higher magnification image of a damage zone beneath a 5 Kg indentation site in glass S4 and clearly shows the fine shear-flow lines which formed in this glass. Other cross-sections revealed the presence of a single full-penny median-radial crack in the fracture plane, bisected by another median-radial crack, similar to the cross-section seen in Figure 7, and which eventually broke through to the surface forming a final half-penny shape.

Figure 12 is a cross-section of a 5 Kg indentation site in glass S5. Two in-plane median-radial cracks had formed, as evident by the arrest lines on the fracture surface. These cracks appear to have been bisected by another median-radial crack extending from the base of the damage zone. Deep and shallow lateral cracks are present. The damage zone consists of a well-defined core of broken apart material that appears dark, and which is surrounded by a well-defined network of shear-flow lines. This is more clearly shown in Figure 13. Other cross-sections revealed the presence of a single full-penny median-radial crack in the fracture plane. Glass S6 exhibited similar cracking and damage zone characteristics as glass S5. This is not surprising since these glasses had similar composition.

A cross-section of a 1 Kg indentation site in glass NS1 is shown in Figure 14, and Figure 15 shows a transmitted light image of the same cross-section. A full-penny median-radial crack had initiated as evident by the arrest lines, then subsequently broke through to the surface. The transmitted light image reveals another median-radial crack perpendicular to the fracture surface. A well-defined region of shear-flow lines exists outside a core region of broken material, as seen in Figure 14.

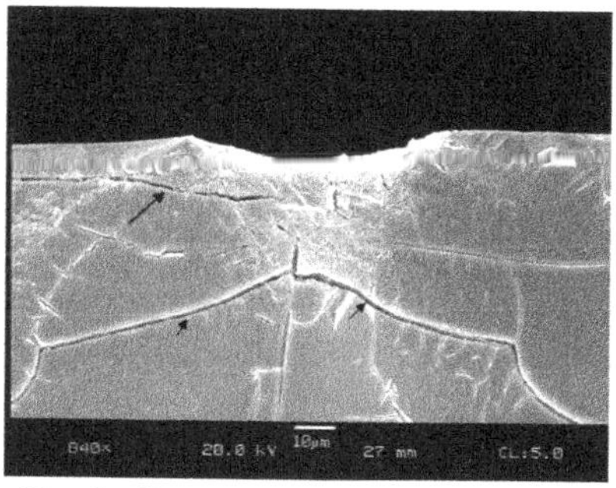

Fig. 8. SEM image of cross-section from 1 Kg indentation site in glass S3. See text for arrow descriptions. Scale bar is 10 μm long.

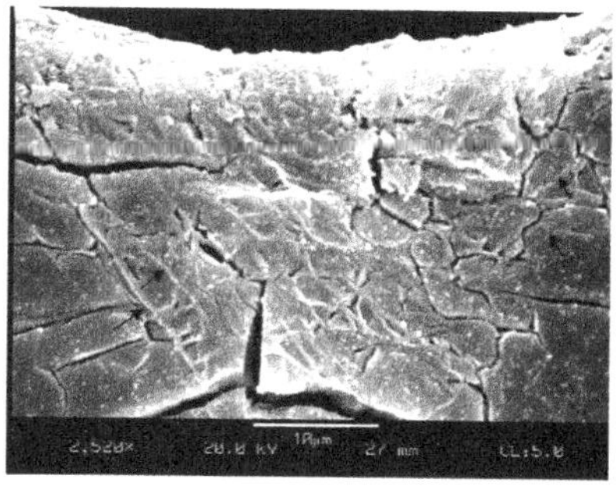

Fig. 9. Higher magnification image of Figure 8 illustrating in more detail the damage zone. Scale bar is 10 μm long.

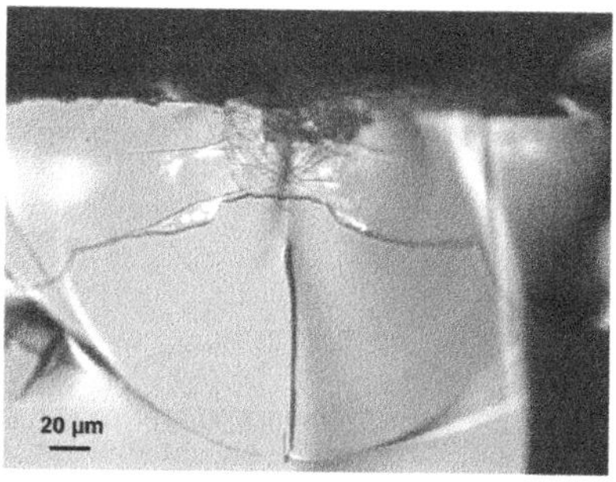

Fig. 10. Cross-section of a 1 Kg indentation site in glass S4. Refl. light, BF.

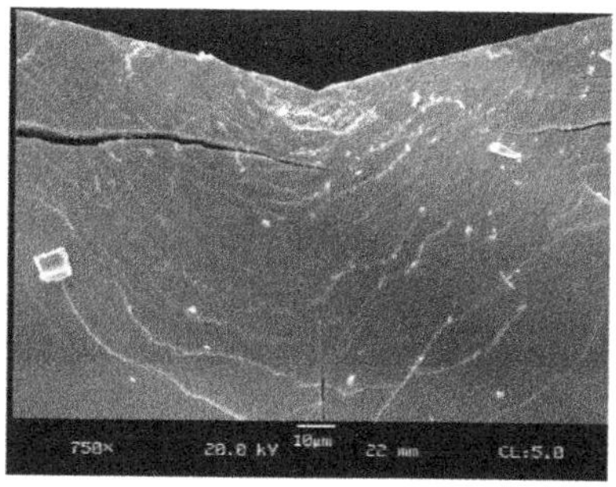

Fig. 11. SEM image of damage zone beneath a 5 Kg indentation site in glass S4. Scale bar is 10 μm long.

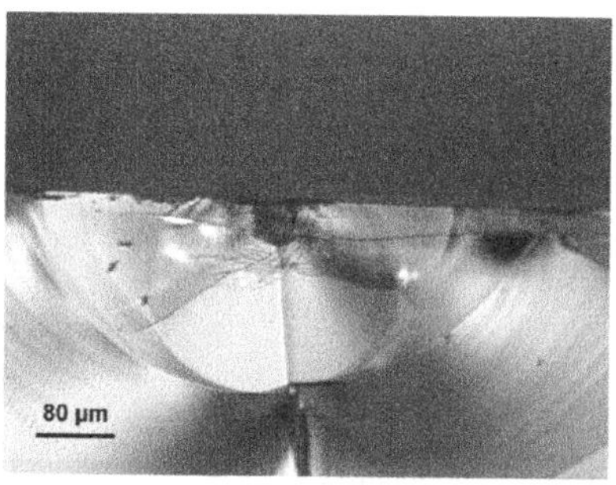

Fig. 12. Cross-section of a 5 Kg indentation site in glass S5. Refl. light, BF.

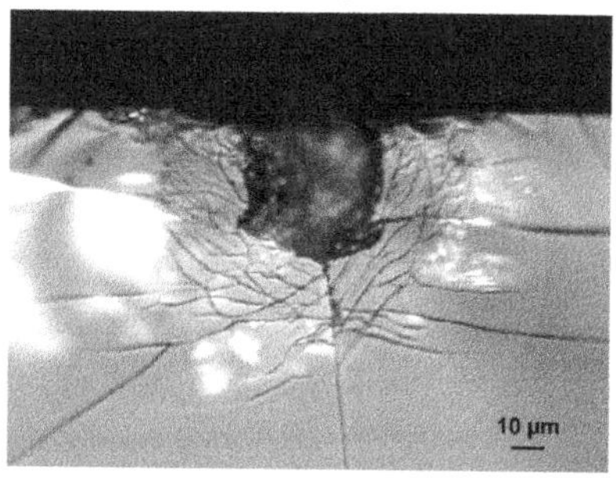

Fig. 13. Higher magnification image of the damage zone seen in Figure 12. Refl. light, BF.

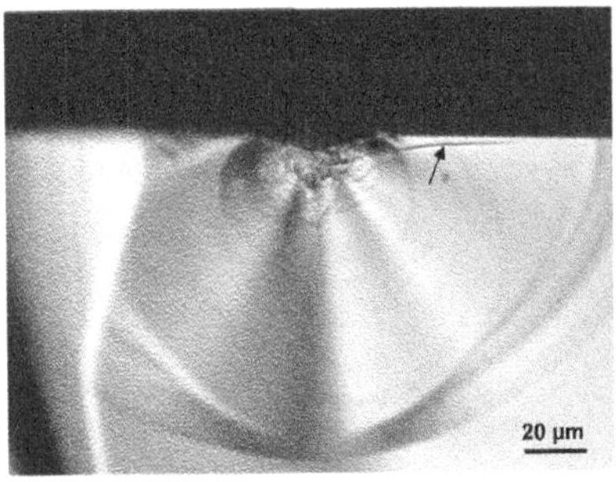

Fig. 14. Cross-section of a 1 Kg indentation site in glass NS1. Black arrow points to a shallow lateral crack. Refl. light, BF.

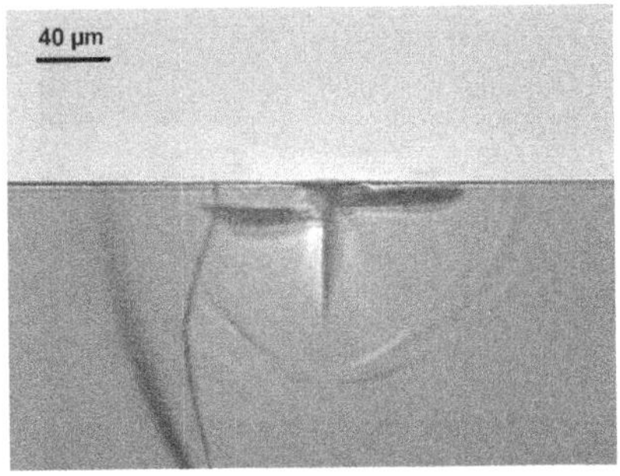

Fig. 15. Same indentation cross-section as in Figure 14, but viewed in transmitted light. Trans. light, BF.

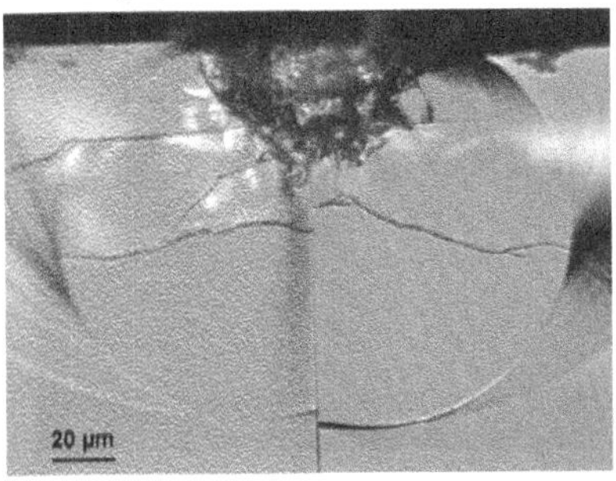

Fig. 16. Cross-section of a 1 Kg indentation site in glass NS2. Refl. light, BF.

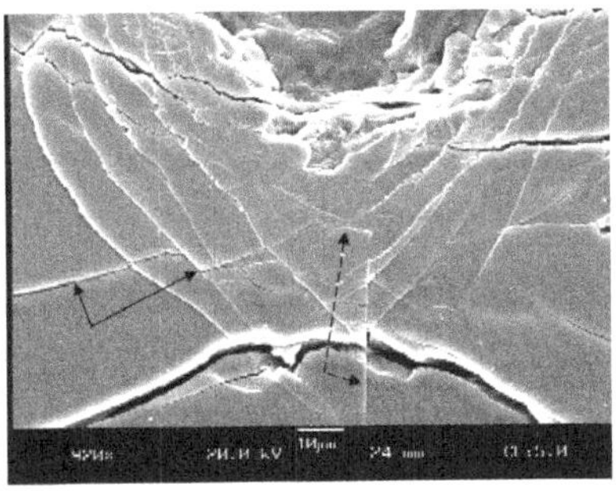

Fig. 17. SEM image of the damage zone beneath a 5 Kg indentation site in glass NS2. See text for arrow descriptions. Scale bar is 10 µm long.

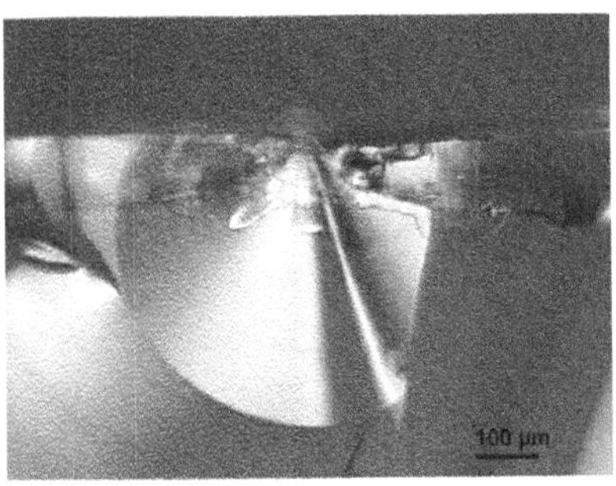

Fig. 18. Cross-section of a 5 Kg indentation site in glass NS3. Refl. light, BF.

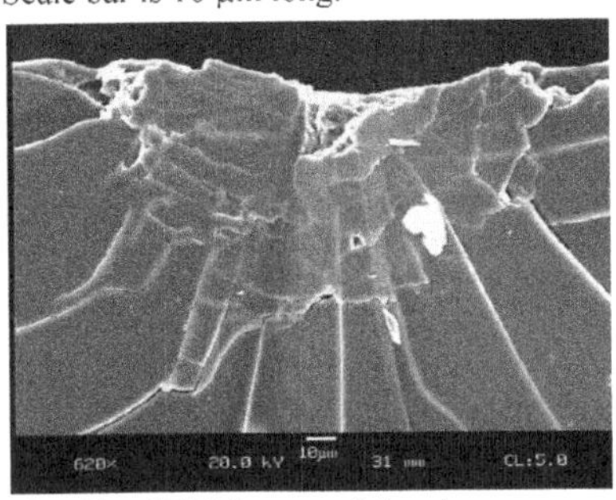

Fig. 19. SEM image of the damage zone beneath a 5 Kg indentation in glass NS3. Scale bar is 10 µm long.

Figure 16 is a cross-section of a 1 Kg indentation site in glass NS2. Two in-plane median-radial cracks initiated which were initially constrained at the surface, then subsequently broke-through and formed a final half-penny shape, the extent of which is not fully seen in this image. A damaged zone is apparent and appears broken apart. Other 1 Kg damage zones showed more shear-flow lines outside the damaged core. A 5 Kg indentation site shown in Figure 17 illustrates the fine network of shear-flow lines in this glass nicely. The double solid arrow indicates a shear-flow line/lateral crack pair while the double broken arrow indicates a median-radial crack/shear-flow line pair. The central core of broken apart material is also visible.

The damage below a 5 Kg indentation site in glass NS3 is shown in Figure 18. A full-penny median-radial crack developed and which broke through to the surface early in its development as shown by the arrest lines which intersect the surface early in the development of the crack. A shallow lateral crack on the top left side of the indentation is seen intersecting the surface, and the top right portion of the indentation site has been chipped away from shallow lateral cracks. Figure 19 shows in detail the damage zone beneath another 5 Kg indentation site, revealing the network of fine shear-flow lines which formed in this glass. A large number of cracks can be seen radiating outward from the damage zone. This type of cracking behavior was only seen in this glass. These radiating cracks were also present beneath the 1 Kg indentation sites, as were the shear-flow lines in the damage zone.

In Figure 20 a cross-section of a 1 Kg indentation site in the float glass is shown. The two small broken arrows show one of several arrest lines for the two in-plane median-radial cracks which initiated. The initial shapes of these median-radial cracks were quarter-penny and showed no signs of constraint at the surface after initiation. This indicates these cracks most likely initiated at a low load on the unloading cycle, when the surface elastic compressive stresses were small, allowing the cracks to breach the surface immediately after initiation. The large solid arrow points to a lateral crack, and the small solid arrow points to a median-radial crack extending down from the base of the damage zone, which then turned to the right and propagated more-or-less parallel to the surface. The damage zone appears to be heavily cracked. A cross-section of a 1 Kg indentation site made with the recording microindenter is shown in Figure 21. In-situ observation showed the median-radial cracks to initiate on unloading for this test. A full-penny median-radial crack had formed in the plane of fracture, as evident by the arrest lines. The arrest lines indicate this crack was initially constrained at the surface, indicating the compressive stress was still high at the time of initiation, which is consistent with the relatively high load at which median-radial cracks were first observed to initiate in this particular test. A SEM image of the damage zone beneath a 1 Kg indentation site is shown in Figure 22. Many microcracks are apparent, but a few shear-flow lines can be seen as well. A SEM image of the damage zone beneath a 5 Kg indentation site in float glass is shown in Figure 23. A small ring-cone crack is visible. A higher magnification image of this damage zone is shown in Figure 24. The region of the damage zone immediately below the surface appears smooth and virtually crack-free. Surrounding this region the material is heavily cracked, however some shear-flow lines are visible. However, it is apparent that the shear-flow behavior in this glass is not nearly as extensive or as well-defined as in optical glasses S3, S4, S5, S6, NS1, NS2, and NS3.

In Figure 25, a 1 Kg indentation site in the Pyrex™ borosilicate glass is shown. Numerous surface ring cracks are visible, indicative of this glass's 'anomalous' behavior. Figure 26 is a 1 Kg indentation site in Suprasil 312 vitreous silica, where numerous surface ring cracks are seen. A cross-section of a 1 Kg indentation site, from a vitreous silica microscope slide glass, is shown in Figure 27. A sub-surface cone crack is apparent and the damage zone appears to be devoid

Fig. 20. Cross-section of a 1 Kg indentation site in float glass. See text for arrow descriptions. Refl. light, DIC.

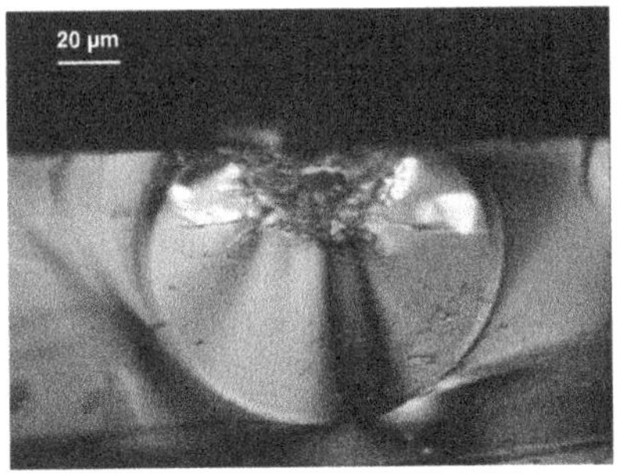

Fig. 21. Cross-section of another 1 Kg indentation site in float glass produced with the recording microindenter. Refl. light, DIC.

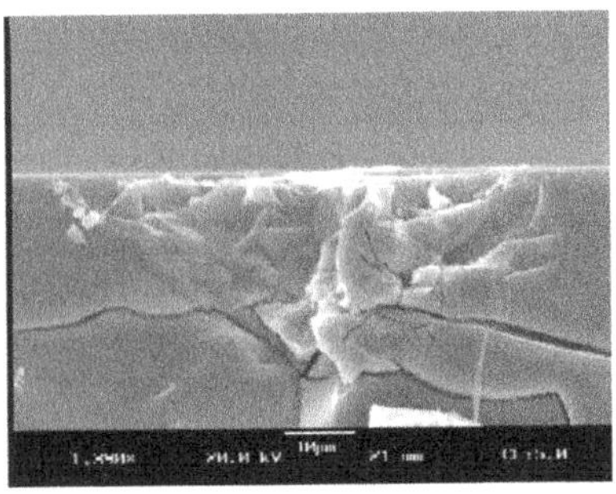

Fig. 22. SEM image of the damage zone beneath a 1 Kg indentation site in float glass. Scale bar is 10 µm long.

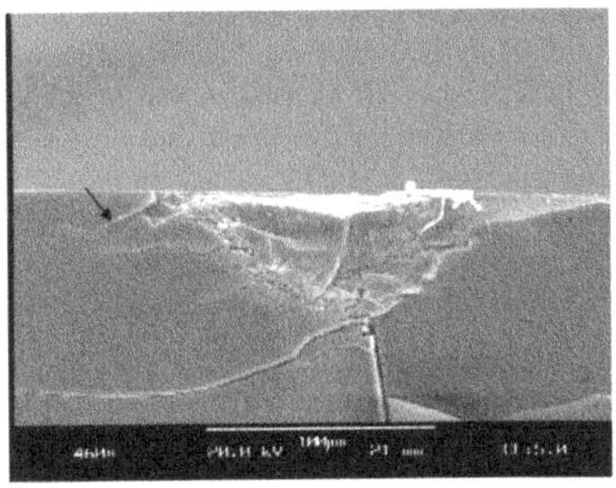

Fig. 23. SEM image of the damage zone beneath a 5 Kg indentation site in float glass. Arrow indicates a small ring-cone crack. Scale bar is 100 µm long.

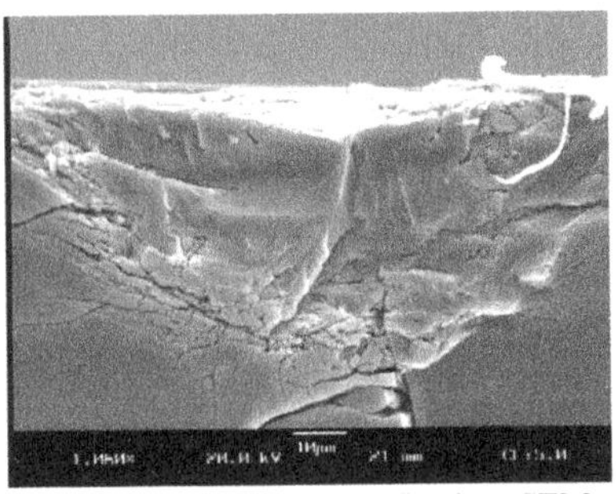

Fig. 24. Higher magnification SEM image of damage zone shown in Figure 23. Scale bar is 10 µm long.

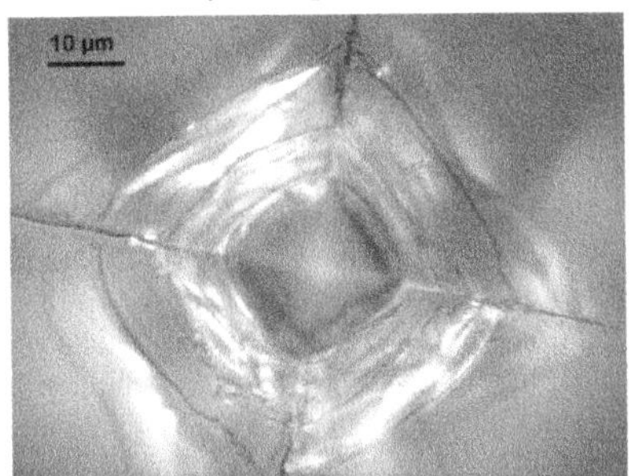

Fig. 25. 1 Kg indentation site in 7740 Pyrex™ borosilicate glass. Refl. light, DIC.

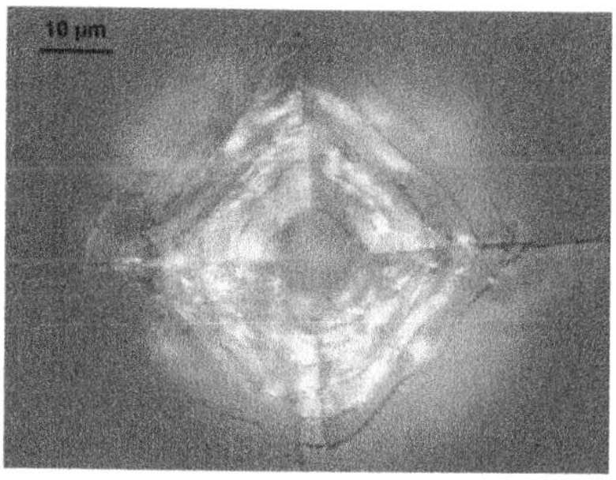

Fig. 26. 1 Kg indentation site in Suprasil 312 vitreous silica. Refl. light, DIC.

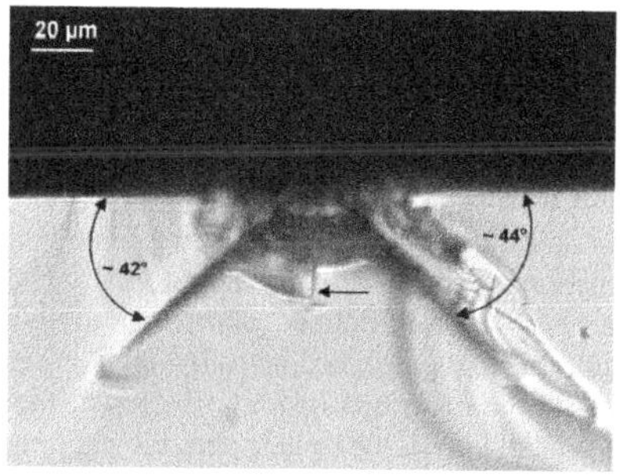

Fig. 27. Cross-section of a 1 Kg indentation site in vitreous silica microscope slide glass. Trans. light, BF.

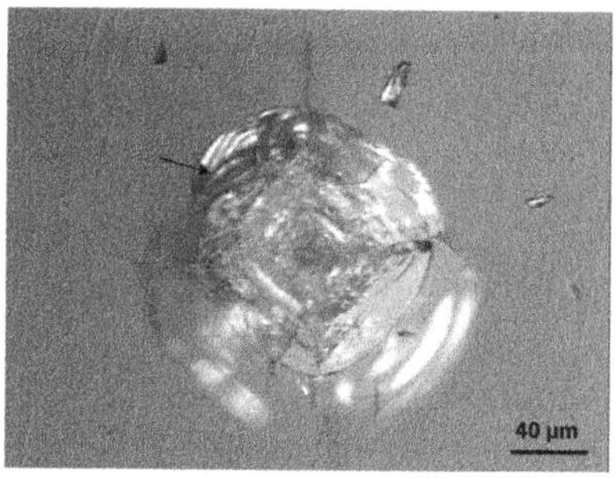

Fig. 28. 4 Kg indentation site in Suprasil 312 vitreous silica. Refl. light, DIC.

of any shear-flow lines, but does appear to contain some microcracks. A 4 Kg indentation site in Suprasil 312 vitreous silica is shown in Figure 28, where some surface chipping has occurred.

DISCUSSION

Damage Zone Behavior

Optical glasses S3, S4, S5, S6, NS1, NS2, and NS3 initiated median-radial cracks at relatively low loads on the loading cycle. The usual notion that 'normal' glasses median-radial crack only on indenter withdrawal in the low-load regime, does not hold for these seven glasses. Glasses S4, S5, S6, NS1, NS2, and NS3 exhibited median-radial cracking at an average load of less than 500 g on loading, with glass S3 showing a slightly higher average initiation load, but which was < 1 Kg. In glass NS3 median-radial cracks were found to initiate instantaneously on loading. All seven of the aforementioned glasses had damage zones characterized by extensive shear-flow line formation. This is indicative of 'normal' i.e. volume conserving, behavior. The greater the shear-flow component of deformation the more volume conserving the behavior should be, since pure shear deformation should only produce a shape change in the material. Glasses S1, S2, vitreous silica, and 7740 Pyrex™ borosilicate behaved 'anomalous.' The lack of shear-flow lines in the damage zones in glasses S1, S2 and vitreous silica suggest that densification (compaction) was the primary deformation mode. Although the sub-surface damage zones beneath indentations in the Pyrex™ glass were not examined, given the high silica content and the easy formation of ring-cone cracks, it is likely that densification was the major deformation response in this glass as well. The shear-flow lines in the float glass were not as extensive or as well-defined compared to those in the seven 'normal' optical glasses. In addition, for float glass median-radial cracking primarily took place on unloading for loads up to 4 Kg, with the exception of only two tests. However, even for these two tests the initiation of median-radial cracks took place near the maximum load (4 Kg). Recall that 'anomalous' glasses S1, S2, vitreous silica and 7740 Pyrex™ borosilicate also initiated median-radial cracks on unloading for loads up to 1 Kg (glasses S1 and vitreous silica also initiated some median-radial cracks on loading). In addition, small partial ring-cone cracks were seen around indentations at 4 Kg and higher in the float glass (see Fig. 23). It is clear the float glass did not behave as 'normal' as the seven optical glasses, or as 'anomalous' as glasses S1, S2, vitreous silica, or 7740 Pyrex™ borosilicate, but had a behavior somewhere in between, and thus could be classified as 'normal/anomalous.' Ernsberger[19] found evidence of densification beneath Vickers indentations in soda-lime sheet glass, which suggests a portion of the deformation behavior is 'anomalous.'

G/K Ratio and Deformation Response of Glass

The ratio of shear modulus to bulk modulus (G/K) may provide additional insight into the deformation response of all the glasses examined here, since the permanent indentations arise from the shear and hydrostatic-like components of the stress field. All of the 'anomalous' glasses had relatively high values of G/K (see table III) i.e., $G/K \geq 0.750$. The seven 'normal' optical glasses had $G/K \leq 0.601$. The float glass had an intermediate G/K value of 0.687. Since G measures elastic resistance to shear deformation, and K elastic resistance to hydrostatic compression (compaction), the ratio should indicate which deformation mode is more dominant, and consequently whether the glass behaves more 'normal' or more 'anomalous.' Although G and K are elastic properties, it is believed that since the volume of the surrounding (confining) elastic material in an indentation test is much greater than the inelastically deformed region below the indenter, it is the properties of the elastic region that primarily controls the inelastic response of glass in the damage zone. When G/K is low, the glass structure will undergo shear-flow, since this mode of deformation offers the least resistance, and is thus favored. When G/K increases, shear-flow becomes more difficult, and compaction (densification) assumes a more dominant role, resulting in 'anomalous' behavior. The structures and mechanical properties of the glasses are consistent with this idea. Glasses S3, S4, S5, S6, NS1, NS2, and NS3 contain little or no silica, and hence lack or have limited connectivity in all three spatial dimensions and presumably have little interstitial free volume, with the possible exception of glass NS2. In addition these glasses, with the exception of glass NS2, also have relatively high densities. The lack of silica and relatively high density suggest these glasses, with the possible exception of glass NS2, have tightly packed structures. These glasses will therefore show little densification since there is limited open space for the glass to compact into, and also because of their tendency to shear readily (low G/K). Although glass NS2 presumably has a more open structure than some of the other 'normal' glasses, its low value of G/K allows this glass to shear readily and thus behave 'normal' nonetheless. The lack of open space in most of these glasses, and their tendency to shear readily and want to change shape beneath a sharp indenter, is believed to result in large strains and stresses, and the formation of median-radial cracks early on the indentation loading cycle. The experimental results are consistent with this view. For the more 'anomalous' glasses with higher G/K, the tendency to shear-flow is diminished, resulting in less strain incompatibility between the elastic matrix and surrounding damage zone. This tends to delay the initiation of median-radial cracks to higher loads. Even though glass S1 had median-radial cracks initiate on loading for the 1 Kg maximum load tests, one or more median-radial cracks initiated on unloading in 80% of the tests as well.

Most of the glasses classified as 'normal' by Wagner[21] had $G/K \geq 0.65$. Recall the glasses he classified as 'normal' median-radial cracked on unloading (as did the glasses he classified as 'anomalous') for maximum loads up to 1.4 Kg. In addition, the sub-surface damage zones did not have as well-defined or as extensive a shear-flow line network as did the 'normal' glasses in the current work. Therefore, it is believed these glasses are best classified as 'normal/anomalous.'

Crack Initiation Behavior

Based on observations of the cross-sections of the indentations, it is concluded that what appeared to be the initiation of two separate median-radial cracks from opposite indentation corners on loading in glasses S3, S4, S5, S6, NS1, NS2, and NS3 was in fact a single full-penny median-radial crack. Since the observation of cracking was from below the contact site the initiation of a full-penny median-radial crack could appear as two separate median-radial cracks. Additional median-radial cracks then initiated as separate entities at different times.

CONCLUSIONS

The usual notion that 'normal' glasses median-radial crack only on indenter unloading, at least for relatively small maximum loads, does not hold for the 'normal' optical glasses examined here. Evidence of extensive shear-flow beneath Vickers indentation sites in seven of the optical glasses suggest that little, if any, densification occurred. Median-radial cracks initiated at relatively low loads (< 1 Kg) on the loading cycle for these seven glasses. These loads were significantly less than those required to initiate median-radial cracks in silicate-based glasses investigated by other researchers. The first median-radial cracks to initiate had a full-penny morphology. The final morphology of all median-radial cracks was half-penny.

A new classification scheme of glass behavior under a Vickers indenter emerged. Three broad categories of deformation/cracking behavior in Vickers indentation of glass can be distinguished: 'normal', 'normal/anomalous', and 'anomalous.' 'Normal' (non-densifying) glasses median-radial crack readily on indenter loading i.e., they exhibit a very low threshold load for median-radial cracking, whereas 'normal/anomalous' glasses have a considerably higher threshold load for median-radial cracking on indenter loading. 'Anomalous' (densifying) glasses will tend to begin initiating ring-cone cracks early on the indentation loading cycle (< 1 Kg). The ring-cone crack system is the only crack system particular to 'anomalous' (densifying) glasses in Vickers indentation testing of glass. In addition, threshold loads for median-radial crack initiation in 'anomalous' glasses appear to be higher than those in 'normal' glasses, but not as high as the threshold load observed in 'normal-anomalous' glasses. In this respect, the 'normal-anomalous' glasses have the best resistance to median-radial crack initiation.

Several other important points need to be mentioned. Since glass composition can be made to change gradually, a gradual transition from one category of crack initiation behavior to the next with changing glass composition is to be expected, and as such the three categories described above are to be viewed with this in mind. In the current work the specimens were tested under temperature and relative humidity which did not vary significantly from one specimen to the next. Thus, the specimens were examined on a more-or-less equal environmental basis. Future experiments could be aimed at testing all specimens in exactly the same atmosphere, i.e., one in which the variations are next to nil. The test environment, specifically the amount of water vapor present, is known to affect crack initiation in glass. Thus, the threshold loads for cracking would also be expected to change as well, but in a manner preserving the relationship between the glass types observed in the current work e.g., 'normal' glasses would still be expected to have a lower median-radial crack threshold load than 'normal-anomalous' and 'anomalous' glasses.

REFERENCES

[1]B.R. Whittle and R.J. Hand, "Morphology of Vickers Indent Flaws in Soda-Lime-Silica Glass," *J. Am. Ceram. Soc.*, 84 [10] 2361-2365 (2001).

[2]T. Haranoh, H. Ishikawa, N. Shinkai, and M. Mizuhashi, "Crack Evolution in Vickers Indentation for Soda-Lime-Silica Glass," *J. Mater. Sci.*, 17 [5] 1493-1500 (1982).

[3]H. Ishikawa and N. Shinkai, "Critical Load for Median Crack Initiation in Vickers Indentation of Glasses," *J. Am. Ceram. Soc.* 85 [8] C124-C127 (1982).

[4]J.T. Hagan and M.V. Swain, "The Origin of Median and Lateral Cracks Around Plastic Indents in Brittle Materials," *J. Phys. D: Appl. Phys.*, 11 2091-2102.

[5]B.V. Tanikella and R.O. Scattergood, "Acoustic Emission During Indentation Fracture," *J. Am. Ceram. Soc.*, 78 [6] 1698-1702 (1995).

[6]A.G. Evans and T.R. Wilshaw, "Quasi-Static Solid Particle Damage in Brittle Solids-I. Observations, Analysis, and Implications," *Acta. Metallurgica*, 24 [10] 939-956 (1976).

[7]B.R. Lawn, T.P. Dabbs, and C.J. Fairbanks, "Kinetics of Shear-Activated Indentation Crack Initiation in Soda-Lime Glass," *J. Mater. Sci.*, 18 [9] 2785-2797 (1983).

[8]J.T. Hagan and S. Van Der Zwaag, "Plastic Processes in a Range of Soda-Lime-Silica Glasses," *J. Non-Cryst. Solids*, 64 [1-2] 249-268 (1984).

[9]J.T. Hagan, "Shear Deformation Under Pyramidal Indentations in Soda-Lime Glass," *J. Mater. Sci.*, 15 [6] 1417-1424 (1980).

[10]D.B. Marshall and B.R. Lawn, "Residual Stress Effects in Sharp Contact Cracking," *J. Mater. Sci.*, 14 [8] 2001-2012 (1979).

[11]J. Lankford, "Threshold Microfracture During Elastic-Plastic Indentation of Ceramics," *J. Mater. Sci.*, 16 [5] 1177-1182 (1981).

[12]B.R. Lawn and M.V. Swain, "Microfracture Beneath Point Indentations in Brittle Solids," *J. Mater. Sci.*, 10 [1] 113-122 (1975).

[13]A. Arora, D.B. Marshall, and B.R. Lawn, "Indentation Deformation/Fracture of Normal and Anomalous Glasses," *J. Non-Cryst. Solids*, 31 [3] 415-428 (1979).

[14]J.T. Hagan, "Cone Cracks Around Vickers Indentations in Fused Silica Glass," *J. Mater. Sci.*, 14 [2] 462-466 (1979).

[15]B.R. Lawn and E.R. Fuller, "Equilibrium Penny-Like Cracks in Indentation Fracture," *J. Mater. Sci.*, 10 [12] 2016-2024 (1975).

[16]B.R. Lawn, A.G. Evans, and D.B. Marshall, "Elastic/Plastic Damage in Ceramics: The Median/Radial Crack System," *J. Am. Ceram. Soc.*, 63 [9-10] 574-581 (1980).

[17]R.F. Cook and G.M. Pharr, "Direct Observation and Analysis of Indentation Cracking in Glasses and Ceramics," *J. Am. Ceram. Soc.*, 73 [4] 787-817 (1990).

[18]C.R. Kurkjian and G.W. Kammlott, "Indentation Behavior of Soda-Lime Silica Glass, Fused Silica, and Single-Crystal Quartz at Liquid Nitrogen Temperature," *J. Am. Ceram. Soc.*, 78 [3] 737-744 (1995).

[19]F.M. Ernsberger, "Mechanical Properties of Glass," *J. Non-Cryst. Solids,* 25 [1-3] 293-321 (1977).

[20]K.W. Peter, "Densification and Flow Phenomena of Glass in Indentation Experiments," *J. Non-Cryst. Solids,* 5 [2] 103-115 (1970).

[21]C.J. Wagner, "Influence of Composition on Crack Initiation Behavior of Glasses," M.S. Thesis. Alfred University, Alfred, NY, 1998.

[22]D.K. Shetty, I.G. Wright, P.N. Mincer, and A.H. Clauer, "Indentation Fracture of WC-Co Cermets," *J. Mater. Sci.,* 20 [5] 1873-1882 (1985).

[23]G.R. Anstis, P. Chantikul, B.R. Lawn, and D.B. Marshall, "A Critical Evaluation of Indentation Techniques for Measuring Fracture Toughness: I, Direct Crack Measurements," *J. Am. Ceram. Soc.,* 64 [9] 533-538 (1981).

[24]Engineered Materials Handbook, Vol. 4, *Ceramics and Glasses.* Edited by Samuel J. Schneider. American Ceramic Society, Westerville, OH, 1991.

[25]S.M. Wiederhorn, "Fracture Surface Energy of Glass," *J. Am. Ceram. Soc.,* 52 [2] 99-105 (1969).

IMPACT DAMAGE EVALUATION FROM REMAINED SURFACE DAMAGE BEHAVIOR

Yoshio Akimune
National Institute of Advanced Industrial Science and Technology (AIST)
1-1-1Umezono,
Tsukuba, 305-8568 Japan

Takao Izumi
Nissan Motor Co., Ltd.
1, Natsushima-cho,
Yokosuka-city, 237-8523 Japan

ABSTRACT

The present study concerns a model experiment in which oxide scales striking at the turbine blade and/or ceramic components are conducted by zirconia particles shot to Si_3N_4 specimens at room temperature, 1100°C and, 1200°C. In this way, the temperature dependence of impact damage is evaluated by comparing detected surface damage morphology and inside damage at room temperature with that at 1100 and 1200°C.

When a zirconia particle was shot into a target of silicon nitride at room temperature, the surface damages on the specimens were ring cracks and radial cracks with a shallow crater and the inside crack was mainly Hertzian crack. However, at elevated temperature, the surface damages were ring cracks and radial cracks with a deep crater, and inside crack turned into median crack. This may be attributed to change in response behavior to impact from elastic to elastic-plastic, owing to reduced hardness caused by softening of glassy phase at the grain boundary of silicon nitride at elevated temperature. From the surface damage measurements, supposed inside crack is suggested for visual inspection.

INTRODUCTION

It has been known that in ceramics, owing to its brittle nature, cracks extended under localized stress may become a trigger for instantaneous fracture. A typical example of this event is a fracture in a ceramic blade of gas turbine caused by impact of oxide scales, and it has been recognized as one of serious problems.[1] Therefore, application to a machine part does not expand most apart from a turbocharger.[2]

However, ceramic application such as coating material of gas turbine for power generator advances, depending on superior anti-oxidation characteristics of ceramics at an elevated temperature.[3] A nondestructive evaluation technique to discover internal cracks and experimental study for understanding a crack become important. Before going to understand a coating films of ceramics, fracture behavior of monolith ceramics should be examined well. However, there are a few studies of a fracture caused by particle impact on monolith ceramics in a high-temperature. Shockey et. Al,[4] Song et. Al,[5] Akimune et. al[6] reported similar results about gas turbine, but a damage process just under impact site and inference of fracture process were not well understood because ceramics components fractured completely.

The present study concerns a model experiment in which oxide scales striking at the turbine blade. In this way, the temperature dependence of impact damage is evaluated by comparing surface and internal damage morphology at the room temperature with that at 1200°C.

MATERIALS AND PROCEDURES

Commercially available gas-pressure-sintered silicon nitride (EC152: NGK Spark Plug Co., Ltd.: Komaki, Japan) was used in this study. Partially stabilized zirconia (PSZ: Toso Co.,Ltd., Tokyo, Japan) spheres of 1.0 mm in diameter were used as impactors. A disk-shaped specimen of 14 mm diameter and 2.0 mm thickness was cut for impact test. And its surface was polished with diamond paste (6μm and 3μm) to eliminate machining damage.

Table I. Mechanical Properties of Si3N4 Target

	RT	1200 ℃
Density (g/cm^3)	3.26	3.26
Poisson s Ratio	0.25	0.25
Vickers Hardness(GPa)300	15.4 ± 0.4	9.0
Youngs Modulus(GPa)	305	296
Bending Strength(MPa)	912 ± 110	500
Fracture Toughness(MPa m)	7.0 ± 0.3	----

Table II. Properties of PSZ Impactors

Density (g/cm^3)	6.05
Poisson s Ratio	0.3
VickersHardness(GPa)300	12.5
Youngs Modulus(GPa)	200

A schematic of the impact apparatus is shown in Figure 1, which consists of a particle launcher of gas-gun type and a specimen holder installed in a furnace. The experimental apparatus used a helium gas cylinder and the test procedure was almost the same as that described in the literature.[1,6] A target, Si_3N_4 disk, is mounted in the specimen holder within the furnace and the same impact test was conducted at 1200°C. As an acceleration sensor cannot be used in the high temperature experiment, the impact velocity is to be read from a working curve made at room temperature with velocity calibrated against helium gas pressure. In the present experiment, the impact velocity was available in 400, 500, and 600 m/s.

The impact damage was examined by means of the fluorescence flaw detection, followed by morphological observation of surface craters and internal cracks under an optical microscope. A measurement of crater diameter and depth by use of a needle-probe profilometer was conducted.

RESULTS AND DISCUSSION

Surface damage measurement

Surface damages caused by a zirconia sphere shot onto a specimen at a velocity of 400, 500, and 600 m/s at room temperature consisted of concentric ring cracks and radial cracks. (Figures 2-A, 2-B, 2-C).) It is assumed that, within an interval between the impact of particle and the onset of maximal load, Hertzian stress induces ring cracks on the surface. A fracture process is thought that a sphere was pushed down to a specimen and the radial stress occurred and it exceeded material strength level, so ring crack was caused. Stresses increases further, and the next ring crack was produced, but a location is outside of the initial ring crack. It is supposed that this events(ring crack initiation and extension of Hertzian crack) occurred successively from inside to outside.[1,6]

When a sphere was crushed during loading, the stress from a sphere becomes zero, then stress condition in the specimen varies from elastic to elastic/plastic. As a result, it is thought that radial crack occurred so that a stress direction varies in an orthogonal direction in the concentric circle. A sphere of room temperature was collided in the specimen that was heated to 1,200°C, and a crack number increased and a similar damage was remained (Figures 3-A and 3-B).

Diameter of contact point

The morphology of impact damage at the surface of the target was measured by using a needle-probe profilometer in two orientation of X, Y and a mean value was taken. Under a presumption that a crater is formed by the particle impact in a time span from its contact to collapse, it is assumed that the contact diameter is very close to the crater diameter. It is necessary to find contact diameter from theory of Hertzian crack.[7,8] Contact diameter can be calculated that is equal to the diameter of contact point from Hertzian theory.[9]

$$Da=2R(5\pi k \rho /4)^{1/5} V^{2/5} \quad (1)$$

where Da stands for contact diameter, k is given by $k = (1 - \nu_{12})/E_1 + (1 - \nu_{22})/E_2$ [GPa^{-1}], ν_{12} is Poisson's ratio of target material, ν_{22} Poisson's ratio of sphere, R is sphere radii (= 500μm), E_1 is Young's modulus of target material, E_2 Young's modulus of sphere (Table 1), ρ is density of sphere (ρ= 6.05 x 10^3 kg/m^3), and V is impact velocity of sphere (V = 400 to 600 m/s).

Theoretical contact diameter and crater diameter were written jointly in Figure 4. It was near value on theoretical line at room temperature, and it was about 10% bigger than a theoretical diameter caused by a impact at 1,200 °C. It is supposed that elastic recovery can not be occurred by softening of the material, as a result, a diameter of crater remained bigger that that of theory.

Depth of residual crater

Crater depth was also measured by a profilometer in two orientation of X, Y and the mean value was taken. It was 1-4μm with room temperature, however it was confirmed that it spreads to 8-13μm with a high-temperature (Figure 5). There is no plastic deformation of material in Hertzian theory, however only a slight deformation (several micron) was recognized at room temperature. It is thought that the stress field was affected by the plastic deformation. In a high-temperature, it was measured ten several micron for softening of grain boundary glassy phase of material until a ball does collapse after a contact, and just at the impact site, plastic deformation and porous zone that micro-cracks intersect will be formed resulting in a residual crater.[10]

Internal crack

Disc was cut with near center of the crater, and it was polished from a side to the crater center, and internal crack was observed (Figure 6). Hertzian cone cracks were observed in the case of three kinds of velocity of 400, 500, and 600m/s, a damage form of room temperature shows elastic response. Apex angle of a cone is about 90 degrees which is smaller than that of theoretical value (136 degrees), and it is supposed that stress field is a little bit affected by crater deformation, but it can be similar to elastic stress field. Internal crack is median type crack at 1,200°C, an elastic-plasticity response is shown. It can be supposed that generated stress field is affected by sphere trespassing, which is revealed by residual crater depth at the surface.

Median crack system and Hertzian crack system

Radial cracks were observed about a different pressureless sintered silicon nitride with much glassy phase, and median crack is displayed. In Figures 7 with residual strength with median crack was illustrated.[10] The standard relationship between strength and fracture toughness value was,

$$\sigma_f = K_{IC}/Y (C_f)^{1/2} \quad (2)$$

Here, as for σ_{f},; as for residual strength, K_{IC}; the fracture toughness value (SSN, 5MPa $\sqrt{m}$), Y; the stress concentration factor ($\sqrt{\pi}$), C_f; crack length for fracture. Radial crack length and median crack length are come off from a theory contour. Median crack length is converted into C_f as penny crack here. When Lawn[11,12] plotted the value that was converted in $C_f = \Omega D$ ($=4/\Omega\pi^2$), the effect which was near in a theory contour was provided. On this account the median crack form that fractured surface can be handled as Penny crack. A same account was performed about the material which produced Hertzian crack (Figure 7 Right) that C_f was found from cone geometry in a gas-pressure-sintered silicon nitride, and was plotted (K_{IC} = 6MPa $\sqrt{m}$).[10] As a result, apex of cone angle is smaller than that of theoretical value, and appearing is come off from a standard theory contour.

That it becomes hard to clarify strength/crack length relationship from cone geometry, it is because there is some plastic deformation not pure elastic stress field although crack shape can be similar to that for elastic stress field and Hertzian crack was produced.

Observation by fluoroscopic flaw detection

Surface damage and an internal cracks examined by fluoroscopic flaw detection method were shown in Figures 8A and B. When a sample was impacted at 1,100°C, and surface damage and internal damage after impact was observed by fluoroscopic flaw detection method. The Hertzian cone crack which was clear by interior observation of Figure 8B and a ring crack around impression and seven radial crack were observed from Figure 8A.

A sample was impacted at 1,200°C, and internal damage after a puncture was observed by fluoroscopic flaw detection method. By internal crack observation shown in Figure 9B, median crack and small internal cracks were observed, and ring crack around impression and five radial cracks were observed from Figure 9A. When it is to 1,100°C, the stress field in this experiment is still elastic resulting in Hertzian crack, although temperature becomes 1,200°C, stress field becomes an elastic-plastic stress field, then median crack was formed.

Prediction internal damage by observation of crater

With the damage that small particle collides with ceramics, and there are two case, one is immediate fracture with generating large crack and the other is erosion which accumulates small size of fracture. That damage like internal crack is assumed from surface damage on a component by nondestructive testing is concerned. Response behavior and internal crack can be assumed comparing the theoretical contact diameter of Hertzian contact and residual crater diameter from this study. When crater diameter exceeds theoretical value calculated by elastic response, elastic-plastic deformation will be caused resulting in a median crack. It is very likely that median crack occurs just under crater, and in the first case reaches it in fatal crack. The relationship between crater depth and residual strength were shown in Figure 10, so that residual strength was related with a crack form about crater depth. When crater depth is small, generated crack type is Hertzian and size is small, however, crater depth is big, generated crack type is median crack and its size becomes critical.

In Figure 10, the depth of the silicon nitride was 1-2μm, however, it was 5μm for a silicon nitride to show elastic/plasticity behavior and was 1-3μm for a silicon nitride to show elastic behavior. In this way it understands a difference of extremely minute regional impression depth to influence response behavior with a thing by a characteristic of material. When we evaluate the components, damage behavior and residual strength of damage are measured in pretest to assume damage behavior with the same series of production material, and it is thought that crater diameter and depth are measured for a components from market, then real damage behavior can be guessed.

SUMMARY

It is necessary for ceramic material applying to structural part for gas turbine for power generation to give enough strength for damage by flying objectives, however, fracture behavior is complicated, therefore plenty of study is necessary. In this experiment, damage behavior was understood by connecting response behavior and internal crack formation through generated stress field. This method will be useful to examine damage in ceramic mechanical parts in commercial use.

REFERENCES

[1]K. C. Dao, D. A. Shokey, L. Seaman, D. R. Curran, and D. J. Rowcliffe,, "Particle Impact Damage in Silicon Nitride," *Annual Report, Part I, Office of Naval Research*, Contract No. N00014-76-057, Menlo Park, CA, May 1979.

[2]Y. Akimune, Y. Katano, and K. Matoba, "Spherical-Impact Damage and Strength Degradation in Silicon Nitride for Automobile Turbocharger Rotors," *J. Am. Ceram. Soc.*, 72 [8] 1422-28 (1989).

[3]S. Bose and J. DeMasl-Marcin, "Thermal Barrier Coating Experience in Gas Turbine Engines at Pratt & Whitney," *J. of Thermal Spray Technology*, 6(1)March, pp.99-104(1997).

[4]D. A. Shockey, D.C.Erlich, K.C. Dao,"Particle Impact Damage in Silicon Nitride at 1400 °C," *J. Mat. Sci.*, 16,477-482(1981).

[5]J. Song, J. Cuccio, H. Kington, " Impact Damage Study of Silicon Nitride," *Proceedings of the 26th Automotive Tech. Development Coordination Meeting, Soc. Automotive Engineers*, Pa (1988).

[6]Y. Akimune, F. Munakata, Y.Okamoto, N.Hirosaki,"Temperature Dependence of impact Damage in a Si_3N_4 Disk by a Spherical particle," *J. of the Ceramic Soc. of Japan*, 105[1]93-96(1997)

[7]H. R. Hertz, Chs 5 and 6, Macmillan, London, England, 1886.

[8]S. P. Timoshenko and J. N. Goodier, *Theory of Elasticity*. McGraw-Hill, New York, 1970.

[9]B. M. Liaw, A. S. Kobayashi, A. F. Emery, and J. J. Du, "An Impact Damage Model of Ceramic Coating" ; pp. 187-96 in *Fracture Mechanics of Ceramics.*, Vol. 7. Edited by R. C. Bradt, A. G. Evans, D. P. H. Hasselman, and F. F. Lange. Plenum Press, New York, 1986.

[10] Y. Akimune, "Development of Ceramic Composite with better tolerance for particle impact," *PH.D Dissertation*, Tokyo University, 1990.

[11]B. R. Lawn, E. R. Fuller, and S. M. Wiederhorn, "Strength Degradation of Brittle Surfaces: Sharp Indenters," *J. Am. Ceram. Soc.*, 59 [5-6] 193-97 (1976).

[12]B. R. Lawn, S. M. Wiederhorn, and H. H. Johnson, "Strength Degradation of Brittle Surfaces: Blunt Indenters," *J. Am. Ceram. Soc.*, 58 [9-10] 428-32 (1975).

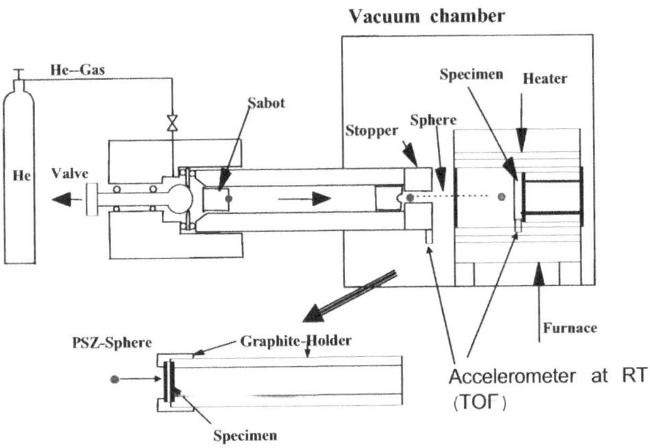

Figure 1 Impact apparatus and specimen holder.

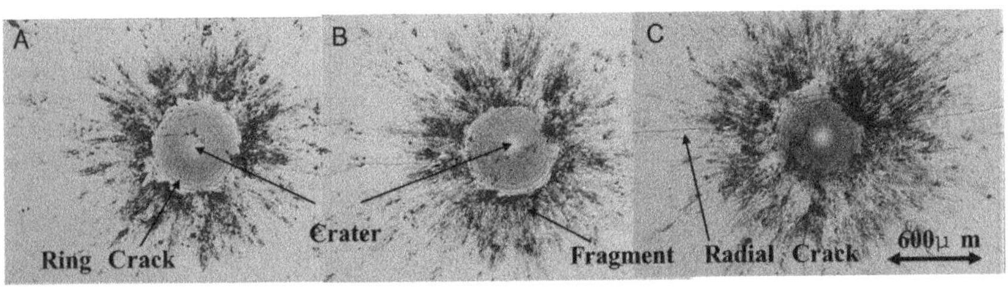

Figure 2A-C Surface impact damage at RT with impact velocities of A; 400, B; 500; C;600m/s.

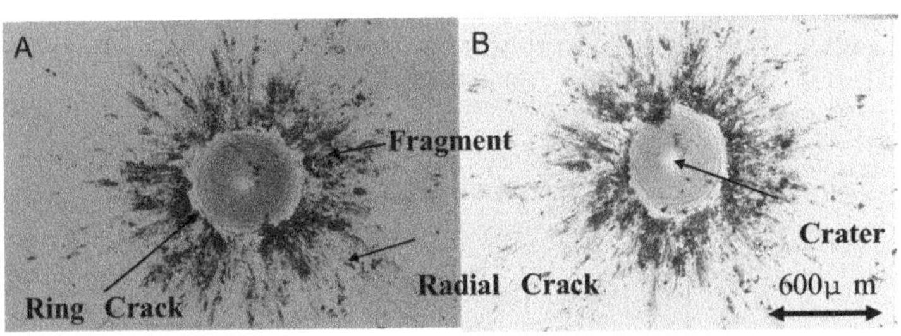

Figure 3A-B Surface impact damage at 1200°C with impact velocities of A; 400, B; 500m/s.

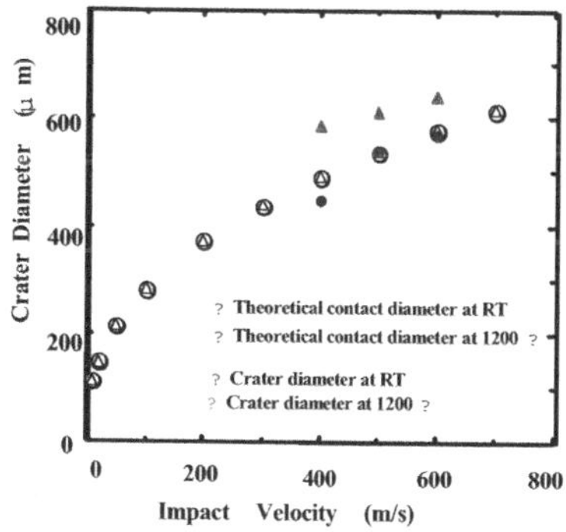

Figure 4 Crater diameter

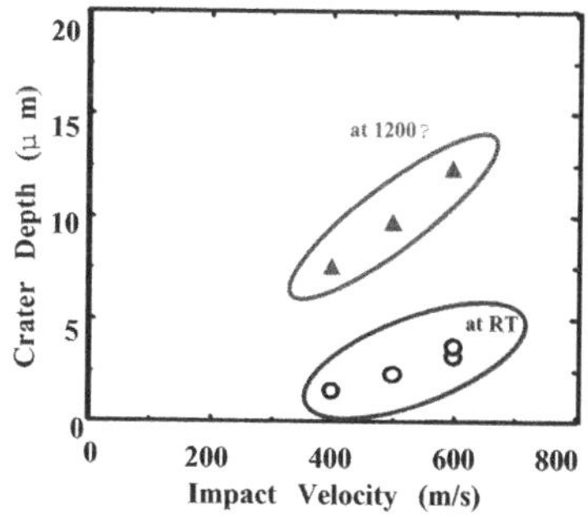

Figure 5 Crater depth

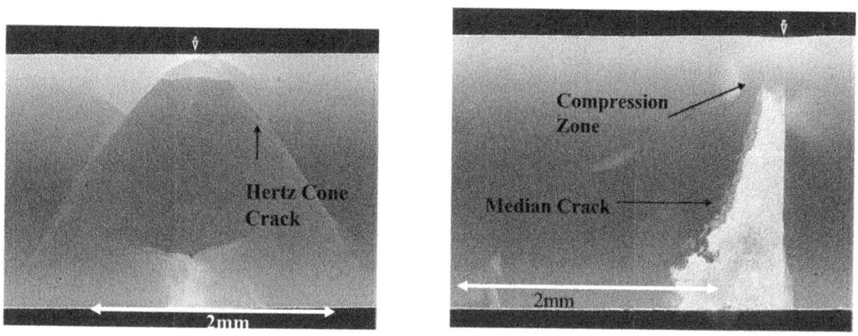

Figure 6 Inside impact damage at RT (left) and 1200°C (right) (V = 600 m/s).

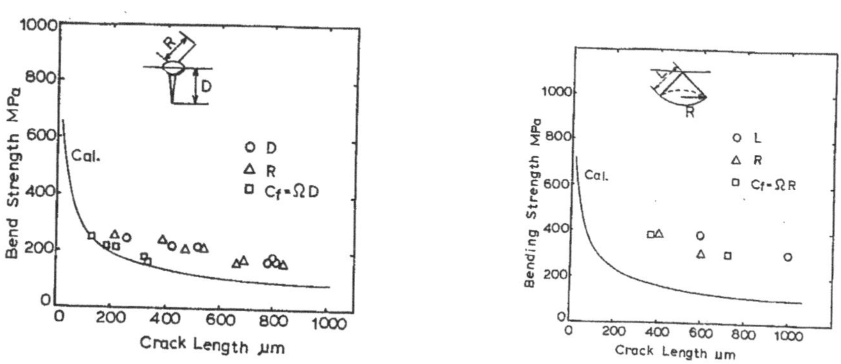

Figure 7 Measurement of residual bending strength (Median crack (left), Hertz crack

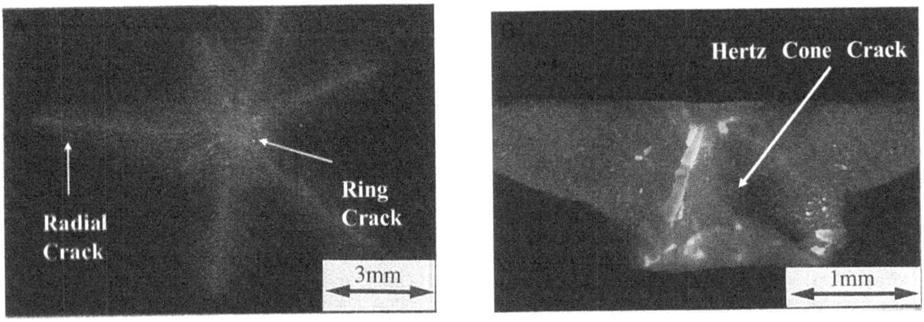

Figure 8 Measurement of damage at 1100°C by fluorescent (A; Surface, B; Cross section).

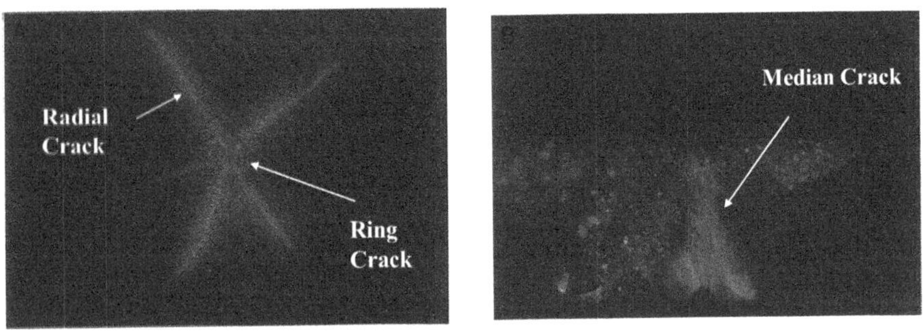

Figure 9 Measurement of damage at 1200° C by

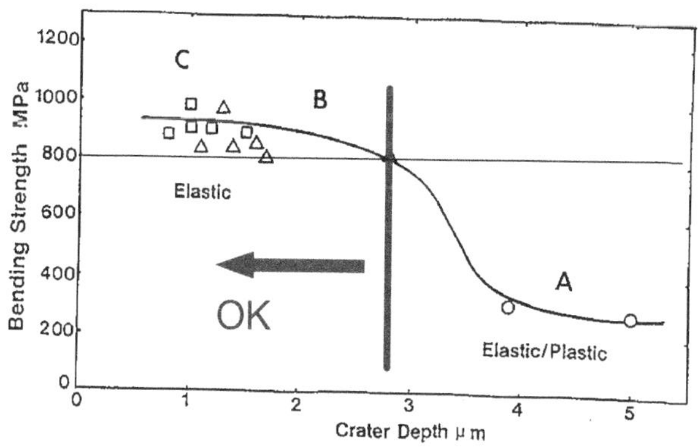

Figure 10 Bending strength when a impact crater is crack origin [10].

GUIDELINES FOR MEASURING FRACTURE MIRRORS

George D. Quinn
National Institute of Standards and Technology
Ceramics Division Stop 8529
Gaithersburg, MD 20899

ABSTRACT

Fracture mirror size analysis is a powerful diagnostic tool for glass and ceramic fractures. There is considerable variability in published fracture mirror constants, however, especially for ceramics. Mirror-mist or mist-hackle boundaries are not sharp, well-defined transitions, and they are subject to interpretation. This paper presents some guidelines for mirror boundary interpretation and analysis that are from a new NIST Guide to Practice for Fractography. Some recommendations are based on common sense, others are based on solid technical grounds, and some are based upon the experience of other experts and the author. The guidelines are intended to help students, novices, engineers, and materials scientists use fracture mirror analysis to solve practical problems.

INTRODUCTION

Fracture mirrors are relatively smooth regions surrounding the fracture origin. In glasses and very fine-grained ceramics, the mirror region is very smooth and highly reflective, hence the name "fracture mirror." In polycrystalline ceramics or composites, the qualifier "relatively" must be used, since there is an inherent roughness from the microstructure even in the area immediately surrounding the origin. Mirrors are circular in some loading conditions such as tension specimens with internal origins, or they are near-semicircular for surface origins in tensile specimens or if they are small mirrors in bending as shown in Figures 1 and 2. Fracture mirrors only form in moderate- to high-strength specimens. Weak specimens do not have fracture mirror boundaries since the crack does not achieve sufficient velocity within the confines of the specimen. Fracture mirror shapes sometimes deviate from the circular or semicircular shapes due to stress gradients or geometrical effects.

Fracture mirrors bring one's attention to an origin, but also give information about the magnitude of the stresses that caused fracture and their distribution. One of the earliest observations in this regard was by Brodmann in 1894:[1] "The radially measured size of these mirrors in general was larger, the smaller was the strength." Smekal[2] also observed a correlation between the overall mirror size and the stress. The relationship was gradually refined so that in the 1950s and 1960s, many authors measured mirror sizes and correlated them with stresses in accordance with the relationship:[3,4,5,6,7,8,9,10]

$$\sigma R^m = A \tag{1}$$

where σ was stress, R was the mirror radius, and m and A were constants. The data were frequently plotted on graphs of log stress versus log radius in order to check for the goodness of fit of equation 1. Leighton Orr[3] recognized that m was 1/2 for annealed glasses and he simplified the equation to:

$$\sigma R^{1/2} = A \tag{2}$$

Figure 1. Fused silica rod broken in flexure (122 MPa). The origin is a surface flaw located at the bottom of the specimen (a) where the stress was a maximum. The mirror-mist boundary is small relative to the cross section size {shown in the insert in (a)} and is approximated by a circle in (b). Close examination of the fracture surface in the vicinity of the flaw (not shown) showed that fracture started from the deepest part of the flaw. The sizes and arcs were judged while viewing through a compound optical microscope and then marked on the photographed image. (Some subtle detail in the mist may be lost during printing.) The mist-hackle boundary is slightly elongated towards the top. (c) is an scanning electron microscope (SEM) image of the same mirror and at the same magnification as (a) and (b). The mist is indistinct in the SEM image. (d) is a composite of two SEM images showing the transition from mirror to mist to hackle. The location of the boundaries as assessed by optical microscopy are marked by dashed lines on the SEM images.

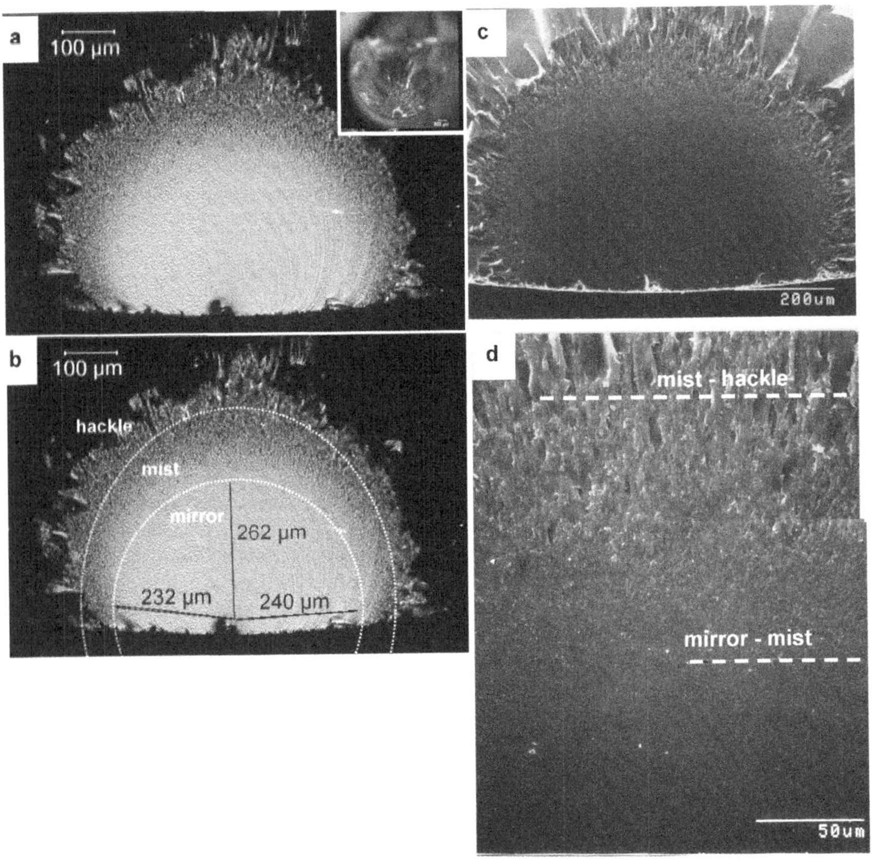

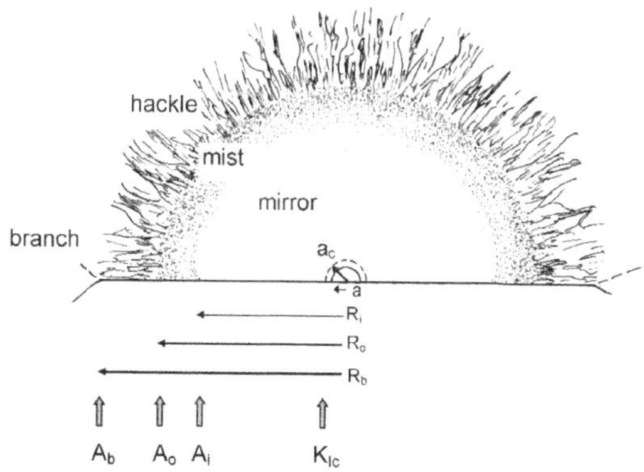

Figure 2. Schematic of a fracture mirror centered on a surface flaw. The various mirror boundaries and their corresponding fracture mirror constants (A) are marked. The initial flaw at the origin may grow in size from a to a_c, at which point the critical fracture toughness, K_{Ic}, was reached.

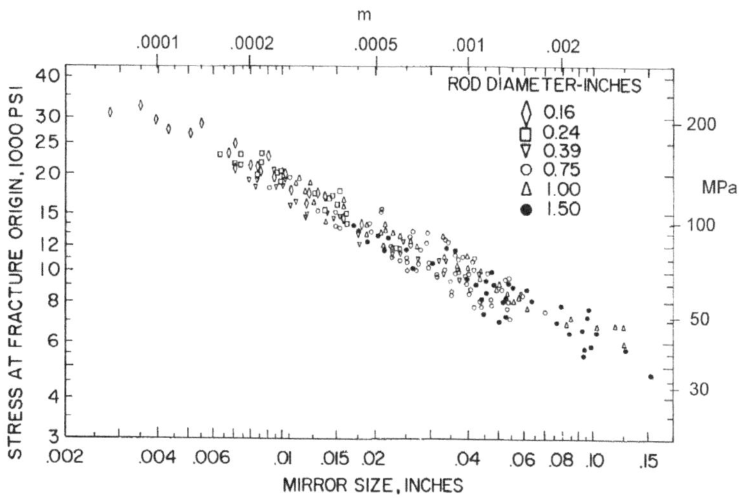

Figure 3. Fracture stress (adjusted for the origin location) versus mirror size for a borosilicate glass from Kerper and Scuderi.[10] This is the original figure from their 1966 paper, and it shows the older system of units then in use with a few SI units added on the top and right side. They fitted the data with eq. 2 for 259 rods broken in the three-point bending and obtained a mirror constant of 2.08 MPa√m (1.89 ksi√in). The rod diameters varied almost by a factor of 10, the mirror sizes by a factor of 50, and the stresses almost by a factor of 10.

In this form, A is the "mirror constant" with units of stress intensity (MPa√m or ksi√in) and is considered by many to be a material property. It is possible to discern separate mist and hackle regions and the apparent boundaries between them in glasses, and each has a corresponding mirror constant, A. The most common notation is to refer to the mirror-mist boundary as the inner mirror boundary, and its mirror constant is designated A_i. The mist-hackle boundary is referred to as the outer mirror, and its mirror constant is designated A_o. The mirror-mist boundary is usually not perceivable in polycrystalline ceramics, and only the mirror-hackle boundary is measured.

Equation 2 has tremendous practical value. It is not necessary to have a-priori information about the how the part was loaded. The smaller the mirror, the larger is the stress at the origin site. A small mirror is proof that the part was strong and had a small strength-limiting flaw. Conversely, large mirrors mean the failure stress was lower and implies a large flaw. In some instances, a part may be so weak that the mirror size is larger than the part cross-section, and hence the mirror markings are not visible. That, in and of itself, is valuable information. The existence of a mirror boundary implies that the part was stressed to a moderate or high level. Alternately, very strong parts (e.g., pristine optical fibers) may have such small mirrors that they are not measurable with optical microscopes. The mirror may not even be found due to excessive fragmentation.

Leighton Orr of the Pittsburg Plate Glass company was the first to use equation 2 in that form and therefore we may refer to equation 2 as "Orr's mirror size equation" or, more concisely, "Orr's equation." Levengood[4], Shand,[7] Kerper and Scuderi,[8,9,10] and later Kirchner[11] credited Orr for equation 2. Orr did not publish his findings until 1972 when he retired,[3] but discussed his empirical equation at a Glass Division meeting of the American Ceramic Society in Bedford, PA in September 1955.[12] Orr told this author in 2001 that equation 2 had been extensively used for glass fracture analysis at PPG in the 1940s.[13] Errol Shand of the Corning Glass works was a leader in the early analysis work.[5-7] His 1959 paper presented an early argument for the 1/2 power in equations 1 and 2 based on stress concentrations at the tip of a sharp crack. This paper incorporated early elements of what would become known later as fracture mechanics. Kerper and Scuderi[8,9,10] at the National Bureau of Standards in Washington performed meticulous experiments on hundreds of glass laths and rods and showed conclusive evidence that equation 2 was applicable over a broad range of specimen and mirror sizes as shown in Figure 3. Some have credited Terao[14] with equation 2, but a careful reading of that paper fails to back the claim. (An exponential form relating mirror size to stress is shown instead.) So although many associate equation 2 with Johnson and Holloway[15] in 1966, the relationship was already in use for over 15 years. Their primary contribution was to use an energy analysis to give some theoretical underpinnings to equation 2.

Congleton and Petch[16] suggested that a fracture mechanics criterion based on a critical stress intensity (K_I) controls crack branching. Henry and James Kirchner and James Conway[17,18] later presented compelling evidence that a fracture mechanics approach was also applicable to mirrors. They used K_{IB} to denote the stress intensity factor at mirror boundaries. It gave the best fit to mirror size data and predicted the exact shape of mirrors in various stress fields. They argued that equation 2 does not take into account the free surface, geometry factors, and non-uniform stress gradients over the crack surface. Their work is considered again later in this paper.

Orr's equation 2 may be used in practice to estimate the stress at an origin. The mirror radius is measured from a fracture surface, the mirror constant found from a data table, and the origin stress calculated. A splendid example of this approach was shown by Morrell et al.[19] for hip balls at the last fractography conference in 2000. The calculated stress is the net tensile stress acting on the flaw and the region around the flaw. It may include several stress sources including mechanical, thermal, and residual stresses.

The author has compiled hundreds of fracture mirror constants for many materials in an American Society for Testing and Materials Standard[20] and more recently in a new National Institute of Standards and Technology (NIST) Guide to Practice for Fractography.[21] It became apparent that the reported A values for identical materials (even for model materials such as fused silica or soda lime silica) varied by at least 20 % (the range) and often much more. For some ceramics, the A values varied by a factor of two. This directly affects accuracy and precision of stress estimates using equation 2. Some of the scatter is due to metrology and judgment issues, but much of it is due to genuine material-to-material variability. Thus, one should not expect all aluminas to have the same mirror constant since there are major differences in their microstructures.

The literature review also revealed that many authors were reluctant to show their mirrors with labeled photographs. In many instances authors did not document how they had measured their mirrors, or critical details were missing. For example, many failed to report whether they corrected stresses for location, or how they regressed the data. Confidence bounds on the A estimates were rarely reported.

There is a strong subjective element to estimating the location of the mirror boundary even for glasses. Johnson and Holloway[15] wrote in 1966:

> "The position assigned to the boundary between mirror and mist zones depends upon the illumination and the magnification at which the fracture is examined, even within the range of the optical microscope. .... However, under given conditions a reproducible position for the boundary can be assigned."

Mecholsky and Freiman[22] said in 1979:

> "While one might think initially that the measurements of a mirror boundary using a microscope is quite a qualitative operation and would vary from observer to observer; in fact, experiments performed over a number of years by a large number of investigators have shown that the values of mirror constants obtained in different laboratories are quite nearly the same."

Nevertheless, they listed some values for a few glasses and ceramics that did in fact vary over a range of as much as 20 % to 30 %. For example, the soda-lime glass values varied by 23 %.

Since many of the early mirror measurements were made while viewing through the optical microscope, it is safe to say that the first perception of a mirror boundary was where the surface roughness was a fraction of the wavelength of light (0.39 μm to 0.80 μm). Threshold levels of detectability have been estimated to be as small as a few tens of nanometers to as much as 0.25 μm. The judgment issue is exacerbated for ceramics, where intrinsic roughness due to the microstructure contributes to interpretation difficulties.

Despite these problems, it cannot be denied that fracture mirror measurements are a powerful diagnostic tool for quantitative analysis. The accuracy and precision of the stress estimate is quite variable and depends on the material, the experience of the fractographer, the microscopy and illumination used, and component geometry and stress gradient effects. Stress estimates may be accurate to within ≈ 20 % with glasses if they form ideal fracture mirrors. The guidelines presented in this paper are intended to bring some long-overdue consistency to fracture mirror analysis and improve the precision of this technique. The next two sections address two fundamental questions:

1. Is there really a boundary that can be measured?
2. Is there a theoretical basis for the fracture mirror constant, A?

IS THERE A BOUNDARY?

The word "boundary" must be used with some caution even with glasses. It is now clear that there probably is not a distinct transition point on the fracture surface corresponding to a mirror boundary. The higher the magnification the smaller the mirror seems to be, since fine detail that was washed out or not resolvable at lower power can be discerned at higher power. The mirror-mist boundary in glasses probably corresponds to surface roughness features that are of the order of 0.1 μm to 0.2 μm, which is a fraction of the wavelength of light and at the threshold of observable features with an optical microscope. Mirror size estimates can vary depending upon the type of microscope used, the illumination, the objective power of the lens, and the judgment criteria of the fractographer.

Careful electron microscopy and atomic force microscopy have shown that the formation of the mirror is a gradual progression of very-localized crack path deviations from the main plane. Superb transmission electron microscope images of various regions in a mirror have been shown by Beauchamp.[23,24] Poncelet[25] showed comparable images in 1958, but Gölz may have been the earliest with extraordinary electron microscope photos published in 1943.[26] Starting well within the mirror, rounded ridges form that are elongated in the direction of crack propagation. These gradually coarsen in amplitude. At some point, they develop slight hackle steps where over running and under running fingerlike crack segments link. This point, where nano scale hackle lines form, could be an important transition point. Close examination of Beauchamp's Figure 2 of ref. 23 shows they are not present inside the mirror, but they are just visible in his Figure 3 near the mirror-mist boundary.

The explanation below closely follows Fréchette's[27] general discussion as well as Beauchamp's[23] analysis. As the crack accelerates away from the origin, micro portions of the crack front begin to twist slightly or tilt up and down out of the main fracture plane. These local deviations occur as consequence of the stress field in front of a fast crack having maxima that are out of plane, unlike the case for a static or slowly moving crack.[28] The momentary tilting or twisting does not persist for very long, since crack plane deviations are restricted by the energetic cost of creating additional crack surface. The slight tilt or twist variations in crack plane quickly rejoin the main propagating crack plane. These tiny local crack perturbations exist well within the mirror region, but are too small to be optically discernable. As the crack advances, they eventually become large enough to be discernable with the optical microscope as the "mist" zone surrounding the origin. The mist has a slight frosty appearance such as when water condenses on a reflecting mirror. As the crack continues to advance, the local perturbations increase and begin to oscillate and form larger tongue-like segments that may deviate from the average fracture plane to the degree that micro steps are generated running parallel to the direction of crack propagation. Some have described the out-of-plane perturbations as "fingers." The perturbations gradually coarsen such that the tongue-like elements can over cut other portions of the crack front thereby generating large "velocity hackle" lines that run parallel to the direction of crack propagation. By this time the crack is running at a velocity that is a high fraction of the terminal velocity. Beauchamp[23,24] pointed out that many of these features are similar in character from within the mirror out to the hackle zone. They differ only in scale. The transitions between the regions are gradual. One intriguing observation by Beauchamp[23] was that tilted or twisted hackle segments that relink with the main crack plane may generate elastic pulses and Wallner lines that could trigger additional hackle along the crack front. This could set off a cascade of hackle that forms a mist-hackle boundary.

Attempts to make objective determinations of the mirror boundary by quantitative surface roughness characterization have been unsuccessful, since there are different scales of roughness for the various features that comprise the mirror and its apparent boundaries. A single threshold roughness value corresponding to the boundaries cannot be specified, although many have tried. Duckworth et al.[29] carefully studied mirror sizes in float glass using optical photographs and a conventional surface profilometer. They obtained a good correlation with optical boundary estimates when the surface roughness reached a level of 0.25 μm for the mirror- mist A_i boundary, and 5 μm for the mist-hackle A_o boundary. Hull,[30,31] Wünsche et al.,[32] and Kuluwansa et al.[33] used atomic force microscopy to show that the mist region in glass or brittle epoxy has a roughness of as small as a few tens of nanometers, which is much less than the wavelength of visible light. The scanned regions were quite small, however, and if the AFM scanned a small region between hackle and river line steps, the measured roughness was much less than if the latter were included. Hull pointed out that the large undulations from Wallner lines needed to be factored out when evaluating the intrinsic mist roughness. His study showed that roughness increased continuously and there were no dramatic jumps in roughness at the boundaries, but the rate of change of roughness did increase significantly at the mirror-mist boundary.

Hull[30] considered the matter in some detail and pointed out that different surface roughness characterization devices such as atomic force microscopes (AFMs), mechanical profilometers, and laser optical profilometers all have different advantages, disadvantages, sensitivities and scanning zone sizes. AFMs can measure tiny regions with very high sensitivities, but may miss large hackle steps in a mist or hackle zone. These can dramatically alter the average or root mean square roughness. A mechanical stylus profilometer or laser profilometer with a 1 μm spot size may miss the small undulations and be more sensitive to larger hackle steps on the fracture surface.

In summary, there are two potential transition points. The mirror–mist boundary in glass may be a transition where nano-micro steps form between ridges that are tens of nanometers tall. The mist-hackle transition boundary may be where a band of microhackle forms that is triggered by self-generated or external elastic pulses and Wallner lines. It is not known whether these possible transition points can be detected or measured accurately and precisely using ordinary microscopy.

IS THERE A THEORETICAL BASIS FOR THE FRACTURE MIRROR CONSTANTS?

The underlying principal that accounts for the micro and macro branching has been attributed to the crack reaching a critical velocity,[28] a critical energy level,[15] a critical stress intensity,[16,17, 18,34] or a critical strain intensity.[35] A velocity criterion for crack branching was discounted by data shown by Congleton and Petch.[16] Richter and Kerkhof, [36] Field,[37] and Döll[38] used ultrasonic fractography to show that cracks approached terminal velocity before the formation of the mist boundary, and there was no pronounced change in velocity when the mirror formed. Congleton and Petch[16] demonstrated that a stress intensity criterion controlled branching and also showed the Johnson and Holloway energy criterion[15] was related to the stress intensity criterion. It is difficult to distinguish between the energy, stress intensity, or strain intensity criteria, since they are all related for isotropic materials.[39]

The two Kirchners and Conway[17,18] showed that a fracture mechanics criterion based on a critical stress intensity gave a better fit to mirror shapes in various stress fields than a simple application of equation 2. A crucial observation was that the stress intensity model fully accounts for the small inward facing cusp in the mirror shape at specimen surfaces, whereas the other models do not. They argued that a more fundamental material parameter might be K_{IB}, the stress intensity factor at branching, rather than the A's in equation 2, since the latter do not take into account the free surface, geometry factors, and non-uniform stress gradients over the crack surface. There is validity to their argument, but in practice the ease of use of equation 2 has led to its widespread utilization. Evaluation of a K_{IB} is more difficult and involves an unusual curve-fitting exercise on a σ versus $1/\sqrt{R}$ graph.[18] One other consideration is that their stress intensity shape factors (the geometry terms, Y) for various surface cracks were based on solutions for static cracks, whereas fracture mirrors are formed by cracks moving at or near terminal velocity. Nevertheless, the size and shape predictions based on A or K_{IB} match very closely for the limiting cases of small mirrors in tension specimens. This is also true for small semicircular mirrors centered on surface flaws in strong flexure specimens. So, at least for some special mirror cases, A should be directly related to a more fundamental parameter K_{IB}.

Mecholsky et al.[40] have shown that A can be correlated to the critical fracture toughness, K_{Ic}. The relationship depends upon the mirror-to-flaw size ratios, which differ for glasses and ceramics. K_{Ic} and K_{IB} may in turn be related, but little work has been done to establish this relationship. K_{Ic} pertains to the resistance to crack propagation of a static flaw in the vicinity of the origin, but K_{IB} pertains to a rapidly moving crack that has had ample opportunity to interact with more of the microstructure and be affected by possible R-curve behavior.

THE GUIDELINES

The boundary criteria for glasses and ceramics adopted in the new NIST Guide[21] are as follows:

The mirror-mist boundary in glasses is the periphery where one can discern the onset of mist. This boundary corresponds to A_i, the inner mirror constant.

The mist-hackle boundary in glasses is the periphery where one can discern the onset of systematic hackle. This boundary corresponds to A_o, the outer mirror constant.

The mirror-hackle boundary in polycrystalline ceramics is the periphery where one can discern the onset of systematic new hackle and there is an obvious roughness change relative to that inside the mirror region. This boundary corresponds to A_o, the outer mirror constant. Ignore premature hackle and/or isolated steps from microstructural irregularities in the mirror or irregularities at the origin.

These are very general criteria and they rely upon perception, but they furnish guidance for the practical steps which follow in the guidelines.

The microstructure of most polycrystalline ceramics obscures the fine details of fracture mirrors. Mist is usually not recognizable. In coarse-grained or porous ceramics, it may be impossible to identify a mirror boundary. In very weak materials, the mirror may be larger than the specimen or component and the boundaries will not be present. Sometimes coarse microstructural elements or flaws in a mirror may trigger early hackle lines that will generate wake hackle within the mirror, prior to the onset of generalized velocity hackle that forms the mirror boundary. The mode of crack propagation, whether it is trans- or intergranular, also affects the mirror markings. Mirrors may be easier to see in materials that start out with transgranular fracture but then develop increasing fractions of intergranular fracture. On the other hand, it may be more difficult to define mirrors when the fracture mode is intergranular from the origin right through the entire mirror. Hence, it is appropriate to add the qualifier "relatively" as in "relatively smooth" when describing the mirror region, since there is an inherent roughness from the microstructure even in the area immediately surrounding the origin.

If the mirror is being measured for a component failure analysis, and if the mirror constant A is known, steps 1 through 9 below are followed and the stress at the origin is calculated in accordance with equation 2. Note that this origin stress may or may not necessarily be the maximum stress in the part. If the fracture mirror constant A is being evaluated by means of testing laboratory specimens, and the origin stresses are known, then additional steps 10 through 12 may be used.

Examples of how to judge the boundaries are shown in Figures 1, 4 and 5. Low-power images are also shown to provide an overview. The boundaries were assessed while looking through an optical microscope and the digital images were then marked. The twelve guidelines are presented below with clarification notes and details. Space limitation precludes a full treatment in this summary paper, but the NIST Guide includes copious notes, examples, and marked illustrations.

1. Use an Optical Microscope Whenever Possible.

A compound optical microscope with bright-field viewing is best for glasses. A stereo optical microscope is best for ceramics. A scanning electron microscope may be used if optical microscopy is not feasible.

Differential interference contrast (DIC, also known as Nomarski) mode viewing with a research compound microscope is not recommended by this Guide. It is not suitable for rough ceramic fracture surfaces. It also creates complications with glass fracture surfaces. There is no question that DIC mode viewing can discern very subtle mist features in glasses, but the threshold of mist detectability is highly dependent upon how the polarizing sliders are positioned. Hence, DIC measured radii are quite variable. DIC measured radii can be substantially smaller than those obtained with conventional viewing modes. It also must be borne in mind that not all users have access to interference contrast microscopes.

Dark-field illumination may be used for glasses, but dark-field images may lose a little resolution with glasses and radii may be slightly larger as a result. It is very effective with highly-reflective mirror surfaces in single crystals.

Confocal optical microscopes and optical interferometers have been occasionally used to examine fracture mirrors, but the author is unaware of any systematic study to correlate apparent mirror sizes to those measured with conventional optical microscopes. Confocal microscopes have a resolution limit of about 0.25 μm, which may be a problem.

Scanning electron microscope images of mirrors are not recommended for glasses, since the mirror-mist boundary is usually indiscernible. SEM images often appear flat and do not have adequate contrast to see the fine mist detail at the ordinary magnifications used to frame the whole mirror. SEM images may be used with very small mirrors that would be difficult to see with optical microscopy, e.g., high- strength optical fibers. Scanning electron microscope images may be used for ceramics if necessary, but contrast and shadowing should be enhanced.

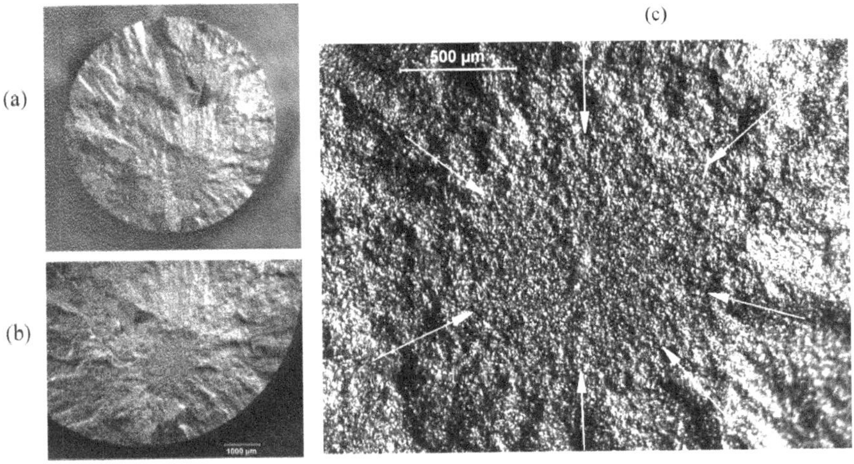

Figure 4. Silicon carbide tension strength specimen (371 MPa) with a mirror centered on a compositional inhomogeneity flaw. Note how clear the mirror is in the low power images in (a,b). The mirror boundary (arrows in c) is where systematic new hackle forms and there is an obvious roughness difference compared to the roughness inside the mirror region.

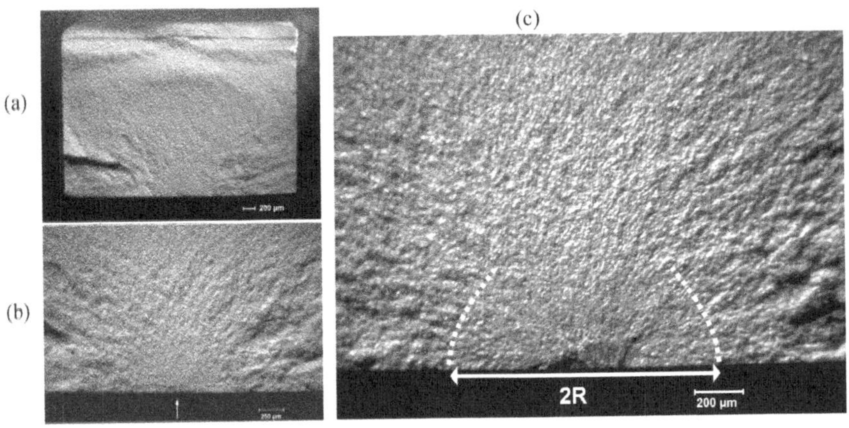

Figure 5. Silicon nitride bend bar with a Knoop surface crack in a silicon nitride (449 MPa). The mirror is incomplete into the stress gradient, but the mirror sides can be used to construct boundary arcs in (c). Radii are measured in the direction of constant stress along the bottom.

2. The Fracture Surface Should be Perpendicular to the Microscope Optical Path or Camera.

This simple and fairly obvious requirement is intended to avoid the foreshortening that can occur if the specimen is tilted. A small amount of tilting is acceptable in order to get a favorable reflection in a glass piece. The requirement poses a small problem if the mirrors are examined with stereo binocular microscopes. These have two different tilted optical paths. If viewing with both eyes in a stereo microscope, the specimen should be flat and facing directly upwards. The observer's brain will interpret the image as though he is facing it directly. Alternatively, if a camera is mounted on one light path of the stereo microscope, and it is used to capture or display the mirror, then the specimen should be tilted so that the camera axis is normal to the fracture surface. For example, tilt the specimen to the right if the camera is attached to the right optical path.

3. Optimize the Illumination to Accentuate Topographical Detail.

The mist and hackle features should be accentuated. Glasses may either be illuminated from directly down onto a fracture surface or by grazing angle, vicinal illumination. Ceramics should not be directly illuminated, since the light will reduce contrast, especially in translucent or transparent materials. Ceramics should be illuminated with vicinal illumination. Thin coatings may be applied to translucent or transparent ceramics as needed. Stereo microscopes are strongly preferred for ceramics. Vicinal illumination is less convenient with compound light microscopes, but the observer should experiment with whatever illumination options are available to accentuate subtle surface roughness and topography features.

4. Use a Magnification Such that the Fracture Mirror Area Occupies about 75 % to 90 % of the Width of the Field of View for Glasses, and Approximately 33 % to 67 % of the Width of the Field of View for Ceramics.

Observers usually mark the mirror boundaries closer to the origin at greater magnifications than they would at lower magnifications. This is because mist or micro hackle markings are easier to see at distances closer to the origin at high magnification. This is particularly the case with glasses. Conversely, at very low magnifications, much detail is lost and observers typically overestimate the mirror size. Fracture mirrors are reasonably easy to see in glasses and magnifications should be used such that they nearly fill the field of view. Mirror interpretation is more problematic with polycrystalline ceramics. Excessive magnification often leads to confusion as to where the boundary is located. Even though a mirror may be obvious at low or moderate magnification, at higher magnification it may be impossible to judge a boundary. It is more practical to view the mirror region and the natural microstructural roughness therein compared to the hackle roughness in the regions outside the mirror. "Stepping back" and using the 33 % to 67 % rule should help an observer better detect the topography differences. Images recorded at these magnifications are also more convincing when shown to other fractographers or engineers. Supplemental lower-magnification images may also be made to aid interpretation, such as shown in Figures 4 and 5. The images should not be more than 5 times lower magnification, otherwise it is difficult to correlate features in one image to another.

The guide has additional information on how to interpret ceramics with microstructures that make judgment difficult. Some self-reinforced silicon nitrides (Figure 4) and yttria-stabilized tetragonal zirconia polycrystals (Y TZP) are difficult (Figure 6). The mirror regions are somewhat bumpy in the self-reinforced silicon nitride. The zirconia has intrinsic micro hackle lines well within the mirror. The mirror boundary was judged to be the point where systematic radiating new hackle commenced and there is an obvious roughness change relative to the inside-mirror region.

The word systematic requires elaboration. Mirror hackle lines are created after the radiating crack reaches terminal velocity. Premature, isolated hackle can in some instances be generated well within a mirror, however, but it should be disregarded. Wake hackle from an isolated obstacle inside the mirror (such as a large grain or agglomerate) can trigger early "premature" hackle lines. Steps in scratches or grinding flaws can trigger hackle lines that emanate from the origin itself.

(a)

(b)

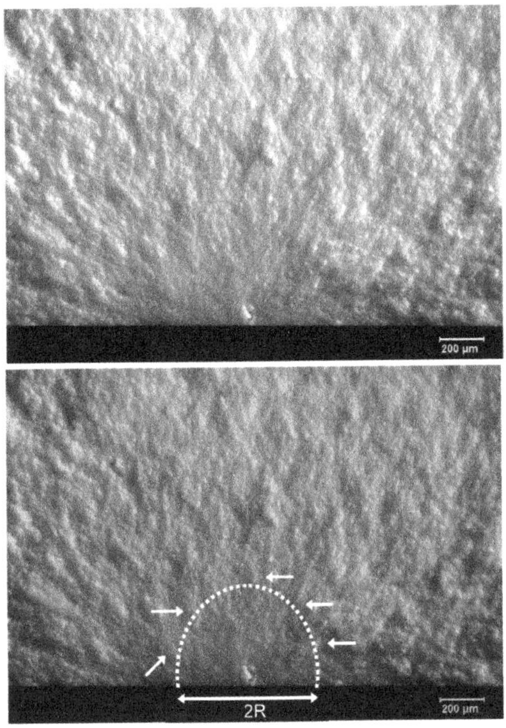

Figure 6. A fine-grained 3 mol % yttria-stabilized tetragonal zirconia (3Y-TZP) polycrystal (Ref. 41). The mirror is difficult to delineate in this material. (a) shows the uncoated fracture surface of a 2.8 mm thick flexural strength specimen (486 MPa) with vicinal illumination. (b) shows an interpretation for a mirror-hackle boundary where systematic new hackle is detected (small white arrows) as compared to the roughness inside the mirror. The marked circle is elongated somewhat into the depth due to the stress gradient. The radius was 345 µm.

5. Measure the Mirror Size While Viewing the Fracture Surface with an Optical Microscope Whenever Possible.

Mirrors should be evaluated while looking in a microscope. Use either calibrated reticules in the eyepieces or traversing stages with micrometer-positioning heads. If a digital camera and high-resolution computer monitor are available, the image can be viewed both through the microscope and measured on the digital monitor, provided that it has been calibrated.

Interpretation from two-dimensional photos alone should only be done as a last resort. Measurements from photos may be necessary for very small specimens or very strong specimens with tiny mirrors, such as in fibers or microelectro-mechanical system (MEMS) devices. Scanning electron microscope images may be used. Again, the fractographer should take an overall framing photograph or image shot in accordance with the 75 % to 90 % rule for glasses and 33 % to 67 % rule for ceramics. Higher- and lower-magnification images may be used to help aid in interpretation. Mirror size measurements from photographs are usually less accurate or precise. They frequently overestimate mirror sizes unless conditions are carefully optimized to accentuate contrast and topographic detail. Two-dimensional photographic renditions of a three-dimensional fracture surface usually lose much of the topographic detail discernable by the eye with a compound optical or stereo microscope. Video cameras should not be used to capture mirror images since they lack adequate resolution. Mirror size measurements made on computer monitor screens are also subject to inaccuracies, also because they are two-dimensional renditions of a three-dimensional fracture surface. Nevertheless, high-resolution cameras and monitors are beginning to match the capabilities and accuracy of an observer peering through the optical microscope.

6. Measure Radii in Directions of Approximately Constant Stress Whenever Possible.

Measurements should be taken from the center of the mirror region, but some judgment may be necessary. A common practice is to judge whether a mirror is approximately semicircular or circular. If it is, then multiple radii measurements may be made in different directions and averaged to obtain the size estimate. The center of the mirror may not necessarily be the center of the flaw at the origin. Careful inspection of tiny localized fracture surface markings (Wallner lines and micro hackle lines) may reveal that fracture started at one spot on a flaw periphery. For example, fracture from grinding or impact surface cracks in glass often starts from the deepest point of the flaw and not at the specimen outer surface. Figure 1 shows an example. Large pores often trigger unstable fracture from one side. If an exact mirror center cannot be determined, measure a mirror diameter and halve the measurement. This is commonly done for semicircular mirrors centered on irregular surface-located flaws. Circular embedded mirrors are easiest to interpret, such as in Figure 4. Small semicircular mirrors on the surface of a part, such as in a bend bar or a flexurally-loaded plate, are also not difficult. The mirror relationship holds up very well in glass optical fibers tested in tension for mirror radii almost as large as the fiber diameter.[42] The mirror radius should simply be measured from the origin to the mirror-mist or mist-hackle boundary on the opposite side of the fiber, R_d as shown in Figure 7.

Mirror shapes are commonly affected by stress gradients in a plate or a beam. Mirror radii are elongated in the direction of decreasing stress. Examples are shown in Figure 8. In such cases, measure the mirror radius along the tensile surface where the stress is constant. Do not measure the mirror radii into the gradient. Even with this precaution, there is considerable evidence that the data begin to depart from the stress-mirror size curves and the relationship in Orr's equation 2 when the sizes approach the cross-section thickness. For mirrors radii larger

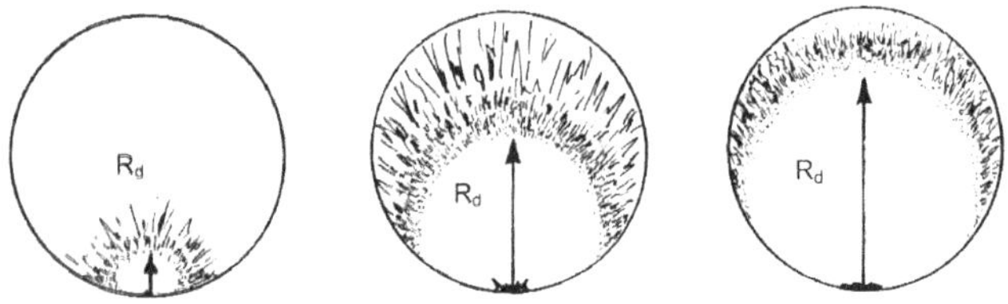

Figure 7. Mirrors surrounding surface origins in rods or fibers loaded in direct tension. (After Ref. 42) Measure both the mirror-mist radius and mist-hackle radii into the depth.

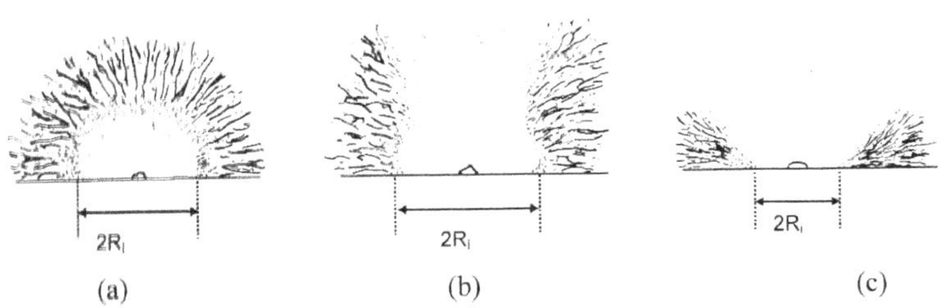

Figure 8. Elongated mirrors in bending stress fields. If the mirror is small relative to the part size, then the mirror may be semicircular, as shown in (a). Weaker parts have larger mirrors that flare into the interior and are incomplete, as shown in (b) and (c). Measure the mirror size (R_i or $2R_i$ for the mirror-mist in the illustrations here) in the direction of constant stress.

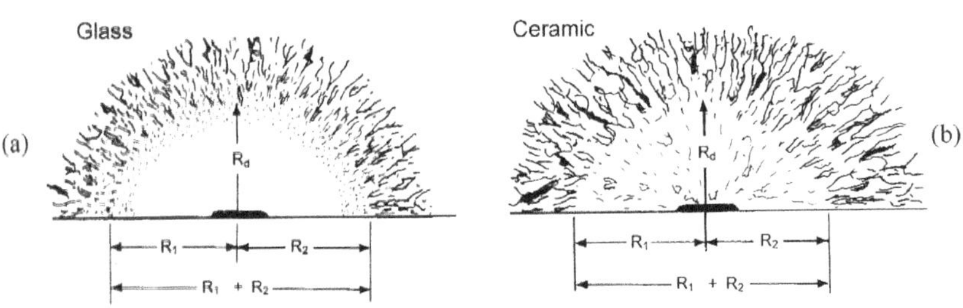

Figure 9. Grinding cracks and scratches can cause mirror elongations along the surface, even in bend bars with stress gradients. (a) shows a schematic of such a mirror with the mist-hackle boundary marked in glass, and (b) shows a comparable image in a polycrystalline ceramic. The ceramic has some intrinsic microstructural roughness inside the mirror. The mirror-hackle boundary is marked. Use an average radius: $R_{avg} = \{(R_1 + R_2 + R_d\}/3$.

than the plate thickness, mirrors are larger than they would otherwise be if they were completely within a large part in uniform tension.

A trend for mirrors to elongate the opposite way, along the external surface of a specimen, was detected by the author in recent work on the fractographic analysis of grinding flaws in structural ceramics.[43,44] Long surface cracks often caused mirrors to have perceptible deviations from a semicircular shape, as shown in Figure 9.

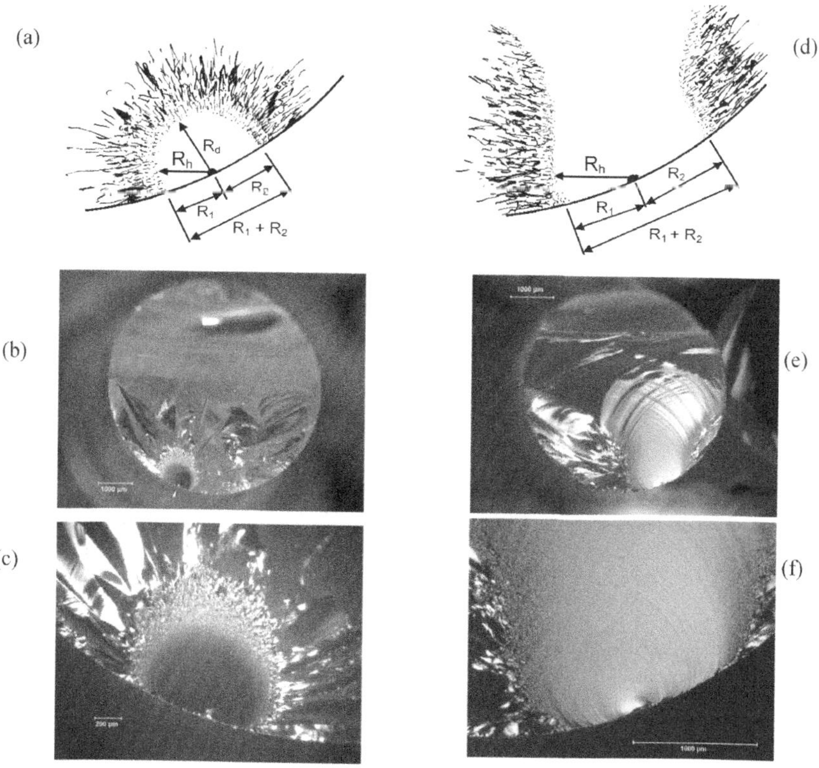

Figure 10. Fracture mirrors in two rods tested in flexure. The maximum tensile stress is at bottom center. The cantilever curls at the top of the rods in (b) and (e) help confirm the location of the maximum tensile stress. Fractures started at flaws part way up the sides of the rods, causing the mirrors to have unequal radii. One rod (a, b, c) was sufficiently strong that a nearly semicircular mirror formed, albeit with unequal radii due to the stress gradient. The second rod (d,e,f) was weaker and has an incomplete mirror boundary. In each case, use the horizontal radius, $R = R_h$ that is in the direction of constant stress if the origin and mirror center is distinct. Otherwise, if the mirror center is not clear, use $R_{avg} = (R_1 + R_2 + R_d) / 3$ if the mirror is nearly semicircular. Use $R_{avg} = (R_1 + R_2) / 2$ if the mirror is elongated into the interior and R_d is large.

In some cases, it may be difficult to measure mirrors in directions of constant stress. The two sides of a mirror may have unequal lengths, since the stresses are different on either side of the mirror. Figure 10 shows examples from the author's research of round rods broken in flexure.[43,44] Many origins were not at the rod bottom where the stresses were a maximum, but part way up the side of the specimen. The specimen orientation was easily determined from observation of the cantilever curl. The maximum tensile stress on the bottom of the specimen was on the rod directly opposite the cantilever curl. The mirror radii had obviously different lengths due to the stress gradient. A radius in the direction of constant stress, R_h, should be measured as shown in Figure 10, if the mirror is centered on a well-defined origin site. If there is any doubt, then an average radius may be computed. Use $R_{avg} = (R_1 + R_2 + R_d) / 3$ if the mirror is nearly semicircular. Use $R_{avg} = (R_1 + R_2) / 2$ if the mirror is elongated into the interior and R_d is large or is incomplete. For origins located in the interior of a rod broken in flexure, only use the radii in the direction of constant stress.

There is one important detail about mirror sizes that warrants discussion. Mirrors located on a specimen external surface have *small cusps* at the intersection with the outer surface, as shown in Figure 11. Cusps are often detected in glass mirrors, but they are rarely if ever discerned in polycrystalline ceramic mirrors. The small cusp is a consequence of fracture mechanics. A small element of material near the tip of a crack at the specimen exterior surface experiences greater stress intensity than a similar element buried in the interior, whereby neighboring

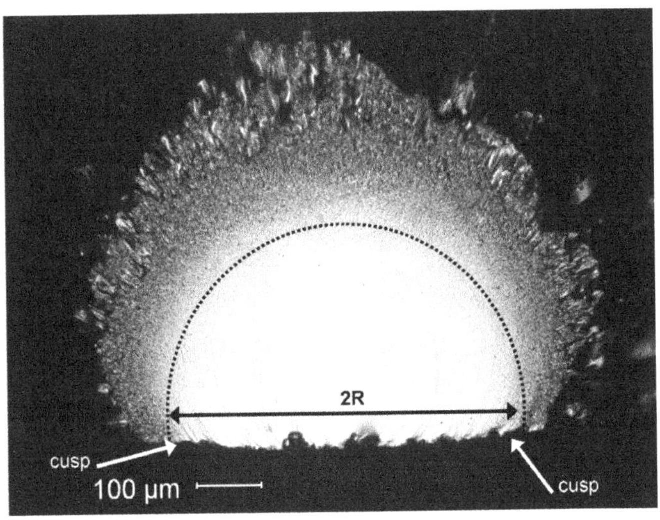

Figure 11. Fracture mirror in a fused silica rod (116 MPa). Mirrors on the surface have inward tilting cusps. These should be truncated and a mirror site measured by using the arc of the overall mirror shape. This mirror is slightly elongated into the interior due to a bending stress gradient. (Remember that some subtle mist detail may not duplicate well in printing.)

elements can "share the load." The slightly-greater stress intensity at the surface triggers the mirror markings a bit sooner than for interior elements. The convention adopted in the NIST Guidelines is to truncate the cusp. Extend the semicircular (or other mirror shape) arcs as shown in Figure 11. Kirchner et al.[17,18] discussed the shapes of fracture mirrors that intersect outer surfaces, and showed that the local enhancement of the stress intensity K_I accounts for the cusps. Another reason to be wary of measurements right along the surface is that surface roughness, machining damage, or other surface irregularities may trigger mist or hackle formation a bit earlier than in the interior. Others have noted that measurements taken right on an exterior surface are slightly different than those taken into the interior. Even Shand[6,7] recommended that size measurements be taken 0.1 mm (.004 in) beneath the exterior surface to avoid "distortions." Mecholsky and Freiman in 1979[22] recommended ignoring the cusp at the surface on the basis of fracture mechanics considerations: "In measuring the mirror-mist and mist-hackle boundaries, these should be projected to the tensile surface to compete a circular arc, since there is curvature at the surface due to free surface effects."

A dilemma occurs when mirrors are large in plates or beams broken in flexure, as shown in Figure 8c. Cusps cannot be detected. In this case the only plausible way to measure a mirror radius is directly along the surface. In such cases, the general warning of step 7 below is applicable.

Residual stresses alter mirror shapes. If the mirror is very small relative to the stress gradient, the mirror shape may remain circular or semicircular if along the surface. On the other hand, if the mirror is larger or the stress gradient is steep, then the gradient alters the mirror shape, as shown in Figure 12. Figure 12a shows an annealed plate that requires an applied stress of $\sigma_a = \sigma_f$ to cause fracture. Figure 12b shows the case where the same plate has residual surface compression stress $\sigma_r = \sigma_c$ from ion exchange or thermal tempering, so that an applied stress to cause fracture is $\sigma_a = \sigma_f + \sigma_c$. In other words, the applied stress must be increased to overcome the residual surface stress. Nevertheless, the net stress at the surface at the moment of

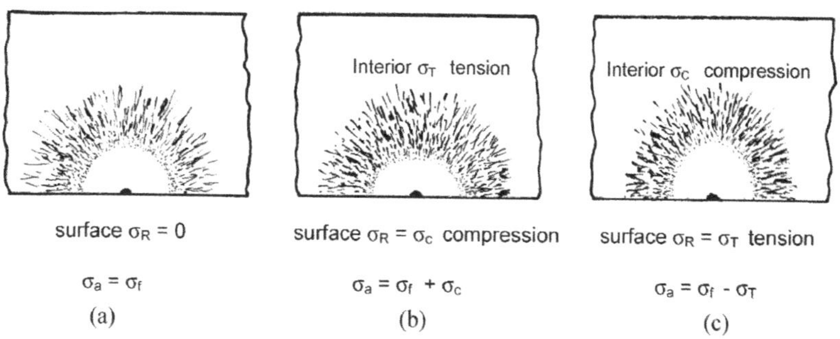

Figure 12. Surface residual stresses also may alter a mirror shape. σ_a is the applied stress to cause fracture, and σ_f is the fracture stress in an annealed plate in tension. (a) shows a surface mirror in an annealed plate. (b) shows the mirror shape in a plate with surface compression stresses that decrease into the interior and become tensile. (c) shows a mirror in a plate with surface tensile stresses that diminish into the interior and become compressive.

fracture is $\sigma = \sigma_a - \sigma_c = (\sigma_f + \sigma_c) - \sigma_c = \sigma_f$, the same stress as in the annealed plate. Hence the mirror radii along the surface are *unchanged* compared to the annealed plate. In contrast, in the direction into the interior, tensile stresses combine with the applied tensile stress to cause the mirror markings to form sooner, at a shorter radius into the interior than in the annealed plate. In this example, the mirror shape is flattened to a semiellipse. Mirror radii should be measured *only along the surface* (or just beneath the surface if there is a cusp) in these cases. Figure 12c shows that surface residual tensile stresses have the opposite effect: mirror radii are elongated into the interior. Mirror radii again should only be measured along the surface, since again the net stress to cause fracture is $\sigma = \sigma_f$. There are two possible paths for analysis if there are residual stresses:

(a) The mirror is measured on a component. The applied stress and the residual stresses are unknown. In this instance the net stress σ at the origin can be evaluated from R and equation 2.

(b) The mirrors are collected in laboratory conditions with multiple specimens such that the apparent origin stresses σ_a, from applied external stresses are known. In this instance, one or more matched pairs of σ_a and R are obtained. Graphical analysis shown below in step 11 reveal the existence and magnitude of the residual stresses.

7. Exercise Caution When Mirrors are Large Relative to the Specimen Cross-Section Size.

At some point, one can expect departures from the stress - mirror size relationship. The point where the departure occurs depends upon the loading geometry and the stress state. Pronounced deviations occur once the mirror size approaches or is greater than the component thickness in plate or beam *bending* fractures. Experimentally measured radii are much greater than predicted by equation 2. Shand recommended that the maximum mirror size should be no more than 15 % of the rod diameter for flexure tests.[5] Kirchner and Conway[18] warned about limitations in the fracture mechanics models for mirror radii exceeding 20 % of a rod diameter tested in flexure. On the other hand, Castilone et al.[42] had success with mirrors that were almost as large as the fiber diameters for fibers tested in direct tension. Mecholsky and Freiman[22] warned that systematic deviations from the mirror size relationship occur at large mirror sizes but also at very small sizes, the latter due to internal stress effects, e.g., from thermal expansion anisotropy of grains in ceramics.

8. Show at least one photo with arrows or lines marking the mirror size.

9. Report How the Mirrors Were Measured.

This simple step is often overlooked or ignored, and the reader is left wondering exactly what has been done. The fractographer should report the microscope used, confirm that interpretation was made while looking through the microscope, whether photos had to be used, and approximately what magnifications were used. The directions in which the mirror radii were measured should be documented. The approximate shape of the mirrors (semicircular, circular, or elliptical) should be noted. It should also be noted whether the mirrors were an appreciable fraction of the size of the cross section or not. Lastly, and most importantly, the judgment criterion used should be reported.

10. Use the Stress at the Origin Site.

If the specimen was broken in controlled conditions where the stress distribution was known (e.g., beams, rods, or plates in flexural loadings) correct the stress for location in specimens with stress gradients. No correction is needed if a part was stressed in uniform tension. On the other hand, many parts or laboratory specimens do have stress gradients. The general principal that should be followed is that the mirror formation is guided by the stresses in the vicinity of the origin. While this may seem obvious, it is probable that some analysts in the past have erroneously used nominal stresses in a specimen or component rather than the actual stress that was acting upon the mirror region in a body. In contrast, some researchers have correlated the stress at every site along the mirror periphery with the mirror radius at that periphery site, but this complex process is not practical on a routine basis.

11. Evaluate the Fracture Mirror Constants

Once a set of matching mirror radii and fracture stresses has been compiled, plot the data as linear stress versus inverse square root of mirror size, as shown in Figure 13. Perform a linear regression analysis and calculated the mirror constant from the regressed line. The older graphical representation of log stress versus log radius may be used if necessary, but is not preferred. The mirror constants should be reported as either MPa$\sqrt{m}$.

Use linear regression methods to obtain A in accordance with equation 2 with a zero intercept on a graph of σ versus $1/\sqrt{R}$. In a typical strength-test experiment in a laboratory, applied stress, σ_a, and mirror radius, R, are independently measured. It is customary to regress σ on R. A is the slope of the regression line. Use some judgment in the regression analysis since fracture mirror data usually have moderate scatter. If the data do not appear to fit a trend that has a zero intercept, regress the data with a non-zero intercept, as shown in Figure 13b. Again use some judgment in the interpretation, since a strict linear regression fit may produce implausible outcomes, particularly if the data are collected over a limited range of mirror sizes and stresses.

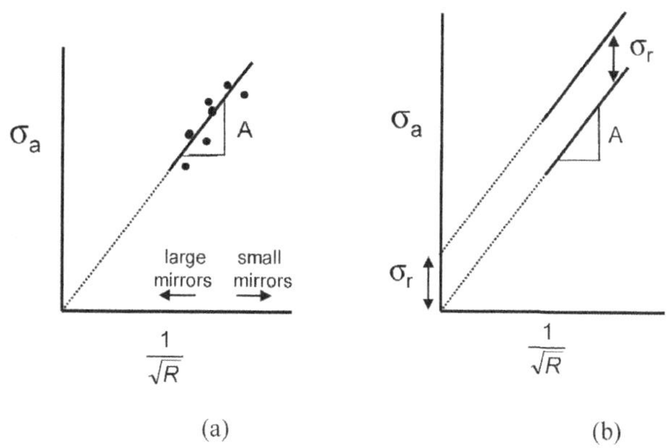

Figure 13. Plot of applied stress σ_a versus $1/\sqrt{R}$. (a) shows the trend for parts with no residual stresses. (b) shows it for parts with residual stresses. Compressive residual stresses move the locus up with a positive intercept σ_r, but with the same slope. Tensile residual stresses shift the data downwards with a negative intercept (not shown).

Report the intercept if it deviates significantly (> 10 MPa for glasses; > 50 to 100 MPa for ceramics) from zero. Investigate possible residual stresses or specimen size or shape issues if the intercept deviates significantly from zero. Consistent units should be used with this approach. That is to say, if the stress axis is MN/m^2 or MPa, then the abscissa (horizontal axis) should be $1/\sqrt{R}$ where R is in units of *meters*. The mirror constant A as a slope is easily visualized. In addition, a nonzero intercept may be conveniently interpreted as an effective residual stress. If residual stresses σ_r are present in addition to the externally applied stress, σ_a, then the net stress acting on the origin site is:

$$\sigma_{net} = (\sigma_a + \sigma_r) = \frac{A}{\sqrt{R}} \tag{3}$$

and:

$$\sigma_a = \frac{A}{\sqrt{R}} - \sigma_r \tag{4}$$

An intercept below the origin corresponds to a net tensile residual stress. A positive intercept corresponds to residual compressive stress, since the usual sign convention is for compressive stresses to have a negative sign.

Some caution is advised, since residual stresses are often nonuniform. The estimate from the intercept is an effective residual stress, which in reality may vary in magnitude through the mirror region. Once again is it prudent to measure mirror radii in directions of constant stress. If the mirror is in a heat-strengthened or tempered piece (where stress may be constant along the surface, but change dramatically through the thickness) the mirrors should only be measured along the surface (or just underneath to avoid the cusp). Residual stresses from an indentation or impact site are very local to the origin and may have very little effect on a mirror size.

Although most researchers have felt that the regressed lines should go through the origin in annealed test pieces, there is evidence by J. Quinn[45] that a small but measurable intercept may exist in even annealed materials. The intrinsic intercept was evaluated as 10 MPa (1,500 psi) for glass, a value that interestingly concurs with Orr's[3] estimate of the minimum stress necessary for

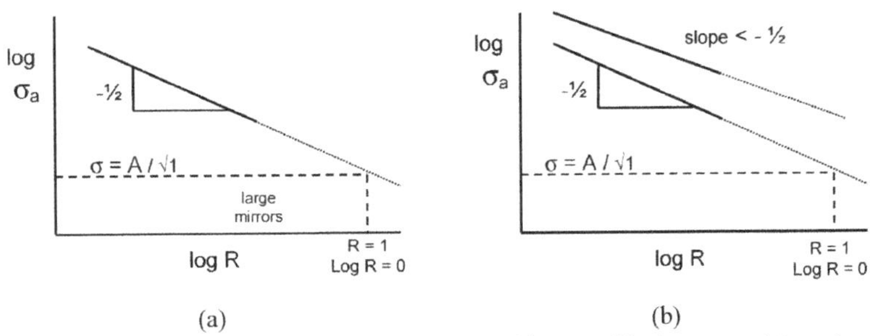

Figure 14. Plot of log σ_a versus log R for parts with no residual stresses (a), and parts with residual stress (b). Compressive residual stresses move the locus upwards, but with a different slope and intercept. Tensile residual stresses move the loci below the baseline curve (not shown).

branching in glass. One popular alternative analysis method is based on plotting the data on a log-stress versus log-radius graph as shown in Figure 14. This method of showing the results and calculating a mirror constant was common in the older technical literature and is occasionally still found today. Graphs of this type were used when researchers were not sure whether equation 2 with the √R relationship was appropriate. Forty years of research have shown it is, so there no longer is a need to test for the trend. From Orr's equation 2:

$$\log \sigma_a = -\frac{1}{2}\log R + \log A \qquad (5)$$

If stresses are in units of MN/m^2 (MPa), and the mirror size is measured in meters, then the mirror constant A has units of $MN/m^{1.5}$ or MPa√m. If the mirror size is 1 m, then log R = 0. Then log σ = log A and hence, σ = A. Hence, the mirror constant A corresponds to the value of stress that would create a mirror of size 1 m. Since most actual mirrors that are measured are usually much smaller than unit size, it is apparent from Figure 14 that the mirror constant (or the stress for R = 1) lies beyond the range of data usually collected. Deviations from the linear relationship on the log - log plot occur when residual stresses are present but unaccounted for, or when the mirror size is large relative to the component size, or when there are stress gradients. The residual-stress deviations cause the line to have a slope other than -½, as shown in Figure 14b. Attempts to compute the residual stresses may then be made by guessing values of the residual stresses σ_r, replotting the data, and checking the goodness of fit of a line of slope –½. This is a fairly cumbersome process and the σ versus 1/√R procedure is simpler. The two analyses put different weights on large and small mirror measurements. In one case the mirror constant is a slope of a line, in the other it is an intercept at R = 1, a rather large mirror size not likely to be realized in practice. Some of the variability in published mirror constants probably is due to the use of the two different curve-fitting schemes. It is also certain that some researchers have evaluated unannealed test pieces and then force fitted regression lines through the data with zero intercepts in the former scheme, or lines of slope – ½ through the data in the latter scheme.

Regression analyses on the log - log graph are more vulnerable to deviations of the data from the correct trends when mirror sizes are large. Upward deviations from the log stress - log radius graphs have been noted in a number of studies (e.g., Shand,[5,7] Orr[3]). Regression lines chase the upward deviations and dramatically alter the estimate of the mirror constant. On the other hand, with the σ versus 1/√R graph, oversized mirror data points are closer to the origin and have less influence on the regression line and less effect upon the slope, A.

Finally, regression analysis with the σ versus 1/√R approach minimizes the deviations of stress σ from the fitted line. Regression analysis for the log – log graph minimizes deviations of log σ from the fitted line. The former is preferred from a mathematical perspective. For all the reasons above, the linear-stress versus inverse square root radius approach was adopted in the NIST Guidelines. Analysis is simple and intuitive. The uncertainty of the slope can be estimated from routine analyses available in many statistical software packages.

12. Mirrors Sizes Should Be Collected Over a Broad Range of Sizes and Fracture Stresses if Possible. Data From Different Specimen Types and Sizes May Be Combined.

This is a fairly obvious conclusion. Superb examples are shown by Kerper and Scuderi[8] for borosilicate glass rods with diameters that varied by a factor of ten and by Mecholsky and Rice[46] for various sized fused silica rods, disks, and fibers. Ideally, data from many small specimens could be complemented by judicious testing of a few large specimens. Another common procedure is to anneal or fine grind/polish some specimens to obtain high strengths, but also abrade or damage others to obtain low strengths. Sometimes the mode of loading can be changed to alter the fracture stress. For example, some studies have generated mirrors with large four-point and small three-point flexure specimens. Some specimens may be tested in inert conditions and others in conditions conducive to slow crack growth.

SOME FINAL THOUGHTS

One is also struck by the fact that nearly all the surface-centered mirrors shown in the literature, even in the classical papers, are not exactly semicircular despite all the schematics that imply that they are. So fractographers should not be alarmed if their mirrors are not perfect.

The steps in this Guide have occasionally appeared sporadically in the past. Shand recommended that stresses be corrected for the origin location,[5] that radii be measured beneath the surface to avoid surface effects,[6,7] that low-angle vicinal illumination be used.[5,7] He also warned about deviations from the trends if the mirror sizes were too large relative to the component thickness.[5] Shand also said that mist could not be discerned in glass ceramics.[5] Morrell et al.[19] agonized over the interpretation of mirrors in Y-TZP zirconia, but settled on stereo optical microscopy at a fixed magnification, with grazing incidence illumination. The specimen sides were masked to block transmitted-light scatter. Matching fracture halves were mounted together to aid the interpretation. The most comprehensive set of recommendations predating the new NIST guidelines were crafted by Mecholsky and Freiman.[22] Six of their recommendations match steps in the guidelines: optical microscopy is preferred over scanning electron microscopy whenever possible; suitable magnifications should be used; mirror boundary arcs should be projected to the outer surface to complete a circular arc to eliminate the surface cusps; lighting should be varied to obtain optimum contrast; radii should be measured in directions of constant stress and not into gradients; and caution should be used with data from large mirrors relative to part thickness.

CONCLUSIONS

A set of guidelines for measurement of fracture mirrors and determinations of fracture mirror constants have been devised. The goal is to bring consistency to the procedures used to measure fracture mirrors. This should facilitate improved data bases and better estimates of failure stresses. Most students of the technique have concluded that consistent readings are possible, provided that defined procedures are used. The new guidelines have been prepared on the basis of a review of sixty years of literature, recommendations by experts, and the author's own experiences. Measurement of the mirror sizes requires interpretation. The perception of the observer and the type of equipment are important factors. Although advanced microscopy and software tools hold considerable future promise, it is unlikely that a simple definitive boundary criterion, such as a specific surface roughness, will emerge.

REFERENCES

[1] C. Brodmann, "A Few Observations on the Strength of Glass Articles," *Nachrichten von der Gessellschaften zu Göttingen, Mathematisch-Physikalische Klasse,* 1, 44 – 58 (1894).

[2] A. Smekal, "Die Festigkeitseigenschaften spröder Körper," *Ergebnisse der Exakten Naturwissenschaften*, 15 (1936) 106 -188 (1936).

[3] L. Orr, "Practical Analysis of Fractures in Glass Windows," *Materials Research and Standards*, 12 [1], 21 - 23, 47 (1972).

[4] W. C. Levengood, "Effect of Origin Flaw Characterization on Glass Strength," *J. Appl. Phys.*, 29 [5], 820 – 826 (1958).
[5] E. B. Shand, "Breaking Stress of Glass Determined from Dimensions of Fracture Mirrors," *J. Am. Ceram. Soc.*, 42 [10], 474 - 477 (1959).
[6] E. B. Shand, "Strength of Glass - The Griffith Method Revised," *ibid.*, 48 [1], 43 - 48 (1965).
[7] E. B. Shand, "Breaking Stresses of Glass Determined from Fracture Surfaces," *The Glass Industry*, April (1967) 190 - 194.
[8] M. J. Kerper and T. G. Scuderi, "Modulus of Rupture of Glass in Relation to Fracture Pattern," *Cer. Bull.*, 43 [9], 622 – 625 (1964).
[9] M. J. Kerper and T. G. Scuderi, "Relation of Strength of Thermally Tempered Glass to Fracture Mirror Size," *ibid*, 44 [12], 953 – 955 (1965).
[10] M. J. Kerper and T. G. Scuderi, "Relation of Fracture Stress to the Fracture Pattern for Glass Rods of Various Diameters," *ibid*, 45 [12], 1065 - 1066 (1966).
[11] H. P. Kirchner, R. M. Gruver, and W. A. Sotter, "Fracture Stress-Mirror Size Relations for Polycrystalline Ceramics, *Phil. Mag.*, 33 [5], 775 – 780 (1976).
[12] Levengood, Ref. 4 above, footnote page 821. See also *Glass Industry*, 36 [10] October, 1955 p. 523.
[13] Private Communication. Letter from L. Orr to G. Quinn on December 15, 2001.
[14] N. Terao, "Sur une Relation entre la Resistance a la Rupture et le Foyer d'Eclatement du Verre," *J. Phys. Soc. Japan.*, 8 [4], 545 – 549 (1953).
[15] J. W. Johnson and D. G. Holloway, "On the Shape and Size of the Fracture Zones on Glass Fracture Surfaces," *Phil. Mag.* 14, 731 - 743 (1966).
[16] J. Congleton and N. J. Petch, "Crack Branching," *Phil. Mag.*, 16 [142], 749 – 760 (1967).
[17] H. P. Kirchner and J. W. Kirchner, "Fracture Mechanics of Fracture Mirrors," *ibid*, 62 [3-4], 198 – 202 (1979).
[18] H. P. Kirchner and J. C. Conway, Jr., "Criteria for Crack Branching in Cylindrical Rods: I, Tension; and II, Flexure," *ibid*, 70 [6], 413 – 418 and 419 – 425 (1987).
[19] R. Morrell, L. Byrne, and M. Murray, "Fractography of Ceramic Femoral Heads," pp. 253 - 266 in *Fractography of Glasses and Ceramics, IV, Ceramic Transactions* Vol. 122, eds., J. R. Varner and G. D. Quinn, American Ceramic Society, Westerville, OH, 2001.
[20] ASTM C 1322-96, "Standard Practice for Fractography and Characterization of Fracture Origins in Advanced Ceramics" *Annual Book of Standards*, Vol. 15.01, ASTM Int., West Conshohocken, PA. 1996.
[21] G. D. Quinn, "Guide to Fractography of Ceramics and Glasses," NIST Special Publication 960-16, Gaithersburg, MD, 2007.
[22] J. J. Mecholsky and S. W. Freiman, "Determination of Fracture Mechanics Parameters Through Fractographic Analysis of Ceramics," pp. 136 -150 in *Fracture Mechanics Applied to Brittle Materials*, ASTM STP 678, S. W. Freiman, ed, ASTM, Int., West Conshohocken, PA, 1979.
[23] E. K. Beauchamp, "Mechanisms of Hackle Formation and Crack Branching," pp. 409 – 446 in *Fractography of Glasses and Ceramics III*, eds. J. Varner V. D. Fréchette, and G. D. Quinn, Ceramic Transactions, Vol. 64, American Ceramic Society, Westerville, OH, 1996.

[24] E. K. Beauchamp, "Fracture Branching and Dicing in Stressed Glass," Sandia Laboratories Research Report, SC-RR-70-766, Jan. 1971.

[25] E. F. Poncelet, "The Markings on Fracture Surfaces," *J. Soc. Glass Technol.*, 42, 279T – 288T (1958).

[26] E. Gölz, "Übermikroskopische Feinstructuren an Glassbruchflächen," (Ultramicroscopic Fine Structures on Glass Fracture Surfaces), *Zeitschrift Physik*, 120, 773 – 777 (1943).

[27] V. D. Fréchette, *Failure Analysis of Brittle Materials,* Advances in Ceramics, Vol. 28, American Ceramic Society, Westerville, OH, 1990.

[28] E. H. Yoffe, "The Moving Griffith Crack," *Phil. Mag.*, 42, 739 – 750 (1951).

[29] W. H. Duckworth, D. K. Shetty, and A. R. Rosenfeld, "Influence of Stress Gradients on the Relationship Between Fracture Stress and Mirror Size for Float Glass," *Glass Technology,* 24 [5], 263 - 273 (1983).

[30] D. Hull, Chapter 5, in *Fractography, Observing, Measuring and Interpreting Fracture Surface Topography*, Cambridge Univ. Press., Cambridge, 1999.

[31] D. Hull, "Influence of Stress Intensity and Crack Speed on Fracture Surface Topography: Mirror to Mist Transition," *J. Mat. Sci.*, 31, 1829 - 1841 (1996).

[32] C. Wünsche, E. Rädlein, and G. H. Frischat, "Morphology of Silica and Borosilicate Glass Fracture Surfaces by Atomic Force Microscopy," *Glastech. Ber. Glass Sci. Technol.,* 72 [2], 4 (1999).

[33] D. M. Kuluwansa, L. C. Jensen, S. C. Langford, and J. T. Dickinson, "Scanning Tunneling Microscope Observations of the Mirror Region of Silicate Glass Fracture Surfaces," *J. Mater. Res.*, 9 [2], 476 – 485 (1994).

[34] A. B. J. Clark and G. R. Irwin, "Crack-Propagation Behaviors," *Exptl. Mech.*, 6, 321 - 330 (1966).

[35] H. P. Kirchner, "Brittleness Dependence of Crack Branching in Ceramics," *J. Am. Ceram. Soc.*, 69 [4], 339 - 342 (1986).

[36] H. G. Richter and F. Kerkhof, "Stress Wave Fractography," pp. 75 - 109 in *Fractography of Glass*, eds., R. C. Bradt and R. E. Tressler, Plenum, NY, 1994.

[37] J. E. Field, "Brittle Fracture: Its Study and Application," *Contemp. Phys*, 12, 1 - 31 (1971).

[38] W. Doll, "Investigations of the Crack Branching Energy," *Int. J. Fract.,* 11, 184 - 186 (1975).

[39] Y. L. Tsai and J. J. Mecholsky, Jr., "Fracture Mechanics Description of Fracture Mirrors Formation in Single Crystals," *Int. J. Fract.,* 57, 167 - 182 (1992).

[40] J. J. Mecholsky, Jr., S. W. Freiman, and R. W. Rice, "Fracture Surface Analysis of Ceramics," *J. Mat. Sci.*, 11, 1310 - 1319 (1976).

[41] G. D. Quinn, J. Eichler, U. Eisele, and J. Rödel, "Fracture Mirrors in a Nanoscale 3Y-TZP," *J. Amer. Ceram. Soc.*, 87 [3] (2004) 513 - 516.

[42] R. J. Castilone, G. S. Glaesemann, and T. A. Hanson, "Relationship Between Mirror Dimensions and Failure Stress for Optical Fibers," pp. 11 - 20 in *Optical Fiber and Fiber Component Mechanical Reliability and Testing II*, eds., M. J. Matthewson and C. R. Kurkjian, Proc. SPIE, 4639 (2002).

[43] G. D. Quinn, L. K, Ives, and S. Jahanmir, "On the Nature of Machining Cracks in Ground Ceramics," *Machining Science and Technology,* 9, 169 – 210 (2005).

44 G. D. Quinn, L. K. Ives, S. Jahanmir, and P. Koshy, "Fractographic Analysis of Machining Cracks in Silicon Nitride Rods and Bars," pp. 343 – 365 in *Fractography of Glasses and Ceramics IV*, , 2001.

45 J. B. Quinn, "Extrapolation of Fracture Mirror and Crack-Branch Sizes to Large Dimensions in Biaxial Strength Tests of Glass," *J. Am. Ceram. Soc.*, 82 [8], 2126 - 2132 (1999).

46 J. J. Mecholsky, Jr. and R. W. Rice, "Fractographic Analysis of Biaxial Failure in Ceramics," pp. 185 - 193 in *Fractography of Ceramic and Metal Failures*, eds. J. J. Mecholsky, Jr. and S. R. Powell, ASTM STP 827 American Society for Testing and Materials, Westerville, OH, 1984.

SECTION IV

IDENTIFYING AND UNDERSTANDING FLAWS IN CERAMICS

INFLUENCE OF BINDER ON STRUCTURE AND STRENGTH OF ALUMINA CERAMICS

Keizo Uematsu and Satoshi Tanaka,
Nagaoka University of Technology
Kamitomioka, Nagaoka Niigata, 9402188 Japan

ABSTRACT

The influence of binder systems on structure and properties of alumina ceramics was examined with a variety of new analytical tools. Outlines of these tools are presented first. Two types of binders were used to prepare alumina granules by spray drying; poly-acryl-acid (PAA) and poly-vinyl-alcohol (PVA). PVA binder segregated at the surface and subsurface of the granules, whereas PAA binder was homogeneously distributed. Compaction tests on single granules showed that the PAA binder provides lower yield stress of granules than PVA for all conditions examined. In the die compaction process, the relative density of the compact body started to increase at lower pressure, and more homogeneous internal structures were observed in the green compact made with PAA binder than with PVA. Homogeneous structures and higher strengths were achieved in the sintered body. Fracture strength as high as 580 MPa was achieved for the PAA binder system.

INTRODUCTION

Recent studies explicitly showed the relevance between macro-defects and the strength of ceramics. Surprisingly good agreement was noted between the measured strength distribution and that calculated using fracture mechanics with the characteristics of macro-defects in alumina ceramics.[1, 2] Other parameters of microstructure, such as grain size, were found to have little relevance with the strength and its distribution. The grain size may affect the size of macro-defect and/or toughness only indirectly. Clearly, the macro-defect is the most important key for producing ceramics with high performance.

An adequate characterization method was indispensable to establish the above relevance between the macro-defect and the strength of ceramics.[3,6] Macro-defects are very difficult to study with conventional characterization methods including SEM. They are very few in number, which makes them very difficult to be found by common analytical tools. Bulk analysis is needed to find and examine them. However, current bulk analysis involving ultrasonics and x-rays, etc. often do not have resolution high enough to examine their detailed structure. Recent optical methods are easily applicable for many systems. They can be even applicable for both green and sintered bodies. They will provide an opportunity to study the relevance between processing, macro-defects, and strength of ceramics.

This paper will review the optical characterization tools first. It will then discuss the problem associated with binders in the processing of alumina ceramics. Finally, it will show a method to avoid the problems. The achievement of high strength will be discussed in terms of structure.

Optical Characterization Tools

Green specimens are made transparent with an adequate immersion liquid, and the internal structure is observed with an optical microscope in the transmission mode.[2] The selection of liquid is the key point of the tool, since it governs the reflection of light at the interface and thus the scattering of light. The reflection of light at the interface is governed by the ratio of refractive indices (RI) of relevant phases, i.e., the refractive indices of solid and liquid. No reflection of light occurs when the refractive index is the same for the solid and the liquid, making the green compact transparent.

Liquids with a variety of refractive indices are available. Liquids with the RI under 1.79 can be made with safe chemicals. For those with high RI, the liquid may be highly toxic. The highest RI attainable in a liquid is around 2.05 at room temperature. In the visible light range, the refractive index of solids should be approximately under this value to examine the structure of their powder compact. In the infrared region, the achievement of transparency is easier as the wavelength increases. Infrared microscopy can extend the range of materials to be examined by the current tool.[8,9] With this tool, there is no practical limitation in the structural examination for systems made of fine powders. A minor disadvantage of IR microscopy is a slightly reduced resolution of the optical image. Nevertheless, IR microscopy has very high potential, since the size of important defects is very large. In the laser scanning microscope, fluorescent dye is dissolved in the immersion liquid, and its distribution is imaged. The image represents the negative image of particles in the green compact at very high resolution; i.e., a dark image corresponds to regions of high packing density of powder particle, and a bright image to regions of low density.

The requirement on immersion liquid varies with the observation method, i.e., the mechanism of image generation. With polarized light microscopy, the best matching is determined by the refractive indices of the liquid and the solid. The quality of the optical image increases with increasing transparency, since the optical contrast is developed by the optical anisotropy of the solid. Liquids having slight mismatching of refractive index are needed to observe pores and cracks with the normal transmission mode. The optical contrast is generated by the residual scattering of light at the interface of liquid and solids. No structure can be seen when the matching of refractive index is perfect. The requirement of refractive index matching is much reduced in the IR microscope. The light with long wavelength reduces the reflection of light at the interface. The best matching is needed in the examination of confocal laser scanning fluorescent microscopy.

Structures of ceramics can be readily examined using thin specimens without immersion liquids. Ceramics are basically made of transparent material. They appear opaque due to the scattering of light mainly at the boundaries of residual pores in the bulk. Thinned and with both surfaces polished, most of them become transparent, and a survey scan at low magnification will easily reveal macro-defects. The typical thickness needed for examining alumina ceramics is 100-200μm. High magnification is not needed for most cases, since the features of interest are big.

Problems Associated with Binders

Conventionally, poly-vinyl-alcohol (PVA) has been widely used for ceramics production through the powder-compaction process. It can provide the granules and their compact adequate strength needed for subsequent processing. However, the binder tends to segregate at the surface of granules. Figure 1 shows a laser scanning confocal fluorescent micrograph for PVA binder segregated at the surface of the powder granules. The segregated binder appears dark, since it expels the immersion liquid that contains the dye. The segregated binder is known to have detrimental effects on the properties of ceramics.

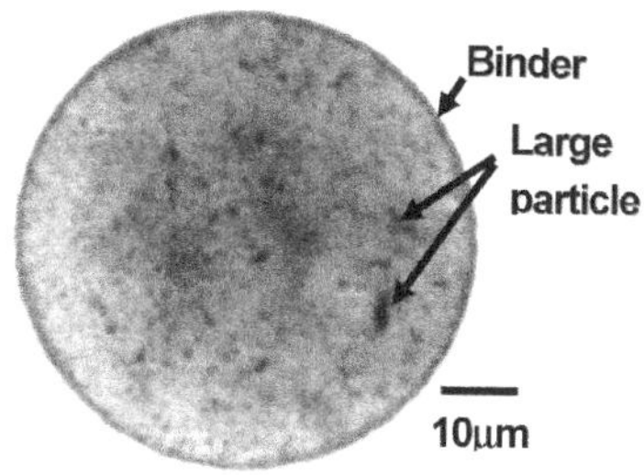

Figure 1. Image of PVA binder segregated at the surfaces of alumina granules; confocal laser scanning fluorescent micrograph.; Y. Saito, K. Uematsu, et al, Am. Ceram. Soc. Bull., 81[1]35-38(2002)

The spray-drying process is responsible for the binder segregation at the surface of granules.[11] In this process, a ceramic slurry containing binder is sprayed as small droplets and dried in a flow of hot air. Figure 2 shows the important point responsible for the segregation process, i.e., the water in the slurry can evaporate only from the surface. The latent heat of evaporation keeps the temperature of water within the droplet below the boiling point. As the surface dries, the water inside the droplet moves to the surface. Binder dissolved in the water also moves to the surface, and is left to segregate at this place as the water evaporates. The migration of PVA with the water is clearly the problem for the binder segregation at the surface of granules.

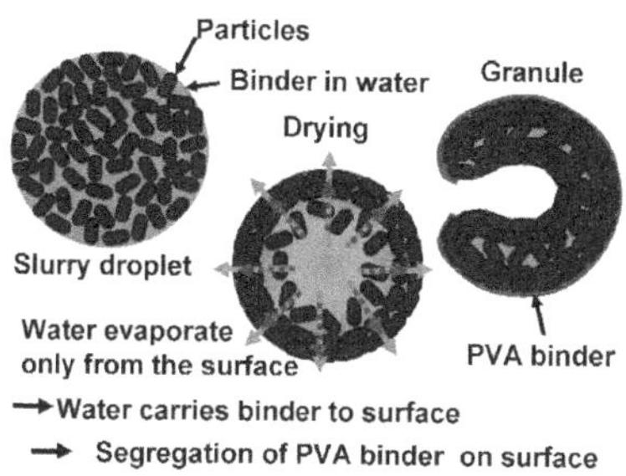

Figure 2. Schematic diagrams showing segregation of PVA binder at the surface of granules during spray drying.

A Method to Avoid the Problem

A binder which does not move with the water may eliminate the segregation. Namely, a binder which bonds strongly on the powder surface appears very promising to avoid the problem of binder segregation. In an alumina system, a strong chemical bonding is expected in a binder containing a functional group in poly-acrylic type.[10,12,14] In fact, the strong bonding of PAA relative to PVA has been reported.[4] Figure 3 shows chemical structures of conventional PVA and novel PAA-type binders. Figure 4 shows the bonding between the PAA-type binder and the positively charged surface near the neutral-to-weakly-alkaline regions of the slurry.[15] This is in clear contrast to PVA, which adsorbs only weakly on the alumina surface.

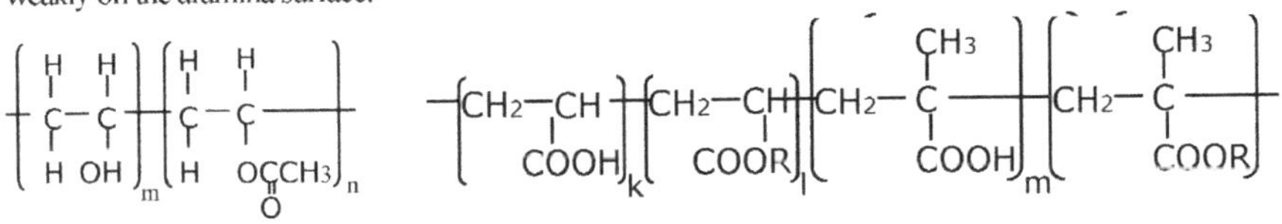

Figure 3. Structure of PVA (left) and PAA (right) binders.

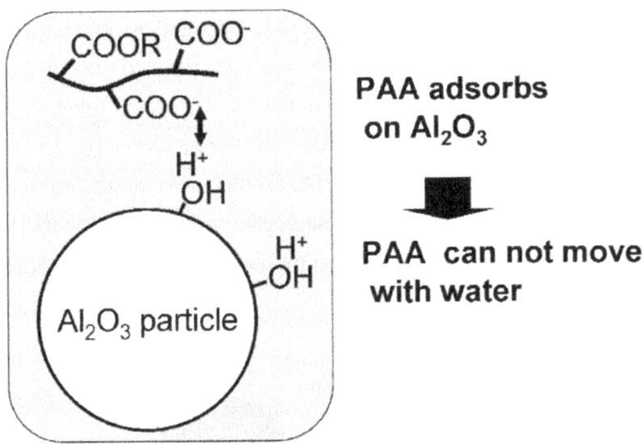

Figure 4. Schematic diagram showing adsorption of PAA on an alumina particle.

EXPERIMENTAL

Fabrication of Granules

The raw materials were an industrial commercial alumina (Al-160SG, Showadenko K.K., Japan) and a poly-acrylic-acid system (PAA) (AQ, Lion, Japan[18]) as a dispersant and binder. The nominal average particle size of the alumina powder is 0.6μm. The components were mixed with distilled water in a ball mill to prepare slurries with binder contents of 2.5 mass%. A slurry was also prepared with polyvinyl alcohol (PVA105, Kurarays, Japan) binder 0.5 mass% and PAA dispersant (D305 Chukyo-yushi Co., Japan) 0.5 mass% as a reference. A spray drier (SD-13, Mitsui-Mining Co., Japan) was used to prepare the granules.

Characterization of Granules

The liquid immersion technique was applied to examine the internal structure of the granules before and after de-binder treatment.[3] In this examination, granules were immersed in diiodomethane to make them transparent, and the internal structure was observed with an optical microscope in transmission mode (OPTIPHOT, Nikon, Japan). Segregated binder can be observed as a dark layer because the refractive index was different from those of alumina and immersion liquid. A scanning electron microscope (JSM-5310L, JEOL, Japan) was used to examine the morphology and structure of granules in detail.

Commercial micro-compaction equipment (MCTE-500, Shimadzu, Japan) was used to examine the compaction behavior of individual granules. To equilibrate with the moisture, the granules were placed in an ambient atmosphere for a day. Granules with a diameter around 50 μm were selected under a microscope for examination. Temperatures were controlled at 16° C for the granules containing PAA binder, and at 25° C for the PVA systems. The moisture contents were 50 and 80% at respective temperatures. These conditions are above the glass transition temperatures for the respective binders. The tensile strength of the granule S_t was determined with the equation proposed by Hiramatsu et al.[19]

$$S_t = 2.8P_f / \pi D^2 \qquad (1)$$

where P_f is the applied load at fracture, and D is the average diameter of the granule.

The spray-dried granules were uniaxially pressed in a die with double action by a universal mechanical testing machine (Autograph AG1, Shimadzu, Japan) at a crosshead speed of 1.0mm/min.[21] Temperatures were controlled at 16° C for granules containing PAA binder system, and at 28° C for the PVA systems with moisture contents of 80%. The compaction curve was constructed from the load-displacement curve recorded automatically for each granule. The relative density (R.D.) of the compact was determined during compression tests using the following equation:

$$R.D = 100\, w/(A\, h\, \rho) \qquad (2)$$

where, w is the weight of filled granules, A is the base area of the compact (cross-sectional area of the pressing punch), h is the height of the compact in the die, and ρ is the theoretical density of alumina: 3.987×10^3 kg/m^3.

Preparation of Specimens for Strength Measurement and Examination of Structure

Granules were placed in a die (45mm x 55mm) and pressed at 20MPa under controlled temperature and moisture content, and then cold isostatically pressed at 200MPa. The compact was heated at 1530°C for 2h and cooled to room temperature.

Characterization of Green and Sintered Bodies

The liquid immersion technique was applied to examine the internal structure of the green bodies.[3] The green bodies, which were sliced and ground to 100-μm thick, were immersed in diiodomethane to make them transparent, and the internal structures were observed with an optical microscope in transmission mode.

The structures of sintered bodies were also observed by optical microscopy in transmission mode. The samples were sliced, ground and polished to 100-μm thick for making samples transparent.

Strength Test of the Sintered Body

The strength of the sintered body was measured by 4-point bending. Cross-head speed was 0.5mm/min. Specimens were 3mm thick x 4mm wide and 40mm long.

RESULTS

Figure 5 shows SEM micrographs of granules prepared in this study. All the granules appear to have spherical shapes with dimples. The absolute densities of granules measured by mercury porosimetry were 1.6 kg/m^3 for PAA binder system and 1.9kg/m^3 for PVA binder system, respectively.

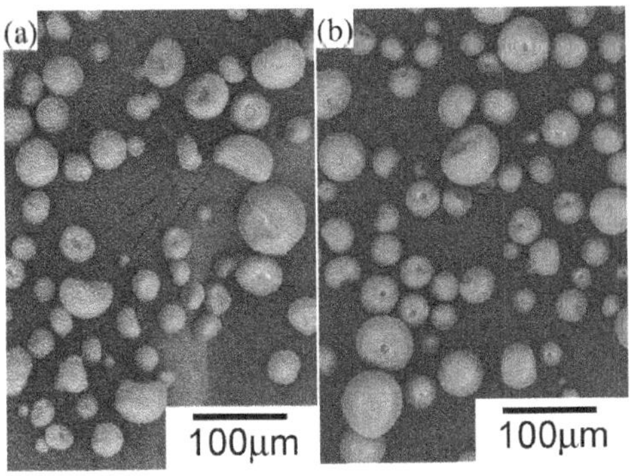

Figure 5. SEM micrographs of granules containing (a) PAA binder system, (b) PVA binder system

Figure 6 shows the liquid immersion photomicrographs of as-spray-dried granules. The granules heat treated at 500°C are shown in Figure 6 (c) and (d). Their internal structure is quite uniform and has a dimple in the center, but not at the rim. The rim of the granules containing the PVA system appears darker than other regions of the granules (Figure 6 (b)). This dark layer at the circumferential region of granules corresponds to the segregated PVA binder. It appears darker since the binder scatters light more than alumina, due to large mismatching of the refractive indices. The dark region disappeared after the granules were heated at high temperature for binder removal. The PAA binder appears to distribute very homogeneously in granules. No structure change was noted before and after the binder removal (Figure 6 (a) and (c)).

Figure 7 shows typical stress-strain curves in the compaction test of a single granule in a humid atmosphere for the PAA system (16°C, 80% Relative Humidity), and the PVA system (25° C, 80%RH). These test temperatures are slightly higher than the glass transition temperatures of the respective binders. With increasing applied stress, the granule deformed gradually until the strain reached approximately 3-5%, and thereafter deformed rapidly. This sudden change corresponds to the fracture of the granules,

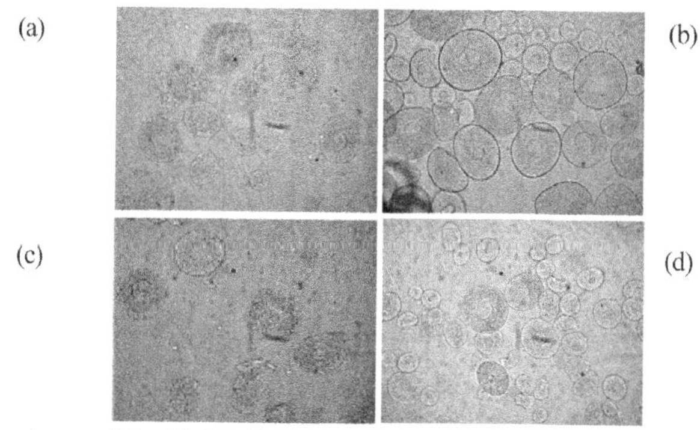

Figure 6. Granules containing binder, (a),(c) PAA system, (b), (d) PVA system (a), (b) before de-binder treatment, (c), (d) after de-binder treatment

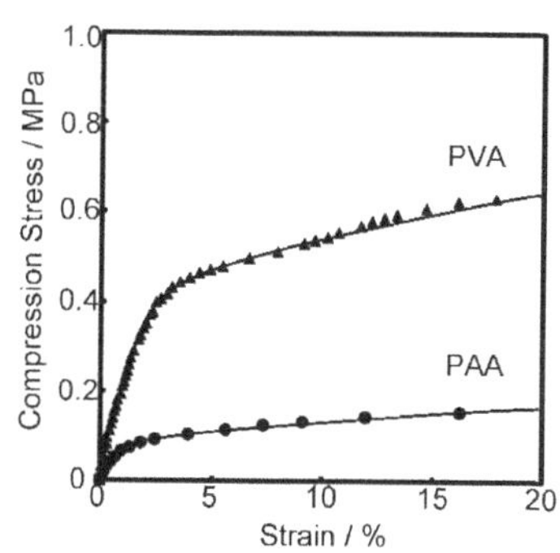

Figure 7. Stress-strain curves of compaction test of single granules

showing the strengths of granules of about 0.1MPa and 0.45MPa for PAA system and PVA systems, respectively.

Figure 8 shows the typical compaction curves for die pressing of these granules. The compaction curves vary with the types of granules. The relative density of the compact increased gradually with increasing stress in both granule systems. Changes of slope appear at two places, about 1MPa and 10 MPa, in the curve for PVA granules. The first change corresponds to the starting point of granule deformation, and the second change is the starting point of particle rearrangement within the granule in the compact body.[20] The curve for PAA granules shows only one change in the slope at 1MPa. Figure 9 shows the structure of the green body examined with the immersion liquid technique in the transmission light mode. The structure depended on the characteristics of the granules. Homogeneous microstructure is noted in the green body made from granules with PAA binder. The shapes of granules are clearly noted in the green body made from granules with PVA binder.

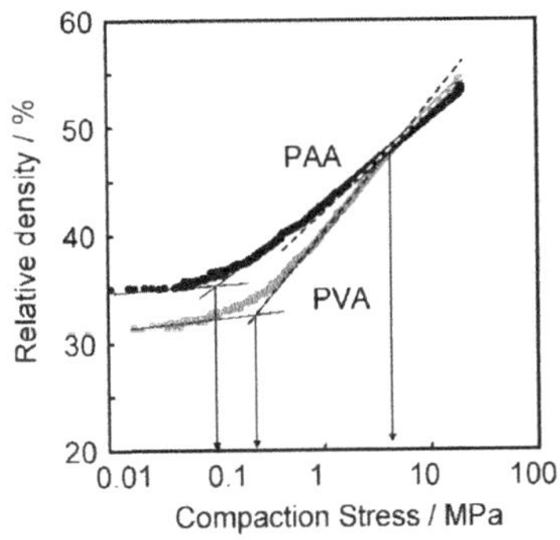

Figure 8. Compaction diagram of die pressing using granules

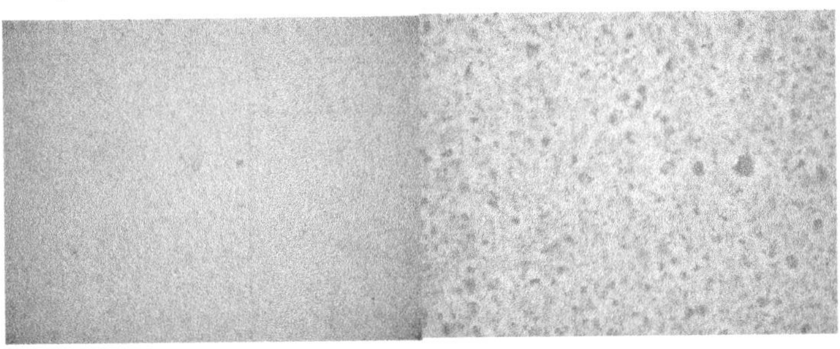

Figure 9. Optical photomicrographs of green bodies.

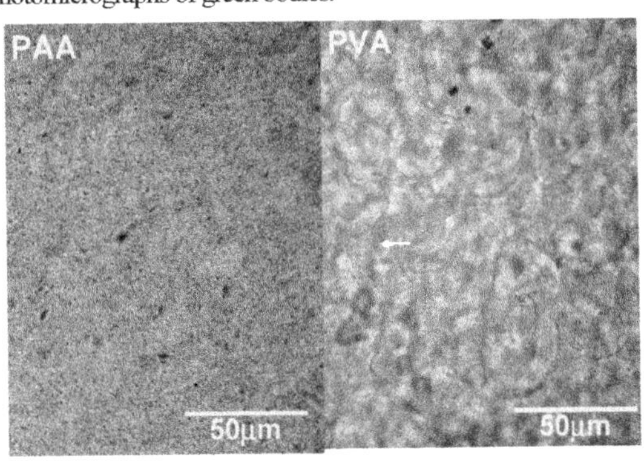

Figure 10. Optical photomicrographs of sintered bodies.

Figure 10 shows the structure of the sintered body observed by optical microscopy in transmission light mode. The large defects were noted as black shadows in the internal structure. The structure was clearly dependent on the characteristics of the granules. Homogeneous microstructure is noted in the sintered body made from the granules with PAA binder. Many large defects are observed in the PVA system.

Figure 11 shows the flexural strengths of these ceramics as Weibull plots. The strengths of ceramics using granules with PAA system were 580 and 540MPa at the relative humidity of 50% and 80% respectively, and were much higher than those of PVA system, 485 and 430 MPa at the relative humidity of 50 and 80% respectively. The Weibull modulus decreased slightly from 21 for the PVA system to 15 for the PAA system at 80% relative humidity. Fracture strength is remarkably improved by changing the binder system. The strength was directly governed by the size of fracture origins (defects). The increased strength is clearly due to the homogenized internal structure in the PAA system.

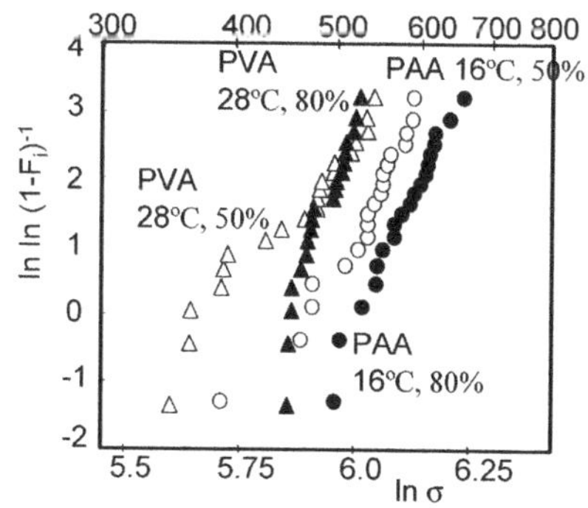

Figure 11. Weibull plots for various ceramics in this study.

DISCUSSION

Segregated binder governs characteristics of the granules and thus the structure and strength of green and sintered bodies as shown in the present results. Structures with no segregated binder are noted in the granules using a PAA binder (Fig.6 (a)). The PAA binder appears to distribute uniformly in the granule, whereas PVA tends to segregate at the surface and the sub-surface of the granules (Fig.6 (b) and (d)). Several granules out of focus appear to have diffuse dark rims due to optical phenomena. The granules with PAA binder deformed easier than PVA granules in the compression tests (Fig.9 and 10). This result is significantly different from what was noted for the granules with the PVA binder. Segregated PVA binder behaves like an eggshell and supports stress.

The differences in the binder distribution and the structure of compacts can be ascribed to the chemical bonding of the binder molecule to the alumina particles. It is well known that the PAA molecules ionize to $R\text{-}COO^-$, H^+ and NH_4^- in the aqueous slurry. Ionized $R\text{-}COO^-$ adsorbs on the positively charged sites on the surface of alumina particles, through the Coulombic attraction force, making the surface negatively charged. The surface charged by $R\text{-}COO^-$ creates a repulsive force to disperse the particles and stabilize the slurry. The adsorbed PAA ions are fixed on particles and do not move during spray drying.

They are uniformly distributed in the granules and behave as a binder after drying. On the other hand, PVA binder ionizes only slightly and adsorbs very weakly on the particles by hydrogen bonding. The PAA molecules added as dispersants in the slurry adsorb preferentially to PVA molecules on the particles, forming free PVA.[21] The free PVA binder can migrate to the drying surface with the evaporation of water during spray drying, causing the surface segregation of PVA binder. Segregated binder is commonly noted on granule surfaces in our past studies, and its origin is ascribed to the flow of water containing PVA in the spray-drying process. The segregated binder hinders the densification under the compaction stress and forms regions of low packing density in the green body. These regions become large defects after sintering and reduce the strength of ceramics.

The binders and their distribution affect the strength and deformation behavior of the granules (Fig. 7 and 8). The granules with PAA binder system showed the onset of density increase at much lower pressure than those with PVA in die pressing. The second change of the slope is noted in the compaction curve for the PVA system, but not for the PAA system. In the PAA system, the rearrangements of the particles within the granules take place simultaneously with the net deformation of the granules during incremental increases of stress. In both cases the relative humidity affects the binder characteristics.[22] Although the compacts were formed above the glass transition temperatures, the deformation of granules appeared to increase with increasing humidity for both cases.

CONCLUSIONS

The influence of binder on structure and properties was examined for compaction processing of alumina ceramics. Alumina granules were prepared with a poly-acryl-acid (PAA) or a poly-vinyl-alcohol (PVA) as binder by a spray-drying method. The following conclusions were obtained:

(1) Segregation of binder can be eliminated by using PAA binder. A strong chemical bonding is expected between the alumina surface and a binder containing a functional group such as poly-acrylic type. It bonds to the positively charged surface in the neutral to weakly alkali regions of the slurry. This is in clear contrast to PVA, which adsorbs only weakly on the alumina surface and often causes non-uniformity in the structure.

(2) The compression strength and deformation behavior were examined on single granules with a micro-compression testing machine. The granule with the PAA binder has a low yield stress. In the forming process, the relative density of the compact body started to increase at low pressure. Homogeneous internal structures were noted in the green compact and sintered body. The number of macro-defects was found to decrease significantly in direct observation.

(3) The strength of the sintered body was improved by using the PAA binder system. Reduction of macro-defects in the sintered body contributed to the increased strength. The measured strength follows the theory of fracture mechanics very well.

ACKNOWLEDGEMENTS

This study was partially supported by a grant from Matsuda foundation. We are grateful to Dr. Kakui and Mr. Ishiguro at Lion Corporation for their courtesy and assistance in preparing the compacts.

REFERENCES

[1]Y. Zhang, M. Inoue, N. Uchida and K. Uematsu, "Characterization of Processing Pores and Their Relevance to the Strength in Alumina Ceramics," *J. Mater. Res.*, 14[8]3370-74(1999).

[2]S. Tanaka, submitted for publication in *Trans. Am. Ceram. Soc.*

[3]Keizo Uematsu, "Immersion Microscopy for Detailed Characterization of Defects in Ceramic Powders and Green Bodies," *Powder Technology*, 88, 291-8(1996).

[4]K. Sato, H. Abe, M. Naito, Y. Saito, K. Uematsu, H. Hotta, "Immersion Microscopy for Ceramics Green Compact," *Am. Ceram. Soc. Bull.*, 85[1], 9301-9310 (2006).

[5]Y. Saito, S. Tanaka, N. Uchida and K. Uematsu, "CSLM for Ceramic Green Microstructure," *Am. Ceram. Soc. Bull.*, 81[1]35-38(2002).

[6]K. Uematsu, and M. Saito, "Liquid Immersion Technique Coupled with IR-Microscopy for Direct Observation of Internal Structure of Ceramic Powder Compact - With Alumina as an Example," *J. Mater. Res.*, 14[12], 4463-65(1999).

[7]K. Uematsu, N. Uchida, Z. Kato, S. Tanaka, T. Hotta and M. Naito, "IR Microscopy for Examination of Particle Packing Structure in Silicon Nitride Powder Granules and a Compact," J. *Am. Ceram. Soc.*, 84[1], 254-56(2001).

[8]S. J. Lukasiewicz, "Spray-Drying Ceramic Powders," J. *Am. Ceram. Soc.*, 72, 617-24 (1989).

[9]Z. Kato, N. Uchida, K. Uematsu and Y. Zhang, "Quantitative Analysis of a Binder Segregation in Granules," *Ceramic Transactions*, 112, 501-6 (2001).

[10]N. Shinohara, M. Okumiya, T. Hotta, K. Nakahira, M. Naito and K. Uematsu, "Formation Mechanism of Processing Defects and their Relevance to the Strength in Alumina Ceramics Made by Powder Compaction Process", *J. Mater. Sci* .34, 4271-77(1999).

[11]Xia Tang, Nozomu Uchida and Keizo Uematsu ,"Binder Surface Segregation During Spray Drying of Ceramic Slurry," J. Mater. Res., 13[7],1881-7(1998).

[12]Yao Zhang, Xia Wu X. K., Whitman D.W., Kaufell W.L., Finch W.C. and Cumbers D. I., "Acrylic Binders for Dry-Pressing Ceramics," *Am. Ceram. Soc. Bull.*, 76[1] 1997, pp.49-52.

[13]J. S. Reed, '*Principles of Ceramics Processing, 2nd ed.*' ch 20, Wiley, New York, 1995.

[14]J. Cerarano and I. A. Aksay, "Processing of Highly Concentrated Aqueous a-Alumina Suspensions Stabilized with Polyelectrolytes," *J. Am. Ceram. Soc.*, 71 ,1062-67 (1988)

[15]L. C. Guo, Y. Zhang, N. Uchida, K. Uematsu, "Adsorption Effects on the Rheological Properties of Aqueous Alumina Suspensions with Polyelectrolyte," *J. Am. Ceram. Soc.*, 81, 549-56 (1998).

[16]K.Yamashita, W. Kabutoya, Y. Zhang, N. Uchida and K. Uematsu, "Thermochemistry of PAA Adsorption on Alumina Surface," *Ceramic Transaction*, 83, 147-54(1998).

[17]W.A.Ducker, Z.Xu, D.R.Clarke and J.N.Israelachivili, "Forces Between Alumina Surface in Salt Solutions: Non-DLVO Forces and the Implications for Colloidal Processing," *J. Am. Ceram. Soc.*, 77, 437-43 (1994).

[18]Kakui T, Additives for Ceramic Green Sheet, *Japan Patent* 07-033533, 28 June 1993.

[19]Hiramatsu Y., Oka Y. and Kiyama H., "Determination of the Tensile Strength of Rock by a Compression Test of an Irregular Test Piece," *J. Min. Eng. Jpn.*, 81, 1024-30, (1965).

[20]A.R.Cooper Jr. and L.E.Eaton, "Compaction Behavior of Several Ceramic Powders," *J. Am. Ceram. Soc.*, 45, 97-100 (1962).

[21]T.Nakashima, H. Mizutani, Y. Zhang, N. Uchida and K. Uematsu, "Interaction Between Binder and Dispersant in Aqueous Slurry of Alumina," *Ceramic Transactions*, 83, 139-46(1998).

[22]Nies C.W. and Messing G.L., "Effect of Glass-Transition Temperature of Polyethylene Glycol-Plasticized Polyvinyl Alcohol on Granule Compaction," *J. Am. Ceram. Soc.*, 67[4] 301-4, (1984).

INFLUENCE OF CONTACT DAMAGE ON STRENGTH AND FRACTOGRAPHY OF BALL ON THREE BALLS TEST SPECIMENS

Robert Danzer[*,§,+], Walter Harrer*, Tanja Lube*, Peter Supancic[*,§]
[*]Institut für Struktur- und Funktionskeramik, Montanuniversität Leoben
A-8700 Leoben, Austria
[§]Materials Center Leoben
A-8700 Leoben, Austria

ABSTRACT

The ball-on-three balls (B3B) test is a new procedure for biaxial strength testing of brittle materials. A disc-shaped specimen is supported by three balls on one side and loaded by a fourth ball on the other side. This mechanically well-defined loading situation enables a quick biaxial strength test even for specimens that are not perfectly flat. Therefore, testing of as-sintered ceramic specimens is possible. In recent investigations of the authors the test has been performed on several ceramic materials, with strengths ranging from low (around 100 MPa) to high (more than 1000 MPa) values. The applicability of the test could be proven. As predicted by Weibull theory, the materials showed a pronounced size effect on strength. If this effect is taken into account, the B3B test results fit well to the ones determined by conventional bending tests.

The load transfer via the balls into the specimens occurs in relatively small areas, which can be determined by an elastic Hertz analysis. It is well known that around the Hertzian contact zones tensile stresses exist, which may cause the occurrence of Hertzian ring cracks. In this paper the question is addressed if such cracks can be come into existence during B3B testing and if contact cracks - if they exist - have an influence on the determined strength values.

INTRODUCTION

Many ceramic components are biaxially loaded. Therefore, biaxial strength testing seems to be appropriate and useful. A good overview of several biaxial tests has been done in the past, as can be found in the papers of Morrell[1] and Godfrey and John.[2] The main disadvantages which were claimed for most of those tests are related to uncertainties in the determination of the stress fields and to the necessity to machine very flat specimens having parallel top and bottom surfaces. To overcome these problems, a new biaxial strength test has been developed in the last years by some of the authors.[3,4]

The ball-on-three balls (B3B) test is an easy and low-cost procedure to perform biaxial strength measurements on brittle materials. In this test, the disc-shaped specimen is supported on one side by three balls and centrically loaded by a fourth ball on its opposite side. This well-defined mechanics situation makes the testing of slightly buckled (as-fired) specimens possible. For linearly elastic materials, the stress field in the specimens has been numerically analysed in a wide range of parameters (geometry of specimen and jig, elastic properties of specimen and jig materials) in a recent paper of Börger et al.[3] A detailed analysis of possible measuring uncertainties has been also performed.[4] The most sensitive parameter for the determination of the highest tensile stress in the specimen ($\sigma_{\max}$) is the thickness (t) of the specimen, which enters with $\sigma_{\max} \propto t^{-\eta}$ in the analysis. η is a number between 2 and 3, which depends on the shape of the specimen. If the uncertainty in the determination of the thickness is 1 %, the uncertainty in the determination of the maximum tensile stress is a little less than ½ %. Other properties, e.g., buckling of the disc, the exact position of the balls, or some friction are much less pronounced,

and thus the uncertainties associated can – in general – be neglected. It could also be shown that the uncertainties in the determination of the maximum tensile stress are much smaller than in uniaxial bending tests. This makes the B3B test very suitable for testing as-fired specimens as well as very small specimens.[4]
Since ceramic components often have the shape of discs or plates, the B3B procedure can be employed to test the strength of components without any further sample preparation, using the component itself as the test specimen.

B3B test results have been published for alumina[5,6] zirconia,[6] silicon nitride[5,6] and zinc oxide[5] ceramics. Specimens of quite different sizes have been tested, with diameters ranging from 2.5 mm to 43 mm and thickness from 0.1 mm to 5.3 mm, yielding a volume range from 0.5 mm^3 to 7.700 mm^3. The strength data showed a clear size effect on strength, which is expected in the framework of the classical Weibull theory.[8-11] For an analysis of this effect the effective volume (if fracture origins are volume flaws) or the effective surface (if fracture origins are surface defects) has to be determined.[10,11] Due to the complex stress fields, the determination of the effective volume (surface) can only be assessed numerically. A paper is in preparation, which gives parameterised solutions for a wide range of specimen and loading geometries[7] Comparisons of B3B tests with uniaxial bending tests have been performed for several materials.[5,6] The B3B test results correspond well to the bending test results, provided that the size effect on strength is taken into account.[5,6]

In general, especially in the case of specimens with a small thickness-to-diameter ratio, the effective volume (surface) is only a small fraction of the specimen's volume. Therefore, the effective volume of "small" specimens may be very small, e.g. a fraction of a cubic millimetre. Measurements on such small specimens are published in references 6 and 7. In that case the effective volume may be too small to contain a large flaw, which may act as a fracture origin. Flaw populations different from those in large specimens may become important.[6,7,12] In recent papers of some of the authors a plateau value of strength has been predicted[13] and determined[6,7,12] for the case of very small specimens.

In this paper strength tests on as-fired discs made of alumina and silicon nitride are performed. The fracture surfaces are investigated, and fracture origins are analysed. Special care is given to answer the question, if the load transfer via the balls into the specimen may cause damage in the specimens and if such a potential damage may have an influence on the strength of the specimen.

EXPERIMENTAL PROCEDURE

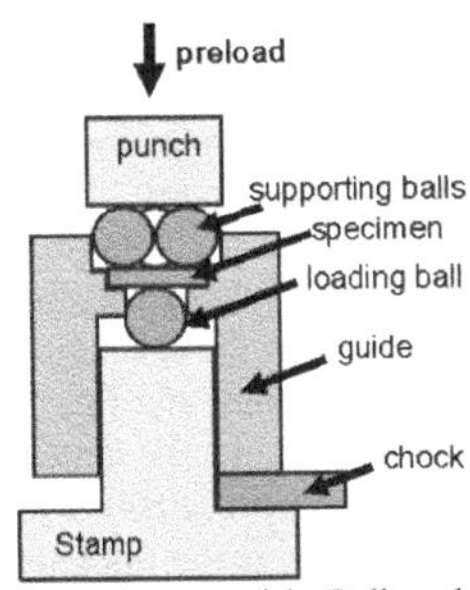

Fig. 1: Experimental set-up of the Ball on three Balls Test

The experimental set-up of the ball-on-three balls test is shown in fig. 1.[14] The specimen is centrally positioned over the loading ball which is placed on the stamp. Specimen and loading ball are positioned by the guide. The specimen is held by three supporting balls, which touch each other. Hence, all four balls and the specimen are exactly positioned by the guide, which has a rotational symmetry and can be machined very accurately with reasonable effort..[14] A preload, which is typically about 10 % of the estimated fracture load, is applied. Then, the chock piece is removed and the positioning aid moves down. The loading ball can freely move and the supporting balls can move outwards. Specimen and balls are only fixed by the friction between balls and specimen. Finally, the load is increased until fracture. The fracture load, $F_{\max}$, is used to determine the maximum tensile stress in the specimen, i. e., the strength of the specimen.

DETERMINATION OF STRENGTH

The center of the top plane of the specimen opposite to the loading ball is the maximum loaded position in the disc, wherein a biaxial tensile stress state occurs. The first principle stress component at this position is the maximum tensile stress in the specimen, called σ_{max}. The stress field has a cloverleaf shape and depends on the applied force, the geometric set-up of the test (dimensions of the disc, especially thickness, and the size and the position of the balls) and on the elastic properties of disc and ball materials.[3] The stress $\sigma_{\max}$ has been analysed by a FE-analysis over a wide range of parameters.[3] It is given by

$$\sigma_{\max} = f(\alpha, \beta, \nu) \cdot \frac{F_{\max}}{t^2} \tag{1}$$

where t is the thickness of the specimen. Due to the biaxial stress state, the stress also depends on the Poisson's ratio ν. $\alpha = t/R$ is the ratio of thickness to the disc radius R, and $\beta = R_a / R$ is the ratio of the support radius R_a to the disc radius. f is a function, which is given in [3]. It is between 1 and 3 for $0.05 \le \alpha \le 0.6$, $0.60 \le \beta \le 95$ and $0.20 \le \nu \le 0.35$. It is interesting to note that the maximum tensile stress strongly depends on the thickness of the specimen. All other geometrical parameters have much less significance. For instance, for a high-strength material (e.g., a silicon nitride of around 1,000 MPa) and for a specimen with a thickness of 2 mm, the fracture load is around 4,000 N or less.

It is also observed that far from the center line of the disc the stresses quickly decrease, with the slope of the stress field strongly dependent on the thickness ratio α.

In addition to the test stress field, significant tensile stresses (Hertzian stresses) also arise around the contact area between the balls and the specimen. The magnitude of such stresses can be estimated by the well-known analysis of Hertz, who investigated the elastic contact between a ball and a plate,[15,16] showing that radial tensile stresses arise in the plate. As a rule of thumb, (for materials having approximately the same Young's modulus and a Poisson's ratio of about $\nu \approx 1/3$) they may be given by[16]:

$$\sigma_{H,t} \approx \frac{F}{6\pi a^2} \tag{2}$$

with the contact area being

$$\pi a^2 \approx \frac{\pi}{2}\left(\frac{FR_b}{E}\right)^{2/3} \tag{3}$$

where F is the applied force and a is the contact radius. The radius of the ball is R_b, and Young's modulus of the tested material is E. The Hertzian stresses can get so high that they generate Hertzian ring cracks. An example is shown in fig. 2.

In the case of the B3B test, this situation is promoted by testing small specimens (this causes a small R_b) of a stiff (high E), high-strength (high fracture load) material. To give an example: a 10 mm diameter disc of a material with a strength of around 1,000 MPa (fracture load around 4,000 N), having a Young's modulus of $2{\cdot}10^{11}$ MPa, is tested using the B3B procedure with a radius of the contact area between loading ball and specimen around 0.4 mm. Hence, the Hertzian tensile stress reaches around 1,200 MPa. If the disc is much smaller, the loading balls as well as the thickness of the disc and the fracture load are smaller too. Then, for F = 500 N, $R_b = 2$ mm and $E = 2{\cdot}10^{11}$ MPa, the contact radius is around 0.125 mm, and the tensile stress increases up to 1,800 MPa. However, although these stresses are very high, it is not expected that they cause Hertzian ring cracks since, as we discussed above, the volume in tension around the contact area is very small.

It must be pointed out that if the contact occurs near the free edges (edge effect) even larger stresses may occur. This may cause breaking out of the edge [17, 18], which is likely to occur when the parameter β is close to one: $\beta > 0.95$. Of course, experiments where fracture starts near a supporting ball are not valid.

INVESTIGATED MATERIALS AND PERFORMED EXPERIMENTS

Tests were made on an alumina and a silicon nitride. For both materials, disc-shaped specimens were delivered by an industrial manufacturer.

The investigated alumina is a typical commercial 99.7 % pure Al_2O_3 produced by Friatec GmbH, Germany (AL23). A more detailed description can be found in [5, 6, 12]. The material was delivered in the form of discs (radius 10 mm, thickness 3 mm). The discs were used as-delivered as B3B specimens, i.e., without any further machining.

The silicon nitride (FSNI) is a commercial gas-pressure-sintered (1800°C, N_2-pressure 10 bar) silicon nitride with Al_2O_3 additions. The material was delivered by FCT Ingenieurkeramik, Germany in the form of discs (radius 10 mm, thickness 5 mm), which were machined out of larger plates by laser drilling. For a better comparison with the alumina, the discs were ground on one side to a thickness of 3 mm. The tensile side of the specimens was always the as-sintered (i.e., not machined) surface. Some mechanical properties for both alumina and silicon nitride specimens are listed in table 1.

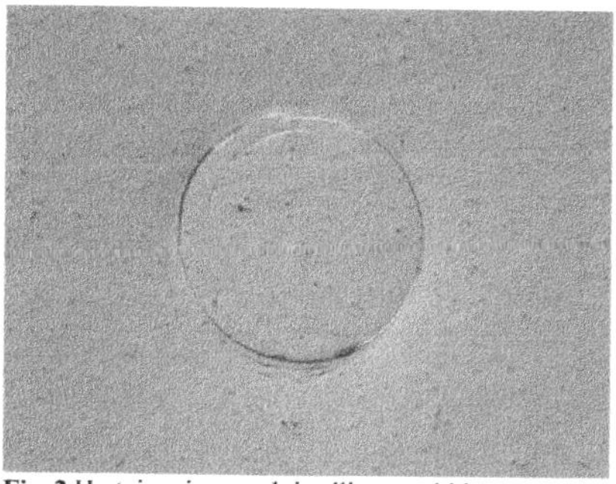

Fig. 2 Hertzian ring crack in silicon carbide.

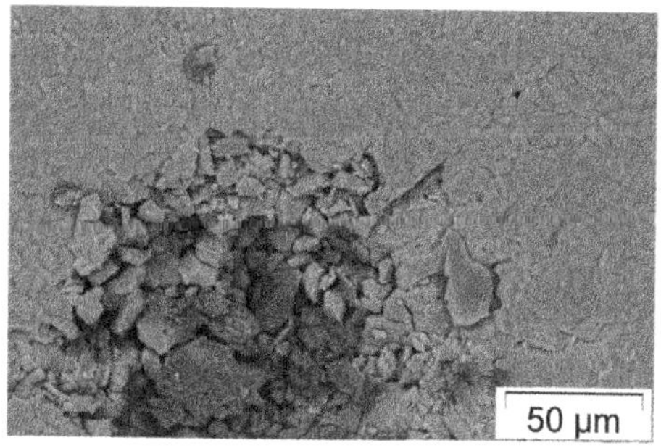

Fig. 3: Typical indentation in the alumina ceramic.

Thirty B3B tests were performed on specimens of each type of material. Then the fracture surfaces were analysed in light and scanning electron microscopes for the identification of critical flaws. In the centre of some specimens, Rockwell C indents were made (force: 200 N) to produce damage and cracks. The depths of the cracks were approximately 50 µm or 2 % of the thickness of the disc. The compressive stresses in the contact area were in any case much larger than in the contact area between ball and specimen in the B3B tests. A typical view of a damaged surface in an alumina ceramic can be seen in fig. 3.

For each material, 10 pre-damaged specimens were B3B strength tested with the damage on the compressive side of the specimen (where the contact during loading in the B3B test occurs) and 3 specimens were tested with the damage at the tensile side.

Table 1: Mechanical properties of investigated materials

	Alumina (AL23)	***Silicon Nitride (FSNI)***
Young´s modulus [GPa]	380[a]	320[a]
Poisson's ratio [-]	0.24[a]	0.28[a]
Characteristic strength [MPa]	>400[a,c]	658[a,c]
Fracture toughness [MPa·m$^{1/2}$]	3.2 ± 0.1[b,d]	5.1 ± 0.1[b,d]

[a)] Data from the supplier; [b)] own measurements, [c)] 4-point bending tests with a 40/20 mm span length according to EN 843-1, [d)] measured with single edge V-notch beam (SEVNB) method [19, 20]

STRENGTH TEST RESULTS

The results of the strength tests are summarised in table 2. Listed are the characteristic strength and the Weibull modulus of the tested samples.

Table 2: Summary of the strength test results.
Investigated are as-sintered specimens (A) and as-sintered specimens with a pre-damage caused by a Rockwell C indent (200 N) made in centre of the compressive-loaded top plane (B) or in the center of the tensile-loaded top plane (C).

	AL 23		FSNI	
Sample number of specimens [-]	characteristic strength [MPa]	Weibull modulus [-]	characteristic strength [MPa]	Weibull modulus [-]
(A) as-sintered 30	357 349-365	15 11-18	688 669-706	12 09-15
(B) with pre-damage on compressive side 10	374 357-391	14 8-19	674 636-715	11 06-15
(C) with pre- damage on tensile side 3	152 145-154	58 -	267 262-269	91 -
Combined evaluation of samples A and B 40	361 354-369	14 11-17	684 668-701	12 9-14

Numbers in the second line refer to the 90 % confidence interval due to the sampling procedure. In the case of the samples C, only 3 experiments were performed, which show very little scatter. In that case the definition of a confidence interval seems not to be appropriate, and the interval given for the strength data is the range of the data (note that the characteristic strength is a little larger as the mean strength).

The strength data follow a Weibull distribution. For both types of materials the characteristic strength as well as the Weibull modulus of the samples A (as-sintered) and B (as sintered and pre-damaged at the compressive side) cannot be distinguished within a 90 % confidence limit, i.e., the samples describe the same population (fig. 4). Therefore, a combined evaluation of both samples seems to be possible. This is shown in Figs. 5a and 5b, where the strength test results of the alumina (a) and the silicon nitride (b) are plotted in a Weibull distribution. It can clearly be recognised that data of sample A and B fit well to the combined Weibull line.

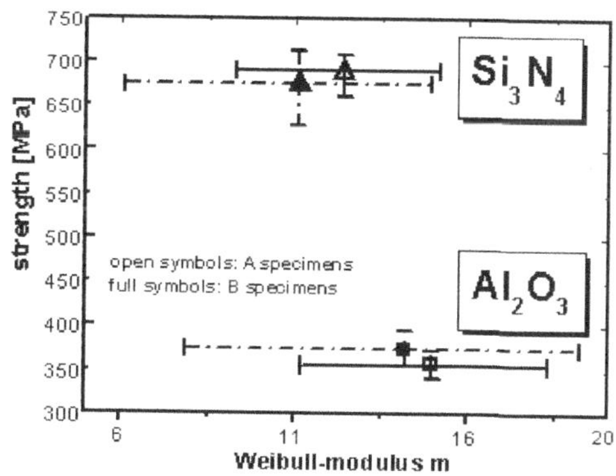

Fig. 4: Strength vs. Weibull modulus. The scatter bars show the 90 % confidence limits.

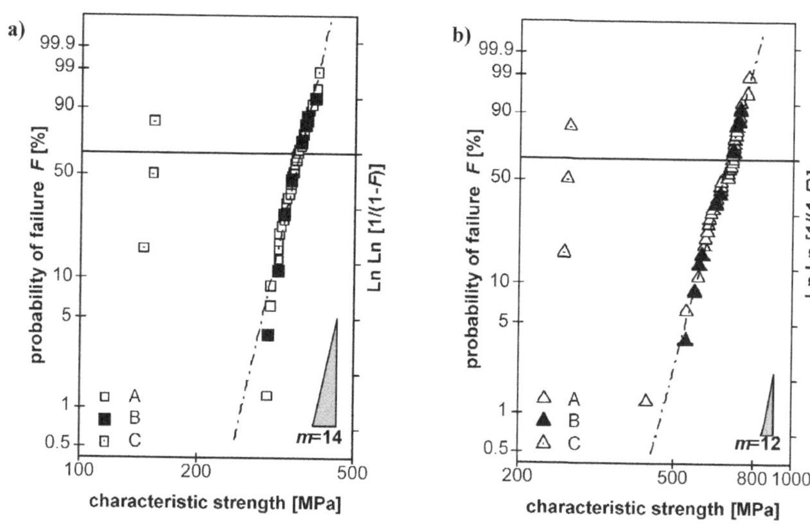

Figs. 5a and 5b: Strength Measurements on AL 23 (Fig. 5a) and FSNI (Fig. 5b). A specimens are not pre-damaged, B specimens were pre-damaged on the compression side, C specimens were pre-damaged on the tensile side.

The test results of the as-sintered specimens with a pre-damage at the tensile loaded side (C) show a strong decrease in strength compared to the results of undamaged specimens. The differences between the test results of the individual samples are very small (± 3% in the case of alumina and ± 2% in the case of the silicon nitride), which results in an extremely high value for the Weibull modulus (in other words, the Weibull distribution is no longer suitable to describe these data). The scatter of data is only slightly higher than the expected measurement uncertainties that result from the bending test [21]. In the case of these experiments the strength was limited by the cracks, which had been created by the indents before strength testing.

In summary it can be stated that the Rockwell indents cause some damage and cracks. If these cracks are positioned on the side of the specimen which is loaded in tension in the B3B test, a significant reduction of strength can be observed. But if the indents are made on the side which is loaded in compression, a reduction of strength cannot be observed. In other words, a crack in or near the contact area of loading ball and specimen is harmless in the B3B testing, in the experiments discussed here.

FRACTOGRAPHY

The figs. 6a and 6b show fractured alumina and silicon nitride specimens. The number of fracture pieces increases with the applied force (with the stored elastic energy). For each of the two pictures (fig. 6a and fig. 6b) the left specimen was damaged on the tensile side of the surface (this mode of loading causes a low stored elastic energy), the specimen in the middle was damaged on the compression side (in that case no influence of the pre-damage on the strength is expected) and the right specimen was not pre-damaged.

a) b)

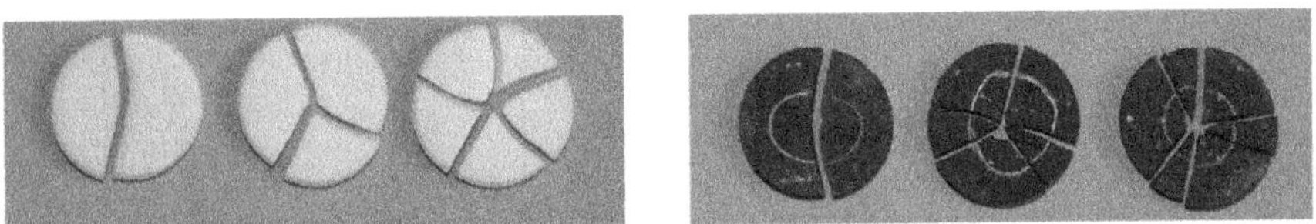

Fig. 6: Fractured B3B test specimens (from left: damaged on tensile side, damaged on compression side, not damaged) a) discs of alumina; b) discs of silicon nitride. The circles indicate the area of the fracture origin.

For the case of alumina (this material has low strength, this means low stored elastic energy) no significant differences between the undamaged (sample A) and the damaged specimens (samples B and C) could be found in their macroscopic fracture behaviour. The specimens broke into two to five pieces. The silicon nitride has a higher strength than the alumina. In this case, the specimens which were pre-damaged on the tensile side broke in two or three pieces, the other specimens into five to eight pieces.

Alumina

In the case of specimens damaged on the tensile side (sample C), the Rockwell indent could be identified as the fracture origin (Fig. 7)

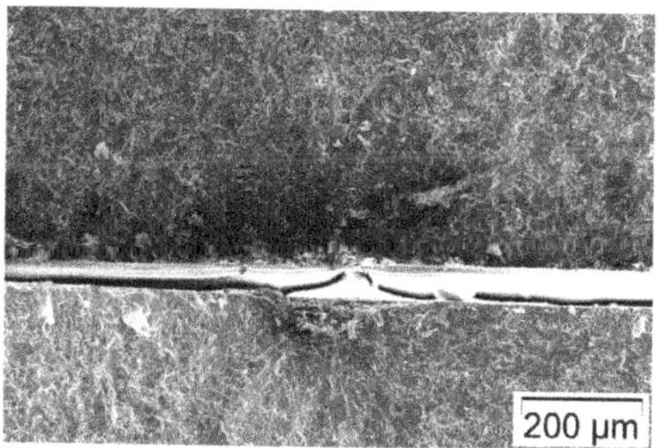

Fig. 7: Specimen with a Rockwell indent (fracture origin) on tensile side.

In all investigated cases, the fracture origins of alumina specimens of sample B and C seem to be at or very near to the loaded tensile top plane of the piece. Only in one case could a pore near the surface be identified as the fracture origin (fig. 9). In the other cases, no microstructural defects could be detected.

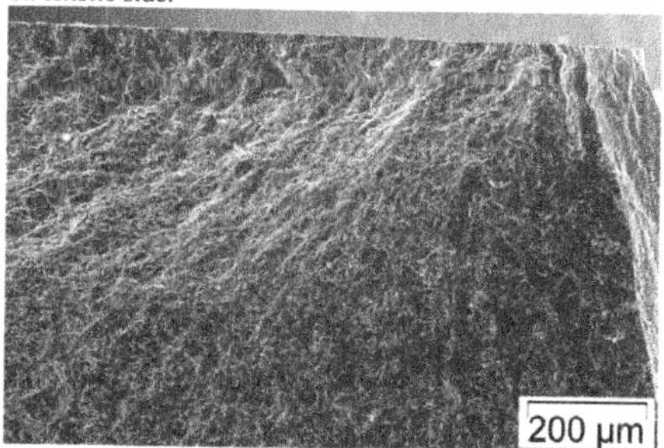

Fig. 8a: SEM photo of a fractured alumina specimen that was not pre-damaged. In this case the fracture origin is a pore near the surface

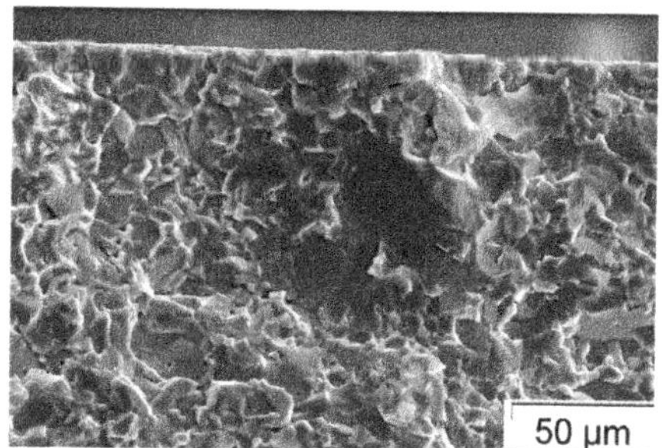

Fig. 8b: Pore-type flaw detected near the surface.

Silicon nitride

In this material glassy regions or clusters of pores near the surface were identified as the fracture origins in the case of samples A and B. Examples are shown in figs. 9 and 10. In the case of sample C the fracture always starts from cracks generated by the Rockwell indent (without illustration).

Fig. 9a: Fracture surface in a silicon nitride specimen.

Fig. 9b: A greater magnification shows that the origin is a glassy region near the surface.

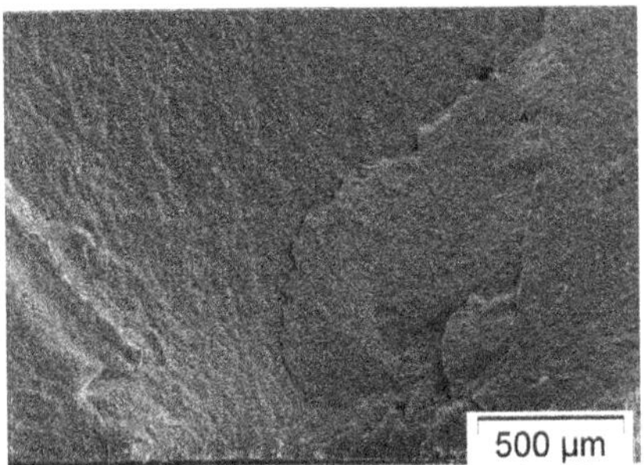

Fig. 10a: Another type of fracture origin in the silicon nitride at a low magnification. The fracture origin is a porous region near the top plane of the B3B specimen.

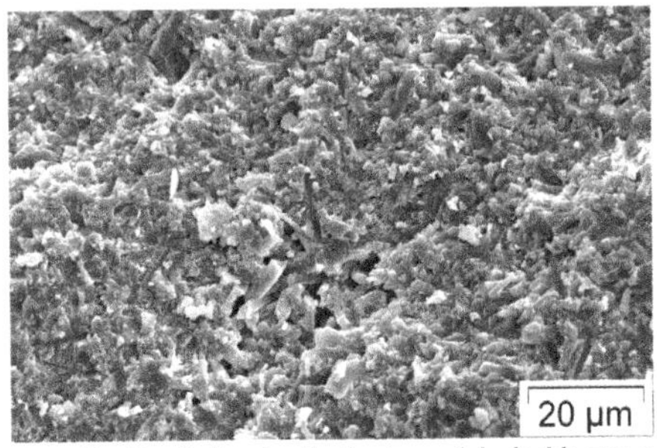

Fig. 10b: Detail showing the fracture origin, in this case a cluster of pores beneath the surface of the specimen.

CONCLUSIONS

The strength of the B3B test specimens with as-sintered surfaces (samples A) was equal to the strength of specimens with as-sintered surfaces and contact damage in the compressive-loaded top plane (samples B). This clearly shows that some contact damage on the compressive surface, which cannot be completely excluded for very small B3B test specimens, has no significance on the strength test results. In contrast, specimens with damage on the tensile-loaded planes (samples C) have a significant loss of strength. The fracture surfaces of specimens of sample B look similar to fracture surfaces of specimens of sample A and the fracture origins look quite similar in both cases. This is a hint again that the contact damage on the compressive surface specimens of sample B has no significance on the fracture in the B3B test. In the case of silicon nitride, volume flaws (glassy regions or clusters of pores) were found to be the fracture origins. It is known from previous investigations on similar materials that a volume effect on strength exists.[5,6] This is consistent with our present observation. In the case of the alumina, only in one test (of 40) could a volume flaw be identified as the fracture origin. In the other cases, no inhomogeneities in the microstructure were found at the position of fracture initiation. Following these ideas, it is assumed that surface defects are responsible for fracture. This agrees with other investigations on the same material, where a pronounced surface effect on strength has been observed.[5,6]

An estimate of the tensile Hertzian stresses around the contact area between loading ball and specimen demonstrates that the stresses increase with increasing fracture load, strength and Young's modulus of the investigated material, and with decreasing size of the loading ball. For high-strength materials and small specimen sizes (diameter of several mm or smaller) tensile stresses of several GPa are likely to occur. In such cases, the development of contact damage becomes possible, but even in this case – which has not been observed in experiments – an influence of this damage on the strength test results in the B3B test is not expected to arise.

REFERENCES

[1]Morrell, R., N.J. McCormick, J. Bevan, M. Lodeiro and J. Margetson, "Biaxial Disc Flexure- Modulus and Strength Testing", Brit. Ceram. Trans., 98, 234-240 (1999).

[2]Godfrey, D.J., S. John, Disc Flexure Tests for the Evaluation of Ceramic Strength, in Proceedings 2nd International Conference of Ceramic Materials and Components for Engines, Verlag Deutsche Keramische Gesellschaft, Lübeck-Travemünde, 657-665 (1986).

[3]Börger, A., P. Supancic and R. Danzer, "The Ball-on-Three-Balls Test for Strength Testing of Brittle Discs – Stress Distribution in the Disc, J. Eur. Ceram. Soc., 22, 1425 – 1436 (2002).

[4]Börger, A., P. Supancic and R. Danzer, "The Ball-on-Three-Balls Test for Strength Testing of Brittle Discs - Part II: Analysis of Possible Errors in the Strength Determination", J. Eur. Ceram. Soc., 24, 2917 – 2928 (2004).

[5]Danzer, R., W. Harrer, P. Supancic, T. Lube, Z. Wang, A. Börger, "The Ball-on-Three-Balls Test-Strength and Failure Analysis of Different Materials, J. Eur. Ceram. Soc., 27, 1481-1485 (2007).

[6]Danzer, R., P. Supancic, W. Harrer, Z. Wang, T. Lube, "Biaxial Strength Testing of Mini Specimens", 16th European Conference of Fracture, July 3-7, 2006, Alexandroupolis, Greece.

[7]Supancic, P., R. Danzer, A. Börger, "The Ball-on-Three-Balls Test for Strength Testing of Brittle Discs – Part III: Determination of the Effective Volume and the Effective Surface", in preparation.

[8]Weibull, W., "A Statistical Theory of Strength Materials". Stockholm, Royal Swedish Institute for Engineering Research (1939).

[9]Weibull, W., "A Statistical Distribution Function of Wide Applicability". J. Appl.Mech., 18, 293-297 (1951).

[10]Munz, D, T. Fett, "Ceramics: Mechanical Properties, Failure Behaviour, Materials Selection," Springer Verlag, Berlin, 1999.

[11]Wachtman, J.B., "Mechanical Properties of Ceramics", John Wiley & Sons, New York, (1996).

[12]A. Börger, R. Danzer, P. Supancic: "Biaxial Strength Test of Discs of Different Size Using the Ball-on-Three-Balls Test," Ceramic Engineering and Science Proceedings 25 [4], 283-289 (2004).

[13]R. Danzer, "Some Notes on the Correlation between Fracture and Defect Statistics: Is the Weibull Statistics Valid for Very Small Specimens?" J. Eur. Cer. Soc., accepted 2005.

[14]Börger A., M. Damani, R. Danzer, T. Lube, P. Supancic, R. Binder: "Verfahren und Einrichtung zur Bestimmung der Bruchfestigkeit von spröden Werkstoffen" Patent No. AT 411 714 B, Österreich, (2004).

[15]Hertz, H., "Über die Berührung fester elastischer Körper." J. reine angewandte Mathematik, 92: p. 165-171 (1882).

[16]Ashby M. F., "Materials Selection in Mechanical Design". Pergamon Press, Oxford, New York, (1992).

[17]R. Danzer, M. Hangl, R. Paar: "Edge Chipping of Brittle Materials", in: James R. Varner, George D. Quinn, Van Derck Fréchette (Eds.), Fractography of Glasses and Ceramics IV, The American Ceramic Society, Westerville, 43-55 (2001).

[18]McCormick N.J., E. A. Almond., "Edge Flaking of Brittle Materials", Journal of Hard Materials. Vol. 1, p. 25-51 (1990).

[19]Damani, R., R. Gstrein, R. Danzer, "Critical Notch Root Radius in SENB-S Fracture Toughness Testing", J. Eur. Ceram. Soc., 16, 695 – 702 (1996).

[20]Damani, R., Ch. Schuster, R. Danzer, "Polished Notch Modification of SENB-S Fracture Toughness Testing", J. Eur. Ceram. Soc., 17, 1685 – 1689 (1997).

[21]Lube T., M. Manner, "Development of a Bending-Test Device for Small Samples", in: Key Engineering Materials 132-136 Part 1, 488-491(1997).

FRACTURE AND FRACTOGRAPHY OF A Si_3N_4-SiC MICRO/NANO COMPOSITE

Monika Kašiarová and Ján Dusza
Institute of Materials Research
Slovak Academy of Sciences
Department of Structural Ceramics
043 53 Košice, Slovak Republic

ABSTRACT

The room-temperature fracture characteristics of a carbon-derived in-situ-reinforced Si_3N_4-SiC nanocomposite have been investigated. Four-point bending strength was measured on specimens with different effective volumes. Weibull statistical analysis has been used for characterization of the strength values in the form of characteristic strength and Weibull modulus. The characteristic strength of the investigated micro/nanocomposite was $\sigma_0 = 675$ MPa and $\sigma_0 = 832$ MPa with the Weibull modulus of 6.4 and 8.6 for effective volume of 18.4 mm^3 and 1.55 mm^3, respectively. Fracture origins of the specimens with effective volume of 18.4 mm^3 tested in four-point bend were mainly volume-located processing flaws with sizes from 5 μm to 460 μm, mainly in the form of clusters of pores and large SiC grains. The fracture mirror sizes, R_0, were measured, and the mirror constant, A_0, was calculated. Good agreement was found between the fracture toughness values measured using single-edge V-notched beam (SEVNB) and estimated by fracture mechanics. Fracture origins in the specimens with effective volume of 1.55 mm^3 tested in four-point bend were mainly sub-surface defects of the same type as in the specimens with a higher effective volume.

INTRODUCTION

Different approaches have been used during the last two decades with the aim to improve the room- and high-temperature properties, reliability, and lifetime of silicon-nitride-based structural ceramics. The nano-particle dispersion-strengthening approach was used to improve the strength values and wear resistance by incorporating nano-sized second-phase particles into the silicon nitride and alumina matrix.[1-2] During the last decade, Si_3N_4 - SiC nanocomposites have been developed in which nano-sized SiC particles are dispersed in the Si_3N_4 matrix.[2] These particles are usually located either intragranularly, with a size of approximately 30-40 nm, or intergranularly, with a size of approximately 150-170 nm.[3] Three main methods have been used for processing such nanocomposites: a) Hot pressing (HP) or gas pressure sintering (GPS) of a mechanical mixture of crystalline sub-micrometer-sized Si_3N_4 and SiC powders; b) HP of a CVD SiCN amorphous powder and a combination of these methods; c) HP or GPS of a mixture of crystalline Si_3N_4 + amorphous SiCN powders. The mechanical properties of Si_3N_4-SiC nanocomposites have been studied by a number of authors during the last fifteen years.[4-6] It was shown that these ceramics exhibit greater high-temperature strength and better creep resistance than the monolithic Si_3N_4. According to Rendtel, et al.[6] the strength of the Si_3N_4-SiC nanocomposite at 1400 °C lies between 80 % and 100 % of the room-temperature strength, with creep rates as low as $1 \times 10^{-9}\ s^{-1}$ at 300 MPa and 1400 °C. This behavior can be explained by a thinner and more refractory intergranular phase of the composites compared to the monolithic materials. This is caused by trapping of a significant amount of intergranular glass in stress-free SiC grain pockets and by the presence of carbon in the intergranular phase.[7] Excellent oxidation resistance for Si_3N_4-SiC nanocomposites compared to monolithic silicon nitride was reported in

Reference 8. This phenomenon was mainly connected with the change in the oxidation and damage mechanisms during operation at high temperatures. D. -S. Cheong, et al.[9] found a very high room-temperature strength of a Si_3N_4–20 vol.% SiC nanocomposite with Y_2O_3+Al_2O_3 as sintering additives. However, strength degradation occurred at temperatures higher than 1000 ^{0}C. A nanocomposite with 4 wt% Y_2O_3 of sintering additive had a lower room-temperature strength (approximately 1 GPa). However, this value of strength remained up to 1400 ^{0}C due to the direct bond of the intergranularly located SiC particles to the Si_3N_4 matrix and due to the inhibition of grain-boundary sliding and cavity formation.

The room-temperature flexural strength and fracture toughness of the Si_3N_4-SiC nanocomposites has been the subject of a number of investigation during recent years.[2, 9-11] It is very difficult to compare the strength and fracture toughness of nanocomposites fabricated and measured by different authors. Firstly, researchers used different processing routes for the material preparation. Secondly, they used different testing methods, which results in significant differences in the measured strength and fracture toughness values. Therefore, it is possible only to establish whether after adding SiC nanoparticles to the matrix toughening/strengthening occurs, or not.

Dusza and Šajgalík reported an increase in strength and Weibull modulus from 987 MPa to 1203 MPa and from 6.7 to 19, respectively, by addition of 10% SiC to silicon nitride in a SiCN-derived nanocomposite.[12] The indentation fracture toughness slightly decreased from 7.3 MPa.m$^{1/2}$ to 6.9 MPa.m$^{1/2}$. The good strength characteristics were associated with a fine and defect-free microstructure and the relatively high fracture toughness caused by crack-bridging mechanisms. It was proposed that activation of these mechanisms was promoted by an intergranular fracture mode and probably by the residual tensile stresses on the Si_3N_4/Si_3N_4 boundaries originating from the intergranularly-located SiC particles. Pezzotti, et al.[13] reported significantly lower fracture toughness (SENB specimens) for a Si_3N_4-SiC nanocomposite, accompanied by transgranular fracture. This is probably connected with the grain-boundary strengthening effect of the intergranularly-located SiC particles, which promotes transgranular fracture and hence eliminates bridging mechanisms.

During recent years an inexpensive in-situ method utilizing carbothermal reduction has been used for preparation of the Si_3N_4-SiC micro/nanocomposite of this study, which was characterized mainly with respect to its high-temperature properties.[14]

The aim of the present contribution is to study the room-temperature fracture of a Si_3N_4-SiC micro/nanocomposite, to characterize the strength-limiting flaws/fracture origins, and to find relations between the strength values, fracture toughness and fracture origins.

EVALUATION OF THE STRENGTH-LIMITING FLAWS – THEORETICAL CONSIDERATION

The strength of ceramics is degraded by the flaws which are a priori present in the material as the result of an inadequate processing route or flaws that are introduced during the surface finishing of the specimens/parts. Typical flaws include volume or sub-surface flaws, such as pores, voids, inclusions, large grains, and surface cracks. These can occur isolated, in close proximity to each other, or forming clusters (e.g., cluster of large grains and pores). Due to the brittle character of ceramics, fracture occurs suddenly when the stress intensity factor at the crack tip, K_I, reaches the critical value K_{IC}. The relationship between failure stress or strength σ_f, flaw size a, and fracture toughness K_{IC} is given by the well-known Griffith-based relationship[15]:

$$\sigma_f = \frac{K_{IC}}{Y\sqrt{a}} \tag{1}$$

where Y is a numerical constant dependent on the loading geometry and crack/flaw depth. The size of the fracture origin can be estimated from equation 1 based on the measured strength and fracture toughness:

$$a = \left(\frac{K_{IC}}{Y\sigma_f}\right)^2 \tag{2}$$

Flaws in the ceramic materials can be treated (according to their character) in different ways as isolated cracks in two-dimensional configuration, isolated three-dimensional cracks, and as multiple flaws. Internal flaws are often modeled as two-dimensional flat elliptical cracks with minor and major half axis a and c, respectively. The stress intensity factor for such a crack located in an infinite body and loaded in tensile mode can be expressed for any point of crack periphery in the form:

$$K_I = \frac{\sigma(\pi a)^{1/2}}{\Phi}\left(\frac{a^2}{c^2}\cos^2\Phi + \sin^2\Phi\right)^{1/4} \tag{3}$$

where σ is the applied stress and Φ is complete elliptical integral of the second kind, which can be approximated [16] by:

$$\Phi \cong \left[1 + 1.464\left(\frac{a}{c}\right)^{1.65}\right]^{1/2} \tag{4}$$

The stress intensity factor at the end of the minor axis of crack can be expressed as:

$$K_I = \frac{\sigma(\pi a)^{1/2}}{\Phi} \tag{5}$$

Surface flaws can be modeled as flat semi-elliptical cracks, and the stress-intensity factor is:

$$K_I = \frac{\sigma M(\pi a)^{1/2}}{\Phi} \tag{6}$$

where M is the surface correction factor. Depending on the shape and size of the crack, the minimum value of K_I occurs at the deepest point or at the surface. From the results of the numerical analysis of Raju and Newman[16] and of Kobayashi[17] for shallow nearly semi elliptical flaws loaded in tension, it was shown in reference 18 that the value of M is from 1.03 to 1.2.

In the case of irregular-shaped (but also regular) cracks the ratio a/c can be replaced by the crack area A.[19] The stress-intensity factor in this case has a form:

$$K_I = \frac{1}{1.68} Y A^{1/4} \sigma \tag{7}$$

Fracture origins such as voids or inclusions can be modeled as isolated cracks in three dimensions. Voids do not cause fracture directly, because usually the stress intensity at a void is low, but fracture occurs when the stress field of the void interacts with the surrounding defects e.g., surface cracks. Voids accompanied by circumferential cracks have been modeled by Baratta[20] and Evans[21] for uniaxial tension. According to the results, the stress intensity factor will increase due to the presence of the void surface, which allows a larger crack opening. This effect decreases as the crack length approaches the size of the void. The effect of inclusions on the fracture strength has been summarized by Evans,[22] and it is influenced by the thermal-expansion coefficient mismatch between inclusion and matrix and by the toughness as well as by modulus of the inclusion with respect to the matrix.

The fracture surface of the ceramics after bending/tensile testing has some reproducible features, which can be quantitatively related to stress at the fracture. The following features typically appear on the fracture surface: fracture origin, mirror, mist and hackle regions. The fracture mirror is a relatively smooth area that surrounds fracture origins.[23,24] The empirical relation between the fracture mirror size and the stress at the origin, σ_f, is given by:

$$A_j = \sigma_f \sqrt{R_j} \tag{8}$$

where R_j is the mirror radius representing the boundaries at which mist (inner mirror constant R_i) or hackle (outer mirror constant R_o) occur, and A_j is the corresponding mirror constant.

EXPERIMENTAL PROCEDURE

The C-derived Si_3N_4-SiC micro/nanocomposite of this study, with the composition listed in Table I, was prepared at the Institute of Inorganic Chemistry, Slovak Academy of Sciences in Bratislava. The sample contained amounts of SiO_2 and C with the aim of achieving 5 wt % of SiC by carbothermal reduction of SiO_2 after densification. The starting mixtures were homogenized in

Table I. Composition of the starting mixture

Compound	Si_3N_4	Y_2O_3	C	SiO_2
Content (wt %)	84.1	4.4	4.1	7.4

polyethylene bottles with Si_3N_4 spheres in isopropanol for 24 h. The dried mixture was sieved through a 25-μm sieve in order to eliminate large hard agglomerates. Green discs with a diameter of 48 mm and 5 mm thick were die pressed under a pressure of 30 MPa. Green discs were then embedded into a BN powder bed and positioned into a graphite uniaxial die. Samples were hot-pressed under a specific atmosphere, mechanical pressure, and heating regime at 1750 °C for 2 h.

The microstructure of the micro/nanocomposite was characterized by X-ray diffractometry (XRD) and scanning electron microscopy (SEM). XRD analysis was carried out using a Philips X-part diffractometer equipped with a CuK_2 radiation source. Polished and plasma–etched sections of the bulk materials were examined in an SEM.

Thirty specimens with an effective volume of 18.4 mm^3 and 1.55 mm^3 were tested in four-point bending on fixture with spans 20 mm x 40 mm and 9 mm x 18 mm, respectively. Before testing, the specimens were ground using a 15-μm by a diamond grinding wheel. Both edges on the tensile surface were rounded with a radius of about 0.15 mm in order to eliminate failure from the edges of the specimens. The specimens were broken at a cross-head speed of 0.5 mm/min, and the test environment was ambient air. The characteristic strength and Weibull modulus were computed using the linear regression method of two-parameter Weibull theory.

The value of fracture toughness, K_{IC}, was measured using the single-edge V-notch beam method (SEVNB). A sharp notch was introduced into specimens (3 mm x 4 mm x 45 mm) using a razor blade.[25] In all cases the notch-tip radius was less than 10 μm. The values of K_{IC} were computed using the formula:

$$K_{IC} = \sigma_f \sqrt{a} Y = \frac{F(S_1 - S_2)}{BW^{3/2}} \frac{3\alpha}{2(1-\alpha)^{3/2}} Y \tag{9}$$

$$Y = 1.9887 - 1.326\alpha - \frac{(3.49 - 0.68\alpha + 1.35\alpha^2)\alpha(1-\alpha)}{(1+\alpha)^2} \tag{10}$$

where σ_f is the fracture stress, F is the fracture load, B and W are the thickness and width of the specimen, respectively, S_1 and S_2 are the outer and inner support spans and $\alpha = \frac{a}{W}$ where a is the crack depth.

SEM and EDX analyses were used to localize and to characterize the fracture origin.[26-30] Fractographic analysis was performed in accordance with ASTM C 1322 and CEN EN 843-6. All fracture surfaces were examined with a stereographic microscope at magnifications up to x 90, to locate fracture origins. Fracture origin size and the mirror radii were measured at magnifications of about x 200 on SEM micrographs. The shape of the fracture origin was approximated by an ellipse, the minor *2a* and major *2c* axis length were measured for the volume-located flaws, and the depth *a* and the width *2c* were measured for the surface-located flaws. In the case of equiaxed flaws, the diameter was measured.

Values of flaw shape factor, Y, were estimated from a table in ASTM C 1322 and in the work of Raju and Newman.[31] Values of the characteristic origin dimension (depth or radius) and flaw shape factor were used to estimate the value of fracture toughness from equation 2. Outer mirror radii, R_0, were measured as a distance from fracture origin to the point where hackle commenced and there was an obvious change in roughness relative to the mirror (mist-hackle boundary). This boundary was more easily distinguishable in comparison to the mirror-mist boundary, which was not clear enough. For surface defects the overall diameter of the mirror was measured on the tensile surface, $2r_1$, as well as the radius into the depth, r. The average radius of the mirror was calculated as $(r_1+r)/2$. For origins located in the bulk of the material the average of two mirror diameters was used to compute the mirror radii. The value of stress at fracture origin was calculated knowing the distance of fracture origin from the tensile surface of the specimen. The fracture mirror size as a function of fracture stress was plotted in order to determine the mirror constant.

RESULTS AND DISCUSSION

The characteristic microstructure of the experimental material is illustrated in Fig. 1a, 1b. The Si_3N_4-SiC nanocomposite consists of a very fine, homogeneously distributed Si_3N_4 grains with a low aspect ratio. The composite additionally contains globular nano- and submicron-sized SiC particles located intragranularly in the Si_3N_4 grains or intergranularly between the Si_3N_4

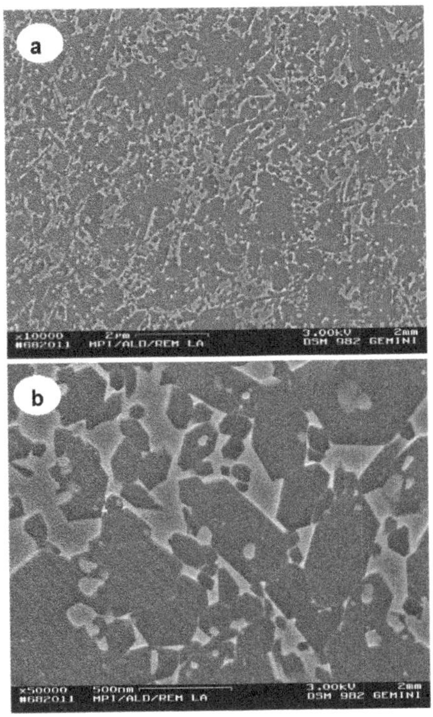

Figure 1. Characteristic microstructures of the investigated material at lower (a) and higher (b) magnification.

grains. It was difficult to distinguish between the SiC particles located intergranularly and the grain-boundary phase, because they are similarly affected by plasma etching. The mean diameter of Si_3N_4 grains is 140 nm, and grains with a diameter larger than 500 nm were observed in the microstructure only occasionally. The volume fraction of the SiC nanoparticles can be estimated approximately as 5 vol%. X-ray analysis revealed that the main phase in this material is β-Si_3N_4

with a small amount of β-SiC. Besides the Si_3N_4 and SiC, some additional crystalline phases were detected in the composite, mainly Y_2SiO_7 and Si_2N_2 O. The Weibull distribution of the measured four-point flexure-strength values of the investigated nanocomposite with the specimen effective volume of 18.4 mm^3 is shown in Fig. 2. Using two-parameter Weibull statistics the characteristic strength of the material was σ_0 = 675 MPa, and its Weibull modulus was m = 6.4. Slightly higher values of characteristic strength σ_0 = 832 MPa and Weibull modulus m = 8.6, Fig. 3, were obtained when testing the specimens with an effective volume of 1.55 mm.[3]

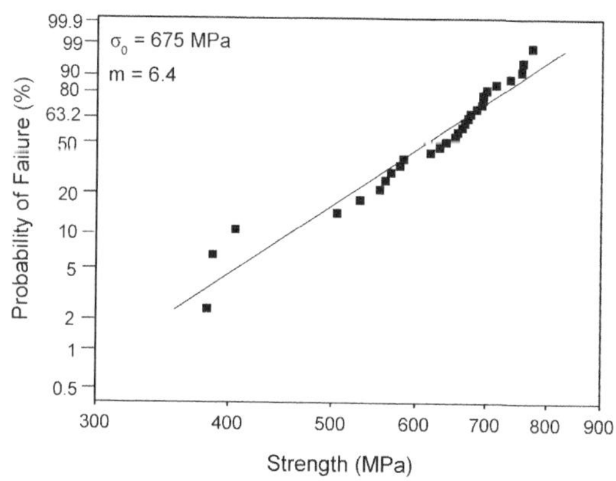

Figure 2. Weibull distribution of the four-point bending-strength values obtained by testing specimens with an effective volume of 18.4 mm^3.

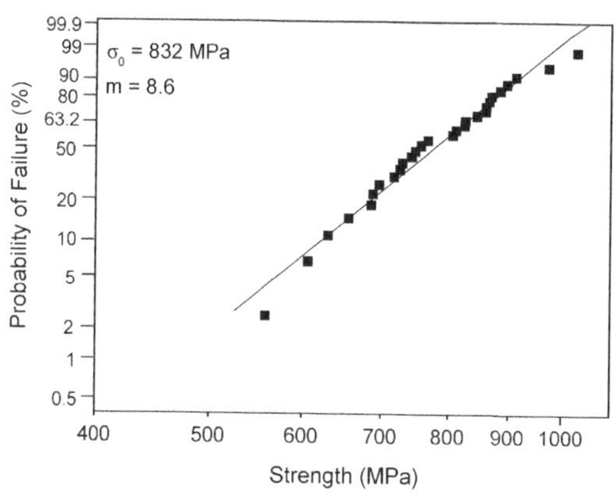

Figure 3. Weibull distribution of the four-point bending-strength values of the carbon-derived Si_3N_4-SiC nanocomposite with an effective volume of 1.55 mm^3.

Weibull parameters, the characteristic strength and Weibull modulus, are lower in comparison with the parameters of the SiCN-derived Si_3N_4-SiC micro/nanocomposite. Of course, specimens with the same effective volume were compared.[12] Bending tests and subsequent fractographic analysis of the fracture surface of failed specimens revealed that a reason for the lower characteristic strength compared to the strength of the SiCN-derived materials is the presence of processing flaws in the investigated materials. The most common location of the fracture origins/flaws was in the volume of the specimens (42 %), but origins were also located near the surface (23 %), on the surface (20 %), and at the edges (15%), Fig. 4a, 4b. Fracture origins were

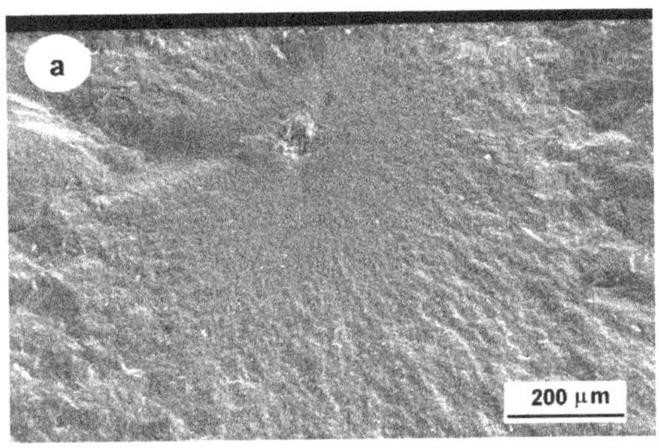

Figure 4a. Volume-located fracture origin of a specimen tested in four-bending mode with an effective volume of 18.4 mm^3.

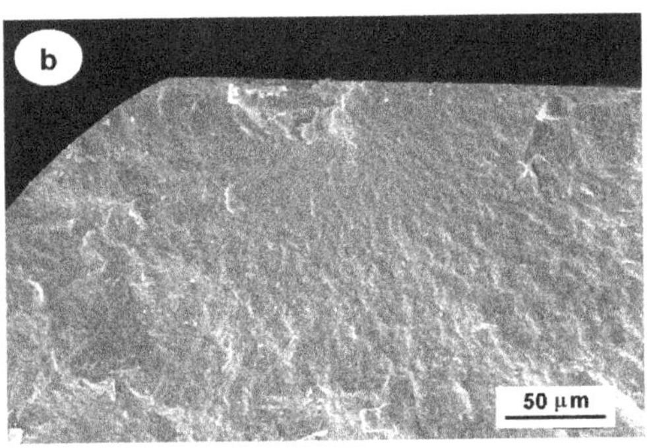

Figure 4b. Surface-located fracture origin of a specimen tested in four-bending mode with an effective volume of 18.4 mm^3.

in all cases processing flaws, and machining-induced flaws were not found. It is necessary to mention that some of the flaws stay undefined or fracture origins were lost by impact between the test piece and testing jig. A study of the size and shape of the fracture origins revealed that their sizes were in the range from 5 µm to 460 µm, with the mean value of half-minor axis length of about 35 µm. Their shape was mostly elliptical, but circular-shaped fracture origins were also observed. In the majority of cases it was easy to define the shape of the defects in at least a two – dimension form. Fractographic analysis of the fracture surfaces of failed specimens combined with EDX analysis revealed two main types of processing flaws acting as failure-initiating flaws in the studied material. The first type of fracture origin was a porous region, often connected with another type of flaw, Fig. 5a. The second type of fracture origin was a cluster of large SiC grains; Fig. 5b. Areas of non-reacted carbon and SiO_2 were found, too. However, this third type of fracture origin occurred rarely compared to the other two types.

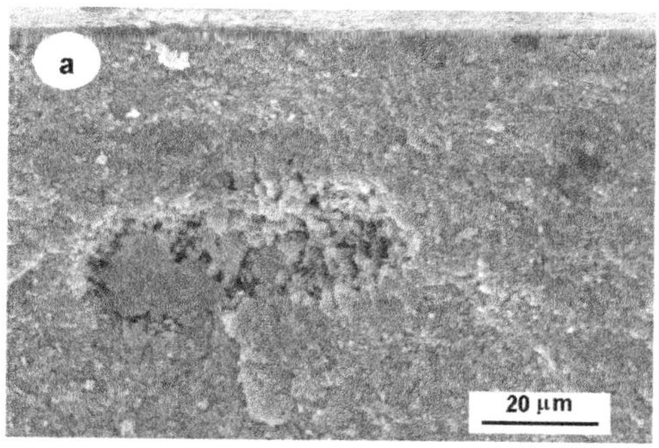

Figure 5a. Porous region connected with a large SiC grains in a specimen tested in four-bending mode with an effective volume of 18.4 mm^3.

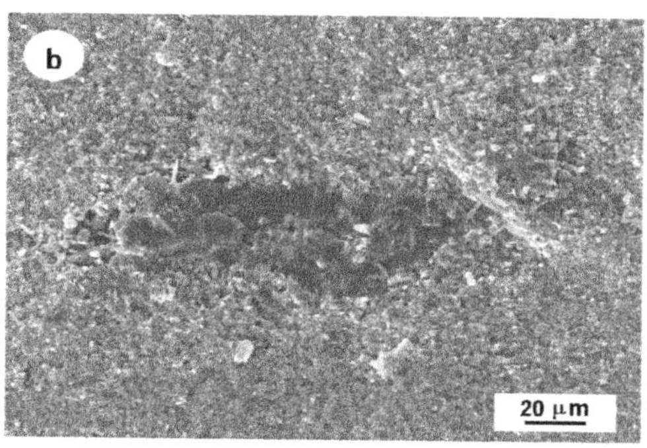

Figure 5b. Large SiC grains in a specimen tested in four-bending mode with an effective volume of 18.4 mm^3.

The type of fracture origins of the specimens with the effective volume of 1.55 mm^3 were the same as described above, but they were located mainly close to the surface of the specimens, Fig. 6. For the specimens with the lower effective volume, the fracture origins were found in almost all fractured specimens. Only in a few cases was the origin lost by impact between the test-piece and the testing jig.

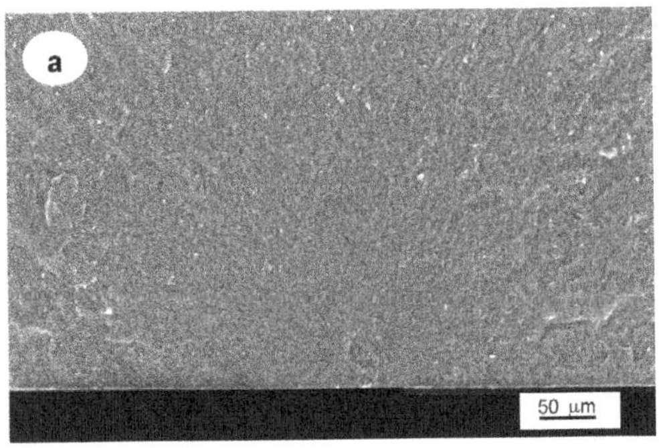

Figure 6a. Characteristic location of the fracture origin in the specimens with effective volume of 1.55 mm^3.

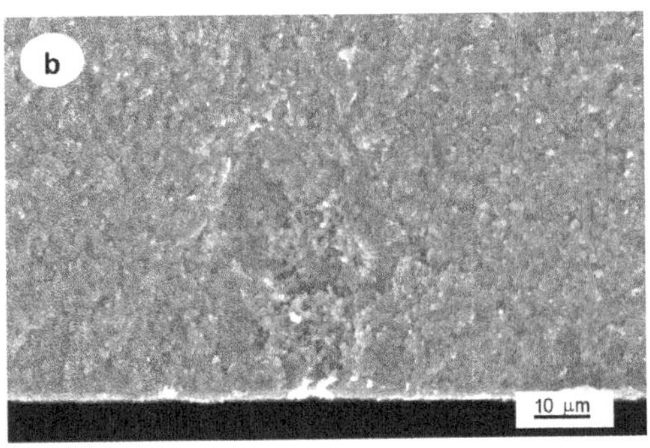

Figure 6b. Characteristic flaw (porous region) in the specimens with effective volume of 1.55 mm^3.

Equation 2 can be used for calculation of the flaw size, c, assuming that the stress at the origin at fracture and fracture toughness of the material are known. Fracture toughness was measured using the single-edge V-notch beam method. It was found to have a value of 3.75 ± 0.53 $MPam^{1/2}$. In Fig.7 the calculated and measured (on the fracture surface) origin sizes for the specimens with an effective volume of 18.4 mm^3 were compared. In most cases the measured size of an origin is larger compared to the calculated origin size. According to Quinn and Swab [32] there are three main reasons which can cause $c_{calc} < c_{meas}$: crack-tip blunting, the use of 2-

dimensional crack models, and stress gradients. Stress gradients and crack blunting could be excluded.

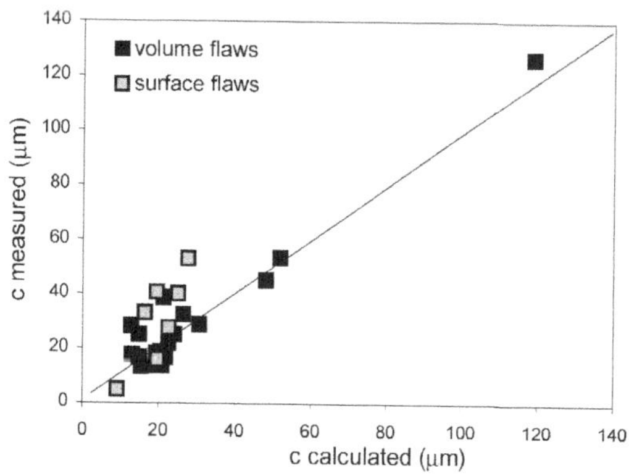

Figure 7. Comparison of the measured and calculated flaw size in specimens with effective volume of 18.4 mm^3.

The stress at the origin was used, not the maximum nominal stress. The SiC nanoparticles theoretically could affect crack propagation and cause a similar effect as crack blunting, but according to the K_{IC} value, this is not the case. The common practice to model origins with circumscribed circles or ellipses, or semicircles and semi-ellipses, as was done in this contribution can be an oversimplification. Such a model may be suitable for some type of origins, such as machining damage; however, the use of a two-dimensional model for the three-dimensional origin can lead to the discrepancies. Generally it is valid that the value of Y for a three-dimensional model is less than or equal to the Y for a penny-shaped origin of the same cross-sectional area. The utilization of the two-dimensional model and overlarge Y factor caused the calculated origin size to be underestimated.

The value of fracture toughness K_{IC}=3.75 ± 0.53 $MPam^{1/2}$ measured by the SEVNB method is significantly lower than the value for a similar material using an indentation technique.[12] However, it is well known that results of the indentation methods are only approximate, and the SEVNB technique gives a more realistic value of fracture toughness. In this paper another method to estimate the fracture toughness of the investigated material is presented, using the Griffith relation and experimentally determined origin sizes and shapes. Fig. 8 shows the relationship between the flaw size and fracture stress at fracture origin. The value of fracture toughness estimated from this relation was found to be 3.74 ± 0.75 $MPam^{1/2}$. This value is surprisingly consistent with the value measured using the SEVNB technique. The difficulties in the estimation of Y factor for the individual defects and the fact that the origins have different composition/microstructure than the surrounding area caused the high scatter of the points around the fit line. But the slope of the fit line which represents the fracture toughness value, is equal to the value from the SEVNB method. Using the empirical relation for the fracture mirror size and the fracture stress at the fracture origin, equation 8, it is possible to obtain information concerning the mirror constant of the material, Fig. 9. In all cases the roughness

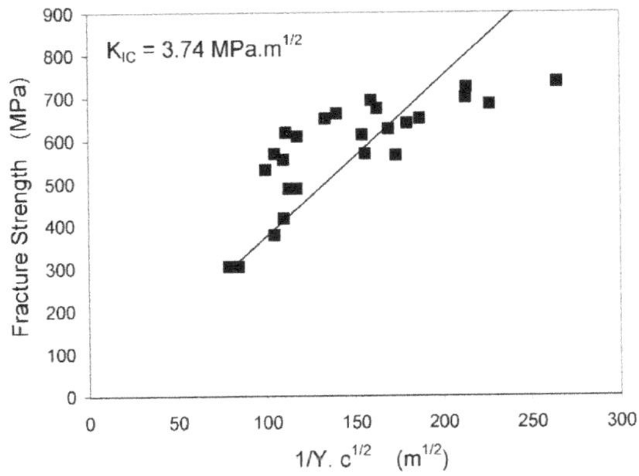

Figure 8. Relationship between the fracture strength and flaw size in specimens with effective volume of 18.4 mm^3.

increases progressively, from within the mirror to the hackle region. The slope of the line is the outer mirror constant (determined from mist – hackle boundary), and the value of this constant

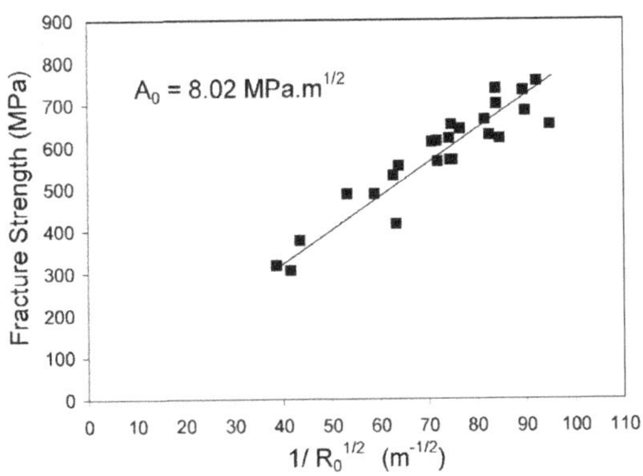

Figure 9. Plot of fracture stress versus the square root of the mirror size in specimens with effective volume of 18.4 mm^3.

is 8.02 MPa$\sqrt{m}$ ± 0.61 MPa$\sqrt{m}$. This value is in good agreement with the measurement of Quinn [33] who determined the value of the mirror constant for silicon nitride bars with machining cracks to be 7.8 MPa$\sqrt{m}$. The fracture toughness of this material determined by the single-edge precracked-beam method was 5.6 MPa$\sqrt{m}$ ± 0.66 MPa$\sqrt{m}$. Choi and Gyekenyesi[34] obtained a value of 9.4 ± 1.19 MPa$\sqrt{m}$ when tested at 1100°C in air and for another monolithic Si_3N_4 indented by Vickers indentor the fracture mirror constant was 5.88 ± 0.14 MPa$\sqrt{m}$. Mecholsky, *et al.*[35] showed that the ratio of outer mirror constant to fracture toughness for a variety of materials was 2.8. The micro/nanocomposite of this study, the mirror constant is 2.1 times larger than the value of fracture toughness (SEVNB).

Although enough results of the value of mirror constants were found for monolithic silicon nitride there is only limited information concerning the fracture origins and the surrounding areas of the nanostructured ceramics. Recently Quinn, *et al.*[36] investigated the relationship between the strength and fracture mirror size in nanoscaled yttria-stabilized tetragonal zirconia polycrystal. The characteristic fracture origins in this material were agglomerates, porous regions and shrinkage cracks, and the corresponding mirror constants were 9.6 $MPam^{1/2}$ for biaxial disc and 8.6 $MPam^{1/2}$ for bending bars. The mirror constant was independent of grain size over a range of 100–500 nm. These results are in a very good agreement with the results of our investigation.

Figure 10a. Character of fracture surface close to the origin of Si_3N_4-SiC micro/nanocomposite.

Figure 10b. Morphology of fracture surface in the hackle region.

As with other polycrystalline materials, a change in the fracture surface topography was observed as the fracture made the transition from slow to rapid crack growth. The fracture mode changed from predominantly intergranular in the slow-crack regime close to the fracture origin (Fig. 10a), to predominantly transgranular in the fast crack regime in the area further from the fracture origin (hackle region), Fig. 10b.

CONCLUSIONS

The room-temperature fracture characteristics of a carbon-derived Si_3N_4 – SiC micro/nano-composite have been investigated. The following main results were found:

- the characteristic strength of the investigated micro/nanocomposite was σ_0 = 675 MPa with the Weibull modulus of 6.4, and σ_0 = 832 MPa with the Weibull modulus of 8.6 using bending bars with an effective volume of 18.4 and 1.55 mm^3, respectively;
- using fractographic methods, in almost all of the specimens, processing-related fracture origins were found with dimensions in the range of 5 μm to 460 μm;
- three types of processing flaws were identified: porous areas, agglomerates of large SiC grains, and areas of non-reacted carbon and SiO_2;
- a good agreement has been found comparing the measured (SEVNB method) and calculated values of fracture toughness;
- the measured outer mirror constant of 8 $MPam^{1/2}$ is in a good agreement with the constants of similar monolithic Si_3N_4 materials;
- a change of the fracture mode was observed from predominantly intergranular close to the fracture origin to predominantly transgranular in the hackle region.

Acknowledgments

Work was supported by the Slovak Grant Agency project No. 2/4173/04, APVT-51-049702, Centrum of Excellence NANOSMART and grant ŠO 51/03R80600/03R0603.

REFERENCES

[1]K. Niihara, “New Design Concept of Structural Ceramics – Ceramics Nanocomposites,” *J. Ceram. Soc. Jpn.*, 99 [10] 974-82 (1991).

[2]M. Herrmann, C. Schubert, A. Rendtel, and H. Hübner, “Silicon Nitride/Silicon Carbide Nanocomposite Materials: I, Fabrication and Mechanical Properties at Room Temperature,” *J. Am. Ceram. Soc.*, 81 [5] 1094-1108 (1998).

[3]P. Šajgalík, M. Hnatko, F. Lofaj, P. Hvizdoš, J. Dusza, P. Warbicher, F. Hofer, R. Riedel, E. Lecomte, and K. Rajan, “SiC/Si3N4 Nano/Micro-Composite - Processing, RT and HT Mechanical Properties, ” *J. Eur. Ceram. Soc.*, 20, 453-462 (2000).

[4]K. Niihara, K. Suganuma, A. Nakahira, and K. Ikazi, “Interfaces in Si_3N_4-SiC Nano-composite,” *J. Mater. Sci. Lett.*, 9, 598-599 (1990).

[5]K. Niihara, K. Ikazi, and T. Kawakami, “Hot-pressed Si_3N_4-32% SiC Nanocomposite from Amorphous Si-C-N Powder with Improved Strength Above 1200°C,” *J. Mater. Sci. Lett.*, 10, 112-114 (1990).

[6]A. Rendtel, H. Hübner, M. Herrmann, and C. Schubert, “Silicon Nitride/Silicon Carbide Nanocomposite Materials: II, Hot Strength, Creep, and Oxidation Resistance,” *J. Am. Ceram. Soc.*, 81 [5] 1109-1120 (1998).

[7]J.-L. Besson, M. Mayne, D. Bahloul-Hourlier, and P. Goursat, "Si_3N_4/SiCN Nanocomposites: Influence of SiC Nanoprecipitates on the Creep Behaviour," *J. Eur. Ceram. Soc.*, 18, 1893-1904 (1998).

[8]M. Herrmann, H. Klemm, B. Göbel, C. Schubert, and W. Hermel, "SiC/Si_3N_4 Nanocomposites with Excellent High-Temperature Long-Term Behaviour," *Key Eng. Mat.*, 161-163, 377-380 (1999).

[9]D.-S. Cheong, K, T. Hwang and Ch.-S Kim, "High-Temperature Strength and Microstructural Analysis in Si_3N_4/20-vol%-SiC Nanocomposites," *J. Am. Ceram. Soc.*, 82 [4] 676-679 (1999).

[10]J.-F. Yang, T. Sekino, Y-H. Choa, and K. Niihara, "Microstructure and Mechanical Properties of Sinter-Post-HIPed Si_3N_4-SiC Composites," *J. Am. Ceram. Soc.*, 84 [2] 406-12 (2001).

[11]S. Sato, M.-C. Chu, J.-H. Kim, Y. Kobayashi, and K. Ando, "Influence of SiC Particle Size on Microstructure and Mechanical Properties of Si_3N_4/SiC Composite Ceramics," *J. Ceram. Soc. Jpn.*, 103 [7] 676-679 (1995).

[12]J. Dusza and P. Šajgalík, "Fracture Characteristics of Layered and Nano-Particle Reinforced Si_3N_4," pp. 187-205 in *Advanced Multilayered and Fibre Reinforced Composites*, Edited by Y. M. Haddad. Kluwer Academic Publishers, Dordrecht, 1998.

[13]G. Pezzotti and M. Sakai, "Effect of a Silicon Carbide Nano-Dispersion on the Mechanical Properties of Silicon Nitride," *J. Am. Ceram. Soc.*, 77 [11] 3039-3041 (1994).

[14]J. Dusza, J.Kovalčík, P. Hvizdoš, P. Šajgalík, M. Hnatko, and M. Reece, "Enhanced Creep Resistant Silicon-Nitride-Based Nanocomposite" *J. Am. Ceram. Soc.*, 88 [6] 1500-1503 (2005).

[15]B. R. Lawn and T. R. Wilshaw, "Fracture of Brittle Solids," Cambridge University Press, 1975.

[16]J. C. Newman Jr. and I S. Raju, "An Empirical Stress-Intensity Factor Equation for the Surface Crack," *Eng. Frac. Mech.*, 15 [1-2] 185-192 (1981).

[17]A. S. Kobayashi, M. Ziv, and L. R. Hall, "Aproximate Stress Intensity Factor for an Embedded Elliptical Crack Near Parallel Free Surfaces," *Int. J. Fract. Mech.*, 1, 81-95 (1965).

[18]I. Bar-On, "Applied Fracture Mechanics," pp. 645-51 in *Engineered Materials Handbook*, Volume 4: Ceramics and Glasses, ASM International, 1991.

[19]G.K. Basal, "Effect of Flaw Shape on Strength of Ceramics," *J. Am. Ceram. Soc.*, 59 87-88 (1976).

[20]F.I. Baratta, "Stress Intensity Factor Estimates for a Peripherally Cracked Spherical Voids and Hemispherical Surface Pit," *J. Am. Ceram. Soc.*, 61, 490-93 (1979).

[21]A.G. Evans, D.R. Biswas, and R.M. Fulrath, "Some Effects of Cavities on the Fracture of Ceramics: II Spherical Cavities," *J. Am. Ceram. Soc.*, 62, 101-106 (1979).

[22]A. G. Evans,"Structural Reliability: A Processing-Dependent Phenomenon, " *J. Am. Ceram. Soc.*, 65, 127-137 (1982).

[23]R. Morrell, "Fractography of Brittle Materials," Measurement Good Practice Guide, No. 14, HMSO, National Physical Laboratory, UK, ISSN, 1368-6550, (1999).

[24]ASTM Designation C1322-02a Annual Book of Standards, Vol. 15.01. American Society for Testing and Materials, West Conshohocken, PA, 2002.

[25]J. Kübler, "Fracture Toughness of Ceramics Using the SEVNB Methods: Preliminary Results," *Ceram. Eng. Sci. Proc.*, 18 [4] 155-62 (1997).

[26]R. Danzer, "Mechanical Failure of Advanced Ceramics: The Value of Fractography," *Key Eng. Mat.*, 223, 1-18 (2002).

[27]J. Dusza and M. Steen, "Fractography and Fracture Mechanics Property Assessment of Advanced Structural Ceramics," *Int. Mater. Rev.*, 44, 165-216 (1999).

[28]G. D. Quinn and J. J. Swab, "Fractography and Estimates of Fracture Origin Size from Fracture Mechanics," *Ceram. Eng. and Sci. Proc.*, 17 [3] 51-58 (1996).

[29]prEN 843-6 Advanced Technical Ceramics - Monolithic Ceramics - Mechanical Properties at Room Temperature - Part 6: Guidelines for Fractographic Investigation, 2001

[30]R. Morrell and J. Kubler, "Fractography of Advanced Ceramics – The CEN Pre-Standard," *Key Eng. Mat.*, 223, 18-8 (2002).

[31]I. S. Raju and J. C. Newman Jr., "Stress-Intensity Factors for a Wide Range of Semi-Elliptical Surface Cracks in Finite-Thickness Plates," *Eng. Fract. Mech.*, 11, 817-829 (1979).

[32]G. D. Quinn and J. J. Swab, "Comparisons of Calculated and Measured Flaw Sizes," pp. 175-92 in Ceramic Transactions, Vol. 122, Fractography of Glasses and Ceramics IV. Edited by J. R. Varner and G. D. Quinn. American Ceramic Society, Westerville, OH, 2001.

[33]G. D. Quinn, L. Ives, S. Jahanmir, and P. Koshy, "Fractographic Analysis of Machining Cracks in Silicon Nitride Rods and Bars," pp. 343-65 in Ceramic Transactions, Vol. 122, Fractography of Glasses and Ceramics IV. Edited by J. R. Varner and G. D. Quinn. American Ceramic Society, Westerville, OH, 2001.

[34]S. R. Choi and J. P. Gyekenyesi, "Crack Branching and Fracture Mirror Data of Glasses and Advanced Ceramics," NASA Tech. Report. TM 1998-206536. NASA Glenn Research Center, Cleveland, OH, 1998.

[35]J. J. Mecholsky Jr., S. W. Freiman, and R. W. Rice, "Fracture Surface Analysis of Ceramics," *J. Mater. Sci.*, 11, 1310-19 (1976).

[36]G. Quinn, J. Eichler, U. Eisele, and J. Rödel, "Fracture Mirrors in Nanoscale 3Y – TZP," *J. Am. Ceram. Soc.*, 87 [3] 513-16 (2004).

FRACTOGRAPHY OF THERMISTORS

Robert Danzer[*,§], Alexander Platzer[§], Peter Supancic[*,§], and Zhonghua Wang[*]
[*]Institut für Struktur- und Funktionskeramik
Montanuniversität Leoben, A-8700 Leoben, Austria
[§]Materials Center Leoben, A-8700 Leoben, Austria

ABSTRACT

In service ceramic thermistor components are exposed to serious temperature changes, which cause large tensile stresses in the components. These stresses limit the tolerable voltage and can even destroy the components.

Destructive tests on the components were made by increasing the voltage up to the failure of the components and a failure statistic (probability of failure in dependence of the applied voltage) has been determined. The "electric" failure statistics turned out to be a Weibull distribution. The experimentally observed fracture modes were analyzed.

An aspect, which has influence on the fracture initiation, is the stress concentration, which arises from some inhomogeneities in the microstructure. But even more important is the change in the local Joule heat production rate, which is influenced by local inhomogeneities. Examples for such inhomogeneities, which could be identified, are large pores, clusters of pores and regions with variations in the chemical doping level.

INTRODUCTION

Thermistors have a wide range of applications e.g., overload protection, degaussing control, temperature sensors and others. In active service ("switching process") the thermistor is connected in series to the electrical load. Induced by the current the thermistor produces Joule heat. This causes a rise of temperature in the ceramic. Connected to this increase of temperature is a rise of the resistivity. Therefore these electrical components are often referred to as Positive-Temperature-Coefficient-Resistors (PTC-R), meaning that their resistivity goes up with a rise of temperature.

These resistors are usually made out of an n-type doped $BaTiO_3$-ceramic (Fig. 1). The material crystallizes in the Perovskite structure. In a temperature range around the so-called Curie-temperature, a dramatic change in the electrical properties occurs.[1-4] At high temperatures the high electric resistance is caused by electric charges, which are accumulated at and near the grain boundaries (double Schottky barriers), and which trap the electrons from the conduction band. From high to low temperatures a ferroelectric transition from the cubic to the tetragonal phase occurs, which causes a loss of the inversion symmetry in the ionic crystal: the unit cell has a dipole moment. The polarization charges compensate locally the action of the double Schottky barriers at the grain boundaries. At low temperatures the PTC-R is semiconducting. The change of resistance can reach up to seven orders of magnitude within a temperature change of about 50 °C. Thermistors come in various shapes and sizes. The most common geometry is a cylinder with

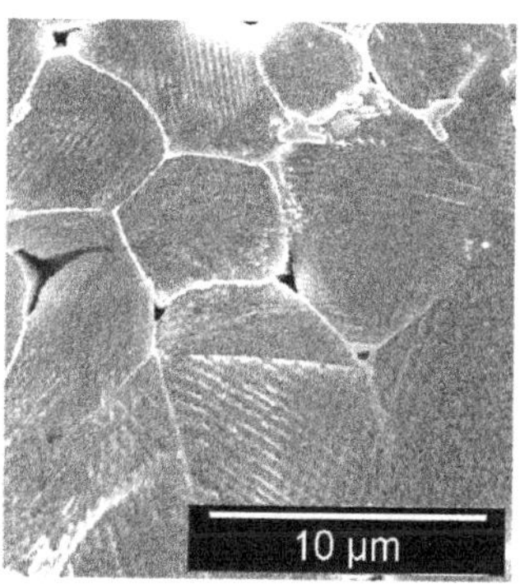

Fig. 1: Microstructure of $BaTiO_3$-based PTC.

typical diameters ranging from 2 to 20 mm and heights ranging from 1 to 5 mm. The circular surfaces are coated with a metal electrode for electrical contact. During the switching process the ceramic has to heat up over the Curie-temperature. Heat is transferred via the surfaces of the component into the environment. Some metallic parts (solder, wires) act like heat sinks and an inhomogeneous temperature distribution develops in the PTC-R. Temperature differences in the ceramic (see Fig. 3 left side) may be boosted by its non-linear thermo electric behaviour and reach more than 100 °C within a fraction of a second. For a component made of a homogeneous ceramic material the highest temperatures occur in the middle region between the top and bottom plane of the resistor. In comparison the regions at and near these planes are much cooler.

The result of this temperature field in the part is thermally induced stress. The thermo-mechanical stresses σ_{th} scales with the factor

$$\sigma_{th} \propto \frac{\alpha E}{1-\mu} \cdot \Delta T \tag{1}$$

where α is the coefficient of thermal expansion, E is the Young's modulus, μ the Poisson's ratio and ΔT the temperature difference in the component.[5] The equation is written for a biaxial stress state (stress in the surface). In the case of the PTC ceramics the material dependent factor $\alpha E/(1-\mu)$ strongly depends on the crystal structure. Fig. 2 shows the coefficient of thermal expansion (a) and the Young's modulus (b) versus the temperature. The sudden change of the properties at the Curie temperature (around 130 °C) can clearly be recognized. Below the Curie temperature $\alpha E/(1-\mu) \cong$ 0.7 MPa/°C and above the Curie temperature it is around 3.5 MPa/°C. Therefore the same temperature difference is 5 times more dangerous above the Curie temperature than below it. This point has been highlighted in a recent paper of some of the authors.[6] The strength of PTC ceramics is relatively low, i.e. it is around 100 MPa or less. It has been shown in thermo-shock test that several tens of degrees may cause the propagation of cracks. Therefore – in a macroscopic stress field caused by macroscopic temperature gradients, local temperature variations of several Kelvin may cause the initiation of fracture.

A typical mechanical stress field is shown in Fig. 3 on the right side. The maximum first principle stress is found on the equator between the main surfaces of the cylinder. This high stress results from the deformation (bending) of the surface associated with the temperature

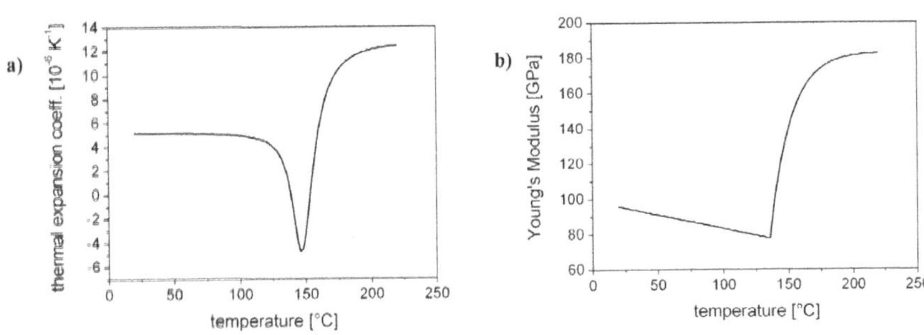

Fig. 2: The thermal expansion coefficient (a) and the Young's modulus (b) of a typical PTC ceramic [13] versus the temperature. The sudden change of the properties at the Curie temperature (~ 130 °C) can clearly be recognized.

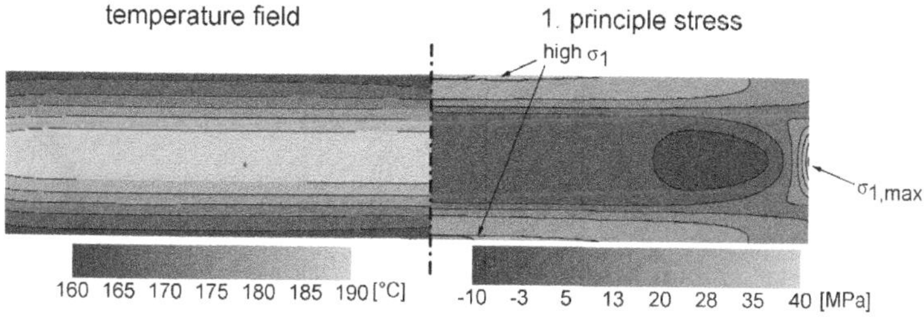

Fig. 3: PTC-R cross-section. On the left side the temperature distribution during the switching process is shown (from Finite Element calculation) for a homogeneous PTC ceramic. The mechanical stress field caused by the temperature field is shown on the right side (also from Finite Element calculation). Electrical conditions: 550 V, consumer resistance 40 Ω, voltage applied for 1 second. For more details of the FE simulation see [9].

difference. This stress may cause the thermistor to break into two parts perpendicular to the electrical field. Dewitte et al.[7] referred to this failure type as a delamination fracture. A second area of high tensile stresses is at the center of the top and bottom planes of the component, where heat is extracted out of the ceramic by the connected metallic conductors. These stresses can cause a so-called longitudinal crack,[8] if a sufficiently large defect is present. Note that the qualitative appearance of the temperature and stress fields shown in Fig. 3 is very typical for PTC components, but the amplitude of the fields sensibly depends on material properties. For example it has been shown in several papers of the authors[8,9,12] that slight changes in the electric behavior of the PTC ceramic may have a significant influence on the mechanical stress amplitudes. The tensile stresses of the simulation reported in Fig. 3 are significant (they reach 40 MPa and more), the mechanical strength is around 80 MPa, see next chapter.

STRENGTH TEST RESULTS

In addition to the electrical measurements, the mechanical strength of the thermistors was tested using the ball-on-three-balls-test (B3B, see also [10, 11]). In this test most of the specimens failed due to pores. The Weibull modulus m of this sample (30 specimens) was surprisingly high: $m = 37$. This indicates that failure originates at defects of quite similar size. This is supported by the fractographic analysis.

A quantitative ceramographic analysis of the pores in the material (not reported here) shows that a bimodal distribution of pores exists in the material. A pore population with a wide distribution of pore sizes (the frequency strongly decreases with size) is superposed by a narrow peaked distribution having a radius of 50 to 70 μm, which is responsible for the fracture events described in Fig. 5. A typical fracture origin is shown in Fig. 4.

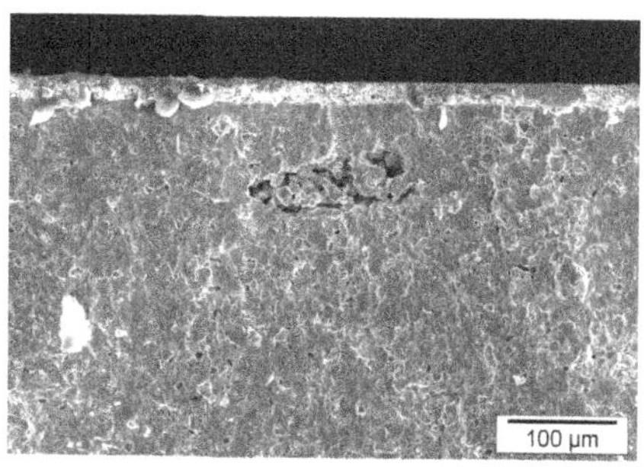

Fig. 4: Typical fracture origin from the mechanical test.

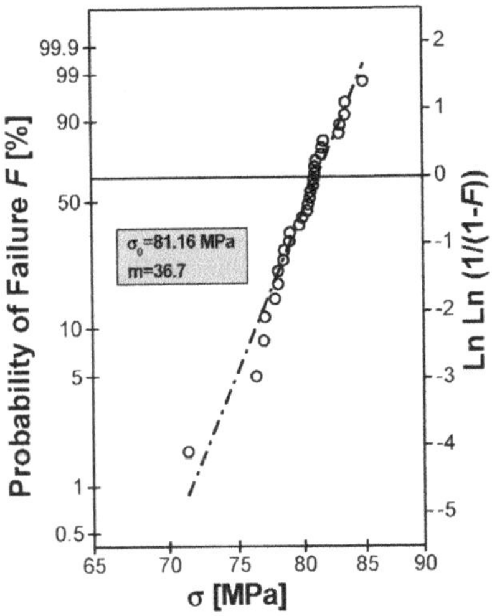

Fig. 5: Weibull-distribution of the mechanical B3B-strength test for thermistor ceramic. For experimental details see [10, 11].

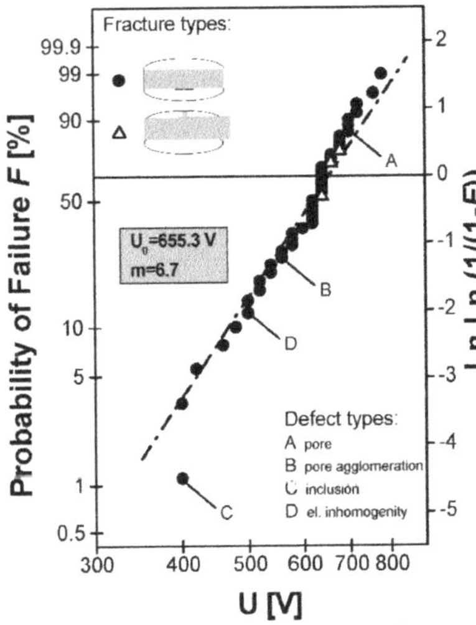

Fig. 6: Electrical Weibull-evaluation. (For fracture types see below). Experiments labeled with letters are discussed later in more detail.

ELECTRICAL TEST RESULTS

For this experiment 45 specimens were submitted to a so-called "step-stress-test." The investigated parts were commercial PTC components used for motor-start applications. They have a diameter of 20 mm and a height of 5 mm. The thermistors have an electrical resistance of about 30 Ω at room temperature. Starting with an electrical load of 400 V, the voltage was increased in steps of 20 V up to a maximum value of 800 V. During this test the components failed at a certain applied voltage. Each switching process was filmed with an infrared camera in order to measure the change of the surface temperature. Analogous to the mechanical test, a Weibull-evaluation of failure depending on the electrical load (voltage) was made. The result of this evaluation is shown in Fig. 6.

The "characteristic voltage" (i.e. the voltage where 63 % of all components has failed) has a value of 655 V, the electrical Weibull modulus a value of 6.7. This indicates a broad distribution in the electrical "strength". The reason for this wide variation is the existence of different types of defects, which cause failure at different levels of applied voltage.

During the experiment two different fracture modes and four different defect types were observed. They are all discussed in detail below.

FRACTURE TYPES

During the electrical experiments two different fracture modes occurred. Most of the parts (over 93%) failed by delamination fracture (see Fig. 7b). The rest of the specimens broke in a combination of longitudinal and delamination fracture (see Fig. 7a and cd), starting closely below the circular surface of the cylinder. This fracture type will referred to as mixed fracture in the following.

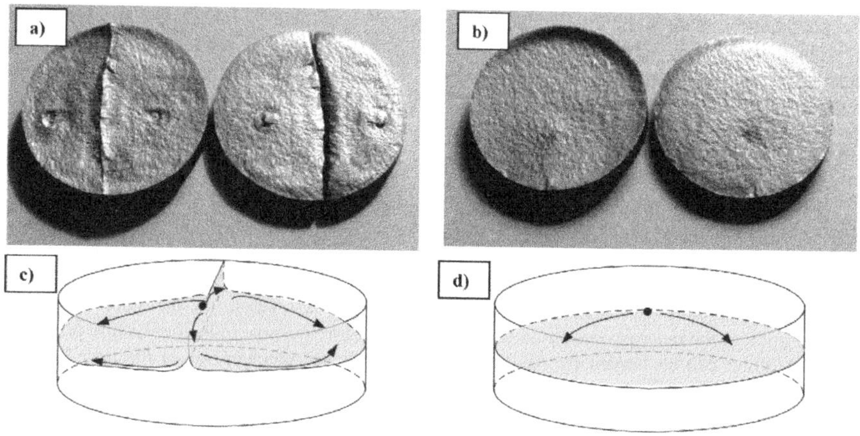

Fig. 7: Examples for fracture types. a) shows the rarely occurring mixed fracture, b) the mostly occurring delamination fracture. In c) and d) sketches of the fracture propagation are shown. The black dots marks the fracture origins, the arrows show the propagation.

Mixed Fracture

This type of fracture occurred in 3 out of 45 parts. Fig. 7a shows the part after failure. In Fig. 8a the fracture surface close to the fracture origin is visible. Arrows mark the crack propagation dissection. Fig. 8b shows the failure-causing defect in detail. This is a large pore positioned closely underneath the circular surface of the cylindrical resistor.

All three parts that broke in this fracture mode had pores as fracture origins. The failure voltage was in the range of 620 to 680 V, which is in the middle to upper region of load spectrum (see Fig. 6). As mentioned above, a large defect must be in the region below the circular surface to cause this fracture; because the tensile stresses in this region are a little smaller than in the region of the equator.

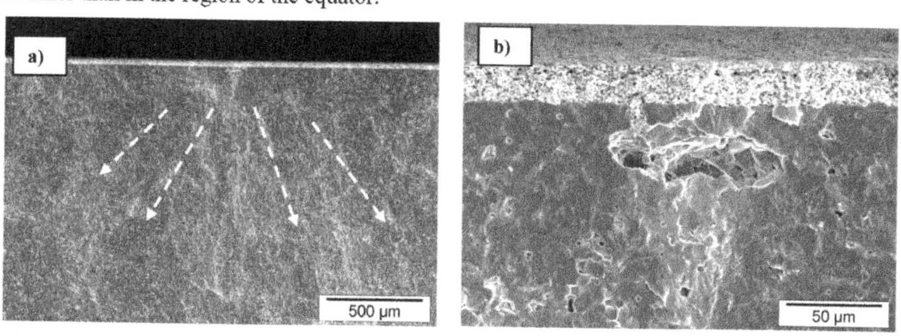

Fig. 8: Mixed fracture. a) Fracture surface close to fracture origin. Arrows mark the crack propagation direction.
b) Detailed picture of pore, with a length of more than 50 µm.

Delamination Fracture

The delamination fracture occurred 42 times during the 45 electrical tests. This is by far the most common type of fracture. It is quite evident, that the PTC-R will break along its equator because of the maximum stress occurring in this area. But as shown below, it is not only a single defect population, which is causing failure, but several different types could be identified.

Failure by a single pore

This failure origin has been detected in about 40 % of all parts. The component presented here failed at a voltage of 700 V. This is within the upper sixth of all tested parts. The failure origin is a pore near the surface in the equator region (Fig. 9a and b). The pore has a diameter of approximately 80 μm (Fig. 9b). The crack propagation is indicated in Fig. 9a.

The pore affects the stresses in the component in two ways: (i) the pore causes a concentration in the macroscopic stress field and can therefore become to be the fracture origin.

This effect can be treated by the generally used fracture mechanics. (ii) the pore has an influence on the local heat generation and the resulting the local temperature distribution and to the local thermo mechanical strains and stresses. An accurate assessment of this effect needs a meso-mechanic description.[8,14] It can be expected that (ii) becomes large if these inhomogeneities of the local temperature field cannot immediately be compensated by heat conduction, i.e., in quick switching processes. The influence of the pore on the temperature field can experimentally be seen in Fig. 9c. The temperature around the pore is lower than in the rest of the equator due to the local lack of Joule heating. Note that the failure occurs above the Curie temperature, where small temperature differences may cause large tensile stresses.

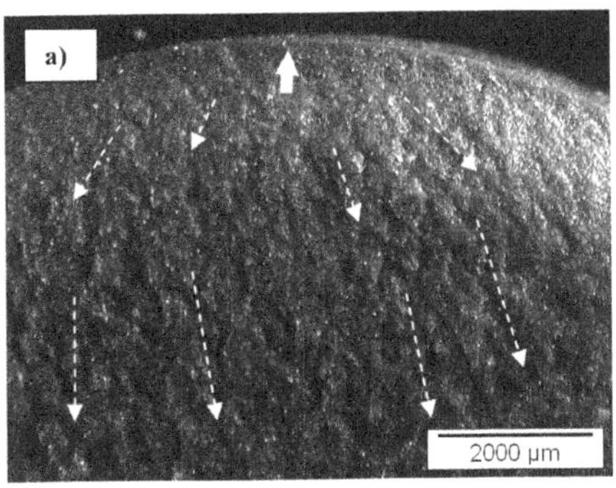

Fig. 9. a) Fracture surface, with fracture origin (thick arrow) and crack propagation (see dashed arrows).

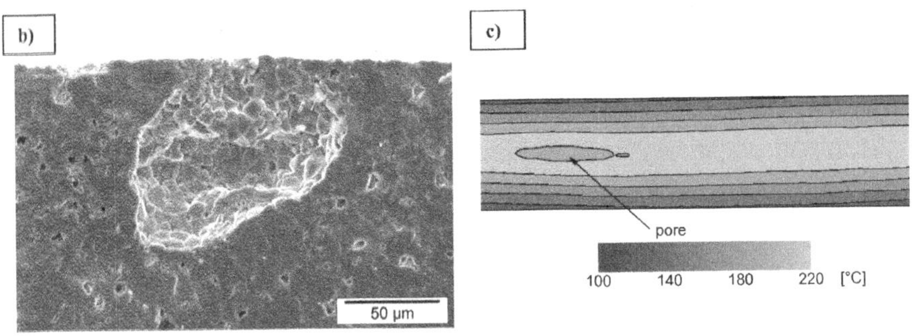

Fig. 9 b) Failure causing pore and d) failure causing pore in detail.
c) Thermographic picture of circumferential surface of the thermistor during failure after 0.56 s of applied voltage (700 V). The region around the pore is cooler.

Failure by an Agglomeration of Pores

The described PTC-R part failed at an applied electrical voltage of 560 V. This is within the lower third of all failed parts. The reason for this failure is an agglomeration of over ten pores of 10 to 50 µm with a total diameter of 500 µm (Fig. 10a and b). Following our experience none of these pores alone would cause a failure initiation. If it would be assumed, that cracks having the size of the pore are distributed in the material, they would be sub-critical and not cause failure in the macroscopic stress field. But these pores cause a reduction of electric current density. Therefore less Joule heat is produced and the volume around the pore agglomeration remains cold. This is shown in Fig. 10c, where a local temperature reduction of almost 40 °C can be recognised. This results in a high local tensile stress around the pores and explains the failure of the component at a rather low voltage.

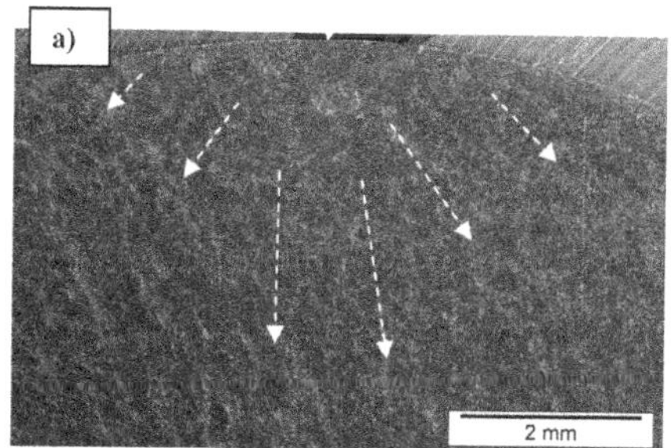

Fig. 10: a) Fracture surface, with fracture origin (thick arrow) and crack propagation (dashed arrows).

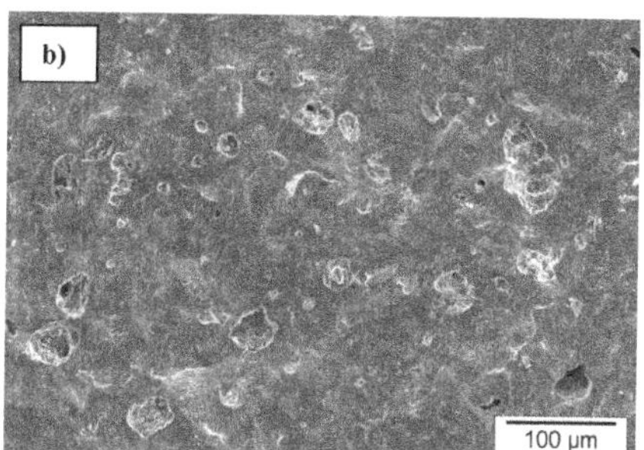

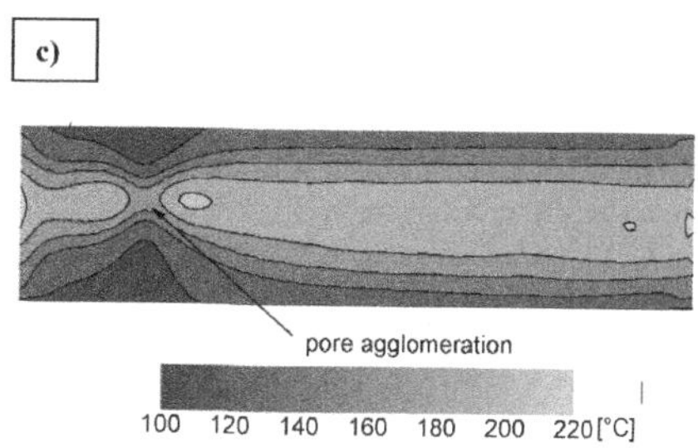

Fig. 10: b) Failure causing pore agglomeration.
c) Thermographic picture of circumferential surface of the thermistor during failure after 0.56 s of applied voltage (560 V).

Failure by a Local Enrichment of Iron

The thermistor component, presented here, failed at a voltage of 400 V. This is the lowest applied voltage during the experiments. The reason for failure is the occurrence of a very large region, which has another color than the rest of the resistor and which was situated at the height of the equator (Fig. 10a). This region has a diameter of about 800 µm and a small dent in its center (see Fig. 10a to c). It can easily be detected by the naked eye. In the thermographic picture the region remains cold during the switching process (compared to the rest of the equator, Fig. 10d), which indicates that it is electrically insulating. An elemental analysis of the region revealed the existence of iron (around one weight percent). It is known that iron ions turn the doped semiconducting barium titanate crystals into an insulator or – depending on the concentration – at least increases the resistivity in a large region. Most likely the iron was incorporated during the powder processing in steel coated mills. In this case the mechanical action of the "inclusion" is negligible, because the mechanical properties are almost not affected by the solute iron. But the large change in the electrical resistance causes a large region to remain much colder than the surrounding material. The thermographic analysis indicates a local drop below the equator temperature of about 30 °C. This cold region is strained by the surrounding extending regions and large tensile stresses occur as a result.

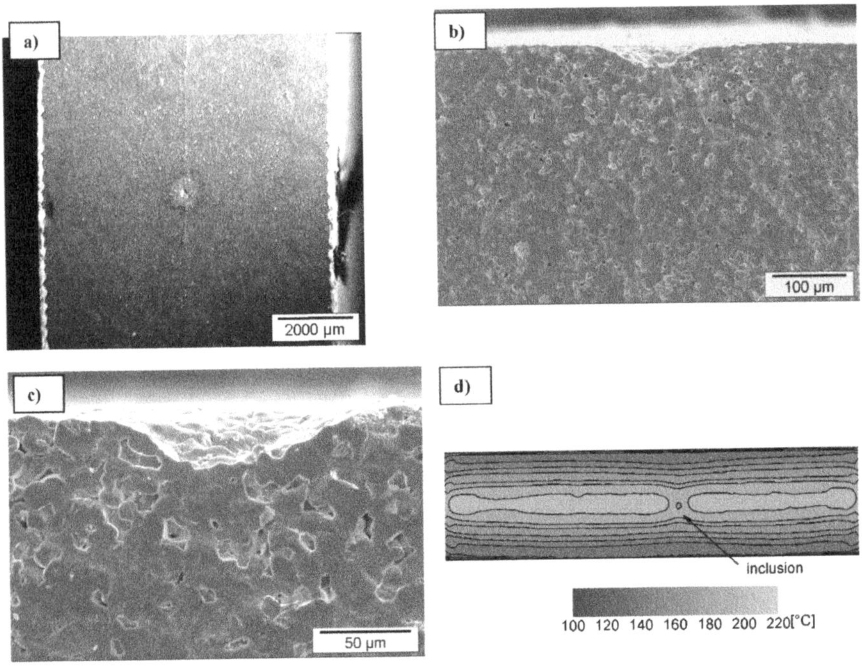

Fig. 10: a) The surface of the disc-shaped PTC-R showing the inclusion and the crack going through it.
b) Fracture origin (dent in the center of the inclusion), and
c) enlarged picture of the dent of the origin.
d) Thermographic picture of the circumferential surface of the thermistor during failure after 1 second of applied voltage (400 V): the cold spot at the inclusion can clearly be recognised.

Failure by a Resistance Inhomogeneity

This case is quite similar to the case discussed above. The presented part failed at a voltage of 480 V. Microscopic investigations (Fig. 11b to d) show neither pores nor inclusions as the failure-causing defect. The location of the fracture origin could be identified (Fig. 11 b and c). There the infrared measurement revealed a local temperature minimum. The (small) temperature difference on the equator (app. 7 °C, Fig. 11a) induced additional thermal stresses, which caused the failure of the part. The local temperature difference indicated a local change of the resistivity, probably caused by a local inhomogeneity of donor elements.

It has been mentioned in the introduction that a local temperature of about 7 °C may cause a tensile stress of up to 30 MPa which has to be added to the macroscopic stress field. Obviously these additional stresses increased the tensile stresses in such a way that fracture could initiate.

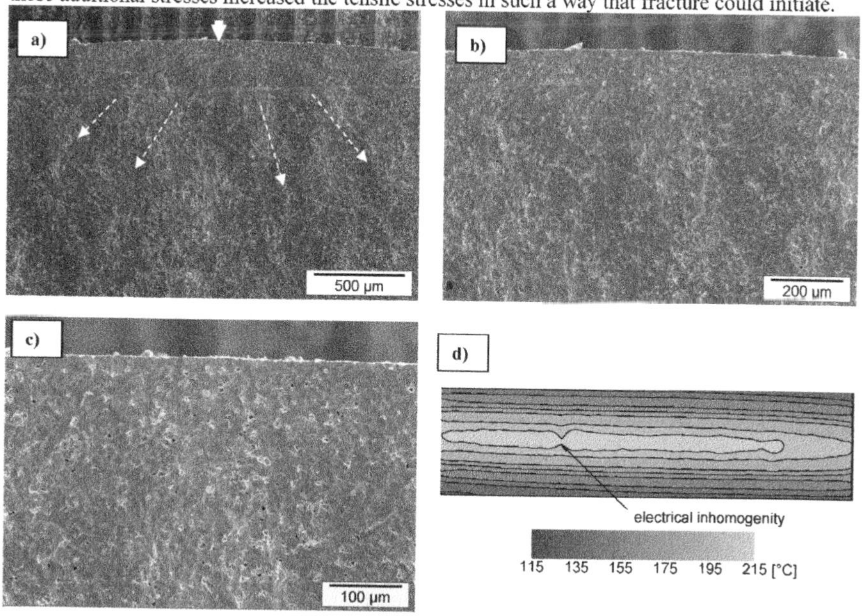

Fig. 11: a) Fracture surface showing the crack propagation but no defect at the origin.
b) Fracture origin, and c) Fracture origin enlarged.
d) Thermographic picture of the surface of the thermistor during failure after 1 second of applied voltage (480 V).

SUMMARY

In most of the mechanical tests in this study, failure of PTC components starts from pores. The existence of these pores is related to hollow agglomerates resulting from the spray drying process.[8] The high Weibull modulus of the strength distribution indicates a quite similar size of the fracture initiating pores. Such pores with a radius larger than 50 µm could be detected by ceramographic analysis.

In electrical tests tensile stresses occur due to temperature differences, which result from Joule heating and heat conduction. In homogeneous ceramic parts two regions with high tensile stresses occur: The first one is around the center of the top (bottom) plane of the component. In this area the ceramic is colder than in the surrounding material due to heat transfer into the metallic contacts and the wire. The second one is the equator of the component. Here the component's surface develops tensile stresses due to the faster heating of its mid-plane.

The macroscopic temperature and stress field is of great significance for the fracture of the components. Fracture occurs if large enough flaws can be found in the regions of high tensile stresses. If the switching process is slow, distributions of the macroscopic temperature field can be compensated by heat conduction. In the other cases additional local minima or maxima in the temperature field occur, which also cause additional tensile stresses.

In the first case, the fracture of the PTC components is quite similar to that in a strength test. The mechanical stress concentrations of microstructural flaws trigger the initiation of fracture. In the second case the local temperature and stress fields produced by the flaws becomes important. Especially at temperatures over the Curie temperature very small temperature differences may cause significant tensile stresses. These stresses may even be more important than stress concentrations due to mechanical reasons (pores, changes in elastic properties). In fact examples were shown were local slight changes in the chemical concentration of dopants caused cold spots in the switching component, and these serve as the fracture origins.

ACKNOWLEDGMENTS

Financial support by the Österreichische Forschungsförderungsgesellschaft mbH, the Province of Styria, the Steirische Wirtschaftsförderungsgesellschaft mbH and the Municipality of Leoben within research activities of the Materials Center Leoben under the framework of the Austrian Kplus Competence Center Programme is gratefully acknowledged.

The kind support of this work by EPCOS OHG is gratefully acknowledged. Special thanks are given to Mr. F. Aldrian and Dr. K. Reichmann.

REFERENCES

[1] W. Heywang: "Resistivity Anomaly in Doped Barium Titanate," *J. Am. Ceram. Soc.* 47 484-90 (1964).

[2] G. Jonker, "Some Aspects of Semiconducting Barium Titanate," *Solid State Electronics* 7 895-903(1964).

[3] B. Kulwicki, A. Purdes, "Diffusion Potentials in $BaTiO_3$ and the Theory of PTC Materials," *Ferroelectrics* 1 253-63(1970).

[4] W. Heywang, "Semiconducting Barium Titanate," *J. Mat. Sci.* 6 1214-1224 (1971).

[5] H. Parkus, Thermoelasticity, 2nd revised and enlarged edition, Springer-Verlag, Wien, New-York, (1976).

[6] R. Danzer, P. Supancic, "Thermal Shock Behavior of $BaTiO_3$ PTC-Ceramics," Ceramic Engineering and Science Proceedings, 27th International Cocoa Beach Conference on Advanced Ceramics & Composites: B, Cocoa Beach, Florida, Vol. 24 [4], 371-376 (2003).

[7] C. Dewitte, R. Elst, F. Delannay, "On the Mechanism of Delamination Fracture of $BaTiO_3$-Based PTC Thermistors," *J. Eur. Ceram. Soc.* 14, 481-492 (1994).

[8] P. Supancic, "Fracture and Fractography of Electroceramics," *Key Engineering Materials* 223, 69-78 (2002).

[9] A. Platzer, "Quantitative Thermographie von PTCs," doctoral Thesis (Montanuniversität Leoben), (2006).

[10] A. Börger, P. Supancic, R. Danzer, "The Ball on Three Balls Test for Strength Testing of Brittle Discs - Stress Distribution in the Disc," *J. Eur. Ceram. Soc.* 22, 1425-1436 (2002).

[11] A. Börger, P. Supancic, R. Danzer, "The Ball on Three Balls Test for Strength Testing of Brittle Discs - Part II: Analysis of Possible Errors in the Strength Determination," *J. Eur. Ceram. Soc.* 24, 2917-2928 (2004).

[12]A. Platzer, P. Supancic, R. Danzer, "Thermography of Advanced Nonlinear Electroceramic Components," *J. of Advanced Engineering Materials*, to be published (Nov. 2006).

[13]P. Supancic, "Mechanical Stability of $BaTiO_3$-based PTC Thermistor Components: Experimental Investigation and Theoretical Modeling," *J. Eur. Cer. Soc.* 20, 2009 – 2024 (2000).

[14]P. Supancic, R. Danzer, „Mechanische Zuverlässigkeit von keramischen Kaltleiterbau-elementen - Möglichkeiten zur Verbesserung," *Zeitschrift für Metallkunde* 95 [7], 607-617 (2004).

A NEW TECHNIQUE FOR MEASURING THE FRACTURE SURFACE ENERGY OF CERAMICS

William Carty, Brian Pinto, and Katherine Sieg
Kazuo Inamori School of Engineering, Alfred University
Alfred, NY 14802

Mithilesh Kumar
Indian Institute of Technology
Bombay, India

ABSTRACT

A novel technique has been developed to introduce an ideal flaw on the surface of a ceramic material allowing the straightforward measurement of fracture surface energy. When a brittle coating (such as a whiteware glaze) with a greater coefficient of thermal expansion (CTE) is applied to a ceramic substrate with a lesser CTE prior to heat treatment, the CTE mismatch places the coating in tension on cooling. This tension results in cracking of the coating – a defect in glazes commonly referred to as crazing. These flaws are essentially atomically sharp cracks that terminate at the body-glaze interface. Therefore the thickness of the glaze layer dictates the size of the flaw and strength of the sample. Using this technique, fracture surface energy values were measured for silica-based and alumina-based electrical porcelain bodies to be 4.5 J/m^2 and 8.4 J/m^2, respectively. This technique applied to polycrystalline alumina yielded a fracture surface energy value of 47.5 J/m^2 consistent with previously published results.

INTRODUCTION

The initial goal of this work was to identify the strength-controlling defects in porcelain bodies. Fractography was not successful in identifying flaw origins or microstructural features responsible for porcelain strength, being able to locate flaw origins in less than 10% of the samples evaluated.[1] It is globally believed that quartz grains are responsible for failure in porcelains, but this is not supported by the fractography results nor direct strain measurements of quartz grains in porcelain bodies via X-ray diffraction.[1,2] To determine the critical flaw size, it was necessary to measure the fracture surface energy. To measure fracture surface energy required the introduction of a know flaw size – a task that is particularly difficult in the heterogeneous microstructures of porcelains.

Within the traditional ceramics arena, it is commonly understood that a glaze in compression on the surface of a whiteware body increases the strength of the article. The electrical insulator industry has exploited this idea extensively. In addition to having a high-strength porcelain body initially, highly compressed glazes can increase the strength of the article by as much as 50%. Interestingly enough, the converse, that a glaze in tension, and thus crazed, decreases the strength of a whiteware body does not seem to be common knowledge. Glazes on commercially produced ware are almost always in compression thus avoiding crazing and enhancing strength, but mainly to create an impervious surface for food safety. Art pottery, however, even functional ware, is frequently crazed, essentially supporting the idea that the relationship between strength and crazing is not well understood. The state of stress in a glaze, and correspondingly, the body, is dictated by the coefficient of thermal expansion (CTE) mismatch between the body and the glaze. A glaze with a lower CTE than the body is placed in compression during cooling, while a glaze with a higher CTE is placed in tension (and thus crazes). The state of stress in the body and the glaze, and the tendency of the article to fail (in one form or another) is complex and depends on a number of factors in addition to CTE mismatch, such as the strength and elastic modulus of the body, the glaze thickness, the shape of the article, etc.

Recent work on the effect of CTE mismatch on the strength of glazed ceramics indicated that the strength scaled linearly with the CTE difference (ΔCTE), defined as the CTE_B-CTE_G (where the subscripts *B* and *G* refer to the body and glaze, respectively).[3] Strength as a function of ΔCTE is shown in Figure 1 for an alumina porcelain body (in which calcined alumina is used as the filler; designated with a subscript *A*) and a standard strength porcelain body (in which ground quartz is used as a filler; designated with a subscript *S*). The alumina and silica porcelain bodies in Figure 1 were prepared in the same factory using an identical process.

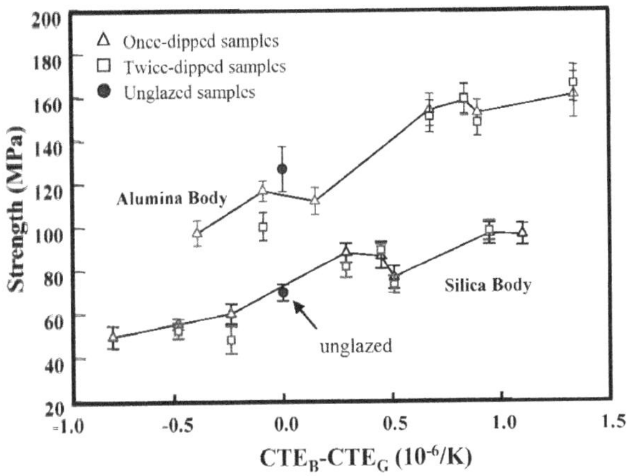

Figure 1. Effect of ΔCTE on the strength of glazed rods prepared from alumina and silica porcelain bodies. ΔCTE < 0 indicates that the glaze is in tension; the glaze is in compression when ΔCTE > 0.

In the cursory evaluation of the data, the strength values were normalized to the elastic modulus of the bodies, and the slope of σ/E versus ΔCTE was identical for the two bodies. If it is assumed that the fracture surface energy values for the two porcelains are equivalent ($\gamma_A=\gamma_S$), the intercepts of the two data sets indicate that the critical flaw size for the two bodies differs by only a factor of 2.23 ($2.23 \cdot c_A=c_S$). The alumina and quartz particle size (D_{50}) used to prepare these bodies differs by approximately a factor of ten ($10 \cdot D_{50,A}=D_{50,S}$). If the literature assertion that the filler particle size controls the strength of porcelain is correct, the critical flaw size difference should reflect the difference in filler particle size.

It was also observed that when the glazes were crazed (ΔCTE<0), the strength decreased further with increasing glaze thickness – an trend that was not commonly observed when the glaze was in compression. It became evident that the crazing cracks were functioning as the strength-controlling flaws (in samples with crazed glazes).

According to Griffith, the failure stress of a brittle material (σ_f) is related to the elastic modulus (E), the energy required to form a new surface (the fracture surface energy, γ), and the critical flaw size (c), by the following relationship:[4]

$$\sigma_F = \sqrt{\frac{2E\gamma}{\pi c}} \quad (1)$$

Rearranging for surface energy yields:

$$\gamma = \frac{\sigma_F^{\ 2}\pi c}{2E} \quad (2)$$

and correcting for Poisson's ratio (ν):[5]

$$\gamma = \frac{(1-\nu^2)\sigma_F^{\ 2}\pi c}{2E} \quad (3)$$

Rearranging Equation 3 gives the linear relationship in which the slope of a reduced strength (σ_R) versus c^{-1} is the surface energy:

$$\sigma_R = \frac{\pi(1-\nu^2)\sigma_F^{\ 2}}{2E} = \gamma c^{-1} \quad (4)$$

Therefore to test this hypothesis it was simply necessary to create a glaze that crazes on cooling; coat porcelain rods with different glaze thicknesses; measure the strength of rods coated with that glaze (in 3- or 4-point MOR); measure the glaze thickness after failure; then calculate the slope of σ_R as a function of c^{-1}.

EXPERIMENTAL PROCEDURES

Three porcelain bodies were evaluated, as listed in Table I. Two of the bodies were industrial electrical insulator bodies (Silica X and Alumina X) and had been processed using exactly the procedure at the factory. The remaining body (Silica Y) was a laboratory body that had been used for several other experiments, including the work done previously to evaluate quartz particle strain. The bodies were slurry processed, filter pressed, then extruded using a vacuum de-aired pug mill to form 13-mm-diameter rods with lengths of approximately 170 mm. These samples were dried in a laboratory oven, bisque fired to 1260°C (with a 3-hour dwell at peak temperature); the rods were fully dense with no open porosity.

Table I. Body compositions listed in oxide weight percent.

Body	SiO_2	Al_2O_3	Na_2O	K_2O	MgO	CaO	TiO_2	Fe_2O_3
Silica X	71.56	21.52	0.96	3.91	0.16	0.35	0.77	0.65
Alumina X	43.84	49.92	0.86	3.3	0.19	0.32	0.74	0.66
Silica Y	64.84	28.15	1.11	4.15	0.15	0.36	0.66	0.57

The glaze was prepared by ball milling a raw glaze (composition listed in Table II) then applied using a dip coating process. The glaze thickness was increased by re-dipping the rods. Some rods were dipped multiple times. The ends of the rod were not glazed to prevent sticking of the rods to the refractory during the firing process – one end of the rod was initially dipped in wax to prevent glaze adherence during the coating process. The rods were fired vertically (to prevent warpage due to pyroplastic deformation) to a temperature of 1160°C (also with a 3-hour dwell). Bulk density and porosity measurements were taken according to an ASTM standard.[6] Elastic modulus (E) and Poisson's ratio were measured using a pulse-echo technique on 3 mm unglazed parallelepipeds cut from the rods.[7] (The elastic modulus of the glaze was not specifically measured for this study, but is typically about 70 GPa). The physical property data are listed in Table III and are consistent with previously reported values for porcelains.[9-12]

Table II. Glaze composition in Seger formula and raw material weight percent.

	Seger
KNaO	0.30
CaO	0.70
Al_2O_3	0.20
SiO_2	1.20

Raw Material	*Wt %*
G-200 Feldspar	47.0
Calcium Carbonate ($CaCO_3$)	35.8
EPK Clay	5.8
Pearl Ash (K_2CO_3)	5.3
Soda Ash (Na_2CO_3)	4.1
Flint (SiO_2)	2.1

Table III. Physical property data for the three porcelain bodies. Standard deviation values are listed in parentheses.

	Bulk Density (g/cm^3)	Apparent Porosity (%)	Elastic Modulus (GPa)	Poisson's Ratio
Silica Body X	2.41 (0.003)	0.25 (0.11)	71 (0.43)	0.18 (0.002)
Alumina Body X	2.72 (0.005)	0.16 (0.06)	127 (1.0)	0.23 (0.008)
Silica Body Y	2.50 (0.003)	0.23 (0.16)	80 (0.81)	0.20 (0.004)

The strengths of coated and uncoated rods were measured using the 3-point MOR test with a 90 mm span following the ASTM standard testing method using a loading rate of 0.1 mm/min.[8] The 90 mm span ensured that glazed portions of the rods rested on the support spans. The rods were marked on the tensile side for the measurement of the glaze thickness after testing.

After strength measurements the fracture surfaces were prepared for examination using scanning electron microscopy (SEM). Using a slow-speed diamond saw a 20-mm-thick section incorporating one side of the fracture surface was obtained from each sample. The fracture surface was then chemically etched using a 20% aqueous solution of HF at 0°C for 10 seconds to help clarify any surface features. The samples were mounted and sputter-coated with Au-Pd. Each sample was then examined in SEM to determine glaze thickness (c). Images at two different magnifications were taken for each sample, and six glaze thickness measurements were taken of each sample. The average and standard deviation of glaze thicknesses were recorded.

RESULTS AND DISCUSSION

Figure 2 illustrates a typical testing sample with a crazed glaze. The rod ends were not glazed to allow the glazed rods to be heat treated vertically. Since the rods were broken in 3-point flexure, the maximum bending moment, and hence, failure occurs at the exact center of the sample, eliminating the necessity for a completely glazed rod. The support span was within the boundaries of the glaze coating.

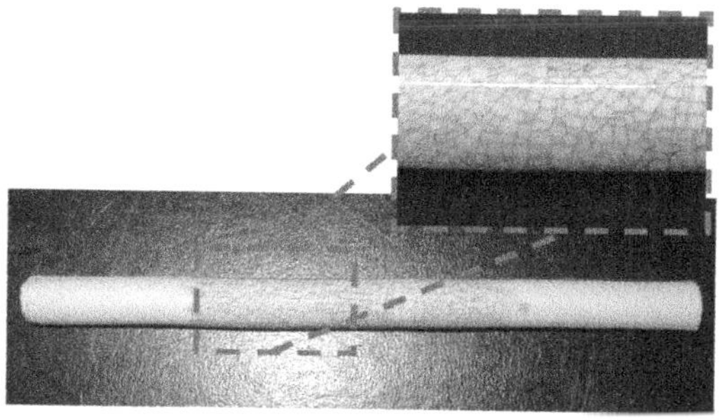

Figure 2. A porcelain rod with a crazed glaze (highlighted with ink).

The interface the cracking characteristics of the crazed glazes were examined using SEM. As illustrated in Figure 3, the glaze-body interface is clearly visible (and the glaze thickness easy to measure). Cracks within the glaze surface, caused by crazing, were also visible in each specimen, spanning the entire glaze thickness and abruptly ending at the glaze/body interface. This phenomenon was observed in all samples, lending support to the hypothesis that the glaze produces a surface flaw of measurable dimension. The significance of these images is they show an abundance of equally sized flaws inhabiting the glaze layer.

Figure 3. SEM photomicrographs illustrating that cracks in the glaze layer stop abruptly at the glaze-body interface. Also evident is the glaze thickness.

Rods with different glaze thicknesses were measured and the results of those measurements were used to calculate the reduced failure stress (σ_R, from Equation 4). The elastic modulus and Poisson's ratio values were measured on un-glazed rods and used for the calculations. (Elastic modulus and Poisson's ratio values were not measured on any glazed rods.) The data are presented in Figure 4: (σ_R) plotted as a function of the inverse of the measured glaze thickness (c^{-1}). The slope is the fracture surface energy.

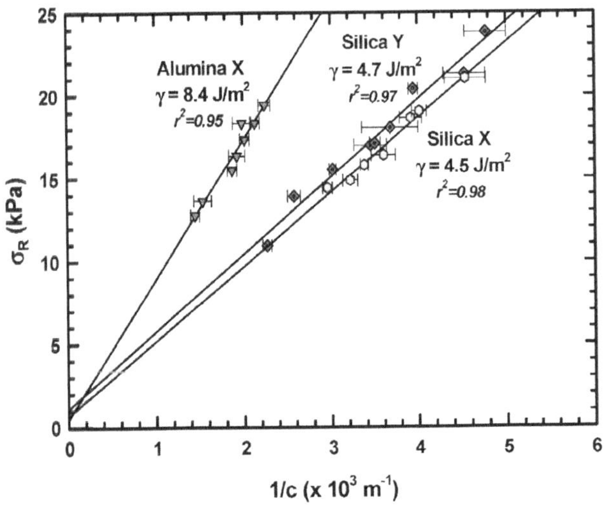

Figure 4. Reduced failure stress, σ_R, versus the inverse of the glaze thickness (c^{-1}). The slope is the fracture surface energy.

As illustrated, the fracture surface energy of the two silica porcelains is nearly identical at 4.5 and 4.7 J/m^2. The alumina body fracture energy is significantly higher at 8.4 J/m^2. This is consistent with a (nearly?) purely elastic failure and the fact that the elastic modulus of the alumina porcelain is significantly higher than that of silica porcelain. Overall this and other data (unreported) indicate that fracture surface energy scales linearly with elastic modulus.

The elastic modulus, Poisson's ratio, measured fracture surface energy (from Figure 4), and the failure strength of unglazed rods were then used to calculate the critical flaw size (listed in Table IV) for these three bodies using a rearrangement of Equation 4:

$$c = \frac{2E\gamma}{(1-\nu^2)\sigma_F^{\ 2}\pi}. \tag{5}$$

Table IV. Critical flaw sizes calculated for the three porcelain body compositions.

Body	c (μm)	E (GPa)	ν	γ (J/m²)	σ_F (MPa)
Silica X	74	71	0.18	4.5	74.4
Alumina X	71	128	0.23	8.4	134.4
Silica Y	89	80	0.20	4.7	78.2

Recall that Silica X and Alumina X had been processed identically in an industrial setting; Silica Y was processed in the lab. The industrial bodies had nearly identical critical flaws, indicating that the flaws are not a function of the filler particle size. It is proposed that the strength-limiting flaws in porcelains are a result of poor mixing, and specifically are clusters of non-plastic particles. It is proposed that these flaws are non-plastic particles (typically quartz, feldspar, or nepheline syenite) because this flaw size would represent small clusters (less than 10 particles) based on the typical size of the non-plastic particles used in modern industrial porcelain production – >98% passing a 325-mesh screen (45 μm opening size). Additional work is now needed to track measured flaw sizes with improved mixing protocols.

APPLICATION TO POLYCRYSTALLINE ALUMINA

To illustrate that this technique has greater applicability than just porcelains, alumina rods were also glazed and tested using a similar technique. The differences were that the glaze was developed to craze on alumina; alumina rods were purchased (rather than fabricated); and the samples were measured using four-point bending (rather than three-point bending). The results were equally as promising.

The measured fracture surface energy for a polycrystalline alumina rod was 47.5 J/m^2. This is consistent with published values for the fracture surface energy of alumina, as listed in Table V.

Table V. Measured and published fracture surface energies for alumina.

Technique (reference)	γ (J/m^2)	Reference
This work (crazed glaze)	47.5	This work
Work of Fracture	41.4 ± 10	13
Warren Analysis	38.5	14
Mouginout & Maugis Analysis	38.4	14
Notch Beam (blunt notch)	31.5 ± 4	15
Notch Beam (sharp crack)	36.3 ± 3	15
Double Torsion	43.0 ± 4	15
Double Cantilever Beam (blunt notch)	41.5 ± 4	15
Double Cantilever Beam (previously propagated)	51.0 ± 18	15

CONCLUSIONS

A new technique, based on the standard notched-beam method, was developed to measure the fracture surface energy of porcelain bodies. By selecting a glaze with a greater CTE than the body, the glaze cracks on cooling, creating a flaw with easily measured dimensions. The glaze thickness, which is then the critical flaw, can be easily altered during glaze application and can be readily measured after fracture. The fracture surface energy is the slope of reduced strength, σ_R (incorporating strength of the glazed rods, elastic modulus, and Poisson's ratio) as a function of inverse glaze thickness (c^{-1}) and was measured to be approximately 4.6 J/m^2 for two silica porcelains and 8.4 J/m^2 for an alumina porcelain. The results indicate that the critical flaw size is not consistent with the size of the filler particle size, and it is proposed that the strength-limiting defects in porcelains are particle clusters that fail to be eliminated during the mixing and processing stages of porcelain fabrication. This technique was also applied to a polycrystalline alumina generating results that were consistent with published fracture surface energy values.

REFERENCES

[1]B. M. Pinto, “Residual Strain in Quartz Particles in a Porcelain Matrix as a Function of Particle Size via X-Ray Diffraction,” M.S. Thesis, Alfred University, 1999.

[2]B. M. Pinto, “Critical Evaluation of Strength-Controlling Variables in Ceramics,” Ph.D. Thesis, Alfred University, 2005.

[3]J. L. Benson, "Effect of Glaze Variables on the Mechanical Strength of Whitewares"; M.S. Thesis, Alfred University, Alfred, NY, 2003.

[4]A. A. Griffith, "The Phenomena of Rupture and Flow in Solids," *Phil. Trans. Royal Soc. of London*, A221, 163-198, (1920).

[5]R. W. Davidge and G. Tappin, "The Effective Surface Energy of Brittle Materials," *J. Mater. Sci.*, 3, 165-73 (1968).

[6]ASTM Standard C 830-93, “Standard Test Methods for Apparent Porosity, Liquid Absorption, Apparent Specific Gravity, and Bulk Density of Burned Refractory Brick and Shapes by Boiling Water,” *Annual Book of ASTM Standards*, 15.01, ASTM, West Conshohocken, PA (2004).

[7]ASTM Standard C 848-88, “Standard Test Method for Young's Modulus, Shear Modulus, and Poisson's Ratio For Ceramic Whitewares by Resonance,” *Annual Book of ASTM Standards*, 15.01, ASTM, West Conshohocken, PA (2000).

[8]ASTM Standard C 1161-94, “Standard Test Method for Flexural Strength of Advanced Ceramics at Ambient Temperature,” *Annual Book of ASTM Standards*, 15.01, ASTM, West Conshohocken, PA (2004).

[9]O. I. Ece and Z. Nakagawa, "Bending Strength of Porcelains," *Ceram. Int.*, 28, 131-40 (2002).

[10]S. B. Vazquez, J. C. M. Valazquez, and J.R. Gasga, "Alumina Additions Affect Elastic Properties of Electrical Porcelains," *Am. Ceram. Soc. Bull.*, 77 [4], 81-5 (1998).

[11]L. Mattyasovszky-Zsolnay, "Mechanical Strength of Porcelain," *J. Am. Ceram. Soc.*, 40 [9], 299-306 (1957).

[12]R. Harada, N. Sugiyama, and H. Ishida, “Al_2O_3-Strengthened Feldspathic Porcelain Bodies: Effects of the Amount and Particle Size of Alumina,” *Ceram. Eng. Sci. Proc.*, 17 [1], 88-98 (1996).

[13]D. B. Binns and P. Popper, “Mechanical Properties of Some Commercial Alumina Ceramics, “*Proc. British Ceram. Soc.*, 6, 71-82 (1966).

[14]S. Bouras and B. Bouzabata, “Study of Hertzian Indentation on a Transparent Vitroceramic,” *Mater. Chem. Phys.*, 43, 127-134 (1996).

[15]R. W. Rice, S. W. Freiman, and P. F. Becher, “Grain-Size Dependence of Fracture Energy in Ceramics: I, Experiment,” *J. Am. Ceram. Soc.*, 64 [6], 345-350 (1981).

SECTION V

FRACTOGRAPHY OF DENTAL AND BIOMATERIALS

THE INCREASING ROLE OF FRACTOGRAPHY IN THE DENTAL COMMUNITY*

Janet B. Quinn
ADAHF Paffenbarger Research Center, NIST
Mailstop 8546
Gaithersburg, MD 20899-8546

Susanne S. Scherrer
University of Geneva, School of Dental Medicine
Dept. of Prosthodontics-Biomaterials 19, Barthélemy-Menn
1205 Geneva, Switzerland

George D. Quinn
National Institute of Standards and Technology
Mailstop 8520
Gaithersburg, MD 20899-8520

ABSTRACT

The development of new fabrication capabilities along with new alloys, polymers, composites and advanced ceramics has revolutionized many industries – including the manufacture of dental restorations. The clinical testing of dental materials has barely kept pace with these technological advances and new product development. Convincing longevity studies encompass time periods much longer than current market-driven introductions of novel dental material systems can afford. Consequently, greater dependence is being placed on failure analysis, including fractographic examination.

These new dental materials are not all ceramics. While any brittle material could theoretically be analyzed by some of the fractographic techniques used in the advanced ceramic community, there has been little cross-over. The required properties of dental restorations, include aesthetics, biocompatibility and mechanical function, and appear to straddle many classes of materials. This is one likely reason for the current multiplicity of standards and test techniques in the dental literature. The establishment of relevant test techniques, longevity assessment and design criteria are linked by failure analyses and fractography.

In this paper, the increasing importance of fractography in the field of dental restorations in general, and the idea of an application-based approach, is presented. Case study examples of failed ceramic crowns are included, and the establishment of a 'fractographic library' of failed dental restorations is discussed.

INTRODUCTION

Materials to replace lost or damaged human teeth have been researched for thousands of years. The ancient Etruscans were among the first to create dental bridges, using carved ivory or discarded human teeth. Throughout the following centuries, carved teeth from oxen, calves, walrus and elephant tusks, various animal bones and even seashells have been used as tooth replacement materials.[1] It wasn't until the late 1700's that the first modern ceramic achieved widespread use as a dental material - porcelain. This led to the first recorded inventors' patent for artificial teeth. It was awarded, with considerable controversy, to Dubois de Chémant by the French King, Louis XVI, in 1789.[2]

---

*Official contribution of the National Institute of Standards and Technology; not subject to copyright in the U.S.

Since then, improved forms of dental porcelains have continued to be developed, as well as other "high tech" materials for dental restorations. Among modern dental materials, ceramics have come to be regarded as the most aesthetic and biocompatible,[3] and recent trends are to avoid metals entirely. While aesthetics are often improved, non-metallic dental materials are less durable than their more ductile predecessors. This need for improvement is fueling continued development of ceramics, composites and various ceramic combinations, such as layers and unusual or textured microstructures, as well as different processing techniques and additives, such as nanopowders. Research into new dental materials is also spurred by expected increases in the need for dental restorations among a critical and demanding public.

A study of dentate adults over 44 years of age in the United Kingdom revealed that 50 % had teeth that were filled and crowned.[4] The study predicted that, in spite of a general decrease in caries among the young, requirements for dental restorations are likely to increase over the next ten years. There are indications that the United States and other northern European countries are experiencing similar dental demographic profiles.[5,6,7] Older people today can expect longer, healthier life spans, and most want to maintain a functional and aesthetically acceptable dentition. Nonmetallic materials have increased importance in this regard, for the dentate elderly are more likely to be on multiple drug regimes, have reduced saliva flow, have difficulty in maintaining oral hygiene and have compromised immune systems or allergies.

Fashion trends and social concerns are also driving the market for an increased need for restorations and material development. Higher expectations for an enhanced appearance are prompting dental restorative work that may not be driven by esthetics alone. Factors such as peer pressure, magazines and reality make-over shows on television are now leading even young people to the dentist's office for cosmetic tooth restorations. The current popularity of mouth jewelry is particularly damaging to teeth. Chipping is not only common between the anterior teeth of subjects with tongue piercings, but on posterior teeth as well.

The clinical testing of dental materials can no longer keep pace with the increased expectations and new product development. Current clinical statistical tests can take years to determine the lifetime expectancies of the most successful materials. Consequently, greater dependence is necessarily being placed on alternative methods, such as fractography and failure analyses, to predict component survival times.

DENTAL RESTORATION MATERIALS AND LONGEVITY

How long do current, non-metallic dental restorations last? This is a complicated question, for there are many kinds of dental materials, and each application is unique. In general, posterior teeth exert much more force than anterior teeth, and there can be a great deal of variability in the shape and extent of a tooth restoration. Furthermore, clinical longevity data is reported in different ways over dissimilar time periods. With this in mind, a cautious, rough estimate of longevity is presented in Fig. 1 from a variety of sources.[8-26] The data is for failed ceramic crowns, excluding partial coverage crowns, inlays, onlays and bridges. Some of the data is from reported total crown failures that might also include failures from decay, tooth pain, soft tissue irritation or discoloration. However, the majority of ceramic restoration failures are from fracture. The largest source of scatter is likely due to differences in crown placement, for anterior restorations have a higher success rate due to the relatively low stresses generated in the front part of the mouth. If the data were limited to posterior teeth alone, total failure rates would be higher. Posterior applications are the most challenging for dentists and material scientists.

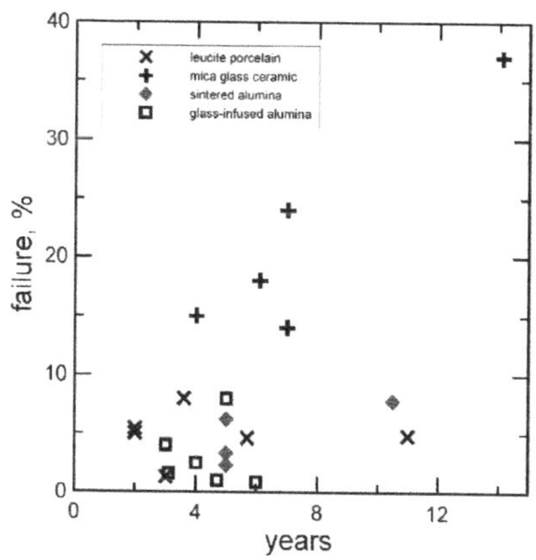

Figure 1. General longevity for some common dental ceramic crown restorations. Many different sources and reporting criteria were used for the same materials, contributing to the scatter.

The materials in Fig. 1 represent different types of ceramics that are typically used in restoring teeth. They are all brittle, but are not as fractographically friendly as failure analysts would wish.

Leucite porcelains might be thought of as composites with particles of the mineral, leucite ($KAlSi_2O_6$), mixed within a matrix of aluminosilicate glass. These "feldspathic porcelains" are of slightly different compositions, with different phases, crystallite sizes and amounts of leucite within the glassy matrix. The leucite porcelain in Fig. 1 (IPS Empress, Ivoclar, Amherst, NY) is typical,[†] with about 0.24 volume fraction crystalline content. The amount and distribution of the crystals affect the ability to distinguish fractographic features as well as affect the properties. The leucite porcelains are among the most difficult materials to analyze fractographically, and many variations of these porcelains are used in dental restorations. Leucite porcelain crowns are usually placed on anterior teeth because of their good aesthetics, so the failure rates in the figure are probably skewed to the low side.

The mica glass ceramics might similarly be considered composites, with a much different chemical composition. In this material, tetrasilicic crystals are precipitated directly from the melt, with the cooling rate and temperature dictating the final crystal size and quantity. The dental glass ceramic in Fig. 1 (Dicor, Corning, NY) has about 0.65 volume fraction mica crystals within the glass matrix. This material is somewhat less difficult to fractographically analyze, in spite of higher crystallinity content, probably because of the small, homogenous size and shape of the mica crystal platelets. This material was the first all-ceramic crown commercially developed for dentistry, and therefore many Dicor crowns were used to restore both anterior and

[†]Certain commercial equipment, instruments, or materials are identified in this paper to foster understanding. Such identification does not imply recommendation or endorsement by the National Institute of Standards and Technology, nor does it imply that the materials or equipment identified are necessarily the best available for the purpose.

posterior teeth for many years. The high failure rates, as shown in the figure, have led to the discontinuation of this material for dental restorations as well as more cautious acceptance of more recent materials.

The sintered alumina in Fig. 1 (Procera, Nobel Biocare, Zurich, Switzerland) is isostatically cold-pressed before sintering. The resulting fracture surfaces are subject to the usual difficulties in discerning fractographic markings in sintered polycrystalline ceramics. Polycrystalline materials such as this are usually glass-coated (veneered) for a more natural appearance in the mouth.

The glass-infused alumina in Fig. 1 (InCeram, Vita Zahnfabrik, Bad Säckingen, Germany) has an unusual microstructure. This dental ceramic consists of a dry-pressed matrix of alumina with 0.31 volume fraction porosity into which a lanthanum-aluminosilicate glass is infused. As expected, the resulting pseudo-composite microstructure results in fracture surfaces on which fractographic features are distinguished with difficulty.

Thus, two main problems are brought out by Fig.1. First, dental materials can be of vastly different ceramic systems, all of which are difficult to fractographically analyze. Rice comments in his review of fracture features that mirrors and other normally observed fracture markings can be totally obliterated with larger grain sizes, lower stress failures, and less than 25 % transgranular fracture.[27] Unfortunately, dental material microstructures generally satisfy several of these criteria. The second point of Fig. 1 is that longevity determinations are difficult and very time-consuming. The low percentage failure rates in the early years constitute only a few of hundreds of restorations. Many restorations have to be studied for many years in order to get a clear, statistically significant picture, rather than general trends. Obviously, some standards would help, such as establishing consistent failure reporting criteria, including location and time frame.

These two problems are likely to be exacerbated in the future. New dental materials are being developed. The most recent include various forms of resin composites, lithium disilicate glass-ceramics and polycrystalline zirconia. Fractographic analyses of these newest materials, presented later in this report, are not less problematic than the materials in fig. 1. The more pressing problem, however, is that clinical durability studies of the newest materials are not commercially feasible, especially as they are hoped to outlast previously available dental materials. As difficult as it may be, fractographic analyses may be necessary if a practicable means of predicting clinical longevity is to be determined. Other approaches, such as strength testing, have not yielded effective correlations with clinical longevity and this further heightens the importance of fractographic failure analyses. Besides aiding in predicting failure by determining failure causes, fractography might also be used to ascertain why strength studies fail to correlate with in vivo failures.

FRACTOGRAPHIC MARKINGS ON DENTAL MATERIALS

As previously mentioned, dental materials encompass many compositions and microstructures, most of which do not facilitate fractographic failure analyses. In difficult cases, it is helpful to study fracture surfaces of broken specimens in which the stress state and approximate crack initiation sites are known. Familiarity with material-specific fracture processes can then be useful for analyses of broken restorations with more complicated geometries. Another approach for fractographic characterization of dental restorations has also yielded good results. This involves focusing on the veneer layer common to many restorations.

While the porcelain restoration materials generally have good aesthetics, the polycrystalline materials such as alumina, zirconia and even composites with high crystalline content are usually coated with glass or porcelain for a more natural appearance. A glassy coating can also aid in avoiding wear of opposing dentition when a hard, polycrystalline restoration is placed opposite relatively soft enamel. Most dental restorations, especially posterior placements, are layered structures. While the core material of the structure might have fracture surfaces difficult to decipher, the glassy coating (veneer) can aid considerably in tracing back to a fracture initiation site. These two approaches, examining flexure bar fracture surfaces for familiarization with fractographic features and utilizing the veneer layer to trace general crack progression, are demonstrated in the next sections by examples.

FRACTOGRAPHIC EXAMINATION OF DENTAL RESIN COMPOSITE FLEXURE BARS

Fractographic analyses of flexure bars for strength tests have long been recognized as crucial to design and failure predictions within the ceramics community.[28] It is significant that the referenced work by Quinn and Morrell give examples where flexure data does correlate with component strength, rather than the reverse. A study of commercial glass-ceramics, a class of materials widespread in dentistry, determined that different batches had identical elastic moduli, fracture toughnesses, densities and microstructures, but very different strengths due to a variation in flaw type.[29] Similar examples can be found for alumina-based ceramics.[30,31] Fractographic examination is thus essential for critical flaw determination and strength predictions as well as general fracture feature recognition.

The previous points can be similarly made for resin composites. Resin composites are somewhat neglected in fractographic analyses,[12] as this class of materials falls "in between" polymers and ceramics. Fractographic specialists in both fields find features difficult to discern when they are masked by rough microstructures that are often typical in failed composite components. These materials are becoming more popular in many fields besides dentistry, however, and they share many fractographic features with both ceramics and polymers.

The following data and photos are of a resin composite (Paradigm, 3M ESPE, St. Paul, MN). The material contains 0.85 mass fraction zirconia-silica particles within a highly crosslinked polymeric matrix. Bars (n = 18) were machined to 3 mm x 4 mm x 18 mm and broken in 3-point flexure in order to provide fracture surfaces for examination. Few dental materials are able to be fabricated into long, standard size "B" bend bars – the manufactured dental blanks only have to be scaled down to be tooth-sized.

Fig. 2 shows the gold-coated fracture surfaces of three specimens with low-incident lighting to bring out smooth areas surrounding the origins (fracture mirrors), and Fig. 3 shows the fracture mirror radius and Weibull graphs. The Weibull graph also contains strength data for

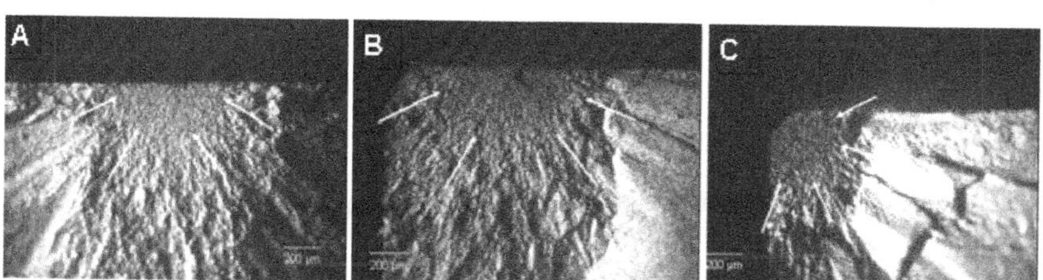

Figure 2. Resin composite flexure bars illuminated to highlight mirrors (arrows). The failure stresses are A) 152 MPa, B) 123 MPa and C) 181 MPa. (The accepted standard uncertainty of the system measurements is 1%.)

the infused alumina mentioned in the first section of this report. The infused alumina bars were 3 mm x 4 mm in cross-section, but available in longer lengths enabling both 3- and 4-point bending. The infused alumina strength difference is exactly what is expected for Weibull size scaling. The infused alumina 3-point data can be compared with the 3-point data for the resin composite. Resin composites are not yet as strong as most dental ceramics, but have some ameliorating characteristics in terms of other properties, fabrication and placement. The measured mean strength of the resin composite bars in this study was 145 MPa with a standard deviation of 17 MPa and a Weibull modulus of 9.9.

From the Fig. 2 photographs, it can be seen that there are indeed smooth-to-rough areas

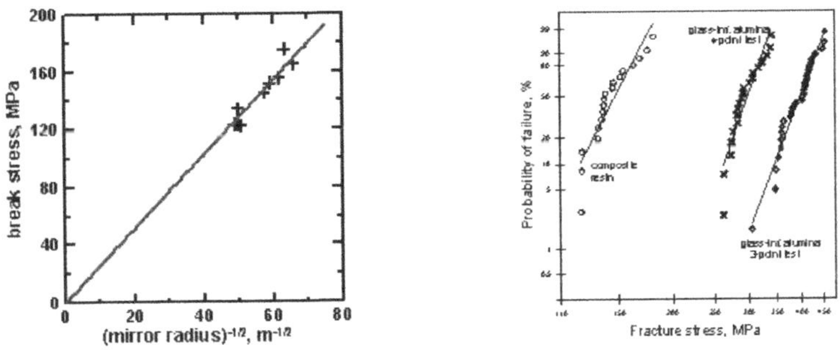

Figure 3. (Left) A mirror constant for the resin composite was measured to be 2.6 MPa·m$^{1/2}$. (Right) Weibull graphs of the composite resin and infused alumina.

resembling mirrors in these resin composites, but they require some subjective judgment to measure. It can be difficult to determine mirror sizes in many materials, but this is especially true of dental materials. Choosing mirror boundaries suggested by the arrows yields a mirror constant of 2.6 MPa·m$^{1/2}$ for a force fit through the origin, as shown in Fig. 3. How to delineate mirror boundaries still is an occasionally contentious subject that is addressed in another paper in this FRACTOGRAPHY V compilation.

The existence of fracture mirrors facilitates finding the fracture origin, but determination of the flaw type is still difficult. Figure 4(a) is an optical close-up of specimen C in Fig. 2, without low-incident lighting, and Fig. 4(b) shows a scanning electron micrograph (SEM) of the

Figure 4. (a) Specimen C in Figure 2, without low-incident lighting, (b) SEM photo of the origin, and (c) Transillumination of the uncoated specimen half.

specimen. The fracture origin appears to be at an inhomogeneous microstructure. Figure 4(c) shows the other, uncoated half of the specimen. Transillumination brings out a shadowed region at the fracture origin, indicating a small circular area where the light is reflected differently. There is a sharp delineation between materials at this point – either a difference in material composition or a subsurface crack. Perhaps an inconsistent microstructure caused the material to separate, or the disparity in the microstructure itself caused the light to reflect differently.

Transillumination was used in examining the fracture surfaces of the other specimens. Figure 5 shows the results for specimens A and B in Fig. 2. Careful manipulation of the light sources brings out dark areas at the fracture origins of these specimens, also. The SEM photos, to the right of the optical pictures, allow closer examination of the fracture origins. It can be seen that there are partial cracks surrounding microstructural inhomogeneities. Examination of all the resin composite specimens reveals such dark areas at the fracture origins, but sometimes these are only visible on one specimen half. This appears to be the critical flaw type for the resin composite specimens (cracking around an inconsistent microstructure), and transillumination was helpful in finding it.

Transillumination has been found to be a powerful tool in fractographic analysis of dental materials in general, which are usually translucent to match the tooth appearance. Dentists often use it when searching for cracks within natural teeth.

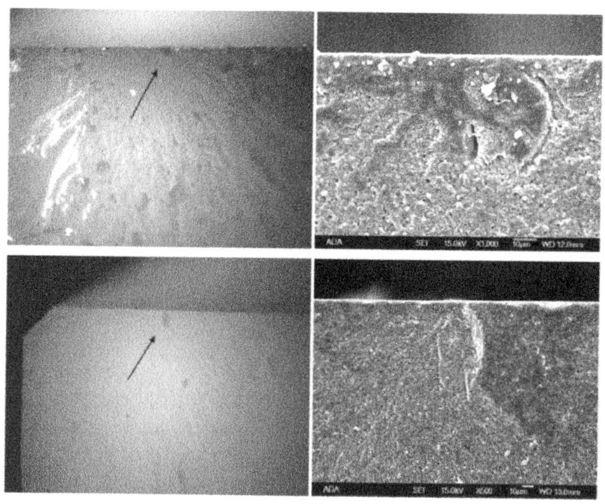

Figure 5. Specimens A and B from Figure 2. Transillumination helps locate the critical flaws.

FRACTOGRAPHIC EXAMINATION OF A RESIN COMPOSITE CROWN

The previous section identified some of the fractographic markings and techniques that might be useful in examination of resin composite failures. This section examines a resin composite crown (Herculite XRV, Kerr, Orange, CA) of similar particle size and slightly less solids loading than the bend bars. The crown failed after three months in vivo after biting on a crusty French fry.

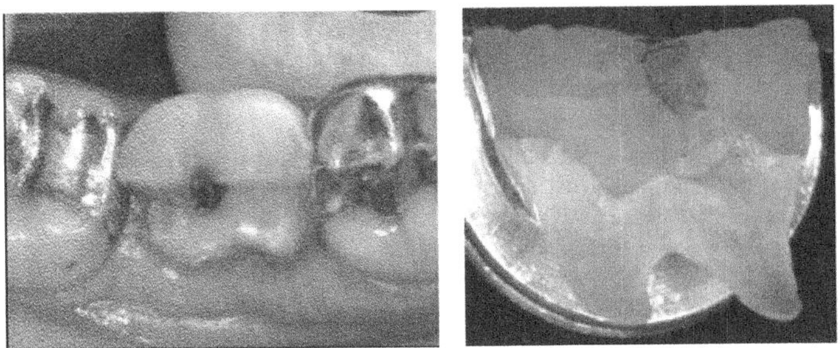

Figure 6. Photographs of a failed resin composite crown, with parts of the tooth structure still adhered to the detached fragment.

Figure 6 shows the two fragments of the broken crown, one still in situ, and the detached piece with parts of the original tooth structure still adhering to it. The dark material in the central part of the restoration is a fiber post placed to reinforce and secure the resin composite.

Figure 7 includes four photos of the detached piece in the right of Fig. 6. The top two photos in Fig. 7 are optical photographs of the Fig. 6 fragment, gold-coated to bring out features. The composite reconstruction can be distinguished from the tooth structure by the change in texture. The optical photograph on the right is a higher magnification of the boxed area, with the lighting adjusted to bring out the rib marks or rings (arrows) surrounding a pore. These marks may be crack arrest lines due to fatigue. Hackle is also apparent, confirming the direction of crack progression away from the pore.

The corresponding SEM photos are just below the optical photos in Fig. 7. The clarity is greater due to the higher depth of field in electron microscopes, but the textural changes distinguishing the tooth structure from the resin composite and the rings are not as easily discerned. The pore at the center of the rings is about 0.2 mm in diameter. It is situated at the

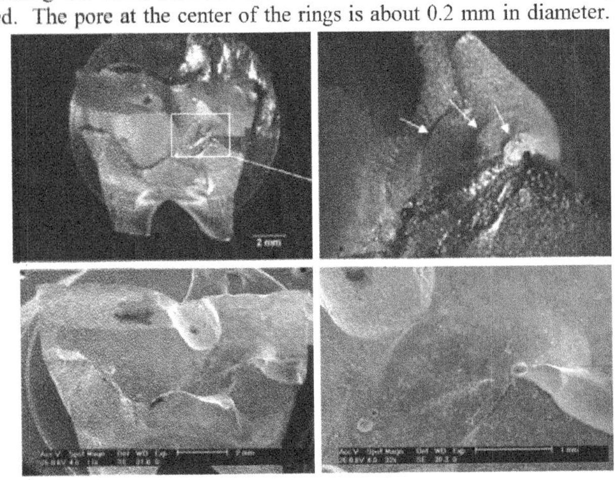

Figure 7. Optical (above) and corresponding SEM photos (below) of the fragment in Figure 6. Clarity is greater in the SEM photos, but some features are more difficult to distinguish.

junction between the tooth and the resin composite at a sharp change in geometry made during cavity preparation.

It is significant that fracture in the resin composite restoration originated at a poorly situated pore while fracture in the flexure tests initiated from discontinuous microstructure and cracks. The strength tests would not be expected to correlate with clinical longevity if the failures arise from different flaw types and populations. Also, the fracture surface of the restoration did not exhibit mirror-like markings, and had less apparent hackle than the flexure test bars. This appears to be typical of restoration failures in general. Almost all failed restorations collected and examined by the authors broke in two pieces, without clear mirrors. Other features on retrieved restorations, such as those found in the previously mentioned glassy veneers, were conducive to fractographic analyses as shown in the examples in the next section.

FRACTOGRAPHIC EXAMINATION OF AN ALUMINA FAILED CROWN

As noted earlier, many of the more robust dental restoration materials are veneered with a glassy layer to acquire the esthetic qualities of natural teeth as well as aid in preventing wear of the opposing dentition. Much work has been done on the relevance of contact stresses to restoration failures.[32] The assumption is that food particles or opposing dentition cause fatigue damage to the exposed restoration or veneer surface, which then becomes the site of eventual fracture initiation. More recent fractographic evidence suggests the failure initiation site is not likely to be the biting surface, but elsewhere, such as the margin areas.[33,34] This is where fractography is of the most practical relevance to the dental community. Choices in the selection, processing and shaping of dental materials are all affected by knowledge of where and how restoration failure initiates.

The sintered alumina of Fig. 1 is used with a veneer, and several such veneered alumina in vivo fractures have been studied by the authors. The fractured restoration in Fig. 8 is typical of brittle whole crown fractures. It appears to be a right first maxillary molar of unknown history. The layered structure of the crown is evident in the photo of the fracture surface on the right, with the veneering layer quite thick on the right side of the restoration. The circle in the photo indicates where fracture initiated. Locating this initiation area, however, was at first quite difficult. At the initial inspection, there were no large obvious pores nor flaws, mirrors nor converging hackle lines. The microstructure of the alumina core material is so rough that any but the most obvious fracture markings are "lost in the noise".

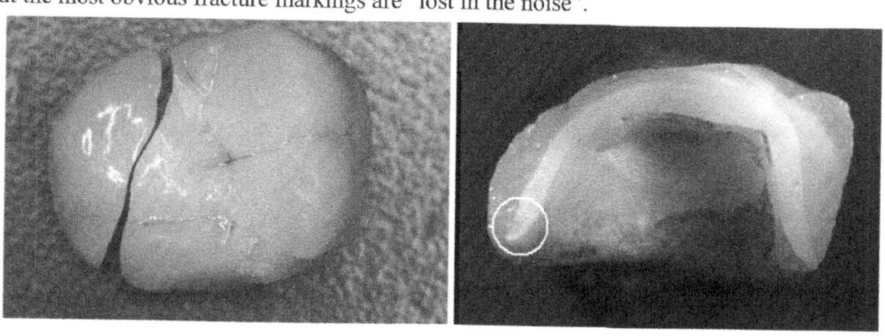

Figure 8. Whole crown failure with a fracture surface shown at right. The layered structure is obvious. Wake hackle in the veneer indicates fracture initiated within the circle.

The SEM composite view in Fig. 9 of the fracture surface shows many features much more clearly than optical microscopes. The individual SEM photos revealed, for example, an abundance of wake hackle from many small pores in the veneer. The SEM images, however, also made the fracture surface appear flat. This camouflaged the compression curl, readily apparent in the optical photo with low incident lighting in the right of Fig. 9. The compression curl somewhat narrowed the search for the region of fracture initiation by indicating the direction of crack growth.

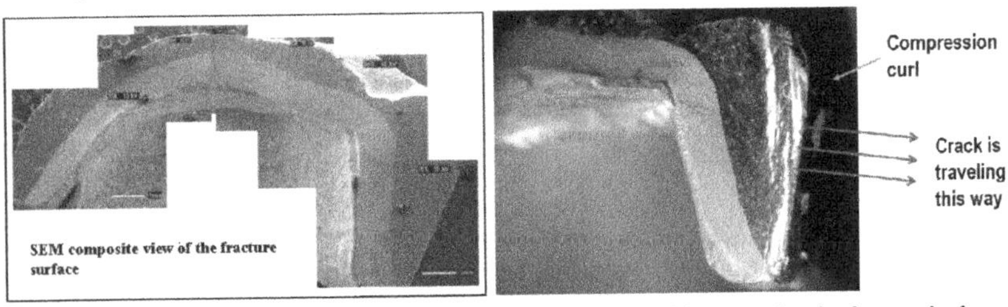

Figure 9. SEM views (left) revealed features such as wake hackle very clearly, but optical viewing (right) allowed for easier specimen manipulation and identification of gross surface irregularities, such as compression curl.

The veneer wake hackle first noted in the SEM photos was also apparent under high magnification optical viewing. The wake hackle wasn't as clear in the optical views, but easy physical manipulation of the specimens without need of a controlled atmosphere dictated the choice of wake hackle mapping under an optical microscope. Figure 10 shows the results of mapping the wake hackle in the sintered alumina veneer. Each arrow indicates one or more wake hackle "tails". The photo in the right of Fig. 10 is an interesting wake hackle pair found near the top of the crown. It is the wake hackle map that enabled determination of the fracture origin location at the restoration margin, where hoop stresses are highest.

Compression curl and wake hackle have thus far proven to be among the most useful tools in characterizing dental crown fractures. Similar case studies can be found in reference 34.

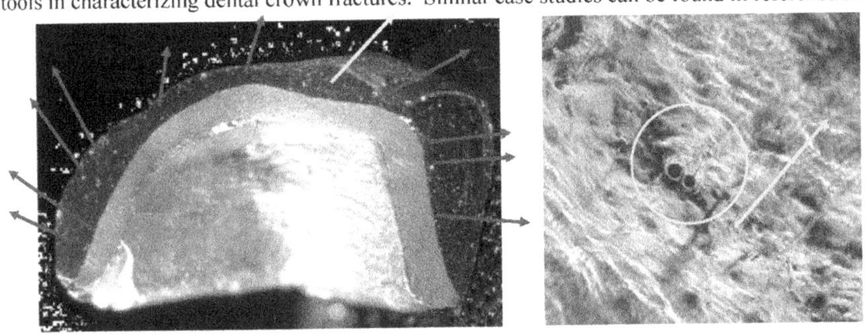

Figure 10. A wake hackle map of the crack propagation direction and an interesting example of wake hackle from a double pore. Fracture initiated at the lower left margin area.

Edge Chipping

Although usually not as serious as whole crown fractures, chipping is a widespread problem for tooth restorations. Chipping might result from chewing hard materials, sports and trauma. This failure mode is greatly exacerbated by current trends in oral jewelry. Tongue studs, in particular, are responsible for many cases of chipped teeth, as shown in the fig. 11 examples.[35,36] In the reference 35 study, tooth chipping was found on molars and premolars in 47% of subjects with a tongue piercing for 4+ years. This does not even include chipping of the central incisors.

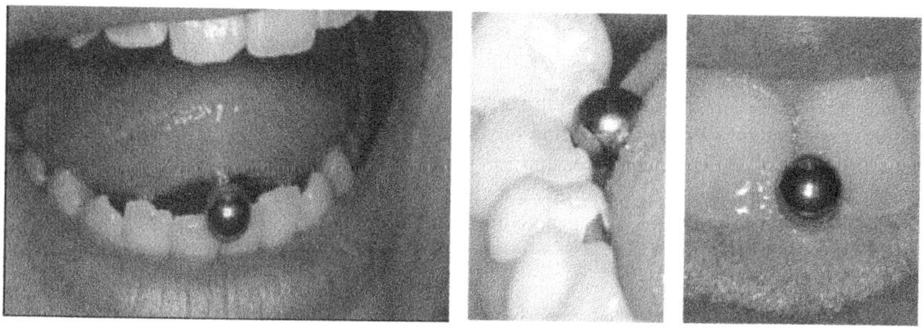

Figure 11. Examples of the adverse effect of tongue studs on teeth.

Modern dental materials, especially porcelains, tend to be brittle and chip easily. This affects not only the clinical longevity due to chipping failures themselves, but restoration shaping techniques that may generate small chips that could initiate eventual crack growth. Thus, a test to determine the susceptibility of a dental material to chipping has obvious clinical relevance.

Such a test exists and has been utilized in diverse applications, including the evaluation of cutting edge integrity in tool steels,[33] ceramic valve design,[20] machinability of brittle materials[37] and heat treatment effects in the flint-knapping of archaeological implements.[38] The edge-chip test is easily performed, it can be used with small amounts of material, and the edge toughness parameter, T_e, has been linked with another important parameter related to clinical longevity, fracture toughness.[19,39] The edge chip test is easily performed, and dedicated equipment has been designed and built (CK10, Nottingham, England) to apply and measure the forces associated with chip formation in different materials. Figure 12 shows a schematic of the test and photos of the test equipment. Further details of this test are given in the references.

Some recent edge toughness studies particularly applicable to the dental field have focused on chips that form when the applied chipping forces are not perpendicular to the specimen surfaces. It was found that the geometry of the chips can be correlated to the angle of the force that made the chips. When the chip force is applied to the specimen top surface at different angles, the shape of the chips as they emerge from the specimen sides change as shown in the glass specimens in Fig.13. Chips formed from forces aimed toward the edge are shallow, while chips formed from forces aimed away from the specimen edge have unusual, rounded-triangular shapes. One way to characterize the differences in chip side geometry is to define a width/depth ratio, w/d.

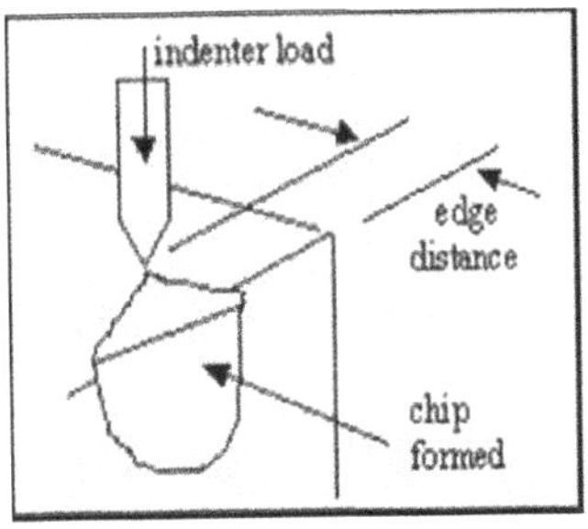

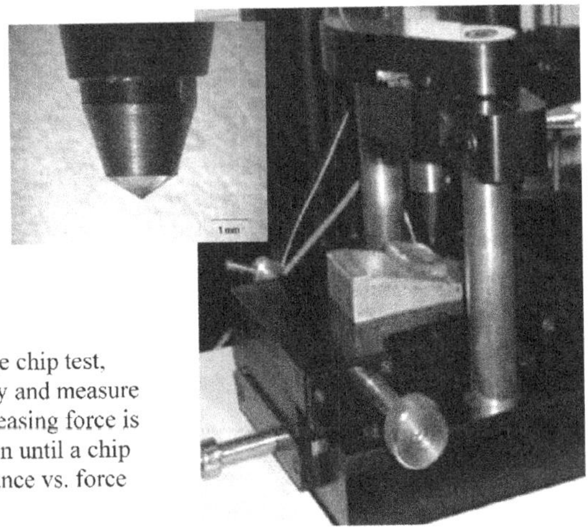

Figure 12. A schematic of the edge chip test, indenter and machine used to apply and measure the chip force. In this test, an increasing force is applied near the edge of a specimen until a chip forms. The slope of the edge distance vs. force is the edge toughness, T.

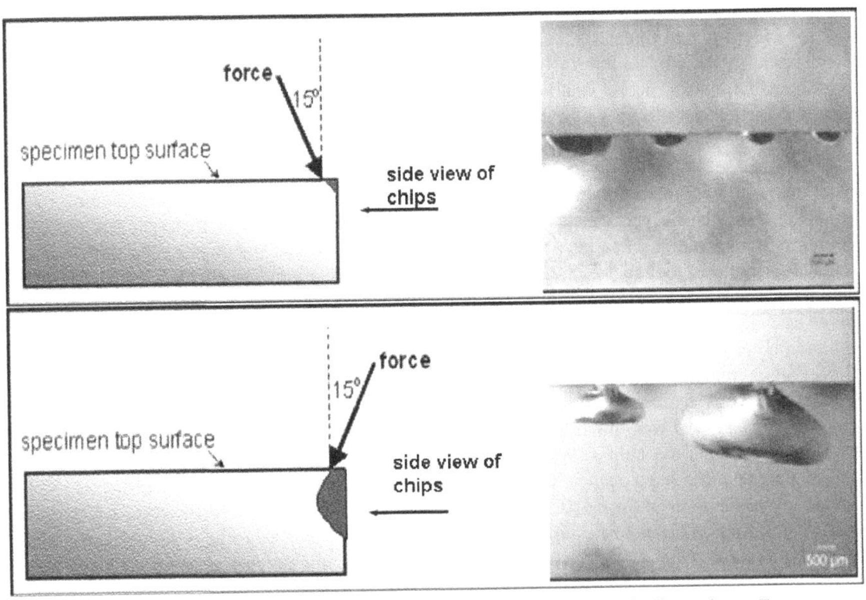

Figure 13. Diagram of force angles and photos of resulting chips in soda-lime glass. Forces aimed towards the edge result in emerging chip shapes that are shallow. Forces aimed away from the edge result in an interesting rounded-triangular shape.

Figure 14 shows how width/depth ratios were defined for specimens with chips resulting from perpendicular forces and for the Fig. 13 specimens, with chips resulting from forces 15° from perpendicular. The measurements were made with the previously described optical microscope system, accurate to the nearest micrometer.

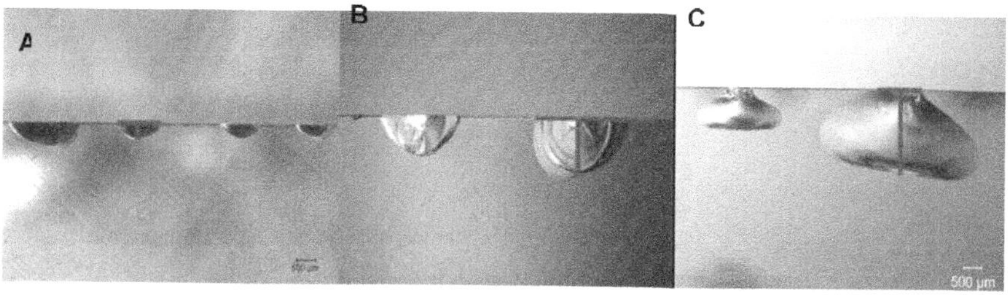

Figure 14. Edge chipping width/depth ratios for chips formed by perpendicular forces and specimens in fig. 13. A) Force aimed toward the edge, w/d = 3.43 ± 0.22. B) Perpendicular force, w/d = 1.66 ± 0.22. C) Force aimed away from the edge, w/d = 0.62 ± 0.05.

Figure 15 is a graph of the data in Fig. 14, indicating the force angle and edge chip geometry relationship for the soda lime glasses in the study. According to the graph, it should be fractographically feasible to approximate the force angles that cause chipping in materials that behave similar to the glass. Theoretically, edge chip shapes are indeed self-similar for all materials, but do all materials, especially dental materials, chip – and are the chips as easily measured as in the glass?

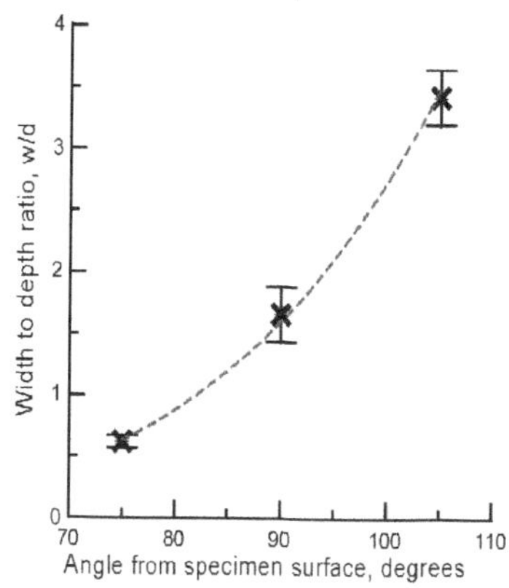

Figure 15. A graph of the Figure 14 data, indicating the feasibility of estimating force angle from edge chip geometry. The dotted line is subjectively added as an aid to visualization.

Figure 16 shows edge chips on glass and two other dental materials at angles 15° from perpendicular. The chips in the leucite porcelain, a very common dental material, appeared very similar to glass, but chips in the dental zirconia were different. Shallow chips in zirconia, from forces toward the specimen edge, often appeared malformed, as shown in the figure, and the material did not chip at all when forces were angled 15° away from the edge. The photo of zirconia is a high optical magnification of indents made right at the zirconia edge. This material plastically deformed rather than chipped. Perhaps a sharper indenter or higher forces at larger edge distances would have resulted in chip formation, but it is clear that this material is more resistant than the others to chipping.

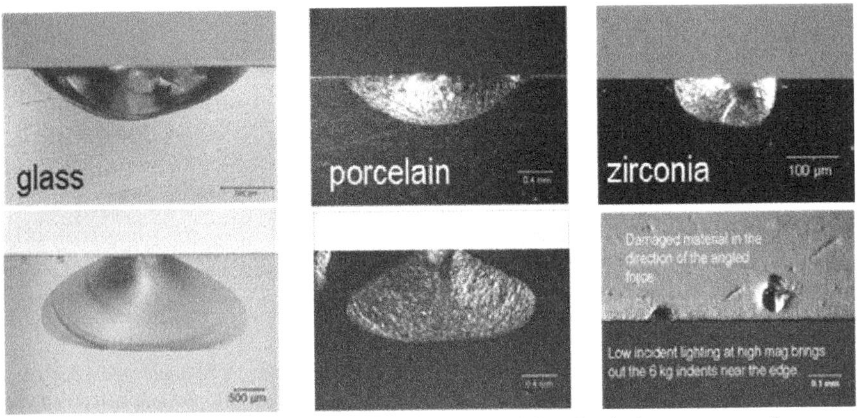

Figure 16. Chips made at 15° angled forces aimed toward the edge in the upper row of photos, and forces aimed away from the edge in the lower row of photos. The zirconia did not form chips at all for forces aimed away from the edge; the material plastically deformed.

ZIRCONIA

Y-TZP zirconia is among the most popular new materials in dentistry. While clinical longevity has thus far had poor correlation with material properties, the strengths and fracture toughness of zirconia is so much higher than other dental ceramics, nearly an order of magnitude, that it is expected that this material will outperform them. Several dental manufacturers currently offer zirconia crowns and bridges, in spite of a lack of optimal esthetics. There are not yet any long term clinical crown studies for this material, but recent dental meetings feature many studies of zirconia mechanical behavior. Figure 17 shows the fracture surfaces of zirconia flexure bars, with arrows delineating the mirrors.

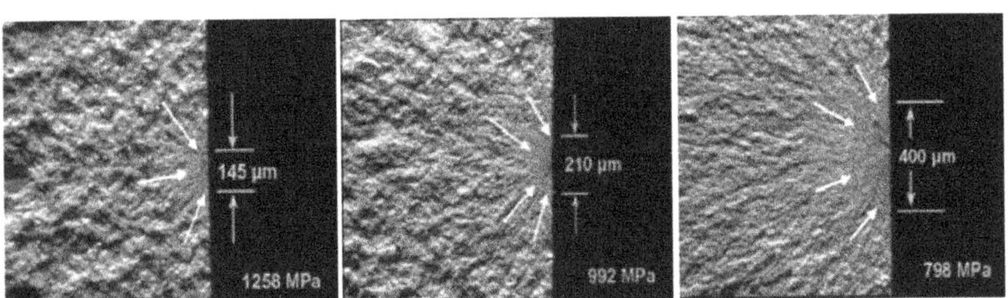

Figure 17. Fracture surfaces of zirconia flexure bars. There was difficulty in identifying the mirror boundaries.

Fracture mirrors are uncommon in dental restorations, as they usually fail under low stresses and are small. In such cases, the mirrors encompass the entire surface. Zirconia restorations may be an exception. This material has high strengths, which often correspond with small mirror sizes. A mirror constant for the zirconia of this study, 10.7 MPa·m$^{1/2}$, was calculated from the subjective mirror measurements, including those in Fig. 17.

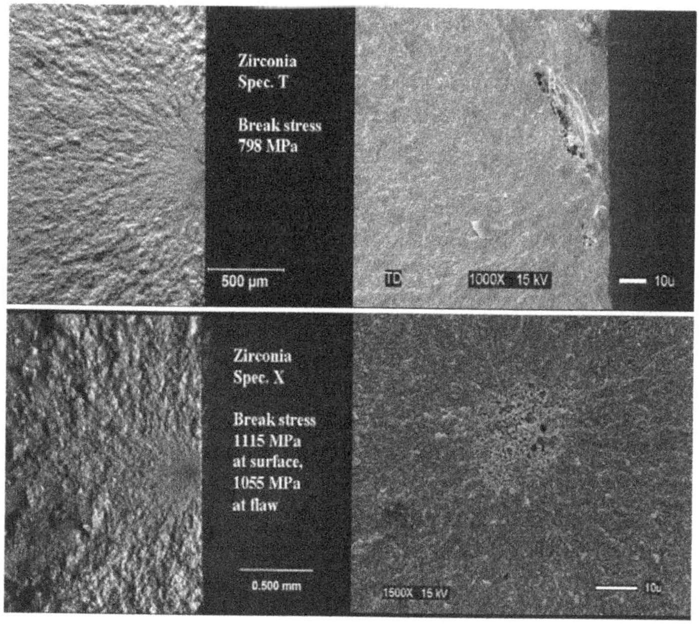

Figure 18. Typical fracture origins in zirconia flexure bars.

The fracture origins in the flexure bars are volume distributed and not much larger than 10 μm. It would be difficult to process this material to decrease or eliminate such small flaws; this material is optimized. To get correspondingly high strengths in zirconia restorations, extreme care would be necessary to avoid stress concentrators and surface discontinuities such as scratches or machining flaws.

SUMMARY AND FUTURE DIRECTIONS

Fractography is starting to be recognized in the dental community for many reasons. Clinical longevity studies can no longer keep pace with the introduction of new materials, and fractographic failure analysis has promise as a possible longevity assessment tool. Also, fractography might be useful in designing dental material strength tests. Many current dental testing methodologies have been shown lacking clinical relevance.

The drive for rapid longevity assessment and relevant strength testing techniques has become stronger as the need for dental restorations is predicted to increase, and market demands for improved esthetics are also increasing. Under these circumstances, fractography is seen as even more promising.

Unfortunately, fractographic evaluation of dental materials is difficult. Dental materials encompass a wide range, including resin composites, glass-infused microstructures, coarse polycrystals and layered structures. Techniques are being developed to overcome difficulties, and some examples were given. Examples were also shown for the edge chipping test, which is promising because of obvious clinical relevance. An important point is that restorations must fail from the same flaw type as in vitro tests in order to be applicable.

Finally, some of the recent fractographic analyses of failed restorations are being collected as part of a dental fractographic 'library'. Dentists and dental researchers are starting to work together to donate failed restorations or replicas obtained in their dental practices. PBRNs (Practice Based Research Networks) have been established by National Iinstitutes of Health (NIH), and is a proposed venue for even more widespread restoration collection. In this continuing effort to augment the 'library', the authors respectfully request the reader to send them any physical restoration fragments or dental fractographic analyses they may have accomplished. It is hoped that large collections and more experience in dental fractography will lead to improved techniques and insight into future analyses as well as statistically significant numbers of failure modes for predictive value.

ACKNOWLEDGEMENTS

This work was accomplished with the appreciated aid of an NIH grant, DE-14534.

REFERENCES

[1] Wynbrandt J, "The Excruciating History of Dentistry," St. Martin's Press, New York, NY (1998).

[2] Ring, ME, "Dentistry, an Illustrated History," Abradale Press, Harry N. Adams, Inc., New York, NY, (1985).

[3] D.W. Jones, "Synthesis and Evaluation of Ceramic, Polymer and Composite Bbiomaterials," 141-219 in Encyclopedic Handbook of Biomaterials and Engineering Part A: Materials, 1, D. L. Wise *et al.*, eds. Marcel Dekker, Inc., New York, NY (1995).

[4] C. M. Pine, N. B. Pitts, J. G. Steele, J. N. Nunn and E. Treasure, "Dental Restorations in Adults in the UK in 1998 and Implications for the Future," *Br. Dent. J.*, 190 (1), 4-8 (2001).

[5] Jones JA, Adelson R, Niesson LC, Gilbert GH, "Issues in Financing Dental Care for the Elderly," *J Public Health Dent* 50:268-275 (1990).

[6] Grytten J, Lund, E, "Future Demand for Dental Care in Norway; A Macro-Economic Perspective." *Comm Dent Oral Epidemiol* 27:321-330 (1999).

[7] Kalsbeek H, Truin GJ, van Rossum GM, van Rijkom HM, Pooterman JH, Verrips GH "Trends in Caries Prevalence in Dutch Adults Between 1983 and 1995." *Caries Res* 32:160-65 (1998).

[8] Fradaeni M and Aquilano A, "Clinical Experience with Empress Crowns," *Int J Prosthodont* 10(3)241-47 (1997).

[9] Fradeani M and Redemagni M, "An 11-Year Clinical Evaluation of Leucite-Reinforced Glass-Ceramic Crowns: A Retrospective Study." *Quint Int* 33(7)503-10 (2002).

[10] Sjogren G, Lantto R, Granberg A, Sundstrom BO and Tillberg A, "Clinical Examination of Leucite-Reinforced Glass-Ceramic Crowns (Empress) in General Practice: A Retrospective Study." *Int J Prosthod* 12(2)122-28 (1999).

[11] Sorensen JA, Choi C, Fanusco MI and Mito WT, "IPS Empress Crown System: Three-Year Clinical Trial Results," *J Cal Dent Assoc* 26(2)130-36 (1998).
[12] Gemalmaz D and Ergin S, "Clinical Evaluation of All-Ceramic Crowns," *J Prosth Dent* 87(2)189-96 (2002).
[13] Lehner C, Studer S, Brodbeck U and Scharer P, "Short Term Results of IPS-Empress Full-Porcelain Crowns," *J Prosthodont* 6(1):20-30 (1997).
[14] Kelsey WP 3rd, Cavel T, Blankenau RJ, Barkmeier WW, Wilwerding TM and Latta MA Kelsey, "4-Year Clinical Study of Castable Ceramic Crowns," *Am J Dent* 8(5)259-62 (1995).
[15] Erpenstein H, Borchard R and Kerschbaum T, "Long-Term Clinical Results of Galvano-Ceramic and Glass-Ceramic Individual Crowns," *J Prosthet Dent* 83(5)530-34 (2000).
[16] Scherrer SS. DeRijk WG, Wiskott HW and Belser UC, "Incidence of Fractures and Lifelong Predictions of All-Ceramic Crown Systems Using Censored Data," *Am J Dent* 14(2)72-80 (2001).
[17] Sjogren G, Lantto R and Tillberg A, "Clinical Evaluation of All-Ceramic Crowns (Dicor) in General Practice," *J Prosthet Dent* 81(3)277-84 (1999).
[18] Malament KA and Socransky SS, "Survival of Dicor Glass-Ceramic Dental Restoration Over 14 Years: Part I. Survival of Dicor Complete Coverage Restorations and Effect of Internal Surface Acid Etching, Tooth Position, Gender and Age," *J Prosthet Dent* 81(1)23-32 (1999).
[19] Probster L, "Four Year Clinical Study of Glass-Infiltrated, Sintered Alumina Ccrowns," *J Oral Rehab* 23(3)147-51 (1996).
[20] Segal BS, "Retrospective Assessment of 546 All-Ceramic Anterior and Posterior Crowns in a General Practice. *J Prosthet Dent* 85(6)544-50 (2001).
[21] Scotti R., Catapano S and D'Elia A, "A Clinical Evaluation of In-Ceram Crowns," *Int J Prosthodont* 8(4)320-23 (1995).
[22] McLaren EA and White SN, "Survival of In-Ceram Crowns in a Private Practice: A Prospective Clinical Trial," *J Prosthet Dent* 83(2)216-22 (2000).
[23] Haselton DR, Diaz-Arnold AM and Hillis SL, "Clinical Assessment of High-Strength All-Ceramic Crowns," *J Prosthet Dent* 83(4)396-40 ((2000).
[24] Oden A, Andersson M, Krystek-Ondracek I and Magnusson D, "Five-Year Clinical Evaluation of Procera AllCeram Crowns. *J Prosthet Dent* 80(4)450-456 (1998).
[25] Fradaeni M, D'Amelio M, Redemagni M and Corrado M, "Five-Year Follow-Up with Procera All-Ceramic Crowns," *Quintessence Int* 36(2)105-13 (2005).
[26] Odman P and Andersson B, "Procera AllCeram Crowns Followed for 5 to 10.5 Years: A Prospective Clinical Study," *Int J Prosthodont* 14(6)504-9 (2001).
[27] Rice RW, "Ceramic Fracture Features, Observations, Mechanisms, and Uses, in Fractography of Ceramic and Metal Failures, ASTM STP 827. J. J. Mecholsky, Jr., and S. R. Powell, Jr., eds., ASTM, West Conshohocken, PA 5-103 (1984).
[28] Quinn GD and Morrell R, "Design Data for Engineering Ceramics: A Review of the Flexure Test," *J Am Ceram Soc* 74(9) 2039-65 (1991).
[29] Lewis D, III, "Observations on the Strength of a Commercial Glass-Ceramic," *Am Ceram Soc Bull* 61(11) 1208-14 (1982).
[30] Taylor D, "Strength-Component Size Relationship for High-Tension Insulator Whiteware," *Br Ceram Trans J* 88 209-12 (1989).

[31] Lewis D, III and Oyler SM, "An Experimental Test of Weibull Scaling Theory," *J Am Ceram Soc* 59(11-12) 507-10 (1976).

[32] Jung YG, Peterson IM, Kim DK, Lawn BR, "Lifetime-Limiting Strength Degradation From Contact Fatigue in Dental Ceramics," *J Dent Res* 79:722-31 (1998).

[33] Kelly, JR, "Approaching Clinical Relevance in Failure Testing of Restorations," *J Prosthet Dent* 81:652-61 (1999).

[34] Quinn JB, Quinn GD, Kelly JR, Susanne SS, "Fractographic Analyses of Three Ceramic Whole Crown Restoration Failures," *Dent Mater* 21(10) 920-29 (2005).

[35] Campbell A, Moore A, Williams E, Stephens J, Tatakis DN, "Tongue Piercing: Impact of Time and Barbell Stem Length on Lingual Gingival Recession and Tooth Chipping," *J Periodont* 73:289-297 (2002).

[36] Levin L, Zadik Y and Becker T, "Oral and Dental Complications of Intra-Oral Piercing," *Dent Traumatology* 21:341-343 (2005).

[37] Quinn J, Su L, Flanders L and Lloyd I, "Edge Toughness" and Material Properties Related to the Machining of Dental Ceramics," Mach *Sci and Tech* 4;2:291-304.

[38] Quinn JB, Hatch JW and Bradt RC, "Edge Chipping and the Thermal Alteration of Lithic Materials," *Advances in Ceramics*, American Ceramic Society, Westerville, OH. (in print).

[39] Almond EA, McCormick NJ, "Constant-Geometry Edge Flaking of Brittle Materials," *Nature* 321;6065:53-55 (1986).

FRACTURE SURFACE ANALYSIS OF DENTAL CERAMICS

Burak Taskonak, Jiahau Yan
Department of Restorative Dentistry
Indiana University
1121 West Michigan St. #DS-112
Indianapolis, IN 46202

John J. Mecholsky, Jr.
Department of Materials Science and Engineering
University of Florida
P. O. Box 116400
Gainesville, FL 32611

ABSTRACT

Ceramic systems have poor long-term fracture resistance, especially when they are used in an oral environment as fixed partial dentures. The objective of this study is to determine the site of crack initiation and the causes of fracture in clinically failed zirconia-based and lithia-disilicate-based ceramic fixed partial dentures. Five fractured three-unit lithia-disilicate-based glass-ceramic fixed partial dentures (Empress 2® all-ceramic system, Ivoclar Vivadent AG, Schaan, Liechtenstein) and five fractured four-unit yttria stabilized tetragonal zirconia (Y-TZP) based fixed partial dentures (Cercon® ceramics, DeguDent GmbH, Hanau, Germany) were retrieved and analyzed. Fractography and fracture mechanics methods were used to estimate the stresses at failure. Delaminations between the glass veneer and zirconia core were observed in Y-TZP based fixed partial dentures and a secondary fracture was initiated from the zirconia core. Secondary fracture dictated the ultimate failure. Failure stresses of the fixed partial dentures that failed due to Zirconia fracture were estimated to be from 325 to 431 MPa. Fractures that had origins on the glass veneer surface had failure stresses between 31 MPa and 35 MPa. Fractures of the lithia-disilicate-based glass-ceramic fixed partial dentures were initiated within the glass veneer layer and fracture initiation sites are controlled primarily by contact damage. Failure stresses of the lithia-disilicate-based glass-ceramic fixed partial dentures were estimated to be from 19 to 68 MPa. We conclude that interfacial delamination in the glass veneer/zirconia core dental ceramic structures controlled the fracture initiation sites and failure stresses in Y-TZP based fixed partial dentures. The design and dimension of the connectors as well as span size of the fixed partial denture can be the key factors in causing fractures at relatively low occlusal loads but high fracture stresses.

INTRODUCTION

Fixed partial dentures (FPDs) are tooth replacements that are fixed permanently in the mouth and cannot be removed for routine cleaning by the patient. These dentures are made to take support from adjacent natural teeth. These dentures can replace only a few missing teeth.[1]

FPDs can be fabricated using metals, ceramics or ceramics (porcelain) fused to metals. However, in recent years ceramics are replacing metals as materials of choice due to their excellent aesthetic properties.[2]

Previous studies showed that ceramic fixed partial dentures fail due to fractures of the connectors.[3-5] It was also reported in the same studies that design of the connector plays an important role in the survival of the FPDs. For example, the connectors with sharp embrasures had stress concentration areas that decreased the life time of the FPD significantly. It was recommended in a previous study that such FPD failures can be prevented by achieving sufficient height and width of connectors as well as reducing the stress concentration areas.[6] The minimum required dimensions for all-ceramic FPD connectors are exclusively dependent on the type of ceramic type used for the core material.[7]

All-ceramic FPDs are fabricated using a core ceramic, *e.g.*, glass infiltrated alumina, lithia disilicate glass ceramic, densely sintered aluminum oxide, veneered with silicate glass to enhance aesthetics. The most recent core ceramic for all-ceramic FPDs is yttria stabilized tetragonal zirconia (Y-TZP). Compared to previously used core ceramics, Y-TZP presents superior fracture toughness (9-10 $MPa{\cdot}m^{1/2}$), and a longer survival time is expected.[8,9] To ensure long-term success, a minimal connector surface area of 7 mm^2 is recommended for FPDs fabricated using Y-TZP.[8,9] However, the long-term clinical data on the survival time of Y-TZP-based FPDs is not available due to their recent introduction to dental therapy. Fractographic analyses of clinically failed Y-TZP FPDs as well as long-term results of clinical studies are critical for the establishment of specific guidelines for their use.

The mechanisms for the failure of dental ceramics have not been studied extensively, although fractographic analysis is a key element in the design and development of dental structural materials.[10] Fractographic analyses have been crucial in efforts to analyze failure mechanisms, identify fracture initiation sites, and determine the probable cause of failure from retrieved clinical specimens. One can determine whether the failure is caused by a processing defect, by contact damage, or by an overload. This paper addresses the application of quantitative fracture surface analysis to ceramic FPDs, material and product development, and "trouble-shooting" in-service failures. The overall objective of this study was to identify, by fractographic techniques, the principal crack initiation sites and the causes of failure in the fractured ceramic fixed partial dentures.

MATERIALS AND METHODS

Five veneered three-unit lithia-disilicate-based glass-ceramic fixed partial dentures (Empress 2® all-ceramic system, Ivoclar Vivadent AG, Schaan, Liechtenstein) and five veneered four unit Y-TZP fixed partial dentures (Cercon® ceramics, DeguDent GmbH, Hanau, Germany) were analyzed in this study (Figure 1). X-ray diffraction analyses showed that the veneering ceramic consisted primarily of amorphous glass. The clinically failed fixed partial dentures were retrieved from two ongoing clinical studies to evaluate each ceramic FPD system separately.

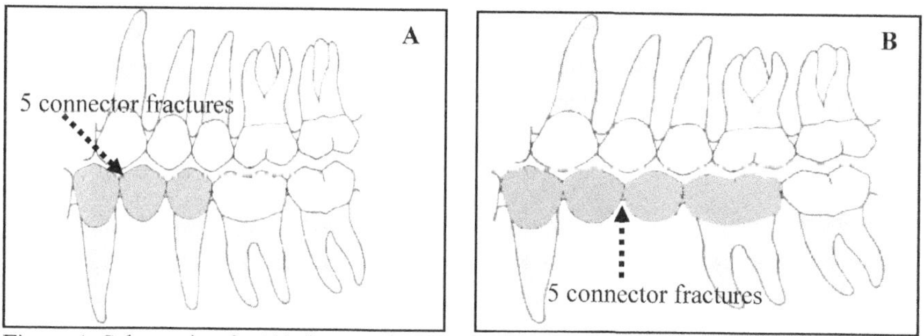

Figure 1. Schematic of a (A) three-unit Lithia-disilicate-based fixed partial denture and (B) four-unit Y-TZP fixed partial denture with fracture initiation sites and numbers.

Sample Preparation

To obtain a clean surface for analysis, we used ethyl alcohol in a sonicated bath to achieve clean fracture surfaces. After removal from the sonicated bath, specimens were rinsed with distilled water. Acetone and methanol were not used because they often leave a film on the surface.[11] After cleaning, fracture surfaces of the specimens were coated with a gold palladium film to accommodate better observation.

Observation

A low magnification magnifier was used as the primary observation tool before any quantitative analysis was attempted. After examination by the aided eye, a low-power microscope was the next step for overall examination. The optical microscope was used for quantitative data recording after preliminary observations. Use of a scanning electron microscope (SEM) was our choice in cases where a better depth of field was required for recording the image.

Analysis

Our first step in failure analysis of FPDs was to locate the origin by looking for a relatively flat region. In general, fracture surface markings can be used to trace back to a fracture origin. The most common fracture markings in dental materials are hackle marks on pores or inclusions (called "wake hackle" and located on the side away from the fracture origin in Fig. 2). These markings fan out away from the origin. Other markings include Wallner lines which are interference marks between the propagating crack and the stress wave and are approximately perpendicular to the general direction of propagation. The origin is observed at the approximate center of these markings. Once the location of the origin is found, quantitative measurements can be made to aid the analysis, and critical crack sizes measured.

We determined the fracture origins by examining the fracture surface and tracing the characteristic markings back to the initiation site.[11] These markings included twist hackle (river marks), wake hackle (fracture tails), cleavage steps, Wallner lines, and branching locations. When characterizing fracture origins, we made scanning electron microscope images (JSM-6400, Jeol, Tokyo, Japan) of the overall sample and of an enlargement of the fracture origin region. The general fractographic procedure is outlined in ASTM standard C1322 (ASTM, 1999).[12] We measured crack-initiating flaws to determine the fracture stress of each specimen.

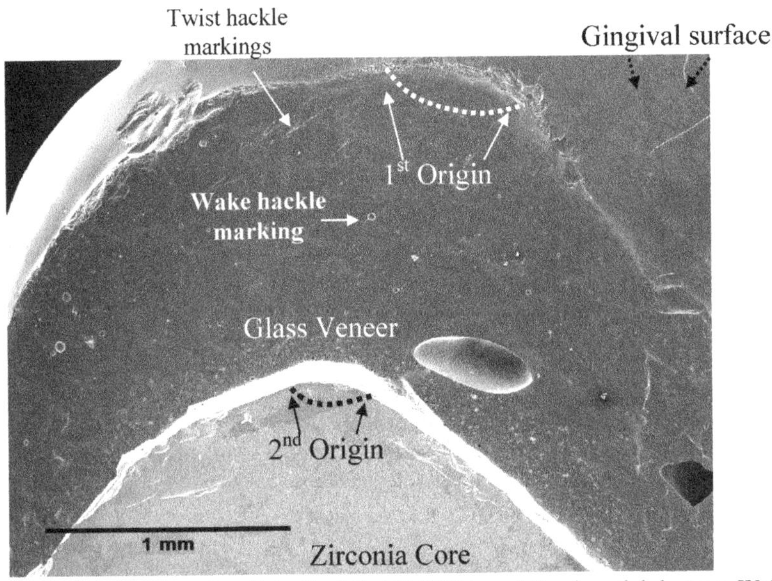

Figure 2. Fracture surface of a clinically failed Y-TZP based fixed partial denture. Wake hackle markings were used to establish the reference points for determining the fracture origin in the glass veneer. The wake hackle markings were often observed on pores as an outcome of fracture passage through or around pores. The markings indicate the direction of the fracture origin in the veneer layer.

Almost all mechanically induced cracks can be idealized as semi-elliptical, sharp cracks of depth, a, and half-width, *b*. The crack sizes are approximated by an equivalent semicircular crack size, *c* [$c = (ab)^{1/2}$]. Stress at failure, σ_f, is calculated based on the known fracture toughness, K_C, and the crack size, *c*:

$$\sigma_f = \frac{K_C}{Y\sqrt{c}} \tag{1}$$

where K_C is the critical stress intensity factor (fracture toughness), and *Y* is a geometric factor, which accounts for the shape of the fracture-initiation crack and loading condition. The quantity *Y* depends on the ratio *a/b*. The equation [$c = (ab)^{1/2}$] allows many irregular crack shapes to be analyzed, and avoids the complications of calculating a geometric factor for each crack[13]. For surface cracks that are small relative to the thickness of the sample, $Y \sim 1.24$. For sharp cracks that are induced by a Vickers or Knoop indentation, $Y \sim 1.65$, and for internal cracks, $Y \sim 1.12$.

RESULTS

Clinically failed ceramic FPDs do not have clear surface fracture markings due to their complex microstructure and multidirectional and repeated intra-oral loading. In addition, not all the clinically retrieved specimens had intact fracture surfaces to perform an analysis. We discarded specimens where the fracture origin was lost. However, we could perform fractography in a total of ten specimens.

Lithia-Disilicate-Based Glass-Ceramic Fixed Partial Dentures

Fracture origins of the 5 veneered lithia-disilicate-based glass-ceramic fixed partial dentures were on the occlusal surface of connectors (Fig 1A). Wake hackle markings from pores were selected as reference points for determination of the location of fracture origins in the glass veneer.[12] The wake hackle markings were observed next to pores on the opposite side, away from the fracture origin in the veneer layer. Porosity existed within the glass veneer because of slurry preparation and the sintering of veneer powders. The thickness of the veneer layers in clinically failed fixed partial dentures varied based on their location. On labial surfaces of fixed partial dentures, the veneer layer was thicker compared with those on the occlusal surface (Fig. 3).

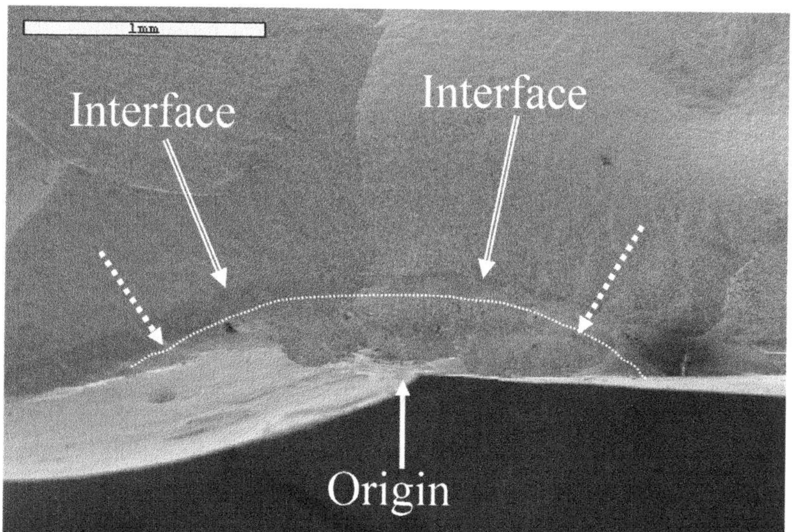

Figure 3. Fracture surface of a clinically failed Lithia-disilicate-based fixed partial denture. Fracture toughness of the glass veneer was used to calculate the stress at failure of samples in which the fracture origin was within the veneer layer. Fracture propagates quickly without any increase in stress from the point where the semi-minor axis of the fracture origin ends.

The mean fracture toughness of the glass veneer was 0.7 ± 0.1 MPa·m$^{1/2}$. We used the fracture toughness value of the glass veneer to calculate the stress at failure (Eq. 1) for the specimens in which fracture initiation occurred within the veneer surface.

Fractographic analysis can determine the stress that would occur from a crack or flaw of a particular size only if the toughness or apparent toughness is known. For fixed partial dentures that are fabricated from a single material, the failure stress can be estimated directly from the crack size, and known toughness. However, for the veneered fixed partial dentures (core/veneer biomaterial), we showed previously that residual compressive stress within the veneer layer is generated because of the elastic/thermal mismatch and visco-elastic relaxation[14]. Therefore, the calculated stress should include an additional term that compensates for the compressive residual stress, which has to be overcome before tensile loading. Veneered lithia-disilicate-based glass-ceramic fixed partial dentures have approximately 22 MPa of compressive residual stresses on the veneer surface; however, this value varies based on the cooling rate, core/veneer thickness and the sintering temperature of the glass veneer[15]. The critical flaw size, *c*, of the lithia-disilicate-based glass-ceramic fixed partial dentures for which an origin could be found ranged from 240 to 939 μm. Estimated failure stresses of the fixed partial dentures that had fracture initiation sites within the veneer layers ranged from 19 to 68 MPa.

Y-TZP Based Fixed Partial Dentures

Unlike the lithia-disilicate-based glass-ceramic fixed partial dentures, fracture origins of Y-TZP based fixed partial dentures were located on the gingival surface of the connectors (Fig 2). Wake hackle markings from pores were the main guiding points that led us to fracture origins (Fig 2). Fracture propagation direction was towards the Y-TZP core after initiation at the veneer surface. However, initial fracture stopped at the interface between the Y-TZP core and glass veneer due to weak interfacial bonding and the fracture toughness difference between the Y-TZP core and glass veneer. Delaminations between the glass veneer and Y-TZP core were observed in all Y-TZP based fixed partial dentures (Fig 4). Y-TZP core has a fracture toughness value of 6.5 $MPa{\cdot}m^{1/2}$ and the glass veneer has a fracture toughness value of 0.7 $MPa{\cdot}m^{1/2}$. A second fracture initiation site was observed within the core layer of each Y-TZP based fixed partial denture (Fig 2). The second fracture that initiated within the core layer dictated the catastrophic failure. Initial fractures that had origins on the glass veneer surface had failure stresses between 31 MPa and 35 MPa. However, failure stresses of the fixed partial dentures that failed due to the Y-TZP core fracture ranged from 325 to 431 MPa. Thus, because of the delamination between the veneer and core, the fracture of the veneer did not result in final fracture of the entire structure.

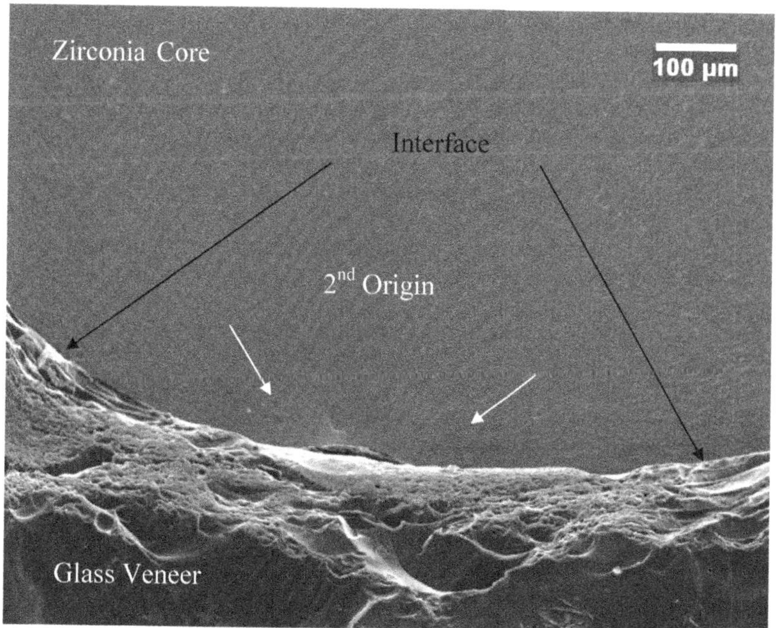

Figure 4. Fracture surface of a clinically failed Y-TZP based fixed partial denture. Delaminations between glass veneer and Y-TZP core were observed in all the Y-TZP based fixed partial dentures.

DISCUSSION

Fracture surface analyses of failed ceramic fixed partial dentures showed that failure origins occurred mostly at surface flaws. Previous investigations have reported that fracture initiated typically along the veneer/core interface of ceramic crowns.[16] However, this was not the case in our study.

In Y-TZP based FPDs, all the connector failures were initiated from the gingival surface where tensile stresses are the highest due to occlusal loading. In contrast, most lithia-disilicate-based FPD failures were initiated at connector sites near occlusal surfaces. Since four-unit Y-TZP based FPDs had longer spans compared to three-unit lithia-disilicate-based FPDs; it is likely that stress concentrations were greater on the gingival surface of middle connectors in Y-TZP based FPDs.

Crack initiating flaws in lithia-disilicate-based FPDs were most likely introduced as a result of repeated contact damage. During the mastication process, the mandible makes lateral, centric, and protrusive movements that allow the opposing cusp tip in the maxilla to exert multidirectional forces on the prosthesis. As a result, fracture can occur in the most vulnerable part of the ceramic fixed partial dentures, i.e., the connector.[3] In addition, mechanical damages resulting from the occlusal adjustment by the dentist or the dental technician can introduce flaws in the fixed partial dentures.

Fracture toughness of the glass veneer was used to calculate the stress at failure of specimens because all the fracture origins occurred within the veneer layer (Fig. 3). In lithia-disilicate-based FPDs the fracture-initiating crack propagates immediately upon reaching the failure stress. Even though the core layer is tougher than the veneer, once crack propagation begins in the veneer, the crack does not arrest or deviate out of plane (Fig.3). Crack progression is not impeded by the core ceramic at the interface between the core and veneer because there is not a great difference in toughness between the core and veneer material and because the core-veneer bond is strong. Thus, the toughness of the veneer is appropriate for calculating the stress from the crack size (Eq. 1). The bilayer materials also include a term for the compressive residual stress generated by the thermal expansion anisotropy (~ 22 MPa) and visco-elastic process.[15]

Residual stresses can be determined by calculating the mirror-to-flaw size ratio and by observing the mirror boundary shape on the fracture surface. However due to low toughness of the glass veneer (0.7 $MPa{\cdot}m^{1/2}$) and small size of the FPD specimens, mirror boundaries were not discernable. Consequently, residual stresses could not be determined using fractographic analyses.

Unlike lithia-disilicate-based FPDs, secondary fracture occurred in Y-TZP based FPDs due to interfacial delamination between the glass veneer and zirconia core (Fig. 4). One can conclude that the failure occurred after veneer fracture. However, fractographic analyses showed that even though there was a fracture in the veneer layer, a FPD was still in one piece until fracture occurred in the zirconia core. Interfacial delamination in the glass veneer/zirconia core dental ceramic structures controlled the fracture initiation sites and failure stresses in the zirconia core. The design and dimension of the connectors as well as span size of the fixed partial denture can be the key factors in causing failure at relatively low occlusal loads but high fracture stresses.

SUMMARY AND CONCLUSIONS

Fractography can be effectively used in the study of clinically failed fixed partial dentures (FPD). Fractographic analysis has been critical in efforts to analyze failure mechanisms, to identify fracture initiation sites, and to determine the probable cause of failure from retrieved clinical specimens. One can determine whether the failure is caused by a processing defect, by contact damage, or by an overload. This paper addressed the application of quantitative fracture surface analysis to ceramic FPDs, to material and product development, and to "trouble-shooting" of in-service failures. Lithia-disilicate-based FPD failures were initiated from connectors and most of the connector failures were associated with fractures that initiated from occlusal surfaces. In Y-TZP based FPDs, all the connector failures were initiated from the gingival surface where tensile stresses are the highest due to occlusal loading. Fracture initiation occurred twice in Y-TZP based FPDs due to interfacial delamination between the glass veneer and zirconia core. One can conclude that the failure occurred after veneer fracture. However, fractographic analyses showed that even though there was a fracture in the veneer layer, a FPD was still in one piece until fracture occurred in the zirconia core.

REFERENCES

[1][Anon], "The Glossary of Prosthodontic Terms." *Journal of Prosthetic Dentistry*, 94(1):10-92 (2005).

[2]P.V. Von Steyern, P. Carlson, K. Nilner, "All-Ceramic Fixed Partial Dentures Designed According to the DC-Zirkon (R) Technique. A 2-Year Clinical Study." *Journal of Oral Rehabilitation*, 32(3):180-187 (2005).

[3]W.S. Oh, N. Gotzen, K.J. Anusavice, "Influence of Connector Design on Fracture Probability of Ceramic Fixed-Partial Dentures." *Journal of Dental Research*, 81(9):623-627 (2002).

[4]W.S. Oh, K.J. Anusavice, "Effect of Connector Design on the Fracture Resistance of All-Ceramic Fixed Partial Dentures," *Journal of Prosthetic Dentistry* 87(5):536-542 (2002).

[5]J.R. Kelly, J.A. Tesk, J.A. Sorensen, "Failure of All-Ceramic Fixed Partial Dentures in-Vitro and in-Vivo - Analysis and Modeling," *Journal of Dental Research*, 74(6):1253-1258 (1995).

[6]A.J. Raigrodski, "Contemporary Materials and Technologies for All-Ceramic Fixed Partial Dentures: A Review of the Literature." *Journal of Prosthetic Dentistry*, 92(6):557-562 (2004).

[7]M. Guazzato, M. Albakry, S.P. Ringer, M.V. Swain, "Strength, Fracture Toughness and Microstructure of a Selection of All-Ceramic Materials. Part II. Zirconia-Based Dental Ceramics." *Dental Materials*, 20(5):449-456 (2004).

[8]P. Christel, A. Meunier, M. Heller, J.P. Torre, C.N. Peille, "Mechanical-Properties and Short-Term Invivo Evaluation of Yttrium-Oxide-Partially-Stabilized Zirconia. *Journal of Biomedical Materials Research*, 23(1):45-61 (1989).

[9]F. Filser, P. Kocher, F. Weibel, H. Luthy, P. Scharer, L.J. Gauckler, "Reliability and Strength of All-Ceramic Dental Restorations Fabricated by Direct Ceramic Machining (DCM)." *Int. J. Comput. Dent.*, 4:89-106 (2001).

[10]J.B. Quinn, G.D. Quinn, J.R. Kelly, S.S. Scherrer, "Fractographic Analyses of Three Ceramic Whole Crown Restoration Failures," *Dental Materials*, 21(10):920-929 (2005).

[11]J.J. Mecholsky, "Fractography - Determining the Sites of Fracture Initiation," *Dental Materials*, 11(2):113-116 (1995).

[12]ASTM, "Standard Practice for Fractography and Characterization of Fracture Origins in Advanced Ceramics, ASTM Designation C1322-96a." (1999) Annual Book of ASTM Standards, Vol. 15.01. American Society for Testing and Materials, West Conshohocken, PA, (1999).

[13]J.J. Mecholsky, "Fracture-Mechanics Principles." *Dental Materials*, 11(2):111-112 (1995).

[14]B. Taskonak, K.J. Anusavice, J.J. Mecholsky, "Role of Investment Interaction Layer on Strength and Toughness of Ceramic Laminates," *Dental Materials*, 20(8):701-708 (2005).

[15]B. Taskonak, J.J. Mecholsky, K.J. Anusavice, "Residual Stresses in Bilayer Dental Ceramics. *Biomaterials*, 26(16):3235-3241 (2005).

[16]J.Y. Thompson, K.J. Anusavice, H. Morris, "Characterization of Fracture Surfaces of Dental Ceramics," *Journal of Dental Research*, 73(196-196 (1994).

STRIATIONS RESULTING FROM FATIGUE CRACK GROWTH IN DENTIN

Devendra Bajaj[1]
[1]Department of Mechanical Engineering
University of Maryland, Baltimore County
1000 Hilltop Circle, Baltimore, MD, 21250

Narayana Sundaram[1]
[1]Department of Mechanical Engineering
University of Maryland, Baltimore County
1000 Hilltop Circle, Baltimore, MD, 21250

Dwayne Arola[1, 2]
[2]Department of Endodontics, Prosthodontics, and Operative Dentistry
Baltimore College of Dental Surgery, University of Maryland
666 West Baltimore Street
Baltimore, MD, 21201

ABSTRACT

The primary objective of this study was to evaluate the striation spacing resulting from fatigue crack growth in the dentin of human teeth. Compact tension (CT) specimens obtained from coronal dentin of patients of two age groups (young < 37 years and old > 50 years) were subjected to cyclic Mode I loads. Fracture surfaces of the CT specimens were examined using a scanning electron microscope (SEM) and contact profilometer. A power spectrum analysis of the surface profiles showed that the striation spacing ranged from 50 to 170 μm. In old dentin (125 ± 23 μm) the spacing was significantly larger than that in young dentin (93 ± 27 μm), suggesting that the mechanisms contributing to crack tip blunting were potentially suppressed by changes in structure of this tissue with age. Fatigue crack growth striations were also identified and examined on cracks that underwent extension in vivo. The striation spacing for in vivo cracks was within the range resulting from the in vitro evaluation. Results of this study suggest that fatigue crack growth contributes to restored tooth failures, and the in vitro approach provides a viable model for evaluating the mechanics and mechanisms of cyclic crack extension in dentin.

INTRODUCTION

The human tooth is comprised of three distinct tissues, namely enamel, dentin and pulp. Dentin forms the bulk of the tooth by both weight and volume. While serving many functions, dentin provides an elastic foundation for the outermost brittle enamel. It also serves as a protective enclosure for the more sensitive pulp and as a conduit for transferring external stimuli to the nerves. Most importantly, dentin acts as an integral foundation for restorative materials which are placed to restore form and function to damaged or diseased tooth structure.

Recent advances in restorative dentistry have led to improvements in the overall oral health of patients. New materials and methods have not only improved the aesthetics of restorations but also their performance. Despite these advances, the failure of restored teeth and tooth fracture is relatively common.[1,2] While it is plausible that fracture in the restored tooth results from a single catastrophic load, it is generally believed that such failures are the result of subcritical cracking induced by repetitive stresses, i.e., fatigue.[3] Damage induced by restorative processes can grow with time due to cyclic stresses resulting from mastication and can enable complete fracture of the tooth.[4-7]

A number of in vitro studies on cyclic crack growth in bovine,[8-10] elephant,[11,12] and human dentin[13-15] have been conducted in the recent past. These studies have been successful in quantifying the fatigue properties of dentin. According to an evaluation of the mechanisms of fatigue crack growth in dentin, cyclic extension is comprised of crack-tip blunting and resharpening.[12]

Fatigue failures in metals are often evaluated in terms of striations on the fractured surfaces. Yet, no study has identified fatigue striations resulting from cyclic crack extension in hard tissues such as bones and dentin and enamel of the tooth. It has been suggested that striations are masked by the underlying structure of dentin.[11] The primary objective of this study was to show that cyclic crack growth in dentin results in fatigue striations and to use these features to further evaluate the mechanisms of crack extension. In addition, a correlation was drawn between the in vitro and in vivo tooth failures.

MATERIALS AND METHODS

Unrestored human molars were obtained from participating clinics within the state of Maryland and the whole body morgue of Stanford University. All methods of acquisition and storage were in compliance with an approved protocol issued by the Institutional Review Board (IRB) of the University of Maryland.

Immediately after extraction, and throughout all aspects of the study, the teeth were stored in Hank's Balanced Salt Solution (HBSS) with record of age and gender of the patient. The extracted molars were molded in a polymer resin and sectioned using a computer controlled slicer/grinder with water based coolant. Diamond slicing wheels were used to obtain primary sections from the molars (Fig. 1(a)). Secondary sections were then introduced to obtain compact tension (CT) specimens (Fig. 1(b)) with overall geometry patterned after ASTM standard E647[16] for fatigue crack growth (Fig 1(c)).

A channel of 1 mm depth was introduced on the back of all the specimens to limit crack curving that occurs with some tubule orientations (Fig. 1(c)). Precision holes (0.7 mm and 1.0 mm diameter) were counterbored to the depth of channel to ensure application of opening mode (Mode I) loads. A sharp razor blade was used to insert a chevron notch in the specimen. Lastly, the front face of each specimen (without backchannel) was darkened using an indelible marker.

A universal testing system[a] was used for fatigue testing of the prepared specimens. This machine has a resolution of ± 0.01 N and the maximum load capacity is 225 N. The specimen was mounted on a specially designed fixture and restrained using two stainless steel pins (Fig. 2(a)). Mode I cyclic loads were applied to the CT specimen within an HBSS hydration bath at room temperature (22°C). To initiate a crack from the notch, a sinusoidal load was applied under load control with peak loads between 8 N and 16 N. A loading frequency of 5 Hz was chosen to balance concerns associated with frequency of mastication and the duration of time required to complete the individual tests. A stress ratio (R) of 0.5 was used for crack initiation.

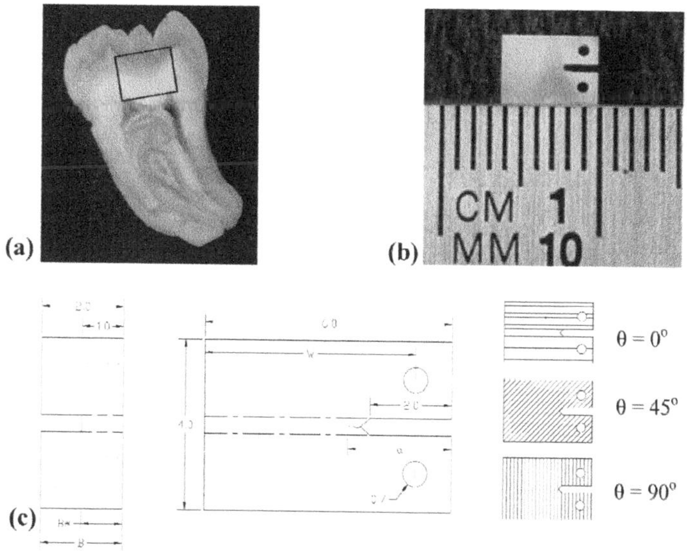

Fig. 1. A dentin CT specimen from a human third molar. (a) A primary section and possible specimen (b) complete dentin CT specimen (c) final geometry of the CT specimen and definition of tubule orientations.

Crack length measurements were conducted visually using a back lighting technique in which the back surface (with channel) was illuminated with white light and inspection of the crack length occurred on the stained face. Due to the partially transparent nature of dentin, the transmitted light underwent diffraction on the crack face and resulted in illumination of the crack tip (Fig. 2(b)). Inspection of the crack length was conducted using an optical microscope (100X) with scaled reticule. Following the initiation process, cyclic loading was resumed using R=0.1 to propagate a crack in the dentin specimen. Measurements of the change in crack length (Δa) were made over specific intervals of fatigue loading until complete specimen fracture. The number of cycles between measurements (ΔN) was chosen according to the observed crack growth rate and typically ranged between 10 and 25 k cycles.

Using the incremental crack length measurements from fatigue testing, the fatigue crack growth rate (da/dN) for the steady state response was modeled using the Paris Law[17] according to

$$\frac{da}{dN} = C(\Delta K)^m \qquad (1)$$

where ΔK is the stress intensity range, and da and dN represent the incremental crack extension (Δa) and number of cycles (ΔN), respectively and the quantities C and m are the fatigue crack growth coefficient and exponent, respectively. The ΔK is determined from the difference in stress intensity at the minimum and maximum loads and can be estimated according to[15] where B

$$\Delta K = \frac{\Delta P}{B^{*}\sqrt{W}}\left(\frac{B^{*}+1}{B+1}\right)(0.131+0.320\alpha+0.211\alpha^{2}) \quad (2)$$

and B* are the specimen thickness and the ligament thickness, respectively and $\alpha = a/W$ as shown in Figure 1(c).

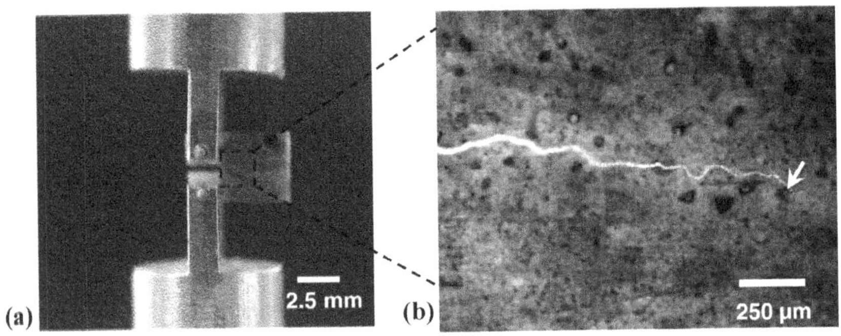

Fig. 2. Experimental setup for fatigue testing of dentin specimen. (a) a CT specimen mounted on special grips of an EnduraTec ELF 3200 system, (b) back illumination of a crack (arrow indicated crack tip).

After completion of the fatigue crack growth test, the fractured CT specimens were removed from the hydration bath and dehydrated in air for at least 72 hours. The fractured surface of the dehydrated dentin specimens were then coated with gold palladium using a cold sputter machine[b]. In order to identify striations on the fractured surface, micrographs were obtained using a scanning electron microscope[c] (SEM) in the secondary electron imaging mode (SEI). A profilometer[d] with 5 μm probe diameter was used to evaluate the surface topography of the fractured specimens. The typical traverse length of the probe ranged from 1 mm to 2 mm depending upon the length of steady state (Region II) crack growth. A cutoff length measuring one fifth of traverse length was used to obtain 8000 data points per profile. An analysis of the power spectrum density (PSD) of profiles was conducted to quantify the primary wavelength of the fatigue striations.

RESULTS

The Paris Law parameters (C and m) were estimated from the fatigue crack growth response. Table I shows the parameters determined for young hydrated specimens with tubule orientation of 0°, 45° and 90° and for old hydrated specimens with tubule orientation of 90°. Note that the fatigue crack growth coefficient (C) and crack growth exponent (m) for old hydrated dentin was significantly different from that of young hydrated dentin. Details of the effect of aging on mechanical properties of dentin are described elsewhere [15,18].

Table I. Paris Law Parameters for the Human Dentin Specimens.

Age group	Tubule Orientation	Paris Law Parameters	
		m	C (mm/cycle).(MPa m$^{0.5}$)$^{-m}$
Young	0°	12.5 ± 1.8	1.57E-05
	45°	11.1 ± 1.8	8.18E-05
	90°	13.3 ± 1.1	1.76E-05
Old	90°	21.2 ± 3.3	6.27E-03

A total of 20 fractured CT specimens were examined in the SEM. Regardless of age, more than 60% of specimens with tubules orientation of 90° showed visible striations. In specimens with tubule orientation of 0° and 45°, striations were less visible and exhibited less definite pattern.

A fracture surface from a CT specimen (male 50 years age) is shown in Figure 3. The image shows two distinct regions of crack growth corresponding to fatigue and fracture. The three regions of cyclic extension are also highlighted. Fatigue striations resulting from cyclic crack growth are clearly visible in Region II. Note that the striation spacing is location dependent and increases with ΔK, particularly near the transition of Region II and Region III. Figure 4 (a) and (b) show SEM images of fracture surfaces from a young (male, age 17 years) molar and old (male, age 50 years) molar. Both images show evidence of striations resulting from cyclic crack extension.

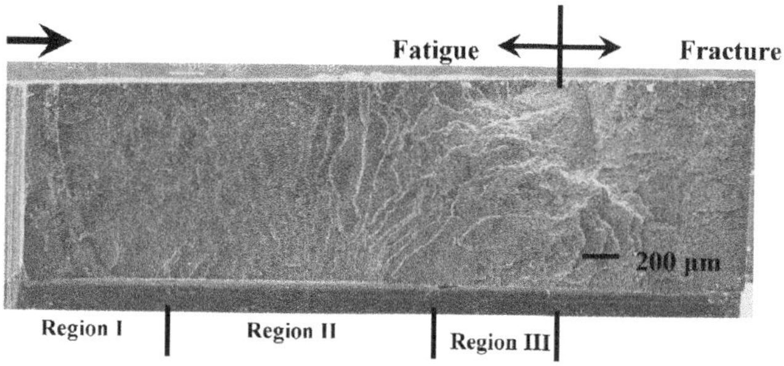

Fig. 3. Fracture surface of a CT specimen (male 50 years age) with 90° tubule orientation. Striations are visible in the Region II of crack growth. The large arrow indicates the direction of crack growth.

Surface profiles were obtained from 20 CT specimens using a contact profilometer. For each specimen, two profiles of the fracture surface were obtained at different locations in Region II. The average surface roughness (Ra) and ten point roughness (Rz) are listed in Table II. Both the Ra and Rz increased from the 90° orientation, to 45° and then to the 0° orientation. For old dentin (θ=90°) the values of Ra and Rz were smaller than those for young dentin.

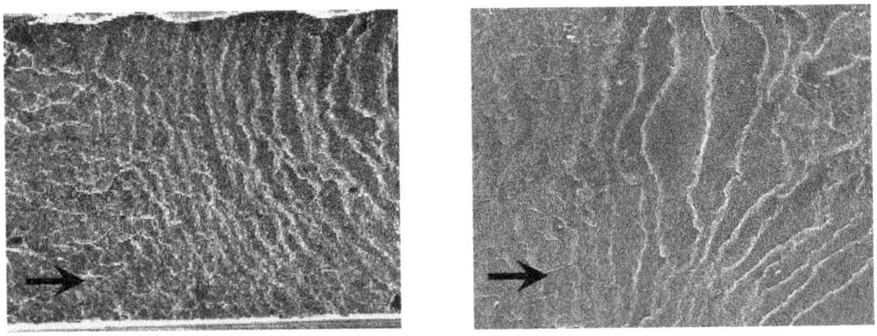

(a) 100 um (b) 100 um

Fig. 4. Fracture surfaces of CT specimen with 90° tubule orientation. Fatigue striations visible on surface of dentin (a) male 17 years age (b) male 50 years age. The arrow indicates the direction of crack growth. Note the difference in average striation spacing between (a) and (b).

Table II. A Comparison of Surface Roughness in Region II Fatigue Crack Growth in Dentin.

Age group	Tubule Orientation	Roughness (μm) Ra	Roughness (μm) Rz
Young	0°	2.5 ± 0.5	12.2 ± 1.7
	45°	2.1 ±1.0	10.6 ± 4.1
	90°	1.6 ±0.4	7.9 ± 2.1
Old	90°	1.5 ± 0.4	7.4 ± 2.2

The power spectrum density (PSD) was generated from the surface profiles using commercial software[e]. Figure 5(a) shows the power spectrum for the fracture surface of a CT specimen from a young (male 17 years age) patient. Two peaks were obtained at 118 μm and 170 μm.

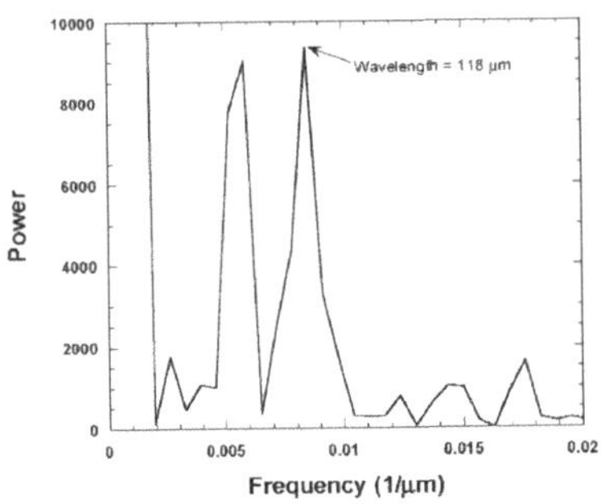

(a)

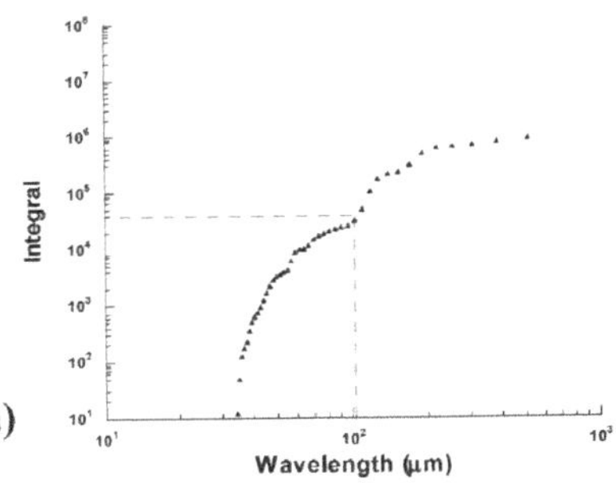

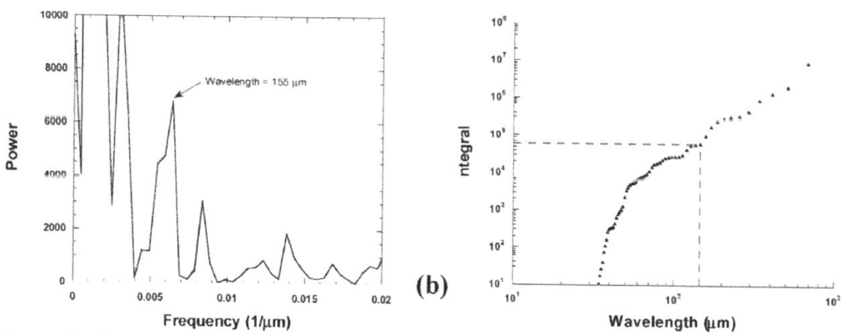

Fig. 5. Results from power spectrum analysis for young and old dentin. (a) Power vs. Frequency for young (male, age 17 years) dentin. A wavelength of 118 µm is visible and gives the largest inflection on the integral vs. wavelength curve. (b) Power vs. Frequency for old (male, age 50 years) dentin. Wavelengths of 120 µm and 155 µm are visible with maximum inflection at 155 µm.

Similarly, the power spectrum for a CT specimen from an old (male, age 50 years) patient (Fig. 5(b)) exhibits two peaks at 120 µm and 155 µm. To obtain an accurate measure of the primary cyclic wavelength, an integral curve was generated for the power as a function of wavelength. The maximum change in slope (inflection) of this curve was identified and used in defining the dominant wavelength (i.e. striation spacing). The integral curve on Figure 5(a) shows a maximum change in slope at 118 µm and for Figure 5(b) gave a maximum change in slope at 155 µm. The results from the power spectrum analysis were then compared with the striation spacing on SEM images of dentin specimens (Fig. 4(a),(b)) to confirm the analysis.

The dominant wavelengths (mean ± S.D.) for young and old dentin specimens are listed in Table III. The primary cyclic wavelength for young dentin with tubule orientations $\theta=0^{\circ}$ and $\theta=45^{\circ}$ was 81 ± 21 µm and 71 ± 17 µm, respectively. According to an ANOVA (4-independent samples) and Tukey's HSD (Honestly Significant Difference) test, the primary cyclic wavelength for old ($\theta=90^{\circ}$) dentin was significantly different (125 ± 23 µm, $p<0.05$) when compared with young ($\theta=90^{\circ}$) dentin (93 ± 27 µm). However, there was no statistically significant difference in wavelength within the young age group.

Table III. Power spectrum analysis of the fatigue crack growth surfaces.

Age Group	Tubule Orientation	Number of Specimens N	Wavelength (µm)	Tukey's HSD test (p)		
				Young		Old
				45°	90°	90°
Young	0°	5	81 ± 21	n/s	n/s	p<0.01
	45°	4	71 ± 17	-	n/s	p<0.01
	90°	5	93 ± 27	-	-	p<0.05
Old	90°	6	125 ± 23	-	-	-

n/s - non significant

Fatigue striations were also identified and examined on cracks that underwent extension in vivo. Extracted restored molars were sectioned axially along the buccal-lingual plane using a slicer/grinder and the sections were then examined under a low magnification microscope (X10) for flaws and cracks. More than 50% of the teeth examined had cracks originating at the line angles along the prepared cavity. In general, they propagated outward into the dentin and perpendicular to the dentinal tubules. Examples are shown in Figure 6.

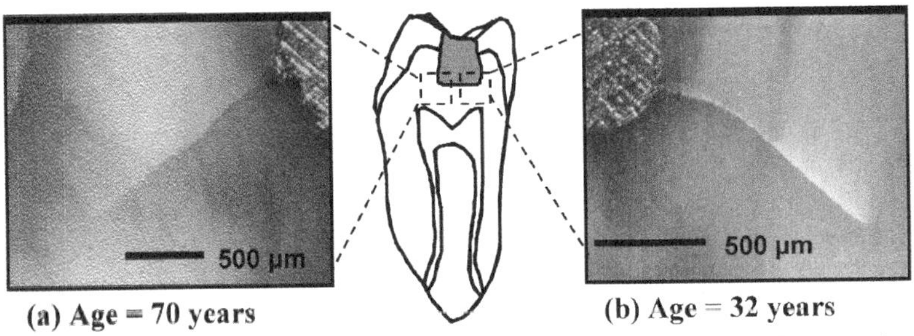

Fig. 6. Cracks in restored teeth originating from the line angle and extending in the coronal dentin.

Similar cracks in the dentin of restored teeth were reported by Arola et al.[9] The sectioned teeth were then broken along the crack and the fracture surface was examined using profilometry and power spectrum analysis. An SEM image of the fracture surface in dentin of a restored tooth obtained from a male, age 32 years, patient is shown in Figure 7(a). The power spectrum for this surface is shown in Figure 7 (b). Two peaks are visible at wavelengths of 85 µm and 113 µm.

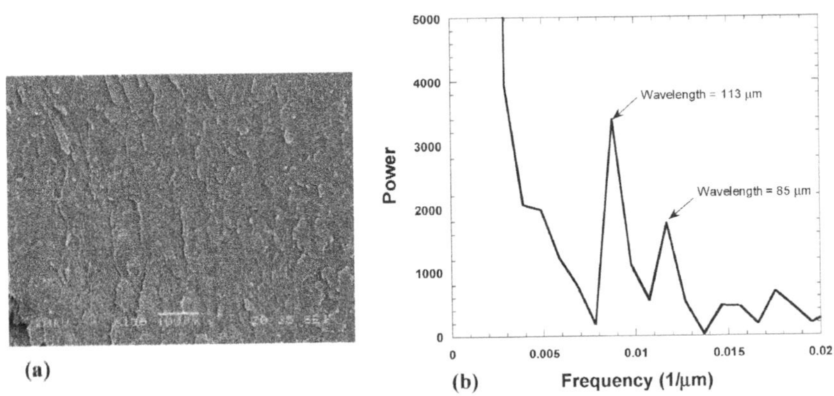

Fig. 7. Striations on fracture surface of restored tooth obtained from a male 32 year old patient. (a) SEM image of surface of restored tooth, (b) Power spectrum analysis of the surface in (a). Note that the wavelengths are within the range obtained from the in vitro analysis.

DISCUSSION

An evaluation of fracture surfaces resulting from fatigue crack growth in extracted human teeth showed that striations are evident. A quantitative analysis of the fracture surface topography was performed to identify the primary cyclic wavelength. The wavelength for old dentin (125 ± 23 μm) was significantly higher ($p<0.05$) than that of young dentin, regardless of tubule orientation. If striations are formed by a cyclic process of extension and blunting, the results suggest that crack tip blunting is less active in old dentin. While striations formed by fatigue crack growth in metals are formed every cycle,[19] the relatively low number of striations on the CT specimens indicates that this is not the case for dentin.

Recent evaluations by the authors have shown that the fatigue crack growth rate in old dentin is significantly faster (X100) than that in young dentin at the same level of driving force.[15] To further understand the changes in cyclic crack extension with aging, an estimate of the number of cycles between each striation was performed using previously reported results for average crack growth rates in dentin.[15] Figure 8 shows the average number of cycles required for striation spacing of 93 ± 27 μm for young and 125 ± 23 μm for old dentin to grow as a function of stress intensity range (ΔK). For a ΔK of 0.83 $MPa{\cdot}m^{0.5}$ (typical of Region II growth), approximately 60,000 cycles are required to extend a crack 93 μm in young dentin. At the same ΔK, a striation length of 125 μm in old dentin required only 1000 cycles of crack extension.

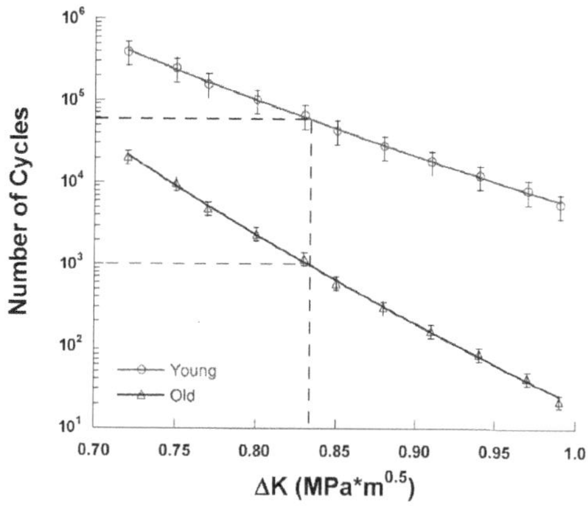

Fig. 8. The number of cycles required for a crack to extend a length equivalent to the dominant striation spacing.

An evaluation of cracks found in a restored tooth and grown in vivo was also conducted. The results from this evaluation showed that the striation spacing found on the fracture surface of restored teeth was within the range found in the in vitro evaluation. Thus, through similarities in fracture surfaces, fatigue and fatigue crack growth is indeed a major contributor to fracture of restored teeth and the in vitro approach used to examine crack extension is a viable model to evaluate both the mechanics and mechanisms of cyclic crack extension in human teeth.

CONCLUSIONS

An experimental evaluation of the influence of tubule orientation and aging on fatigue striations in dentin was conducted. According to a comparison of fatigue striations resulting from in vitro experiments and in vivo extension, the following conclusions were drawn:

(1) Fatigue crack growth striations were identified on fracture surfaces of dentin specimens and restored teeth. The striations are believed to be a result of blunting and subsequent resharpening of the crack tip.

(2) In young hydrated dentin, the striation spacing ranged from 71 ± 17 μm for θ=45° to 93 ± 27 μm for θ=90°. The highest wavelength was found in old dentin (125 ± 23 μm) with θ=90°. The larger striation spacing on fracture surfaces of old dentin suggests that aging suppresses the mechanism of blunting of the crack tip.

(3) The primary wavelength of striations on the fracture surface of restored teeth was consistent with those resulting from in vitro experiments and suggests that tooth fracture is indeed a result of fatigue and fatigue crack growth.

REFERENCES

[1]C.E. Cameron, "The Cracked Tooth Syndrome: Additional Findings," *JADA.*, 93, 971-5 (1976).

[2]W. Geurtsen, R. Schwarze, H Gunay, "Diagnosis, Therapy, and Prevention of the Cracked Tooth Syndrome," *Quintessence. Int.*, 34, 409-17 (2003).

[3]H.W.A. Wislott, J.I. Nicholls, C. Belser, "Stress Fatigue: Basic Principles and Prosthodontic Implication's," *Int. J. Prosthodont.*, 8, 105-16 (1995).

[4]M.E. Gher, R.M. Dunlap, M.H. Anderson, L.V. Kuhl, "Clinical Survey of Fractured Teeth," *JADA*, 114,174-7 (1987).

[5]H. Letzel, "Survival Rates and Reasons for Failure of Posterior Composite Restorations in Multicentre Clinical Trial," *J. Dent.*, 17, S10-7 (1989).

[6]F.J. Burke. "Tooth Fracture in Vivo and in Vitro." *J. Dent.*, 20, 131-9 (1992).

[7]D. Arola, M.P. Huang, M.B. Sultan. "The Failure of Amalgam Dental Restorations Due to Cyclic Fatigue Crack Growth, " *J. Mater. Sci.*, 10, 319-27 (1999).

[8]K Tonami, H. Takahashi, "Effects of Aging on Tensile Fatigue Strength of Bovine Dentin," *Dent. Mater.*, 16, 156-86 (1997).

[9]D. Arola, J.A. Rouland, D. Zhang, "Fatigue and Fracture of Bovine Dentin." *Exp. Mech.*, 42, 380-8 (2002).

[10]D. Arola, J.A. Rouland, "The Effects of Tubule Orientation on Fatigue Crack Growth in Dentin," *J. Biomed. Mater. Res.*, 67, 78-86 (2003).

[11]J.J Kruzic, R.K. Nalla, J.H. Kinney, R.O. Ritchie, "Crack Blunting, Crack Bridging and Resistance-Curve Fracture Mechanics in Dentin: Effect of Hydration," *Biomaterials,* 24, 5209-21 (2003).

[12]J.J. Kruzic, R.K. Nalla, J.H. Kinney, R.O. Ritchie, "Mechanistic Aspects of In Vitro Fatigue-Crack Growth in Dentin," *Biomaterials,* 26, 1195-204 (2005).

[13]R.K. Nalla, V. Imbeni, J.H. Kinney, S.J. Marshall, R.O. Ritchie, "In Vitro Fatigue Behavior of Human Dentin with Implications for Life Prediction," *J. Biomed. Mater. Res.,* 66, 10-20 (2003).

[14]R.K. Nalla, J.H. Kinney, S.J. Marshall, R.O. Ritchie, "On the In Vitro Fatigue Behavior of Human Dentin: Effect of Mean Stress," *J. Dent. Res.*, 83, 211-5 (2004).

[15]D. Bajaj, N. Sundaram, A. Nazari, D. Arola, "Age, Dehydration and Fatigue Crack Growth in Dentin," *Biomaterials*, 27, 2507-17 (2006).
[16]ASTM E647, Standard Test Method for Measurement of Fatigue Crack Growth Rates, In Annual Book of ASTM Standards, Philadelphia PA: *American Society for Testing and Materials*, (2000).
[17]P.C. Paris, F. Erdogan, "A Critical Analysis of Crack Propagation Laws," *J. Basic. Eng.*, 85, 528-32 (1963).
[18]D. Arola, R.K. Reprogel, "Effects of Aging on the Mechanical Behavior of Human Dentin," *Biomaterials*, 26, 4051-61 (2005).
[19]C.R. Brooks, A. Choudhury, Failure Analysis of Engineering Materials, McGraw-Hill (2002).

---

[a] EnduraTEC ELF 3200, BOSE Cooperation, Eden Prairie, MN, USA.
[b] Model LLC Desk II, Denton vacuum, Moorestown, NJ, USA.
[c] Model JSM-5600, JEOL, Peabody, MA, USA.
[d] Model LV-50E, Hommelwerke, Rochester Hills, MI, USA.
[e] Version 7.0.1, MATLAB, Natick, MA, USA.

FATIGUE FRACTURE OF DENTAL RESIN BONDED CERAMIC DISKS

L. H. Mair
University of Liverpool, School of Dental Studies
Pembroke Place
Liverpool, L3 5PS, UK.

P. Padipatvuthikul
University of Liverpool, School of Dental Studies
Pembroke Place
Liverpool, L3 5PS, UK.

ABSTRACT

Many modern dental restorations are bonded onto the teeth using modified epoxy resins and composites. In the mouth these restorations are subjected to cyclic forces that can reach peak loads of 800N. In this study, resin-bonded ceramic specimens were subjected to cyclic loading. Fracture and damage were studied using SEM and incident light microscopy with $AgNO_3$ staining.

Fifty-one ceramic cylinders (Φ=5mm x 4mm) were fabricated by sintering porcelain powder.[1] One surface was etched with 9.5% hydrofluoric acid for 15 minutes. Three bonding systems were used.[2-4] The ceramic surface was treated with a primer (silane) followed by the recommended resin. The enamel of an extracted tooth was etched with orthophosphoric acid for 30 seconds, rinsed and air dried then the resin applied. The cylinders were then bonded together. The tooth was held in a fatigue testing machine and the cylinders subject to a "staircase" fatigue test using 5000 cycles at 80 cycles/ min. The fracture surfaces of the failed specimens were examined in a SEM. Other samples were stained with $AgNO_3$, and prepared for incident light microscopy.

The fatigue limit was Calibra 13.74 $\pm$ 0.1 MPa, Nexus 11.41 $\pm$ 1.44 MPa, Panavia 12.28 $\pm$ 2.66. In all systems the fatigue crack initiated at either the ceramic/resin interface or within the ceramic itself. The crack propagated through the bonding composite and did not fracture the underlying tooth. Staining indicated damage in the ceramic and leakage along the ceramic resin interface.

In this study the specimens failed by mixed mode fracture. Staining indicated that there was subsurface damage prior to failure.

[1]Vitadur, Vivadent, [2]Calibra with Prime and Bond resin, Dentsply, [3]Nexus with Optibond resin, Kerr, [4]Panavia with Clearfil SE Bond resin, Kuraray.

INTRODUCTION

Over the past 20 years, resin bonding has emerged as an important technology for bonding ceramic dental restorations to teeth. The resin bond has two essential functions; it must hold the restoration in place and simultaneously prevent crack initiation and propagation by bonding the entire surface to the underlying tooth. Failure in the first function leads to complete loss of the restoration whereas failure in the second results in cracks within the ceramic. In essence, resin bonding involves impregnating both the etched tooth surface and the etched enamel surface with a modified epoxy resin that is then made to set by the generation of a free radical. This is normally generated by photoinitiation. The resin layer therefore gains a micromechanical attachment to the etched surface. The resin surfaces are then bonded together using a filled resin that can cross polymerize with the unconverted vinyl groups in the resin layer (-C=C). This results in a resin bonded "sandwich" with tooth on one side and ceramic on the other (Figure 1).

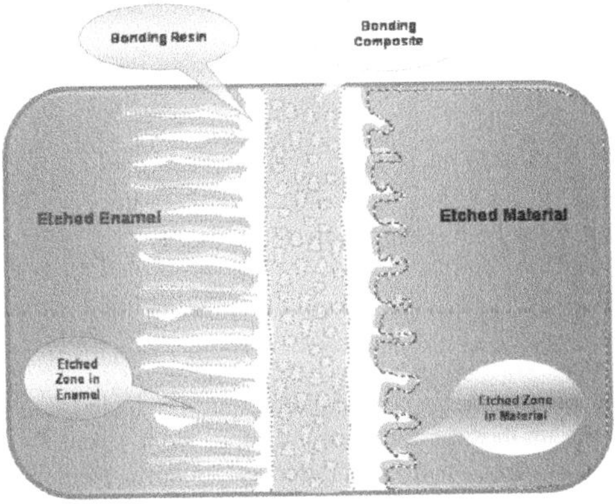

Figure 1: Resin bonded sandwich

Research in this area has focused on improving the strength of the bond between the ceramic and tooth. The majority of laboratory studies have used monotonic tests such as tensile or shear tests to assess the bond strength. Although these tests are quick and expedient they do not replicate the forces that are applied to dental restorations in the mouth. In the oral environment, teeth and restorations are subject to cyclic forces that can result in fatigue failure. A number of studies measuring fatigue forces have set the upper load at between 2 and 12 kg with the majority being under 5 kg.[1-4] Again, this does not represent the situation in the mouth where the chewing force in normal patients is 3 – 16 kg and patients may exert loads as high as 80 kg when grinding their teeth in parafunction.[5-8] The purpose of this investigation was to determine the Fatigue Limit of resin bonded ceramic disks and determine features of fatigue fracture. In addition to scanning electron microscopy the fatigue zones were investigated using silver nitrate staining. Silver ions (Ag^+) penetrate into cracks and crazes in the materials. These are subsequently precipitated to metallic silver (Ag) within microspaces in the materials. The color is caused by the scattering of light by colloidal silver.[9,10]

MATERIALS AND METHODS

Preparation of Ceramic Surface

Fifty-one ceramic disks were fabricated by sintering ceramic power (Vitadur Alpha, VITA Zahnfabrik, Germany) in a vacuum furnace (Multimat MC II, DeTrey, Dentsply, Germany). The disks dimensions were Φ5mm × 4mm, built up in layers to minimize contraction shrinkage. One surface of each disk was etched with porcelain etching-gel (Hydrofluoric Acid 9.5%) for 15 minutes and rinsed with water. The etched surface was sandblasted with 50μm alumina and then cleaned in an ultrasonic bath and treated with ceramic silane primer. A layer of bonding agent was then applied to the silanated surface; the carrier solvent evaporated and the resin made to set.

The prepared disks were then divided into three groups of seventeen that were bonded with three commercially available dental bonding systems. These systems are described in Table I.

Table I: Bonding Agents

Dentine Bonding System	Bonding composite	Impregnating resin	Manufacturer	Batch No.
Calibra System	2 paste dual cure* composite based on Bis-GMA.	Prime and Bond NT Single bottle Dual cure resin Solvent: Acetone	Dentsply, USA	0207000948
Panavia System	2 paste tri-cure# system	Clearfill SE Bond Two liquid, dual cured resin Solvent: H_2O	Kuraray, Japan	00013A
Nexus System	2 part dual cure* Base on Bis-GMA	Optibond Solo Plus Single liquid. Light cure resin Solvent: ethanol	Kerr, USA	Base 423969 Catalyst 428992

* light cure + subsequent chemical cure
\# light cure + subsequent anaerobic and chemical cure.

In essence each system has a resin component suspended in a carrier solvent. After evaporation the solvent is evaporated and the resin made to set.

Preparation of Tooth Surface

Fifty-one recently extracted human teeth were collected and embedded in polymethylmethacrylate (PMMA) with one flat surface of enamel protruding above the resin surface. This surface was gently abraded to produce a flat bonding area. The enamel was then etched with 37.5% phosphoric acid for 30 seconds, rinsed in H_20 for 20 seconds and air dried. A layer of bonding agent was then applied to the etched enamel surface, the carrier solvent evaporated and the resin make to set.

Bonding of the Resin Impregnated Surfaces

The two paste composites were mixed and applied to the resin coated ceramic disks. These were then "seated" onto the resin coated enamel. Any gross excess was removed and the conversion initiated by photopolymerization in the first instance. Most of the systems contained an additional chemical initiator. A completed specimen is shown in Figure 2. After preparation the specimens were stored in water for 24 hours at room temperature before testing.

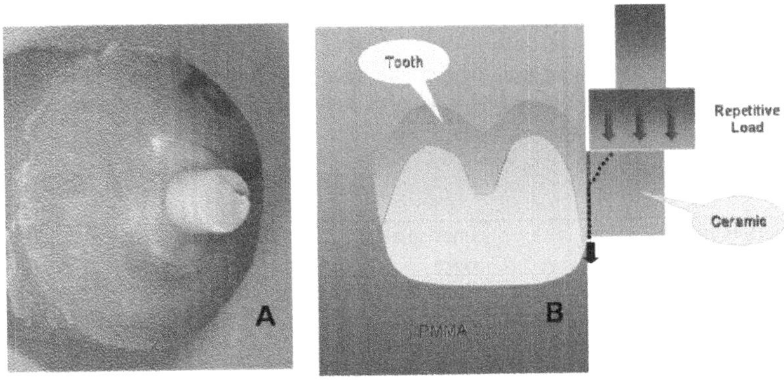

Figure 2:
(A) Photo of prepared specimen
(B) Diagram showing loading point and eventual fracture path

Fatigue Tester

A customized Fatigue Tester was constructed for this study (Figure 3). In essence this consists of a flat anvil that is weighted onto the specimens by means of weights applied through the lever principle. The applied load could be increased from 8 kg to 80 kg in 4 kg steps. The weight is removed by lifting the lever with a mechanical cam. For this experiment, the cycle rate was fixed at 80 cycles/ min.

Figure 3: Fatigue Testing Device with 5 arms capable of delivering 100kg cyclic loads.

Fatigue Test

Three groups of seventeen specimens were subjected to the staircase fatigue test to determine the Fatigue Limit. In essence specimens were subjected to 5,000 cycles at 13 kg. If the sample survived the next specimens was tested at 17 kg, the next one at 21 kg, the next one at 25 kg etc. When a specimen failed the next one was tested at 4 kg less then the load at failure. This "up and down" (staircase method) was continued until all samples had either survived or failed. The analysis of the data is based on the least frequent event (failures or non-failures). To determine the mean fatigue limit and the standard deviation of the mean, the data can be arranged as in Table II. The lowest stress level at which the failure occurs is denoted by i=0, the next i=1, etc. The mean fatigue limit (X), and its standard deviation(s) are given by formulae [1] and [2].[11-15]

$$X = X_o + d(A/N \pm ½) \qquad [1]$$

$$S = 1.62\, d\left(\frac{NB-A^2}{N^2} + 0.029\right) \qquad [2]$$

X_o is the lowest stress level considered in the analysis, and d is the stress increment employed in the tests. In formulae [1], the positive sign is used when the analysis is based on non-failures, and the negative sign is used when failures are considered.

Table II Method for analyzing staircase test data

Stress (MPa)	i	n_i	i n_i	i^2 n_i
16.8	2	4	8	16
14.8	1	3	3	3
12.8	0	1	0	0
		$N = \Sigma n_i$ = 8	$A = \Sigma i n_i$ = 11	$B = \Sigma i^2 ni$ = 19

Statistical difference was determined using the Behrens-Fisher T test.[16-18]

Examination of Fractured Surface

The fractured surfaces were sputter coated with gold and examined in a scanning electron microscope (JEOL JSM-35C , Cambridge, UK).

Examination of "Survivors"

The excess embedding resin was removed with a high power saw and then the specimens were immersed in $AgNO_3$ for 48 hours. They were then removed and sectioned with a diamond disk cutter (Buehler, Isomet 11-1180, Coventry, UK). The sections were examined in an incident light microscope using crossed polars to exclude directly reflected light and reveal the color of silver.

Determination of Shear Bond Strength

Identical specimens were prepared and the shear bond tested determined in a universal testing machine (Nene Instruments, Wellingborough, UK) using a cross head speed of 50 mm/min.

RESULTS

The statistical analysis showed that there were two homogeneous subsets and these are shown in Table III.

Table III: Mean Fatigue Limit ± Standard Deviation (MPa) and homogeneous subsets.

Subsets			
1	Calibra	Panavia	
	(13.74+ 0.1)	(12.28 +2.66)	
2		Panavia	Nexus
		(12.28+2.66)	(11.41+1.44)

For all materials the values of the fatigue limit was significantly lower than the shear values.

Table IV: Fatigue Limits Compared to Shear Values

Bonding systems	Fatigue Limit (MPa)	Shear Bond Strength (MPa)	T Value	Probability(t)
Calibra (Prime&Bond)	13.74 ± 0.1	21.22 ± 2.66	t = 11.75	p < 0.001
Panavia (Clearfill bond)	12.28 ± 2.66	16.14 ± 2.95	t = 3.50	P = 0.002
Nexus (Optibond)	11.41 ± 1.44	17.91 ± 5.09	t = 4.99	p < 0.001

Scanning Electron Microscopy

For most specimens the cracks initiated in the ceramic immediately beneath the area of application of load. The crack then propagated back through the interface between the ceramic and resin and then through the bonding composite. Fracture steps were apparent in the fractured ceramic with interspersed "river patterns" occurring as the fracture passed through the bonding composite. There was evidence of crack bifurcation and division in the composite leading to "edges" of overlapping composite (Figure 4F).

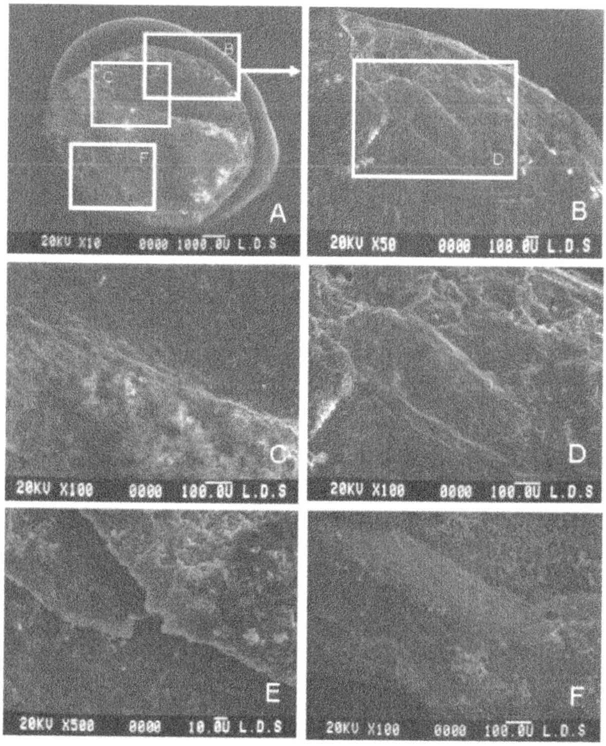

Figure 4: A typical fracture surface of the resin bonded ceramic specimen.
(A) Low power SEM with key to other micrographs. Crack runs from top to bottom.
(B) Steps in fractured ceramic.
(C) Magnified view of (B) showing secondary steps on the fractured surface.
(D) Junction between ceramic, resin and composite.
(E) Junction between composite, resin and tooth.
(F) Layered "edges" in composite.

Silver Nitrate Staining

All specimens had evidence of silver nitrate staining at the junction between the ceramic and tooth. This indicates a potential pathway for the ingress of oral fluids and stain. A number of specimens had staining within the ceramic close to the resin bonded junction with composite (Figure 5C&D). Careful examination of a number of specimens indicated the staining of a delta of cracks in the fatigue zone of the ceramic.

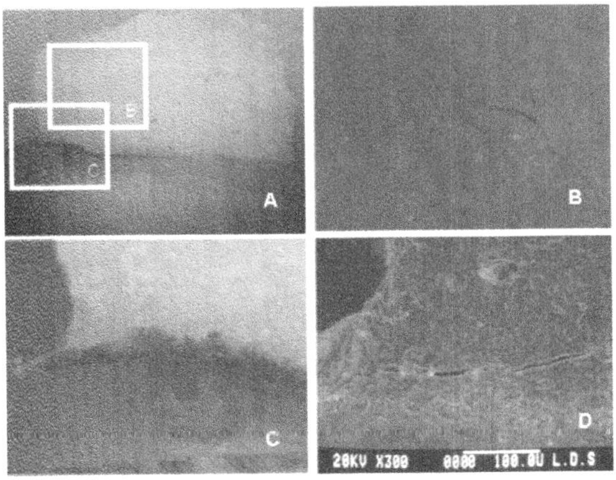

Figure 5: Silver nitrate stained specimen
(A) Low power photomicrograph showing location of B&C.
(B) Silver stained crack delta in fatigue zone of ceramic.
(C) Silver staining of ceramic-composite-tooth interface.
(D) SEM of specimen shown in (C).

DISCUSSION

Although the fatigue limit values were significantly lower than the shear values, and appeared to be around the value of normal human bite force, it has to be remembered that the surface area of the samples was relatively small compared to the surface area of a dental crown or veneer. For a larger surface area the bond strength would be higher. This explains the clinical success of these restorations when used in patients with normal bite forces. However, the bonds do not equate to the forces generated in parafunctional patients. This study indicated that the cracks initiated in the ceramic rather than at the interface. Therefore, when attempting to improve the bond strength of these restorations, for the management of patients with a heavy bite, it would seem important to increase fracture resistance of the ceramic before attempting to improve the bond strength of the bonding system.

The experiment showed that silver nitrate could stain early cracks occurring in the ceramic. Although the fracture resistance of ceramic cores for conventional crowns has been increased by the use of milled alumina or zirconia cores the strength of the overlying veneering ceramic has remained fairly constant. As resin bonded ceramics composite restorations do not have fracture resistance cores, they remain vulnerable to crack initiation by fatigue.

ACKNOWLEDGEMENTS

We would like to thank Mr. H. Babakan, Mr. M. Blance, Mr. L. Blundell, Mr. P. Smith (dental technicians) for their help and advice on constructing the specimens. Mr. G. Abbott for constructing the Fatigue Testing Machine and Mrs. V. Payne for her technical assistance.

REFERENCES

[1]W.P. Saunders, "Effect of Fatigue Upon the Interfacial Bond Strength of Repaired Composite Resins," *J. Dent.*, 18(3), 158-162 (1990).

[2]A. Htang, M. Ohsawa, H. Matsumoto, "Fatigue Resistance of Composite Restorations: Effect of Filler Content," *Dent. Mater.*, 11(1), 7-13 (1995).

[3]K.K. Narva, L.V. Lassila, P.K.Vallittu, "Fatigue Resistance and Stiffness of Glass Fiber-Reinforced Urethane Dimethacrylate Composite," *J. Prosthet. Dent.*, 91(2), 158-163 (2004).

[4]J.F. McCabe, Y.Wang, M. Braem, "Surface Contact Fatigue and Flexural Fatigue of Dental Restorative Materials," *J. Biomed. Mater. Res.* 50(3), 375-80 (2000).

[5]L.D. Garner and N.S. Kotwal, "Correlation Study of Incisive Biting Forces with Age, Sex and Anterior Occlusion." *J. Dent. Res*, 52, 698 (1973).

[6]L.L. Fogle and A.G. Glaros, "Contribution of Facial Morphology, Age and Gender to EMG Activity Under Biting and Resting Condition. A Canonical Correlation Analysis," *J. Dent. Res.*, 74, 1496 (1995).

[7]K. Miyaura, Y. Matsuka, M. Morita, A.Yamashita, T. Watanabe, "Comparison of Biting Forces in Different Age and Sex Groups: A Study of Biting Efficiency with Mobile and Non-Mobile Teeth," *J. Oral. Rehabil.*, 26, 223 (1999).

[8]K. Nishigawa, E. Bando, M. Nakano, (2001), "Quantitative Study of Bite Force During Sleep Associated Bruxism," *J. Oral. Rehabil.*, 28(5), 485 (2001).

[9]L.H. Mair, "The Colors of Silver with Silver Nitrate Staining in Dental Materials," *Dent. Mater.*, 8, 110-117 (1992).

[10]R.H. Doremus, "Optical Properties of Small Silver Particles," *J. Chem. Phys.*, 42, 414-417 (1965).

[11]R.A. Draughn, "Compressive Fatigue Limits of Composite Restorative Materials," *J. Dent. Res.*, 58(3), 1093-1096 (1979).

[12]J.L. Drummond and M.S. Bapna, "Static and Cyclic Loading of Fiber-Reinforced Dental Resin, *Dent. Mater.*, 19, 226-231 (2003).

[13]U. Lohbauer, U., T. von der Horst, R. Frankenberger, N. Kramer, A. Petschelt, "Flexural Fatigue Behavior of Resin Composite Dental Restoratives," *Dent. Mater.*, 19, 435-440 (2003).

[14]J. McCabe, "In Vitro Wear Testing of Composite Resin. In: *Posterior Composite Resin Dental Restorative Materials*, Peter Szulc Publishing (1985).

[15]J. McCabe, T. Carrick, et al., "Alternative Approaches to Evaluating the Fatigue Characteristics of Materials," *Dent. Mater.*, 6, 24-28 (1990).

[16]D.I. Best and C.W. Rayner, "Welch's Approximate Solution for the Behren's-Fisher Problem," *Technometrics*, 29, 205 -210 (1987).

[17]E.L. Lehmann, "*Testing Statistical Hypotheses*," New York: John Wiley & Sons (1986).

[18]W. Chase and F. Bown, "*General Statistics*," New York: John Wiley & Sons (1986).

FRACTURE MIRROR CONSTANTS OF BIOCERAMICS FOR HIP JOINT REPLACEMENT: DETERMINATION AND APPLICATION

S. Hecht-Mijic
CeramTec AG, Central Development
Plochingen, Germany

INTRODUCTION

In a failure analysis on ceramic components of hip joint replacement information on the stress which led to fracture is of outmost interest. An estimate of this stress is possible from a special fracture surface marking which is related to a material characteristic called the "fracture mirror constant."[9,10] The determination of the fracture mirror constants of two bioceramic materials will be described and application of this constant to evaluate the fracture stress in "burst tests" will be discussed.

Mirror/Mist/Hackle Transitions

On the region of the fracture origin of glasses and ceramics areas of different roughness can be observed. These areas are called "mirror, mist and hackle." They extend rotation-symmetrically around the fracture origin. A schematic sample of the region around the fracture origin is given in Figure 1.

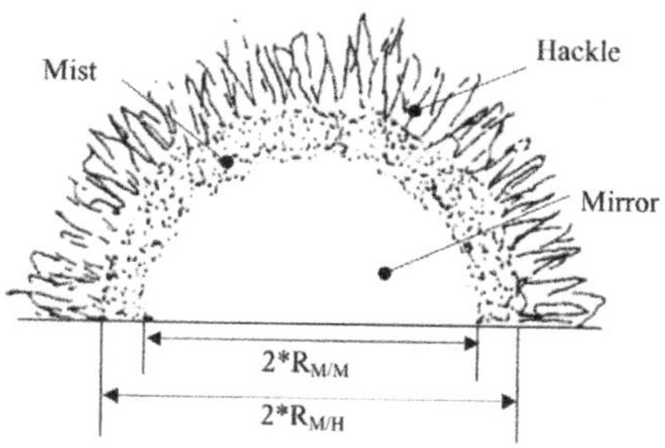

Figure 1: Schematic drawing of the glass or ceramic fracture surface structure, around a fracture origin located at the specimen surface.

In contrast to glasses where the mirror/mist/hackle transitions are relatively obvious, in ceramic materials these transitions are less clearly discernable. This is caused by the fact that ceramic materials are not amorphous materials but exhibit, due to their grain structure, an inherent fracture surface roughness which can especially obscure the mirror/mist transition. Therefore, in the present study the focus was on the transition from mist to hackle; i.e., the radius R_{Hackle} (R_H, in Figure 1 named $R_{M/H}$) was measured and subsequently the fracture mirror constant A_{Hackle} (A_H) was determined.

Experiments

The fracture mirror constants were determined for two ceramic materials which are in use for components of total hip joint replacement, namely Biolox forte, which is a medical grade alumina according to ISO 6474,[6] and ASTM F 603,[3] respectively, and Biolox delta, which is a alumina matrix composite ceramic. The main material constants are given in Table I.

Table I:
Typical material constants of Biolox forte and Biolox delta (according to CeramTec data sheet).

	Biolox forte Alumina ceramic	Biolox delta Alumina matrix composite ceramic
4-point-bending strength [MPa]	> 580	> 1000
Density [g/cm^3]	> 3.98	> 4.36
Fracture Toughness K_{IC} [MPa$\sqrt{m}$] (determined by the SEVNB-method)	3.4	5.7
Youngs Modulus [GPa]	380	350

The 4-point-bending strength of bars was determined and subsequently the radii R_{Hackle} were measured on the fracture surface structure. Fracture surfaces were generated on rectangular bars having the dimensions 3 mm x 4 mm x 45 mm. The upper and lower spans were 20 and 40 mm, respectively. Fourteen Biolox forte specimens and thirty Biolox delta specimens were broken in the 4-point-bending test.

The position of the fracture origin was predefined by setting a Vickers indentation on the polished surface of every specimen. In order to obtain the complete sequence of surface structure of mirror, mist and hackle, as shown in Figure 1, the Vickers indentation forces have to lie within a certain range. This range was determined in preliminary tests.

The radius R_H was measured at the transition from mist to hackle with a light microscope. This was done on both fragments, and parallel as well as perpendicular, to the fracture surface edge containing the fracture origin. Calculation, with R_H and with the 4-point-bending strength, according to equation 1 led to the fracture mirror constant A_H.

$$A_H = \sigma_{4Pkt} \cdot \sqrt{R_H} \quad (1)$$

In a second part of the study the fracture mirror constant A_H was exemplarily applied to make an estimation of fracture stress in burst tests. For this purpose burst tests according to ISO 7206[7] and ASTM F 2345[4] were carried out on hip joint ball heads made of Biolox forte and Biolox delta, respectively, and the fracture surface structures were analyzed. The radii on the fracture surface in the region of the fracture origin were measured by use of a light microscope. Seven ball heads made of Biolox forte and nine ball heads made of Biolox delta, both of an outer diameter of 28 mm, cone diameters 12/14 and neck length "L", were investigated.

RESULTS

Fracture Mirror Constant A_{Hackle} of Biolox Forte and Biolox Delta

Table II contains ranges of values obtained in the experiments on Biolox forte and Biolox delta: The range of Vickers indentation forces F_V which led to the formation of the complete sequence of mirror, mist and hackle on the fracture surface is presented. The resulting range of 4-

point-bending strength obtained on the bending bars is given, too. Finally, Table II contains the range of the measured values R_{Hackle}, whereas ranges of R_{Hackle} of measurement parallel and perpendicular with relation to the fracture surface edge are listed separately.

Table II: Range of Vickers indentation forces F_V which led to the formation of the complete sequence of mirror, mist and hackle on the fracture surface. Resulting range of 4-point-bending strength. Range of the measured single values R_{Hackle}, ranges of R_{Hackle} measured parallel and perpendicular with relation to the fracture surface edge.

		F_V [N]	$\sigma_{4\text{-point}}$ [MPa]	$R_{H,\ parallel}$ [mm]	$R_{H,\ perpendicular}$ [mm]
Biolox forte	Min	0.15	327	0.21	0.16
	Max	5.0	624	0.47	0.51
Biolox delta	Min	2.0	529	0.15	0.14
	Max	7.0	788	0.33	0.31

F_V: Vickers indentation force
$\sigma_{4\text{-point}}$: 4-point-bending strength
$R_{H,\ parallel}$: Radius R_{Hackle} measured parallel to the fracture surface edge
$R_{H,\ perpendicular}$: Radius R_{Hackle} measured perpendicular to the fracture surface edge

Of course, in Table II a high 4-point-bending strength corresponds to a low Vickers indentation force and vice versa. Exemplarily, some details based on the values in Table II shall be discussed for Biolox forte: Vickers indentation below 0.15 N did not lead to a defined position of the fracture origin at the Vickers indentation. This means that F_V = 0.15 N was found as a lower limitation in the experiments presented here. The value $\sigma_{4\text{-point}}$ = 624 MPa corresponds with F_V = 0.15 N and is almost equal to the strength 646 MPa obtained in regular 4-point-bending tests on undamaged bars, i.e., according to ISO 14704[8] and ASTM C 1161.[1] Of course, one should keep in mind that $\sigma_{4\text{-point}}$ = 624 MPa stems from a single measurement while 646 MPa is based on a statistical evaluation of 180 (6x30) strength data.

For illustration, the fracture surface structures of two bending bars of Biolox forte and Biolox delta respectively are shown in Figure 2 and Figure 3. There is evidence of the relation between the level of the 4-point-bending strength σ_{4Point} and the radius R_H of the fracture surface structure, namely a high 4-point-bending strength σ_{4Point} corresponds to a small radius R_H and vice versa.

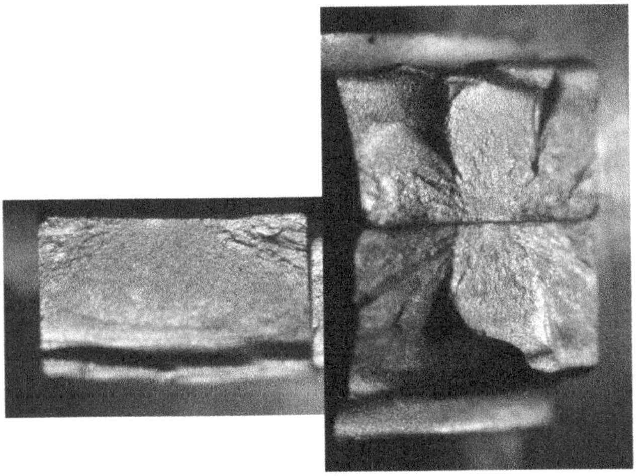

Figure 2: Fracture surfaces of bending bars with σ_{4Point} = 327 MPa (left) and σ_{4Point} = 514 MPa (right) made of Biolox forte.

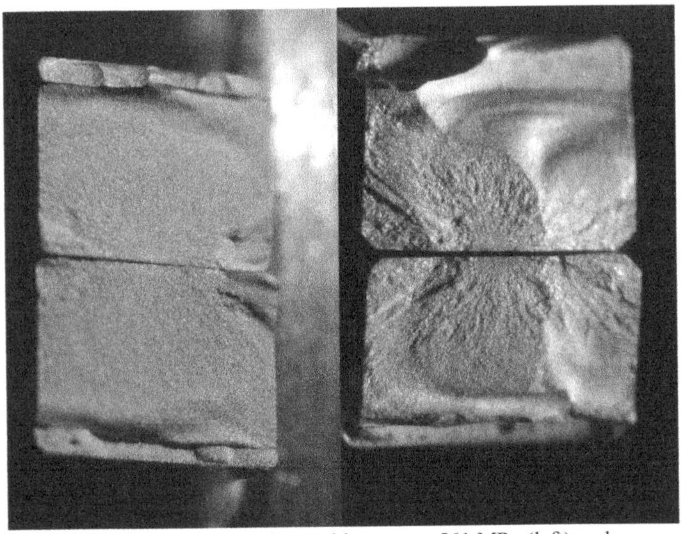

Figure 3: Fracture surfaces of bending bars with σ_{4Point} = 561 MPa (left) and σ_{4Point} = 788 MPa (right) made of Biolox delta.

Transforming (1 into logarithmic form one obtains equation 2 of the type y = mx + n, with an axis intercept n = log A.[2]

$$\log \sigma_{4Point} = -0.5 \log R_H + \log A_H \qquad (2)$$

For Biolox forte and Biolox delta, the according data interpretation for the measured single values parallel and perpendicular to the fracture surface edge are given in Figure 4. Following the method of least squares a straight line was fitted through the given data.

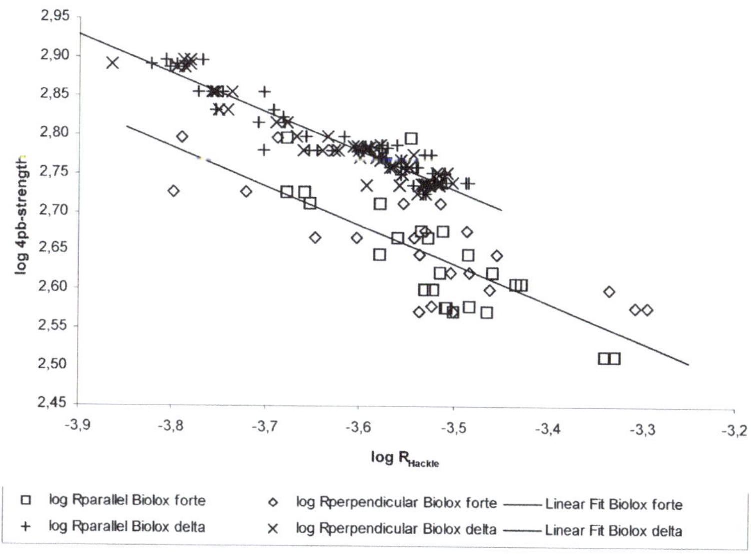

Figure 4: Relation between log stress and log R_H for an assumed slope = - 0.5.

Alternatively A_H can be obtained by plotting the bending strength against $1/\sqrt{R_H}$ (see equation 3).

$$\sigma_{4Point} = \frac{A_H}{\sqrt{R_H}} \qquad (3)$$

Likewise, a straight line of the type y = px + q, with the condition q = 0, was fitted through the experimental data according to the method of least squares. Following this route one also obtains the fracture mirror constant A_H with p = A_H. The graphic representation of the corresponding analysis is given for Biolox forte and Biolox delta in Figure 5.

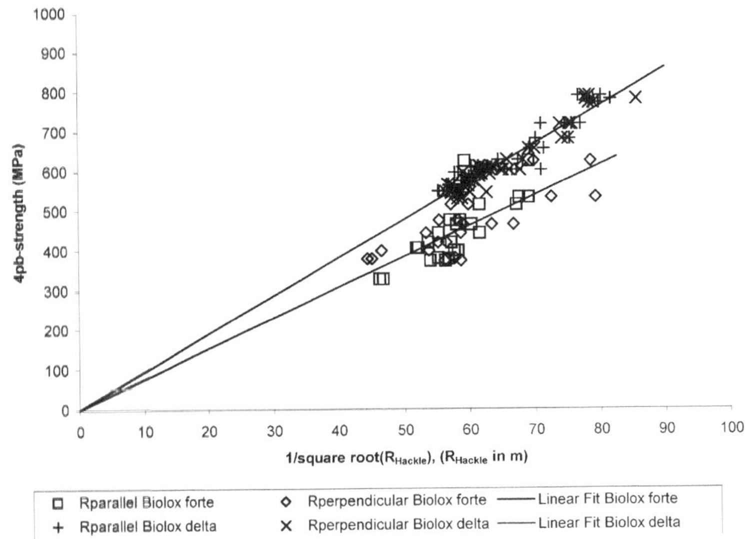

Figure 5: Relation between stress and $1/\sqrt{R_H}$ for a predetermined zero-crossing.

A summary of the values of A_H of Biolox forte and Biolox delta determined by the methods described above is presented in Table III and Table IV.

Table III: Fracture mirror constant A_H for Biolox forte, evaluated according to equation (2) and equation (3) and in dependence on the direction of measurement of R_H.

	Fracture mirror constant A_H [$MPa\sqrt{m}$] of Biolox forte, according to applied equation	
	$\log\sigma_{4Pkt} = -0.5\log R_H + \log A_H$	$\sigma_{4Pkt} = \frac{A_H}{\sqrt{R_H}}$
parallel	7.6	7.7
perpendicular	7.7	7.7
parallel and perpendicular	7.6	7.7

Table IV: Fracture mirror constant A_H for Biolox delta, evaluated according to equation (2) and equation (3) and in dependence on the direction of measurement of R_H.

	Fracture mirror constant A_H [$MPa\sqrt{m}$] of Biolox delta, according to applied equation	
	$\log\sigma_{4Pkt} = -0.5\log R_H + \log A_H$	$\sigma_{4Pkt} = \frac{A_H}{\sqrt{R_H}}$
parallel	9.6	9.6
perpendicular	9.5	9.5
parallel and perpendicular	9.5	9.5

With a statistical analysis of all A_{Hackle} single values of Biolox forte and Biolox delta one obtains, in addition to the mean values of A_{Hackle}, the standard deviations and relative standard deviations given in Table V.

Table V: Standard deviations and relative standard deviations of A_{Hackle} of Biolox forte and Biolox delta.

	Biolox forte	Biolox delta
Standard Deviation (SD)	0.82	0.32
Relative SD [%]	0.11	0.03

Application of the Fracture Mirror Constant A_{Hackle} – Determination of Burst Stress on Hip Joint Ball Heads Made of Biolox Forte and Biolox Delta

The primary fracture surface structures of a series of seven hip joint ball heads made of Biolox forte and a series of nine hip joint ball heads made of Biolox delta were analyzed after burst testing. According to the higher 4-point-bending strength as well as the higher K_{IC} the burst force mean values of ball heads made of Biolox delta are higher than burst force mean values of ball heads made of Biolox forte. For the investigation presented here, and in order to allow a comparison of the features of the fracture surface structure, burst force values in the same range were selected for both materials.

The minimum and maximum burst force values F_{Burst}, radii R_{Hackle} and the estimated burst stresses σ_{Burst} are listed in VI. The single values of burst stresses and of the corresponding burst forces are given in Figure 6.

Table VI: Biolox forte. Ball head burst forces F_{Burst}, radii R_H and according burst stresses σ_{Burst}.

		F_{Burst} [kN]	R_H [mm]	σ_{Burst} [MPa]
Biolox forte	Min	44	0.26	296
	Max	70	0.68	479
Biolox delta	Min	45	0.25	281
	Max	82	1.14	599

F_{Burst}: Burst force
R_H: Radius R_{Hackle}, here R_H is a mean value calculated from 3 independent measurements on each of the two primary fragments (i.e., overall 6 single R_H values)
σ_{Burst}: Burst stress

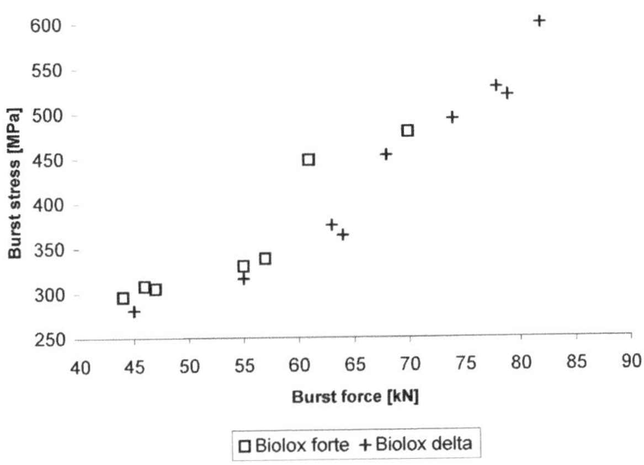

Figure 6: Burst stress versus burst force.

From the overall 6 single R_H values from each ball head the standard deviation s_{RH} as well as the relative standard deviation $rels_{RH}$ were calculated, and the mean values of the relative standard deviation over all ball heads ξ_{relsRH} for Biolox forte and for Biolox delta are obtained, namely 2.3 % and 3.8 %, respectively. The accuracy of the stress values derived from the standard deviations of A_H and of R_H is obtained for Biolox forte and Biolox delta, namely ±13 % of the burst stress and ±7 % of the burst stress, respectively.

At a certain burst force one obtains a significantly smaller radius R_H for Biolox forte compared with Biolox delta (see Figure 7). According to equation 1 the ratio of the square root R_H of two materials should be equal to the ratio of A_H. In the case of Biolox delta/Biolox forte this ratio should be 1.23. The experimental results show that this ratio increases from 1.20 to

1.40 when the burst force increases from 40 to 70 kN. One should bear in mind that in all calculations it was assumed that crack propagation occurs in a constant stress field. This is neither the case in the experiments with ceramic ball heads described above nor in the case of in-vivo fractures. Nevertheless, the fracture mirror constant relation allows useful comparisons with similar components where similar stress distribution can be assumed.

In ball heads the crack propagates in the direction of decreasing stress. Further, the stress is decreasing at a higher rate near the surface of the conical bore than in the bulk (Figure 13 in: Hecht-Mijic/Richter [2006]). This means that for small cracks the error in stress has to be assumed to be larger than for larger cracks. Thus the discrepancy can be explained with a certain plausibility.

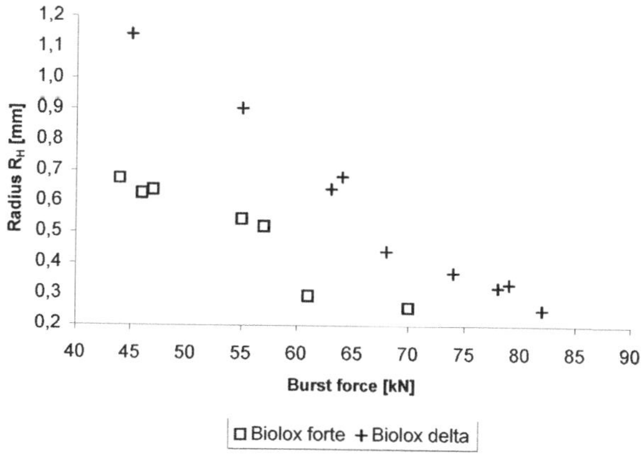

Figure 7: Radius R_H versus burst force.

SUMMARY

Due to their relevance for both component development and failure analysis the fracture mirror constants of the bioceramic materials Biolox forte and Biolox delta were determined experimentally to be 7.7 MPa√m and 9.5 MPa√m, respectively.

The determination of the radius R_H is influenced to a certain degree by subjective judgement. In order to assess the reliability of the stress evaluated on the basis of the fracture mirror constant determined by one operator a statistical analysis of all A_{Hackle} data of Biolox forte and Biolox delta was performed. The standard deviations and the relative standard deviations obtained by this analysis are lower for Biolox delta. It can be assumed that the stress causing fracture can be determined with a precision of ± 13 % for Biolox forte and 7 % for Biolox delta.

Comparison of the stress calculated by the finite element method with the stresses evaluated from the fracture surface structure of ceramic ball heads for hip joint replacement fractured in a standardized "burst test" revealed satisfactory agreement. Even if the stress field in the ball head is not uniaxial and constant, the fracture mirror constant relation allows useful comparisons with similar components where a similar stress distribution can be assumed.

REFERENCES

[1]ASTM C 1161: Standard Test Method for Flexural Strength of Advanced Ceramics at Ambient Temperature.

[2]ASTM C 1322: Standard Practice for Fractography and Characterization of Fracture Origins in Advanced Ceramics.

[3]ASTM F 603: Standard Specification for High-Purity Dense Aluminium Oxide for Medical Application.

[4]ASTM F 2345: Standard Test Methods for Determination of Static and Cyclic Fatigue Strength of Ceramic Modular Femoral Heads.

[5]Hecht-Mijic, S. and Richter, H.G., "Fractography of Bioceramic Components for Hip Joint Replacement." To be published in Proceedings; Fractography of Glasses and Ceramics V Rochester, USA (2006).

[6]ISO 6474: Implants for surgery – Ceramic Materials Based on High Purity Alumina.

[7]ISO 7206: Implants for surgery – Partial and Total Hip-Joint Prostheses. Part 10: Determination of Resistance to Static Load of Modular Femoral Heads.

[8]ISO 14704: Fine Ceramics (Advanced Ceramics, Advanced Technical Ceramics) – Test Method for Flexural Strength of Monolithic Ceramics at Room Temperature.

[9]Salem, J. and Jenkins, M., "Estimating Bounds on Fracture Stresses Determined from Mirror Size Measurements, *J. Am. Ceram. Soc.*, 85 [3] 706-708 (2002).

[10]Varner, J. and Quinn, G., Fracture Analysis of Glasses and Ceramics, information material; summer school at Alfred University, July 12-15 (2005).

FRACTOGRAPHY OF BIOCERAMIC COMPONENTS FOR HIP JOINT REPLACEMENT

S. Hecht-Mijic, H.G. Richter
CeramTec AG, Central Development
Plochingen, Germany

INTRODUCTION

Fractography plays an important role in the pursuit to providing safe ceramic hip joint components. The focus of the paper is on illustrating the use of fractography by presenting fracture surface examples of ceramic hip joint components from the stages of development, production, quality assurance and in-vivo failure analysis.

In an introductory section basics of fractography as applied to advanced ceramics will be presented. Subsequently, a few typical examples of fractography on hip joint components will be reported. Relations between fracture surface features and information on the failure, such as stress situation which led to the fracture event, will also be addressed. Of special note regarding these relations is that scope of fractography of bioceramic components is beyond the main contents of standards like EN 843-6[1] and ASTM C1322;[2] i.e., identification of the fracture origin.

BASIC FEATURES OF FRACTURE SURFACES OF ALUMINA CERAMICS

Remarks on the Crack Velocity/Surface Feature Relationship

Much basic information of fractography of brittle materials was collected by studying fracture processes in glass. A review on this information especially on the relationship between fracture surface patterns and crack velocity in the range between 10^{-9} and 10^{3} m/s is given by Richter.[3] Stress wave fractography, a method for the investigation of this relationship, was developed and applied by Richter and Kerkhof.[4]

Conclusions on the corresponding relationship for ceramics presented below are drawn by analogy. Features of ceramic fracture surfaces were attributed to crack velocities even if it is not known exactly what the K_I-v dependence is in detail at higher crack velocities in alumina ceramics and at what crack velocities mirror, mist and hackle occur; see Figure 1.

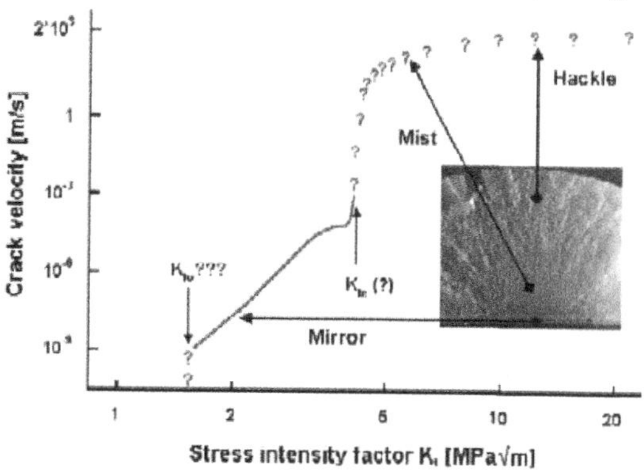

Fig. 1. K_I-v dependence in alumina (schematic) with corresponding fracture surface structure, by Richter.[3]

Recently, the fracture mirror constants of biograde ceramics (Biolox forte and Bioloxe delta) were determined by Hecht-Mijic.[5]

Transgranular and Intergranular Fracture in Bioceramics

The region adjacent to the immediate fracture origin is usually names "fracture mirror." The characteristic of this region is a much smoother surface than regions away from the fracture origin. In alumina ceramics, at least if they are finely grained, a structural change on the fracture surface can be observed in addition to the shiny appearance. In some materials even two "fracture mirror" regions can be distinguished. The inner region has to be attributed to a higher degree of transgranular fracture as opposed to intergranular fracture. This transition from transgranular to intergranular fracture has been reported by Kirchner andGruver.[6]

Fig. 2. Fracture surface of alumina (a), with transgranular and intergranular crack growth regime (b) and (c), by Richter.[3]

Sometimes the inner, shiny area is interpreted as being an area of coarse grains, and the conclusion is drawn that fracture started because of exaggerated grain growth during sintering. However, a closer look reveals grain boundaries within such a flat, shiny zone. The existence of grain boundaries within these flat areas can be proven by a standard thermal etching procedure. In Figure 3 two transgranular fracture surfaces are shown, one without, and one after thermal etching.

Richter[3] defined a particular K_I value at the border of the shiny region, $K_{I\ t\text{-}i}$. He determined $K_{I\ t\text{-}i}$ of the fine-grained "biograde" alumina of around 6 MPa√m. This value is about 50% above the classical fracture toughness of 4.2 MPa√m, and it is around 50% of the K_I value at the onset of the hackly regime. The ratio between the K_I value at the onset of crack branching and the fracture toughness was evaluated by Kerkhof and Richter[7] to the 4 for glass.

Fig. 3. Transgranular fracture surfaces in alumina without (a) and after (b) thermal etching, Richter.[3]

FRACTOGRAPHY OF COMPONENTS

General Remarks

In material and component development for biomechanical applications, such as hip joint arthroplasty, new ceramics and components are characterized to a large part by their strength. Today it is understood that the strength values measured in a strength test provide only about 50% of the information that can be obtained from such a test. The other half consists of findings which one obtains from a fractographic analysis of the fracture surface structures.

Therefore, the following sections contain typical fracture patterns from both ceramic ball heads and ceramic cup inserts as newer findings of interesting fracture surface features. A review on structural ceramics in orthopedics is given by Clarke and Willmann.[8] the role of parameters such as conical press fit, various taper lengths, and taper structure is given in papers by Richter and Willmann et al,[9] Morrell et all,[10] and Richter.[3]

Pre-Sintering Defects Acting as Fracture Origins

If a crack is detected in an experiment through dye-penetrant testing or in proof testing, an extensive analysis of the fragments and the fracture surfaces should be started to clarify the origin of the defect. In order to do so, one tries to assign the defect origin to certain stages of the production process. Amongst others the questions must be answered whether existing crack surfaces occurred prior to or after the sintering of the ceramics. There are two characteristics which allow one to decide between these two kinds of crack surfaces. First, the shape of the ceramic grain edges, and second, the overall roughness of a crack (or fracture) surface.

In an investigation on pressing-technology parameters the origin of cracks in a green body was induced. After sintering of the green body the appearance of the grain edges on the surface existing prior to the sintering was examined and compared to a fracture surface produced on the same but sintered body. As one can see the grain edges on the pre-existing surface appear round and smooth, and the ones from post-sintering fracture appear sharp. Obviously, if these grain-edge characteristics are once shown experimentally for a certain ceramic it is possible to sue them for efficient technical evaluation of components.

Fig. 4. Left: crack induced in a green body by variation of pressing parameters; right / upper half of the picture: grains with sharp edges on a fracture surface which was produced after the sintering cycle, right / lower half of the picture: rounded grain edges on a surface which existed prior to the sintering cycle.

One should keep in mind that the fracture of a ceramic component is an uncommon reason for revision surgery. There are far more common reasons for revision surgery, such as aseptic loosening, dislocation and infection. Clohishi et al.[11] reviewed 439 revision hip surgeries done between 1996 and 2003 in the US. According to this study about 10% of hip arthroplasties have to be revised within a time period of 9 years. 55% of the revision surgeries were for aseptic loosening, 14% was for instability, 13% for osteolysis and 1% for implant fracture. For all implanted Biolox forte ceramic ball heads with a diameter of 28 mm, Willmann[12] reported the fracture rate to be 0.01%.

BALL HEADS

Regular Fracture Pattern from Resistance-to-Load Test ("Burst" Test)

Middleton, Pande et al.[13] calculated the stress distribution in a ceramic ball head in contact with a metal stem under axial loading. The result obtained in this calculation i8s shown in Figure 5. The critical stress is a hoop stress distribution with its maximum in the region of contact between spigot and ball head. An accepted test method for this loading situation is described in ISO 7206 part 10:[14] "Determination of resistance to static load" ("burst" test). More background information can be obtained from the work done by Hinterberger and Ungethüm.[15] Finite element analyses for oblique loading revealed only a slight increase in the stress level and an asymmetric stress distribution.

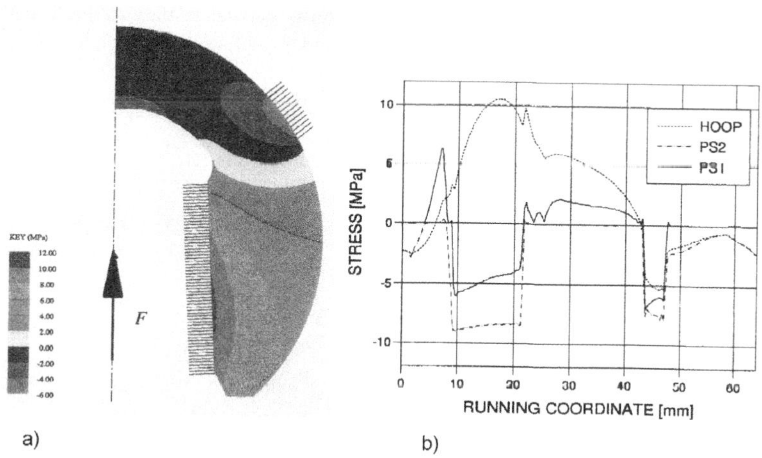

Fig. 5. a) Stress distribution in hip joint head under spigot loading. Numbers indicate distance from dome center in millimeters. b) Stresses along the surface of a hip joint head. Running coordinate is the distance in millimeters measure from the dome center. (Middleton, Pande et al[13]).

IN the test set-up according to ISO 7206 part 10[14] shown in Figure 6 the ball head is positioned on a copper ring in a metal block. A conical metal spigot, representing the conical end of the stem, is pressed into the ball head until the ball head fractures.

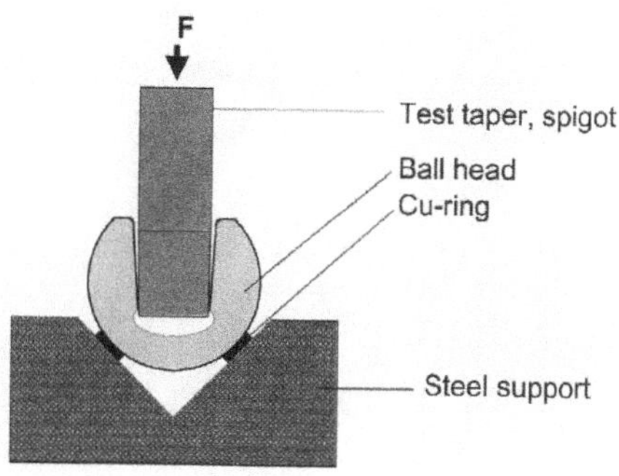

Fig. 6 Set-up for the determination of "resistance to static load" according to ISO 7206-10.[14]

According to two of the fundamental relations of fractography, the position of the fracture origin is at the position of the highest, in other words, critical, tensile stress, namely at the conical bore surface; and the orientation of the primary fracture surface is perpendicular to the critical hoop stress distribution; i.e. this orientation is parallel to the ball head axis. An example of primary fracture surfaces obtained under different load levels is shown in Figure 7.

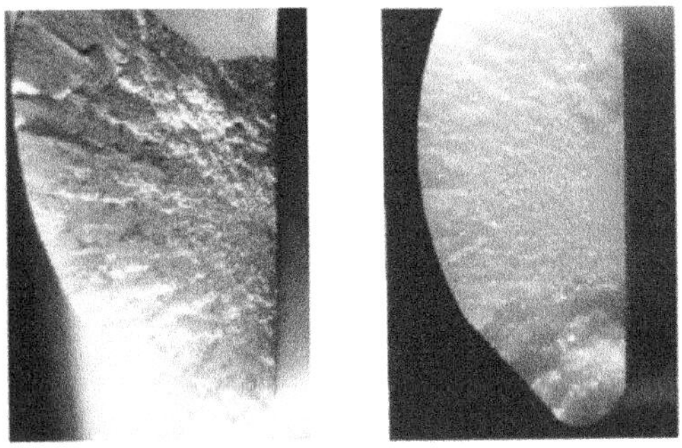

Fig. 7. Primary fracture surfaces of ball heads. (Richter[3])

Another basic principle of fractography is the relation between the magnitude of fracture load and number of fragments; i.e. a high load produces a large number of fragments, and vice versa. Of course, this principle is also true for the number of fragments obtained in the burst test on a ceramic ball head. In Figure 8 ball head fragments resulting from two different burst load levels are shown.

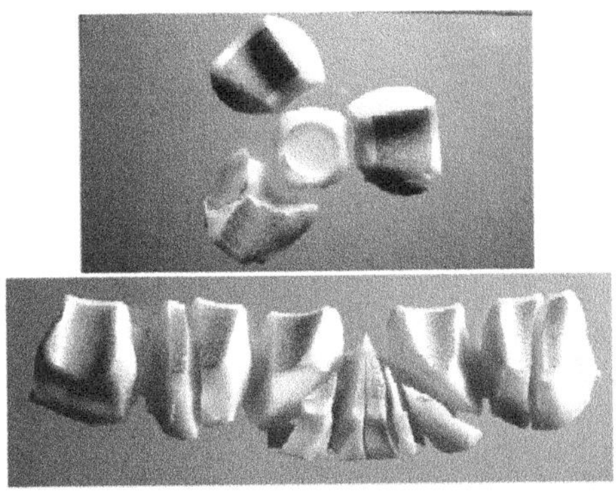

Fig. 8. Fracture pieces of ball heads broken in a test according to ISO 7206 part 10 (Richter[3])

Hard Machining

In the early days of using ceramic hip joints, fractures sometimes occurred with origins on the outside surface of the ball head; i.e. in a region of the ball head where according to finite element analyses only low tensile stresses occur. Two different examples are shown in Figure 9.

The fracture of the example shown in the left half of the figure started in a part of the labeling. One has to bear in mind that in these early days the engraving was done in the green state of the ball head with a gouge. This process led to marks whose depth was up to 0.5 mm. Under unfavorable circumstances these marks could lead to critical stress concentrations. After identification of this weak point, the necessary correcting action was obvious, simple, and very effective: the mechanical labeling was replaced by laser engraving. This change eliminated completely this type of failure.

As a second example, unidirectional grinding was applied to the face of ball heads for demonstration purposes. The fracture origin is on the face side (near the opening of the ball head bore); i.e. not in the region of the contact between metal stem and the conical bore surface of the ball head; see Figure 9. The detailed fractographic analysis revealed the fact that the fracture origin was exactly at the position where the grinding direction was in parallel with the radius direction of the ball head.

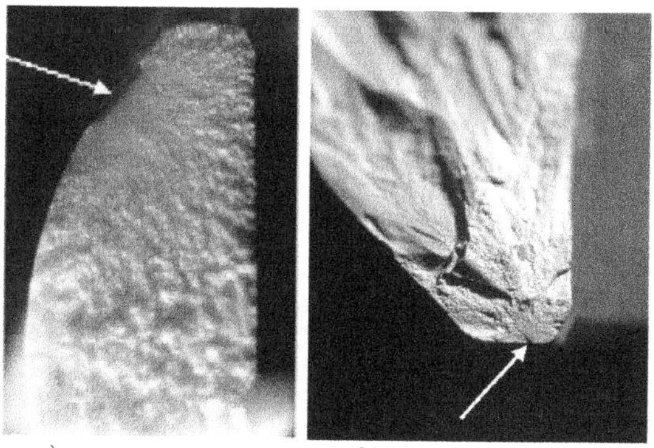

a) b)

Fig. 9. a) Fracture origin in a label, b) Fracture origin on the face of the ball head (Richter[3]).

Irregular Contact Situations and Cyclic Fatigue

Under regular contact conditions between the conical bore of the ball head and the conical end of the metal stem there is a ring-like contact area near the end of the conical bore. Roundness deviations in the stem lead to deviations in the metal transfer layer which can be observed on the conical bore of the ball head. In the example shown in Figure 10 such deviations from the regular contact situation can lead to significant reductions in the load-bearing capability of the ball-head/stem system. It is obvious that not only fracture origins should be he focus in studying features of fragments. Additionally, it can be seen in Figure 10 that the fracture starts at a point at the end of the contact between metal stem and ceramic conical surface. At such transitions additional tensile stresses occur which can cause fracture.

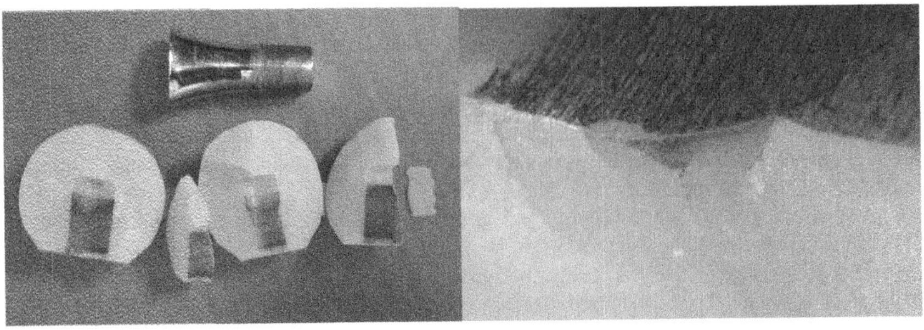

Fig. 10. Ball head fragments and laboratory metal stem after fracture in a cyclic fatigue test. Left: Regions of intensive metal transfer marked by arrows. Right: The continuation of the cone crack on the bore surface is marked by an arrow.

In Figure 11 one finds another feature of basic importance in fractography – arrest lines. They are due to the repeated crack propagation and crack arrest during the cycles of the fatigue test. It is possible to estimate the velocity of crack propagation from the frequency of the cyclic fatigue test and the distance between the arrest lines. In the experiment discussed here the frequency was 10 Hz. From the distance between the arrest lines a rough estimation of the crack propagation velocity reveals $v \approx 3x10^{-2}$ m/s.

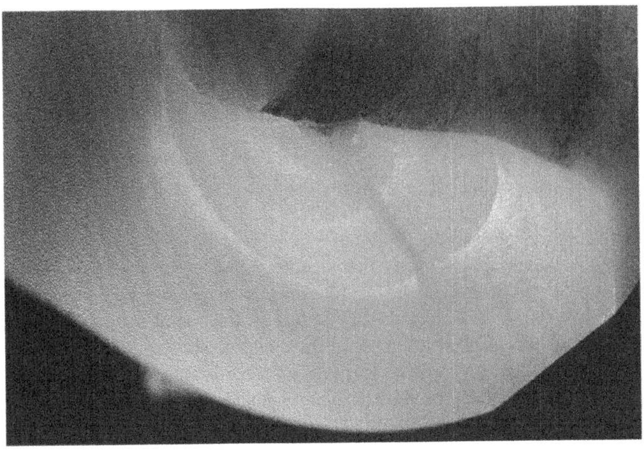

Fig. 11. Fracture surface with arrest lines from cyclic fatigue test.

Fracture Surfaces Obtained in Proof-Testing

It is the goal of proof testing to eliminate weak components by applying a mechanical stress to the component which simulates the service load or an agreed reference load situation. In addition, the mechanical stress has to be at a level at which components fail which would not reach the required time of life. Ceramic ball heads are proof tested in a configuration described by Richter.[16] Since the fracture surfaces contained in the reference loading situation and in the proof test are almost identical, see Figure 12, it can be concluded that the proof-test configuration copies the reference load situation satisfactorily.

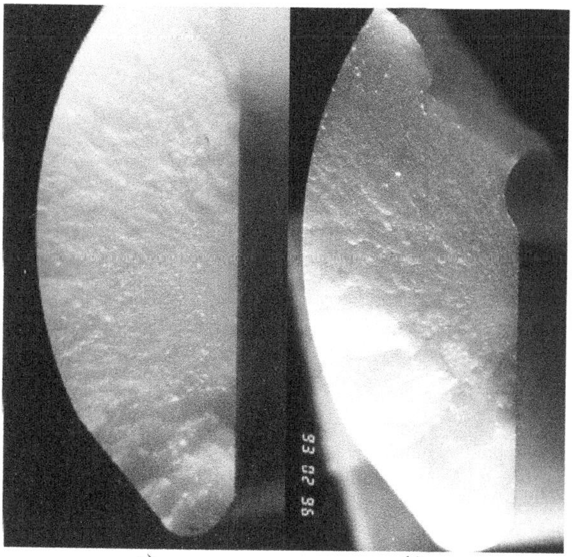

a) b)

Fig. 12. Fracture surfaces of ceramic ball heads. a) obtained in proof test, b) obtained in burst test.

Particle Between Stem and Conical Bore Surface

Any particle between the conical bore surface and the metal stem disturbs the regular stress distribution, leading to stress concentrations which in the end can lead to a reduction in the load-bearing capability. Figure 13 shows a fracture surface obtained with a bone particle between the metal stem and the conical bore surface of the ball head. During the test the bone particle was crushed and its debris was pressed against the conical bore surface. I this way a circular area was left free of metal transfer. The size of the contact area is relatively large compared to the size of the original bone particle. It is obvious that the fracture started near the disturbed contact area. Experimentally, features of the fracture origin and of the contact area on the bore surface were also found with other particles which show plastic or quasi-plastic behavior. The effect of a small hard particle is discussed below.

Fig. 13. Fracture surface and bore surface obtained with a bone particle between ball head and taper.

In-Vivo Failures

As was already pointed out above, the fracture of a ceramic component is a less common reason for revision surgery, compared to other reasons such as aseptic loosening, dislocation and infection. Nevertheless, it is understood that one wants to learn as much as possible about the reasons which led to a fracture event. There is one major cause which complicates the application of fractography as a tool for the analysis of ball heads and cup inserts, namely the very different levels of conservation of in-vivo fracture surfaces. Figure 14 contains three examples of fragments which come from different in-vivo failures.

From the condition of the three examples it is obvious that there is a certain time period between the primary fracture event and the revision surgery. One has to bear in mind that patients often do not realize that the ball head has broken. Because of this, the fragments are held in shape by the cup for a certain time. During this time friction and wear occur between the fragments. Obviously, under certain circumstances there is no possibility left to identify the primary fracture surface, not to mention the identification of the first fracture origin.

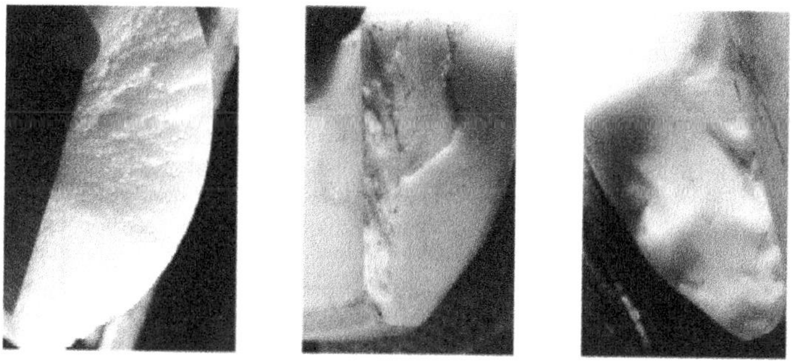

Fig. 14. Fracture surfaces from in-vivo failures (Richter[3])

Another example deals with the contact situation between the ball head and the metal stem. Deviations from the regular contact lead to specific fracture origin features and fracture surface structure features. Such a deviation might be a particle (possibly bone, bone cement or other material) between a metal taper and the ball head. Under this assumption a small region of intensive contact is surrounded by a ring-like region where there is no contact. Further, a so call Hertzian crack can occur in this stress situation and can serve as a starter defect in the regular hoop-stress distribution available in the volume of the ball head. In Figure 15 a fracture surface is shown, the larger part of which occurred under the influence of a hoop-stress distribution. At the position of the fracture origin, the fracture surface in non-planar and is part of a Hertzian crack.

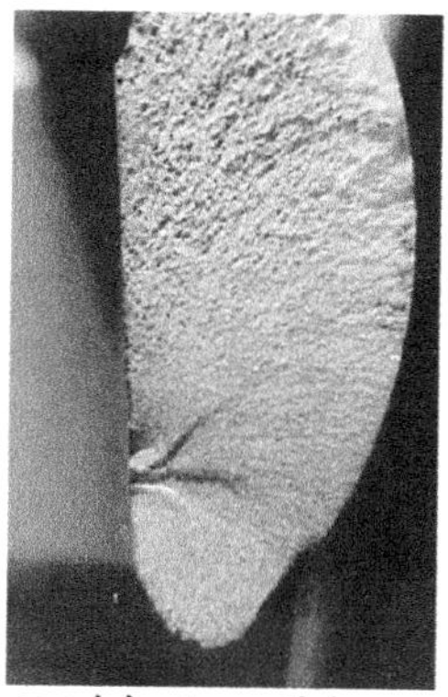

Fig. 15. Fracture origin at a particle contact (Richter[3]).

Richter[3] reports ball head fracture loads obtained under this condition as low as 18-25 kN. This must be compared to fracture loads well above 60 kN in the regular situation.

CUP INSERTS

Regular Fracture Pattern from Resistance-to-Load Test ("Burst" Test)

As with hip joint ball heads, elaborate finite element analysis was carried out on the loading situation of the cup insert. For the calculation as well as for a standard test the established set-up shown in Figure 16 was used. In this set-up the cup insert is in a metal shell(metal back). The contact between these two components is made via a conical press fit. The metal shell is embedded in epoxy resin for simulation of the surrounding bone. The load is applied in axial direction via a ceramic ball head. This test set-up is used for both static (until fracture) and cyclic loads.

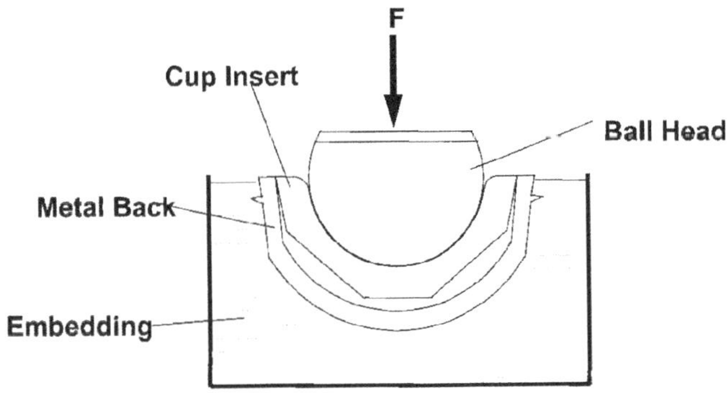

Fig. 16. Standard laboratory test set-up for cup inserts.

In good agreement, both the finite element analysis and the fractographic analysis reveal that the maximum tensile stress is on the outer conical surface which is in contact with the metal shell. Also, tensile stress is on the outer conical surface which is in contact with the metal shell. Also, tensile stresses are on the outside of the bottom. Richter[3] found that the tensile stress on the conical contact area is independent of the ball size within the specified diameter tolerances of the ball head. In contrast to this, it was found that there is a pronounced dependence of the tensile stress on the bottom on the diameter of the ball head used for the load application.

These correlations are reflected in the fracture patterns findings after experiments with the test described above. In figure 17 fragments of cup inserts obtained with different ball head diameters are shown. Ball head diameters in the center of the specified diameter tolerances cause the fracture pattern shown in figure 17a, the shape of the fragment is ring-like and the distance in axial direction between the crack surface and the cup insert rim is from 2 to 6 mm. As expected, the fracture origin is in the outer conical surface. Lower ball head diameters sometimes result in the fracture pattern shown in Figure 17b. In this case the fracture origin was on the outer side of the cup insert bottom. This finding led to tighter tolerances allowed in the cup insert.

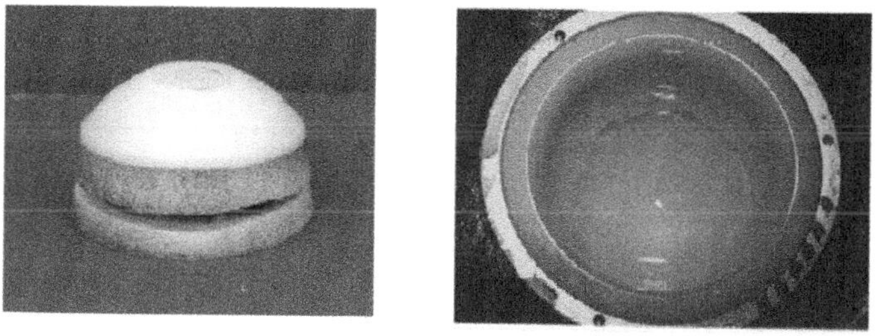

D = 27.99 mm 27.92 mm

a) b)

Fig. 17. Fracture patterns in cup inserts. A) Circular crack, b) Crack at bottom (Richter, Willmann et al[17])

Hard Machining

Part of the challenging development of the hard machining was the necessity of finding appropriate machining parameters for the lapping of the outer surfaces of a ceramic cup insert. It was also shown that under inappropriate machining parameters, in combination with a relatively small test ball head diameter, fracture events can occur during fatigue tests on ceramic cup inserts.

Herewith, an example from an early stage of the hard machining development on cup inserts is presented. The cup insert failed in a fatigue test after 4×10^5 cycles, whereas the overall duration of the test was actually planned to be 10×10^6 cycles without fracture of the cup insert. In the fractographic analysis the symmetric fracture pattern shown in Figure 18 was found.

At a first glance the fracture pattern is very similar to the one shown in the right half of figure 17. Of course, the fracture origin is in the symmetry center of the fracture pattern, too. But in contrast with the example given in Figure 17, the center of symmetry is not in the middle of the cup insert but shifted sideways. From this it must be concluded that there was an additional influence which caused this deviation from the regular, central position of the fracture origin. More precisely, the fracture origin is on the outer surface of the cup insert. It seemed likely that tensile stress maxima on the outer surface of the cup insert during the cyclic loading in combination with an inappropriate quality of the hard machining led to the fracture event. After improvement of the hard machining parameters, fracture origins in this region of a cup insert did not occur again in cyclic loading experiments.

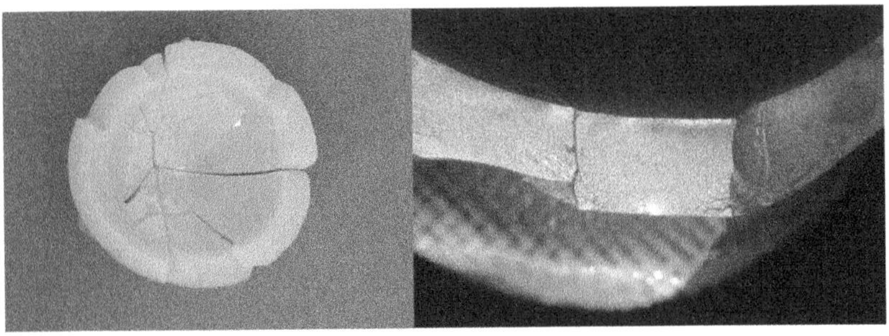

Fig. 18. Left: Fracture pattern, Right: Fracture origin (arrow)

Primary and Secondary Fracture Surfaces

To be able to decide between primary and secondary fracture surfaces is one of the key issues in fractography on bioceramic components. In this regard, knowledge about the correct interpretation of special fracture surface features on fragments is essential. These features help one to clarify the direction of crack propagation, to trace back to the primary fracture origin, and hence, to decide between the fracture surface which originated first (primary fracture surface) and the secondary fracture surfaces. One such type of this special feature is called lances (synonym: twist hackle), Smekal[18] showed that this kind of pattern occurs when torsion is added to a specimen loaded in tension; i.e. when a superposition of mode I and Mode III is present. Under such conditions local directions of crack propagation superpose with the main direction of crack propagation, and, as a consequence, fracture surface structures occur as shown schematically in figure 19.

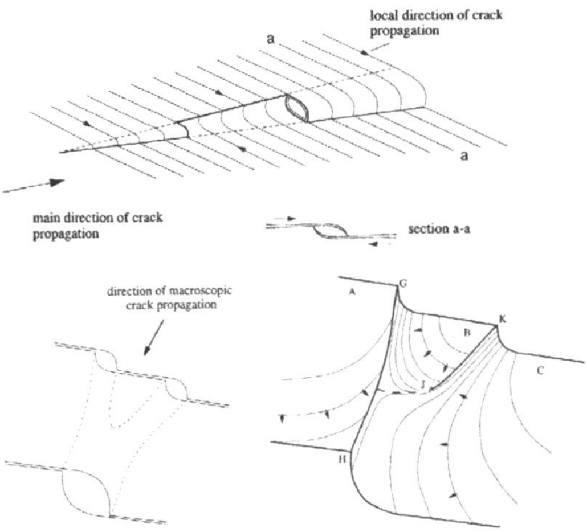

Fig. 19. Directions of crack propagation and resulting fracture surface structure after Smekal.[18]

An example of a secondary fracture surface on a cup insert fragment with lances of an exceptional smooth and wide characteristic is shown in figure 20. This example was achieved in a standard resistance-to-load test at a high burst load (F_{Burst} = 155 kN).

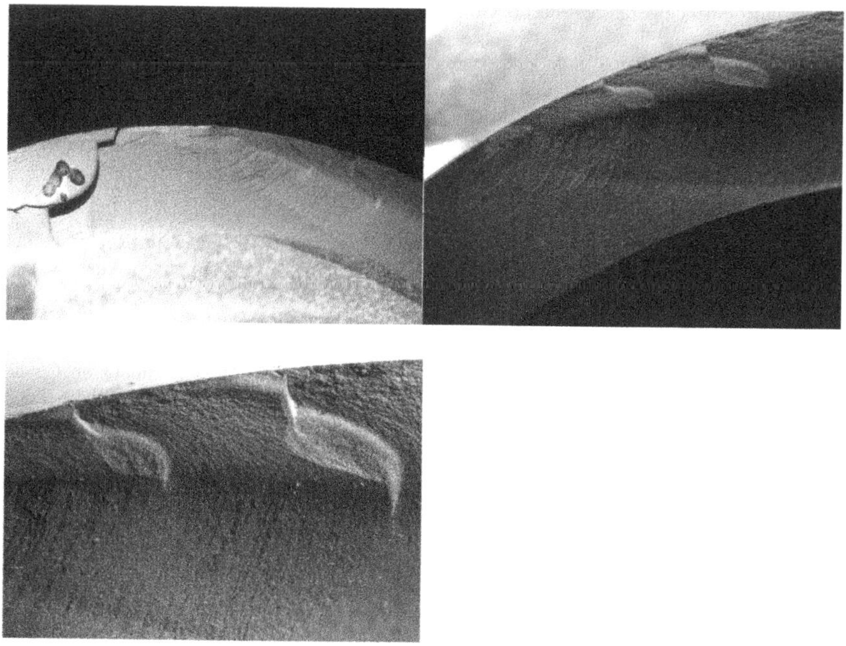

Fig. 20. Lances (twist hackle) on a secondary cup insert fracture surface after standard resistance-to-load test.

SUMMARY

Fractography is a well-established and indispensable analysis tool in the entire process of providing safe ceramic hip joint components, i.e. in development, production, quality assurance, and in-vivo failure analysis. Fractography takes knowledge about the stress distribution in a component into consideration. Fractographic features allow one to decide between the primary fracture surface and secondary fracture surfaces and, finally, to trace back to the fracture origin. Moreover, accompanying experiments, interpretation of metal transfer layers, analysis of number and shape fragments, as well as the analysis of orientation of fracture surfaces are necessary in order to obtain information on the complex conditions which led to a fracture event.

REFERENCES

[1]EN 843-6, "Advanced Technical Ceramics – Monolithic Ceramics – Mechanical Properties at Room Temperature – Part 6: Guidance for Fractographic Investigation," *CEN(DIN)* (2004).

[2]ASTM C 1322-05a, "Standard Practice for Fractography and Characterization of Fracture Origins in Advanced Ceramics," *American Society for Testing and Materials* (2005).

[3]H.G. Richter, "Fractography of Bioceramics," Trans Tech Publications Ltd: Proceedings International Conference on Fractography of Advanced Ceramics in Stara Lesna, May 13-16 (2001).

[4]H.G. Richter, F. Kerkhof, "Stress Wave Fractography" in: Bradt, R.C., Tressler, R.E. [Eds.]: Fractography of Glass, New York, 75-109 (1994).

[5]S. Hecht-Mijic, "Fracture Mirror Constants of Bioceramics for Hip Joint Replacement: Determination and Application," to be published in Proceedings: Fractography of Glasses and Ceramics V, Rochester USA (2006).

[6]H.P. Kirchner, R.M. Gruver, "Fracture Mirrors in Alumina Ceramics," *Phil.Mag.*, 27, 1433-1446 (1973).

[7]F. Kerkhof, H.G. Richter, "Investigation of the Influence of Water Vapour on Crack Velocities in Glass by Ultrasonic Fractography," *Proc. 2nd Intern. Conf. on Fracture*, London, 463-473 (1969).

[8]I. Clarke, G. Willmann, "Structural Ceramics in Orthopaedics," H.A. Cameron [Ed.]: *Bone Implant Interface*, St. Louis, 203-252 (1994).

[9]H.G. Richter, G. Willmann et al, "Influence of the Ball/Stem-Interface on the Load Bearing Capability of Modular Total Hip Endoprostheses," Modularity of Orthopedic Implants, *ASTM STP 1301*, Donald E. Marlowe, Jack E. Parr and Michael B. Mayor, [Eds.], 127-133 (1997).

[10]R. Morrell, L. Byrne, M. Murray, "Fractography of Ceramic Femoral Heads" in: Varner, J.R. and Quinn, G. D. [Eds.]: *Proc. of the Fourth Alfred Conference on Fractography of Glasses and Ceramics*, 253-266 (2000).

[11]J.C. Clohishi, G. Calvert et al, "Reasons for Revision Hip Surgery," *Clinical Orthopaedics and Related Research*, 429, 188-192 (2004).

[12]G. Willmann, "Fiction and Facts Concerning the Reliability of Ceramics in THR," *Proceedings, 8th BIOLOX Symposium*, 193 – 196 (2003).

[13]J. Middleton, G.N. Pande, H.G. Richter et al, "Investigations into the Mechanical Reliability of Ceramic Femoral Heads for Hip Joints," *J. Mater. Sci: Mater. In Medicine*, 5, 503-506 (1994).

[14]ISO 7206-10, "Implants for Surgery - Partial and Total Hip Joint Prostheses, Part 10: Determination of Resistance to Static Load of Modular Femoral Heads" (2003).

[15]J. Hinterberger, M. Ungethüm,"Untersuchungen zur Tribologie und Festigkeit von Aluminiumoxidkeramik-Hüftendoprothesen," *Z. f. Orthopädie und ihre Grenzgebiete,* 116, 294-303 (1978).

[16]H.G. Richter "Application of Proof-Testing to Ceramic Hip Joint Heads" in: Puhl [Ed.]: Bioceramics in Orthopaedics – New applications, *Proc. of the 3rd Intern. Symp. on Ceramic Wear Couple*, Stuttgart, 7-12 (1998).

[17]H.G. Richter, G. Willmann, K. Weick, "Improving the Reliability of the Ceramic-on-Ceramic Wear Couple in THR," Alternative Bearing Surfaces in Total Joint Replacement, *ASTM STP 1346*, Jacobs, J.J. and Craig, T.L [Eds.], 173-185 (1998).

[18]E. Smekal, "Oesterreichische Ingenieur Arch.," 7, 49-70 (1953).

APPLICATION OF FRACTOGRAPHY TO COMPACT BONE

Jiahau Yan, Burak Taskonak
Department of Restorative Dentistry
Indiana University
1121 West Michigan St. #DS-112
Indianapolis, IN 46202

Kari B. Clifton
Department of Physiological Sciences
University of Florida
P. O. Box 100144
Gainesville, FL 32610

John J. Mecholsky, Jr.
Department of Materials Science and Engineering
University of Florida
P. O. Box 116400
Gainesville, FL 32611

ABSTRACT

Most compact bones are highly mineralized (approximately 60–70 wt.% of minerals). They experience quasi-brittle fracture in flexure with fracture strain mostly less than 3%. In this study, we used quantitative fractography to determine the fracture toughness of bovine femur and manatee rib of young adults. Three-point flexure specimens with dimensions 3x3x45 mm were prepared from the mid-sections of bovine femur and manatee rib. Ten specimens were tested wet at room temperature for each animal. In addition, whole ribs were fractured in simulated three point flexure and the fracture surfaces were analyzed in a similar manner. Average fracture toughness of bovine femur and manatee rib bones from small specimens are 4.0±0.4 MPa·m$^{1/2}$ and 3.1±0.4 MPa·m$^{1/2}$, respectively. The fracture toughness of manatee whole ribs, estimated from the fracture surface is ~ 8 MPa·m$^{1/2}$. The differences between manatee and bovine bone will be discussed in terms of microstructure. The differences between small specimens and whole ribs will be discussed in terms of R-curve behavior.

All specimens were then dried, and a Scanning Electron Microscope (SEM) was used to capture SEM images. Most of the fracture origins were found to be at openings for blood vessels and interlayer spaces. SEM images were also used to determine the porosity of the specimens using Image Pro® software. Average porosity of the manatee rib is significantly greater than that of bovine femur. It is found that fractography can be used to identify the fracture origins of compact bone.

INTRODUCTION

Compact bone is the major component of long bones in animals. It provides strength for animals' skeleton. Most compact bones are highly mineralized (approximately 60–70 wt.% of minerals). They usually experience quasi-brittle fracture in flexure with fracture strain mostly less than 3%. Previous studies showed that fracture surfaces of bone revealed fractographic patterns[1-3]. Kimura et al.[1] used fractography to study the relationship between fracture surface and fracture mode of intact human tibia fractured in static bending. Corondan and Haworth[2] used fractography to study the fracture surfaces of accidentally fractured and tensile-tested human long bones. Yan et al.[3] used quantitative fractography to estimate the fracture toughness (K_c) of manatee rib from different age groups. They found that the fracture toughness, K_c, of sub-adult and calf manatee bone material increases as the size of the manatee increases. Figure 1 shows the fracture surface of a manatee rib (Fig. 1A) and the fracture surface of a bovine femur rectangular specimen (Fig. 1B). Both of the fracture origins can be identified from the surface features.

A.

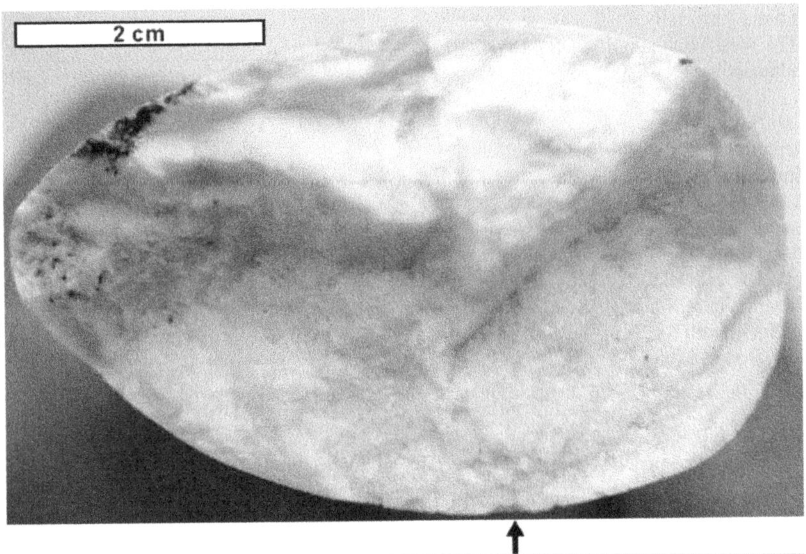

B.

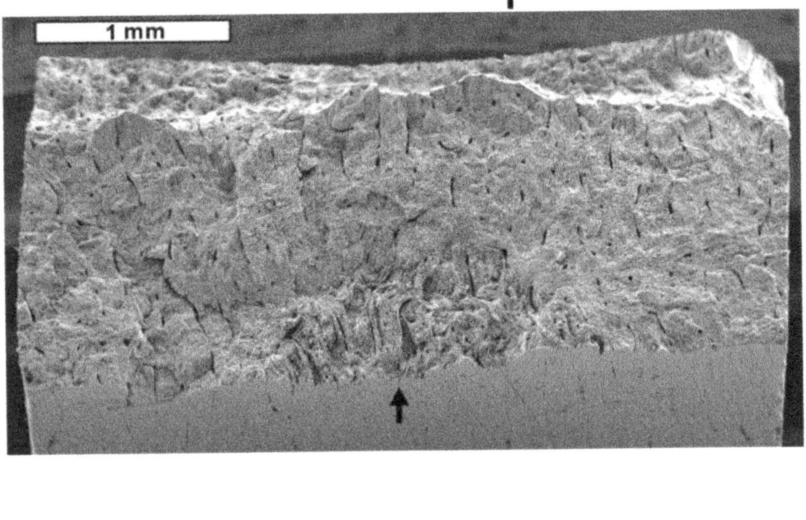

Figure 1. Fracture surface of a manatee whole rib (optical image), A, and a bovine femur sample (SEM), B. Fracture markings can be readily seen on the manatee rib. The fracture origins are indicated by arrows.

Over the last 30 years, many researchers applied linear elastic fracture mechanics (LEFM) to study the fracture behavior and fracture toughness of bone.[4-17] Table 1 lists the testing methods, bone source, fracture direction relative to the long axis of bone, and K_c values of selected studies. Some of the notable testing methods include single-edge notched beam,[4,10,14-15] compact tension,[5-9,12] compact sandwich,[11] chevron-notched short rod,[13] fatigue-precracked beam,[16] chevron-notched beam,[17] and quantitative fractography.[3] The K_c values shown in Table I vary between 1.6 and 12.6 $MPa{\cdot}m^{1/2}$. Though the scatter in data seem large, it may not be surprising considering the difference of bone sources, crack propagation direction, and the natural variability (age, sex, etc.) in animals. One common feature in these techniques (except for quantitative fractography) is that they employed machined cracks as the starter notches. These notches are on the order of 1 to 10 mm and are considered to be much greater than the intrinsic flaws in compact bone (microcracks and vascular channels are usually on the order of 100 μm). Another problem for these techniques is that the collagen phase, which serves as the organic matrix in bone, could deter sharp pre-cracks from forming. Theoretically, fracture mechanics equations were developed based on sharp crack tips. Mukhopadhyay et al.[18] studied the relationship between notch radius and K_c values of alumina (Al_2O_3) and silicon carbide (SiC). They found that the larger the notch radius, the larger the K_c value. A blunt pre-crack may result in a larger apparent K_c value since extra load is needed to break the specimen. One of the advantages of using quantitative fractography is that no artificial notch is needed in the specimens. Therefore, the specimens may fracture in a state more similar to bone fractures in physiological condition. In general, compact bone has microcracks from repeated loading and tiny spacings in between lamellae. The tips of the microcracks and the spacings are mostly in

Table I. Selected Studies Measuring Fracture Toughness of Compact Bone.

Investigators	Testing Method	Bone Source	Fracture Direction	Fracture Toughness ($MPa{\cdot}m^{1/2}$)
Melvin and Evans[4]	Single-edge notched beam	Bovine femur	Longitudinal Transverse	3.21±0.43 5.58±0.52
Behiri and Bonfield[7]	Compact tension	Bovine tibia	Longitudinal	2.8-6.3
Norman et al.[8]	Compact tension	Bovine tibia	Transverse	4.93-12.64
Vashishth et al.[9]	Compact tension	Human tibia	Longitudinal	1.6-2.5
Zioupos and Currey[11]	Single-edge notched beam	Human femur	Longitudinal	6.4[1]
Wang et al.[10]	Compact sandwich	Baboon femur	Longitudinal	2.25±0.18
De Santis et al.[13]	Chevron-notched short rod	Bovine femur	Longitudinal	4.8±0.5
Wang et al.[15]	Single-edge notched beam	Human femur	Transverse	5.09±0.98
Nalla et al.[16]	Fatigue precracked flexure	Human humerus	Longitudinal Transverse	3.53±0.13 5.33±0.41
Yan et al.[17]	Chevron-notched beam	Manatee rib Bovine femur	Transverse Transverse	4.5±0.5 5.8±0.5

Current study	Quantitative Fractography	Manatee rib	Transverse	3.1±0.4
		Bovine femur	Transverse	4.0±0.4

[1]Average value at the age of 35

nanoscale, which is much sharper than most of the notch tips (usually have root radii between several to tens of micro-meters) created by mechanical notching methods. Another advantage is that the origins of the fracture specimens can be identified and related to the causes of fracture initiation.

In this study, quantitative fractography was applied to estimate the fracture toughness of bone specimens from bovine femur and manatee rib of young adults. In addition, whole manatee ribs were fractured in simulated three point flexure and the fracture surfaces were analyzed in a similar manner. Scanning Electron Microscope (SEM) was used to capture SEM images for microstructural study. SEM images were also used to determine the porosity of the specimens using Image Pro® software. The results show that the average K_c of the specimens from bovine femur is significantly greater than that of the specimens from manatee rib.

MATERIALS AND METHODS

Fresh bone specimens were prepared from the femur of an adult cow (approximately 48-month-old) and the rib of an adult manatee. The adult manatee carcass was obtained from the Florida Fish and Wildlife Conservation Commission's Marine Mammal Pathobiology Laboratory. Collection and use of manatee bone for research were conducted under U.S. Fish and Wildlife Service permit #MA067116-0, issued to the University of Florida. All bone tissues and specimens were kept in a freezer or stored in physiological buffered saline solution under 4°C at all times until tested. The specimens were tested within three weeks of sectioning with no difference in storage time between the two tissues.

Bovine femur has a relatively large compact area compared to most common animals. The thick compact bone area makes bovine bone a good fracture-mechanics test candidate since greater thickness can usually assure the tests are under a plane-strain condition, for which the fracture toughness measurement is not a function of specimen thickness.[19] The compact bone in adult bovine femur is mostly secondary osteonal bone with some plexiform bone. In adult manatees, ribs are relatively large compared to other animals. Unlike human ribs, which consist of highly vascular cancellous tissue enclosed in a thin layer of compact bone,[20] manatee ribs are mostly primary plexiform bone with few osteons in adults. They are almost all compact bone, and this makes the sample preparation easier than using other species.

The analysis of the fracture toughness of compact bone using quantitative fractography followed the methods outlined in three publications.[3,21-22] Rectangular bars with dimensions 3x3x45 mm (thickness x height x length) were cut using a diamond band saw from the mid-diaphysis of the bovine femur and mid-body of the manatee rib. Ten specimens were cut from each animal with their long axes parallel to the long axes of the femur and rib. During cutting and polishing, a constant spray of water was supplied to keep the specimens cool and hydrated. Three-point flexure tests were performed with a Mini-Bionix materials testing machine (MTS Systems Corporation, Eden Prairie, MN) at room temperature (22°C) soon after the specimens were taken out of storage. A 10-cycle, 10-N preload was applied immediately to adjust the positioning between the specimen and the supporting rollers before testing. Specimens were loaded to failure in flexure using a standard 3-point fixture under displacement control at 1 mm/min. The span length was 40mm. The load was applied in a direction that caused transverse fracture of the specimens. After each specimen was fractured, the maximum load was recorded and used to calculate the maximum stress.

The fracture surfaces of the specimens were then examined to determine the fracture origins and critical crack sizes. Both matching halves were used for examination. First, the samples were visually inspected. In many cases, apparent twist hackle markings could be seen if a Haversian canal or a vascular channel happened to be near or directly on the tensile side. These fracture patterns can be also easily seen on the fracture surfaces of whole ribs (Fig. 1A). Twist hackle markings are fracture steps in the approximate direction of crack propagation. They occur due to a slight local shear stress in addition to the main tensile stress controlling the overall direction of crack propagation. An optical microscope was then utilized to determine the fracture origin at low magnification (20 to 40X). In most cases, the crack fronts, even for irregular shapes of the fracture origins, tended to grow into a semicircular shape before they reached catastrophic failure (Fig. 1B). After the origins were determined, a higher magnification (100X) was used to measure the crack sizes. For off-center loading for which fracture origins were not at the center line of the tensile surface, adjustments in the critical stress were made.[3] The critical stress (σ_c, in MPa), the critical crack sizes (c, in meters) and stress intensity factor coefficients (Y) were then used to calculate the fracture toughness (K_c) using

$$K_c = Y \cdot \sigma_c \cdot \sqrt{c} \tag{1}$$

Here, critical means the values of the parameters at the time of catastrophic failure. Y values for the three types of crack origin, i.e. internal circular crack, surface semi-circular crack and corner quarter-circular crack, are 1.12, 1.24 and 1.4, respectively. For most specimens, finite width correction factor was also calculated since most of the crack sizes in the specimens were relatively large ($c/W \sim 0.1$-0.2, where W is specimen width).

For the manatee whole rib specimens, two ribs were tested with the loading span (approximately 30 cm) adjusted according to the lengths of the ribs. The fracture stress was estimated using a strain gage attached to the tensile side of the ribs. The fracture origins and crack sizes were estimated using a method similar to the beam specimens. The fracture toughness of the ribs was estimated using Eq. 1.

An SEM was used to characterize the microstructure and confirm the crack origins of the broken specimens. All specimens were air dried after fracture. The sample surfaces were coated with plasma sprayed gold-palladium. No large cracks formed due to the drying process were observed. SEM images were also used to determine the porosity of the fracture specimens. The porosity measurement was carried out using Image Pro® software (Media Cybernetics, Silver Spring, MD).[3] It has been shown that porosity affects the critical strain energy release rate of bone.[24]

RESULTS

Table II shows the results of the fractographic measurements. Average fracture toughness values of bovine femur and manatee rib specimens measured using fractography are 4.0±0.4 and 3.1±0.4 MPa·m$^{1/2}$, respectively. The fracture toughness of the two manatee whole ribs, estimated from the fracture surface is ~ 8 MPa·m$^{1/2}$. Average critical stresses of bovine and manatee rectangular specimens are 219.5±21.4 and 163.2±28.3 MPa, respectively. Average critical crack sizes of bovine and manatee specimens are 183±27 and 219±38 μm, respectively. Using a one-way ANOVA with $\alpha = 0.05$, fracture toughness of bovine femur is significantly greater than that of manatee rib. Average critical stress of bovine specimens is also significantly greater than that of manatee specimens. However, the average critical crack size of bovine specimens is smaller than that of manatee specimens ($p=0.0276$).

Table II. Average fracture toughness, critical stress, critical crack size, and porosity of the bovine femur and manatee rib specimens. Standard deviations of each value are also shown in the table.

	Bovine femur [n=10]	Manatee rib [n=10]
Fracture toughness [MPa·m$^{1/2}$]	4.0±0.4	3.1±0.4
Critical stress [MPa]	219.5±21.4	163.2±28.3
Critical crack size [μm]	183±27	219±38
Porosity [%]	3.5±0.6	7.3±1.8

Figure 2 shows the fracture surfaces and fracture origins of one bovine specimen (Fig. 2A) and one manatee specimen (Fig. 2B). As shown in the figure, the bovine femur specimen has a more uniform microstructure (mostly secondary osteonal bone) and lesser porosity. On the other hand, the manatee specimen shows greater porosity and has more vascular channels. Average porosity of the ten bovine specimens is 3.5±0.6%, which is significantly lesser than that of manatee specimens (7.3±1.8%)

DISCUSSION

The average fracture toughness of bovine femur specimens was found to be about 30% greater than that of manatee rib specimens. This is consistent with a previous study.[17] There, the researchers used a chevron-notched beam (CNB) method and found the fracture toughness of bovine femur and manatee rib to be 5.8 and 4.5 MPa·m$^{1/2}$($p<0.001$), respectively. Bovine femur has a greater fracture toughness value than manatee rib and this could be because of two reasons. First, the bovine specimens exhibit a greater apparent density than the manatee specimens.[23] Wright and Hayes[5] studied the effect of apparent density and found that a five percent increase in

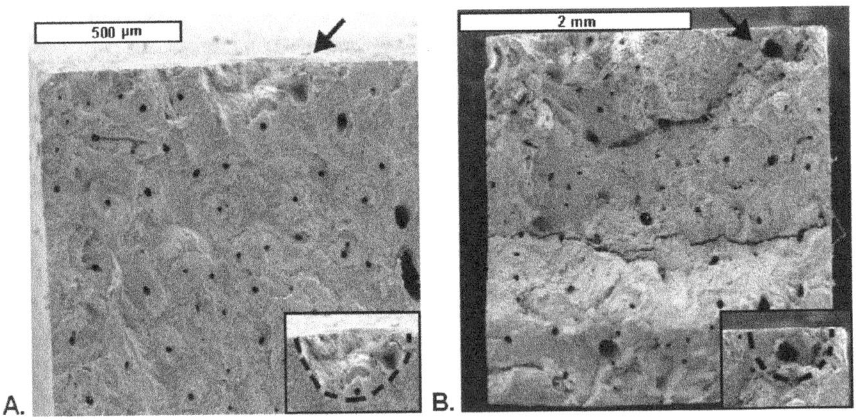

Figure 2. SEM images of the fracture surface of a bovine specimen, A, and a manatee specimen, B. The fracture origins are indicated by arrows. The origin of the manatee specimen is a channel for vascular vessel. Two insets show the crack boundaries. In most cases, the measurement of the crack boundaries has 5-15 μm uncertainty. The uncertainty is considered small since it is the square root of the crack size that is used to determine the fracture toughness of the specimens.

bone density (from 1.92 to 2.02 g/cm^3) resulted in a 30% increase in the fracture toughness. Second, the reason could be that the bovine specimens had a lower porosity (3.5%) than the manatee specimens (7.3%). Yeni et al.[24] studied the influence of bone morphology and found that the critical strain energy release rate (G_c) of human femur decreased as porosity of the bone increased.

The estimated fracture toughness of the manatee rib specimens in the current study is significantly smaller than that estimated using a CNB method.[17] Also, the fracture toughness of manatee whole ribs (with diameters about 5-6 cm), estimated from the fracture surface is ~ 8 MPa·m$^{1/2}$. The fracture origin sizes in the three studies are about 0.22, 2.0, and 3.0 mm, respectively. The reason for the large differences in the measured fracture toughness could be because of R-curve behavior in bone. Recently, several researchers have reported rising R-curve (crack extension resistance) behavior in cortical bone.[9,25-27] For materials with rising R-curve behavior, the greater stability results in additional crack extension and this may lead to greater measured fracture toughness values than the initiation fracture toughness values.[28-29] Vashishth et al.[9] used compact tension specimens to measure R-curves of young bovine and adult human tibiae along longitudinal fracture. They showed that both materials had R-curve behavior. In addition, bovine specimens had greater overall fracture toughness and greater growth toughness, which is defined as ΔK/mm. They demonstrated that the toughening mechanism for the increasing fracture toughness was microcracking. Malik et al.[25] also used compact tension specimens to estimate the R-curve behavior of compact bone. They studied equine bone and tested the specimens along transverse fracture. They found that specimens from lateral cortex and dorsal cortex both exhibited rising R-curve behavior with the fracture mechanisms acting in these specimens to be regionally dependent. They claimed that besides microcracking, mechanisms as osteon pullout, fiber bridging, and crack deflection may also contribute to the increment of fracture toughness. The R-curve behavior in bone may be one of the reasons for the large variation in the reported fracture toughness data (Table I) since the initial notch lengths and critical crack lengths in different techniques could be different.

Fracture origins of the bone specimens were found to be mostly at openings for blood vessels (vascular channels, Haversian and Volkmann's canals) (Fig. 2B) and areas where bone structures were less compact (larger interlayer spacings) (Fig. 1B and 2A).

CONCLUSION

Quantitative fractography was applied to estimate the fracture toughness of bovine femur and manatee rib rectangular specimens. Average fracture toughness of bovine femur and manatee rib specimens are 4.0±0.4 and 3.1±0.4 MPa·m$^{1/2}$, respectively. The reasons for the difference could be that bovine bone has a greater apparent density and lesser porosity. It is demonstrated that fractography can be used to identify the fracture origins of compact bone.

REFERENCES

[1]T. Kimura, K. Ogawa, and M. Kamiya, "Fractography of Human Intact Long Bone by Bending," *J. Legal Med.*, 79, 301-10 (1977).

[2] G. Corondan, and W. L. Haworth, "A Fractographic Study of Human Long Bone," *J. Biomech.*, 19, 207-18 (1986).

[3] J. Yan, K. B. Clifton, R. L. Reep, and J. J. Mecholsky, "Application of Fracture Mechanics to Failure in Manatee Rib Bone," *J. Biomech. Eng.*, 128, 281-9 (2006).

[4] J. W. Melvin and F. G. Evans, "Crack Propagation in Bone." ASME Biomechanics Symposium, New York, NY, 87-8 (1973).

[5] T. M. Wright and W. C. Hayes, "Fracture Mechanics Parameters for Compact Bone - Effects of Density and Specimen Thickness," *J. Biomech.*, 10, 419-30 (1977).

[6] W. Bonfield, M. D. Grynpas and R. J. Young, "Crack Velocity and the Fracture of Bone," *J. Biomech.*, 11, 473-9 (1978).

[7] J. C. Behiri and W. Bonfield, "Fracture Mechanics of Bone - the Effects of Density, Specimen Thickness and Crack Velocity on Longitudinal Fracture," *J. Biomech.*, 17, 25-34 (1984).

[8] T. L. Norman, D. Vashishth, and D. B. Burr, "Effect of Groove on Bone Fracture Toughness," *J Biomech.*, 25, 1489-92 (1992).

[9] D. Vashishth, J. C. Behiri, and W. Bonfield, "Crack Growth Resistance in Cortical Bone: Concept of Microcrack Toughening," *J. Biomech.*, 30, 763-9 (1997).

[10] P. Zioupos and J. D. Currey, "Changes in Stiffness, Strength and Toughness of Human Cortical Bone with Age," *Bone*, 22, 57-66 (1998).

[11] X. Wang, N. S. Masilamani, J. D. Mabrey, M. E. Alder, and C. M. Agrawal, "Changes in the Fracture Toughness of Bone May Not Be Reflected in Its Mineral Density, Porosity, and Tensile Properties," *Bone*, 23, 67-72 (1998).

[12] Z. Feng, J. Rho, S. Han, and I. Ziv, "Orientation and Loading Condition Dependence of Fracture Toughness in Cortical Bone," *Mater. Sci. Eng.* C, 11, 41-6 (2000).

[13] R. De Santis, P. Anderson, K. E. Tanner, L. Ambrosio, L. Nicolais, W. Bonfield, and G. R. Davis, "Bone Fracture Analysis on the Short Rod Chevron-Notch Specimens Using the X-ray Computer Micro-Tomography," *J. Mater. Sci. Mater. Med.*, 11, 629-36 (2000).

[14] P. Lucksanasombool, W. A. Higgs, R. J. Higgs, and M. V. Swain, "Fracture Toughness of bovine Bone: Influence of Orientation and Storage Media," *Biomaterials*, 22, 3127-32 (2001).

[15] X. Wang, X. Shen, X. Li, and C. M. Agrawal, "Age-Related Changes in the Collagen Network and Toughness of Bone," *Bone*, 31, 1-7 (2002).

[16] R. K. Nalla, J. H. Kinney, and R. O. Ritchie, "Mechanistic Fracture Criteria for the Failure of Human Cortical Bone," *Nat. Mater.*, 2, 164-8 (2003).

[17] J. Yan, K. B. Clifton, J. J. Mecholsky, and R. L. Reep, "Fracture Toughness of Manatee Rib and Bovine Femur Using a Chevron-Notched Beam Test," *J. Biomech.* 39, 1066-74 (2006).

[18] A. K. Mukhopadhyay, S. K. Datta, and D. Chakraborty, "Fracture Toughness of Structural Ceramics," *Ceram. Int.*, 25, 447-54 (1999).

[19] R. W. Hertzberg, Deformation and Fracture Mechanics of Engineering Materials, John Wiley and Sons, New York, NY (1996).

[20] H. Gray, Anatomy of the Human Body, Lea and Febiger, Philadelphia, PA (1973).

[21] ASTM International, "ASTM C1322: Standard Practice for Fractography and Characterization of Fracture Origins in Advanced Ceramics," West Conshohocken, PA (2002).

[22] J. J. Mecholsky, "Quantitative Fracture Surface Analysis of Glass Materials," In: Experimental Techniques of Glass Science, C. J. Simmons and O. H. El-Bayoumi, eds, *The American Ceramic Society*, Westerville, OH, 483-90 (1993).

[23]J. Yan, "Elastic-Plastic Fracture Mechanics on Compact Bone," Ph. D. Dissertation, University of Florida, Gainesville, FL, p. 51 (2005).

[24]Y. N. Yeni, C. U. Brown, Z. Wang, and T. L. Norman, "The Influence of Bone Morphology on Fracture Toughness of the Human Femur and Tibia," *Bone*, 21, 453-9 (1997).

[25]C. L. Malik, S. M. Stover, R. B. Martin, and J. C. Gibeling, "Equine Cortical Bone Exhibits Rising R-curve Fracture Mechanics," *J. Biomech.*, 36, 191-8 (2003).

[26]D. Vashishth, "Rising Crack-Growth-Resistance Behavior in Cortical Bone: Implications for Toughness Measurements," *J. Biomech.*, 37, 943-6 (2004).

[27] R. K. Nalla, J. J. Kruzic, J. H. Kinney, and R. O. Ritchie, "Effect of Aging on the Toughness of Human Cortical Bone: Evaluation by R-curves," *Bone*, 35, 1240-6 (2004).

[28]J. A. Salem, J. L. Shannon, and M. G. Jenkins, "Some Observations in Fracture Toughness and Fatigue Testing with Chevron-Notched Specimens," In: Chevron-Notch Fracture Test Experience: Metals and Non-Metals, K. R. Brown and F. I. Baratta, eds., *American Society for Testing Materials*, West Conshohocken, PA, 9-25 (1992).

[29]A. T. Zehnder, C. Hui, and E. D. Rodeghiero, "Chevron-Notched Toughness of Materials with Rising Fracture Resistance Curves," *J. Am. Ceram. Soc.*, 80, 1319-22 (1997).

DESCRIPTIVE FRACTOGRAPHY ON ALL CERAMIC DENTAL CROWN FAILURES

Susanne S. Scherrer, H.W. Anselm Wiskott
Univ. of Geneva, School of Dental Medicine
Dept. of Prosthodontics-Biomaterials
Geneva, Switzerland

Janet B. Quinn
[2]American Dental Association Foundation
Paffenbarger Research Center
Gaithersburg, MD 20899, USA

George D. Quinn
National Institute of Standards and Technology
Gaithersburg, MD 20899, USA

ABSTRACT

Descriptive fractography has been applied to three dental ceramic crown failures for understanding the fracture process and assessing possible design inadequacies. In two cases, the original broken parts could be recovered for stereo and SEM analysis. In the third case, the replica method was used for reproduction of the still in-situ remaining fractured ceramic crown part. The dental replica technique consists in the injection of a high precision silicone impression material onto the fractured surface and pouring of a positive epoxy model which is then gold-coated for SEM evaluation.

The recovered ceramic pieces as well as the replica model were inspected for fractographic features such as hackle, wake hackle, twist hackle, compression curl and arrest lines for determination of the direction of crack propagation and location of the origin.

A general map of the fracture events could be reconstructed for all three cases based on recognition of wake hackle and compression curl. The replica technique proved to be easy to use and allowed an excellent reproduction of the fractured surface. Case specific reasons for failure are discussed.

INTRODUCTION

The dental community is using a variety of ceramic restorative materials such as porcelains (leucite or alumina based), glass-ceramics (leucite, mica, lithium disilicates), alumina-glass infiltrated, and CAD-CAM ceramics including pure alumina and zirconia (3Y-TZP) core materials.[1] The polycrystalline high strength ceramics (alumina and zirconia) serve as substructure materials (i.e. framework or core) upon which glassy ceramics are veneered for an improved appearance. When using all-ceramic restorations in posterior teeth where high masticatory loads are generated, sudden fractures of the full-thickness restoration or of the veneering ceramic (chips) may be expected. Fracture such as large chips or through the full core thickness are considered as complete failures and end up in removal of the restoration. Understanding the reasons for such events is of key importance to the dental community as many parameters may be involved such as design deficiency, material deficiency (fabrication process) or in situ stress-induced conditions. Clinical failure analysis utilizing fractographic techniques are for that purpose gaining more attention within the dental community.[2-8] However, recovery of the clinically failed ceramic part may not always be possible. The main reasons for this are that the broken part has not been recovered by the patient and the still in-situ strongly cemented other half of the restoration cannot be taken out without damaging its fractured surface. In these instances the replica technique is applied to obtain a perfect reproduction of the remaining fracture surface by performing an intra-oral impression of the exposed fractured surface using a low viscosity silicon material. The impression (i.e. the negative of the fractured surface) is poured with cold mounting epoxy resin thus producing a positive of the fractured surface which is then gold-sputtered for SEM evaluation. In dentistry, the term replica is used for both the impression and the poured epoxy model. The microscopic surface analysis is performed on the gold-sputtered epoxy model.

In this report qualitative (descriptive) fractography has been applied on recovered full-thickness all-ceramic crown parts as well as on a replica of a veneering ceramic crown chip. The goal was to identify the characteristic crack features (arrest lines, hackle, wake hackle, twist hackle, compression curl) for mapping the crack direction and determination of the origin from which reasons for failure could be drawn.

MATERIALS AND METHODS

Three clinical ceramic crown failures are studied (Fig.1). Case One is a Procera®AllCeram (Nobel Biocare, Sweden)[a] upper premolar crown that fractured through its full-thickness after 4 years of intra-oral function. The patient remembered biting on something hard. The Procera®AllCeram uses a computer aided system for the core design and manufacturing which is made of 99.9 vol % of alumina and veneered with an alumina based porcelain. Case Two is a fracture of an upper premolar restored by a feldspath-based porcelain

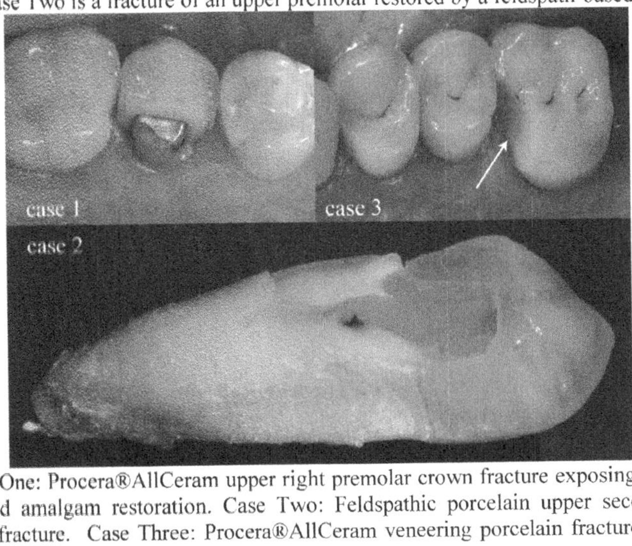

Fig.1: Case One: Procera®AllCeram upper right premolar crown fracture exposing part of tooth structure and amalgam restoration. Case Two: Feldspathic porcelain upper second premolar crown-root fracture. Case Three: Procera®AllCeram veneering porcelain fracture of an upper left molar.

crown which is resin-bonded to a composite resin core-build-up of a root-canal treated tooth. The crown-tooth entity fractured catastrophically during mastication after 9 months of intra-oral use. Case three is a veneering porcelain fracture of a Procera®AllCeram molar that occurred after 4.2 years of intra-oral use. The patient remembers biting on something hard. In this case the replica method was used which consists in taking an intra-oral impression of the in-situ remaining fractured part by injecting an elastomeric low viscosity silicone material (Aquasil Ultra XLV, Dentsply DeTrey, Germany).[9] The impression (negative of the fractured surface) is poured with a cold mounting epoxy resin (Epofix Resin, Struers, Denmark) reproducing a positive of the fractured surface which is then gold-sputtered for SEM analysis. Identification and interpretation of the fracture patterns were based on reference textbooks[10-13] and an ASTM standard[14]. The main fractographic features searched for were arrest lines, hackle, wake hackle and compression curl using a combination of stereo microscope and SEM analyses.

FRACTOGRAPHIC DESCRIPTIONS

Case One

The recovered part of the fractured Procera crown was examined under stereo and the SEM at different sites of interest (zone 1 to 5) starting at the mesial (left) crown margin and moving to the right (distal) side of the restoration (Fig.2). A porcelain veneer chip (1.5 x 2 mm) (zone 1) was visible on the gingival mesial margin. Higher magnifications of that chip (Fig.3) showed arrest lines and wake hackle indicating that the crack started at the margin and propagated upwards. The fact that the chipped surface had considerable staining before it was cleaned with ethanol indicated that the fracture of the chip must have occurred early in the clinical age of the crown, probably during cementation. Primary fracture of the crown occurred on a different plane (zones 2 – 5) that split the crown into two. Fractographic examination showed that this cleavage plane started from the bottom of the chip at the margin. Higher magnifications of zone 2 are shown in Fig. 4. On the stereo image (left) series of parallel running twist hackle lines are observed. Under the SEM at 475x magnification (right) wake hackle in the veneer indicated the crack ran upwards towards the occlusal surface of the crown. Figure 5 (zone 3) shows a series of edge chips on the occlusal (chewing) surface with arrest lines and twist hackle within the veneering porcelain not reaching the alumina core. These edge chips are secondary incidental events not interfering with the general crack propagation direction which continues to be followed in zone 4 and 5 (Fig.6 and 7). Zone 4 (Fig.6) viewed under the stereo microscope (left) shows an array of shear hackle within the core and radiating into the veneering porcelain. These hackle were generated when the crack emerged from the inside corner of the crown (bottom left). The direction of crack propagation is again confirmed by the presence of wake hackle spotted under the SEM (right) within the veneering porcelain indicating the crack moved from left to right (mesial to distal).

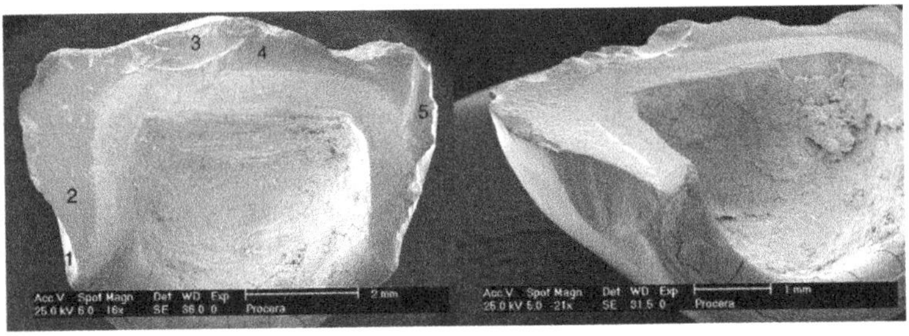

Fig.2: Recovered fractured Procera crown part (Case One) viewed at 16x under the SEM. Zones of interest for detailed analysis are numbered 1 to 5 starting at the mesial (left) margin to the other side of the crown.

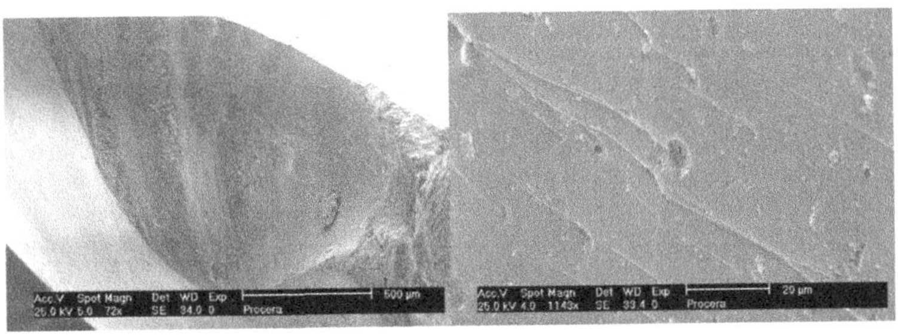

Fig.3: SEM magnifications of zone 1 chip. Arrest lines and wake hackle indicate that the crack ran from the margin upwards. Round pores are commonly found in veneer glasses and are harmless. They are convenient since they trigger wake hackle.

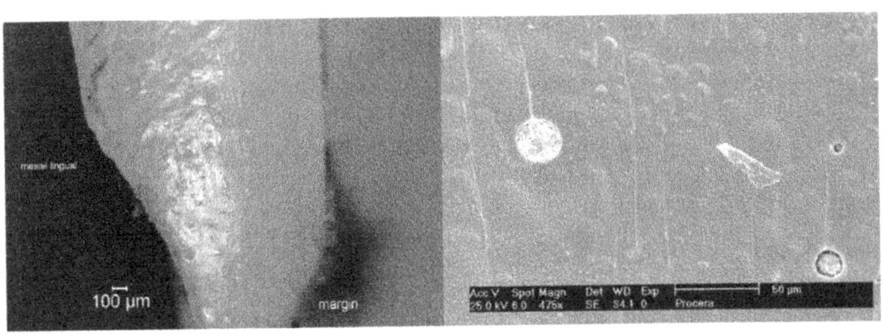

Fig.4: Stereo image of zone 2 showing parallel running twist hackle. SEM (right) confirmation of the presence of wake hackle in the veneering porcelain indicating the crack ran upwards.

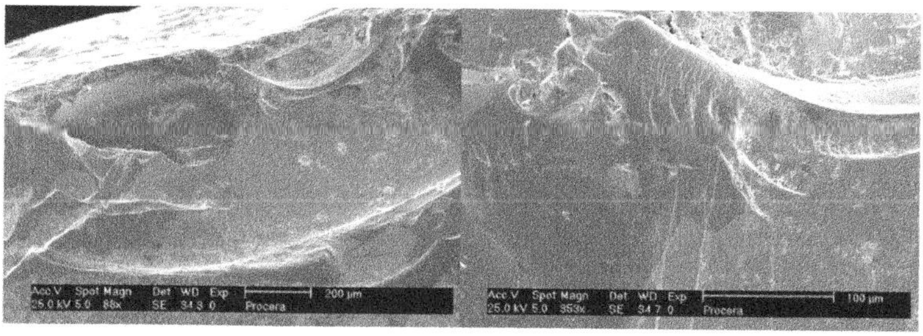

Fig.5: SEM of zone 3 representing secondary edge chips on the occlusal surface. Arrest lines and twist hackle remain confined within the veneering porcelain. The direction of propagation within these edge chips is running from top to bottom.

Finally, zone 5 (Fig.7) contains a classic compression curl (shear lip or cantilever curl) on the crowns outer distal side which marks the end of the fracture event. The search for wake hackle in this area was successful with the SEM confirming the direction of cracking from left to right within the compression curl corresponding to the final breakthrough on this side of the crown. In summary, fracture in this crown occurred from a chip at the margin that must have occurred early in the clinical age of the crown, probably during cementation. The main fracture plane started from the chip, but propagated on a different plane in response to hoop stresses in the crown walls. Additional interpretation is presented in the conclusion section.

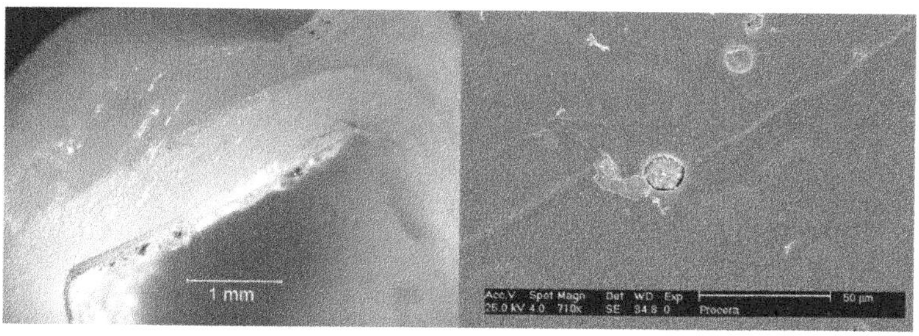

Fig.6: Stereo and SEM of zone 4 below the occlusal surface of the crown. An array of shear hackle radiating from the core towards the veneer confirmed by wake hackle in the veneer.

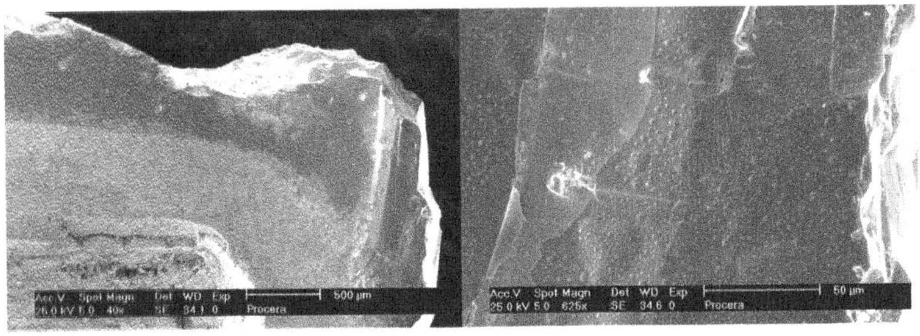

Fig.7: Compression curl and wake hackle confirming the final breakthrough in zone 5.

Case Two

This restored upper second premolar tooth broke suddenly in two parts during mastication ending up with its extraction (Fig.1). The tooth had been reconstructed 9 months earlier with a feldspathic porcelain crown which was cemented adhesively to an internal resin composite core build-up itself adhesively cemented to dentin tooth structure. The recovered tooth half was viewed under the SEM and several zones of interest (1 to 5) selected for detailed microscopic analysis (Fig.8). The left image provides an overlook of the different layers (porcelain crown, resin cement, resin composite core build-up and root dentin) whereas the right image shows a wake hackle identified in zone 1 indicating a direction of crack propagation (dcp) towards the internal margin angle of the crown.

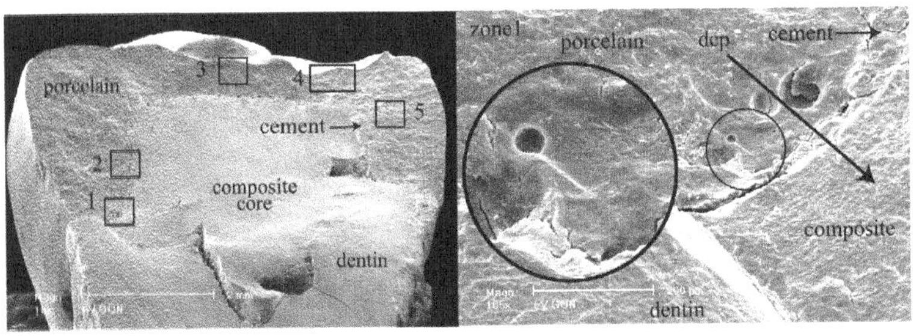

Fig.8: Feldspathic porcelain crown fracture with underlayers of resin core and tooth structure. Zones of interest are marked 1 to 5. Zone 1 is magnified on the right image showing the direction of crack propagation (dcp) based on the presence of a wake hackle.

Figure 9 provides views of zone 2 (left image) and zone 3 (right image) with respectively wake hackle and hackle oriented in the same general direction as in zone 1 which is at approximately 45° angle from upper left of the crown to lower right.

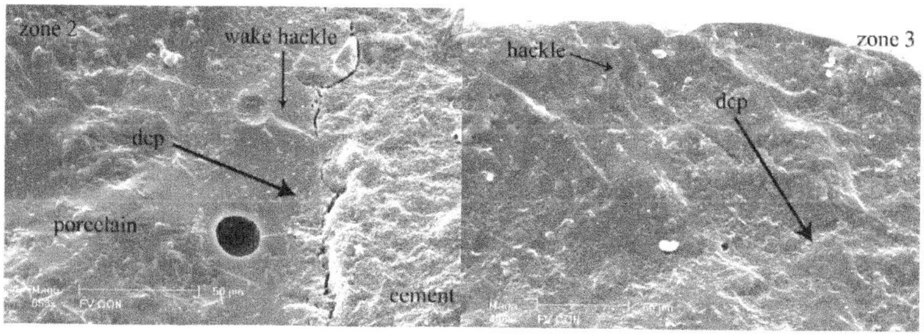

Fig.9: The direction of crack propagation (dcp) for zone 2 (left) and 3 (right) is marked with an arrow based on the identification of wake hackle (left) and hackle (right).

Figure 10 shows selected fractographic features within zones 4 and 5. Zone 4 shows wake hackle from pores within the porcelain near the occlusal surface and oriented locally at a 45° angle from upper right to lower left, whereas in zone 5 the wake hackle indicates again a direction of crack propagation from top to bottom in a similar orientation as in zones 1, 2, and 4.

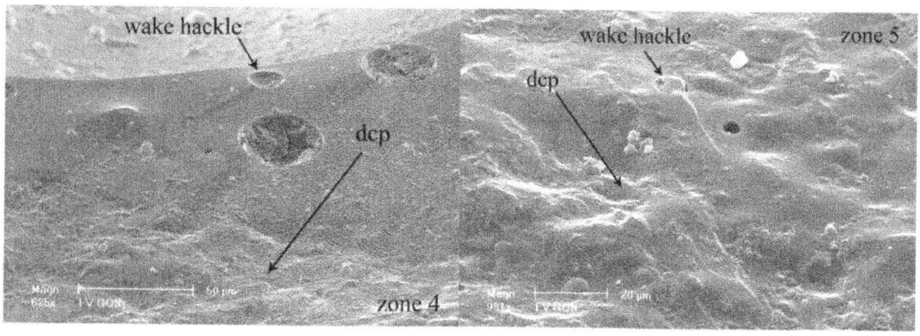

Fig.10: Zone 4 (left) shows wake hackle in the upper feldspathic porcelain layer just below the occlusal surface. In zone 5 (right) hackle and wake hackle indicate the direction of crack propagation (dcp).

The summary SEM map of the crack direction is shown in Figure 11. From this information it can be concluded that high chewing forces acted on the palatal crown cusp and wedge opened a crack on the occlusal surface which then propagated down and through the crown-root structure at high velocity until complete fracture of the tooth in two parts (right image of Fig.11). Additional interpretation is presented in the conclusion section.

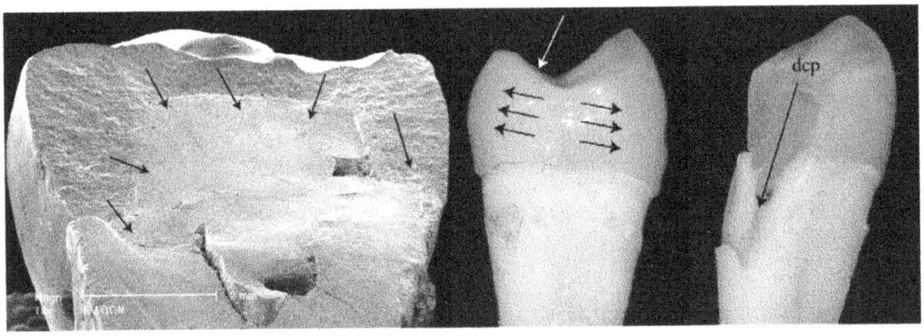

Fig.11: Summary map (left) of the crack direction within the recovered tooth-crown fragment. On the right image, the white arrow shows the direction of the occlusal forces applied to the porcelain palatal crown cusp and the generation of tensile stresses wedge-opening the crack. A cantilever curl at mid root level indicates the end of the fracture event.

Case Three

The gold sputter-coated resin replica of the veneering porcelain failure of a Procera® AllCeram molar crown was inspected under the SEM. Figure 12 (left) shows a top (occlusal) view of the mastication surface of the crown on which major erosive wear patterns are apparent on all four cusps. On the fractured surface (right image) viewed at low magnification hackle and arrest lines are visible. Three zones of interest (labeled 1 to 3) were selected for further microscopic analysis. Zone 1 corresponds to a curved area looking similar to a compression curl and located near the interproximal mesial gingival margin. Zone 2 corresponds to the occlusal edge of the palatal cusp with some surface damage visible. Zone 3 located in the middle of the fracture surface contains a concentration of hackle lines. Figure 13 is a higher magnification of zone 1. Many hackle and wake hackle are oriented from right to left on the images, indicating that the crack was moving towards the gingival margin shortly before break off.

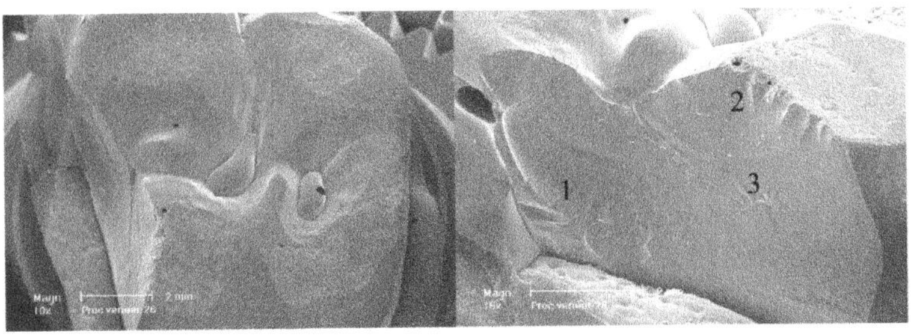

Fig. 12: Resin-poured replica reproduction showing a top (occlusal) and side view of the fractured crown. Major abrasion is visible on the mastication (occlusal) surface (left image). Arrest lines and hackle are visible on the right image. Zones of interest are labeled 1 to 3.

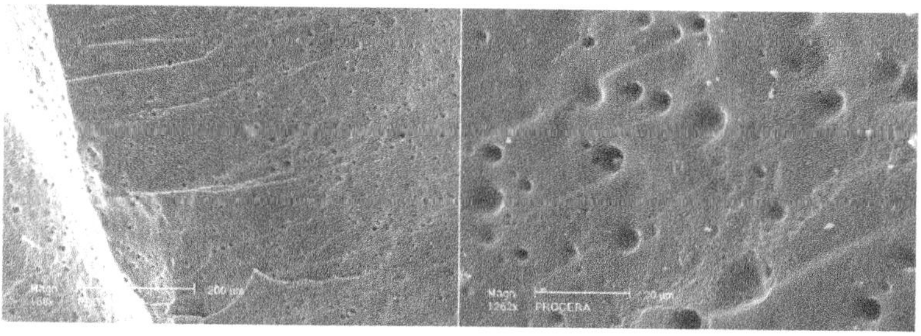

Fig. 13: Low and high magnification of zone 1 showing hackle and wake hackle indicating the crack was moving from right to left towards the interproximal gingival crown margin.

Figure 14 is a high magnification of zone 2. Two deeper surface damages shaped as indents (craters) of about 300 μm in diameter are identified as the origins of the fracture process. They are followed by several arrest lines (Fig.12, right) indicating the position of the crack front at the time of an arrest or disturbance. Wake hackle inside these craters indicate that the crack as moving from top (chewing side) to bottom (gingival margin). Figure 15 illustrates details within zone 3 containing hackle and wake hackle trails of the crack moving downwards. Figure 16 (left) is a synopsis of the failure process. The origin of the fracture is circled on the palatal cusp. The general direction of crack propagation is marked by arrows pointing towards the crown margin.

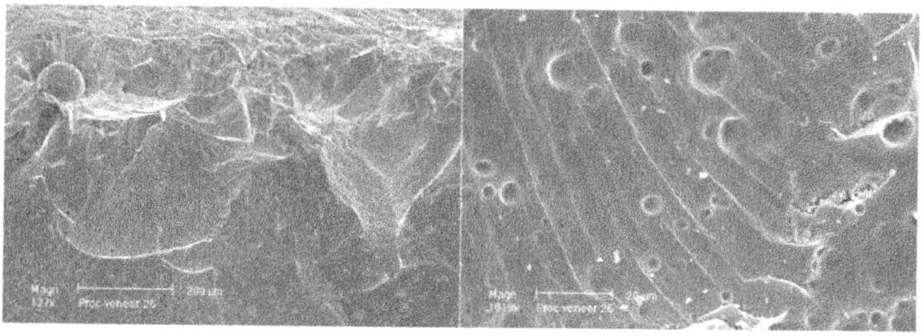

Fig. 14: Illustration of zone 2 showing two contiguous indent-type of surface damage identified as the origin(s). Wake hackle (right image) within the damage zone are followed by larger arrest lines on the left image. The direction of propagation is from top left to bottom right. The round object in the upper left of the left view is a replica artifact from a bubble in the silicone.

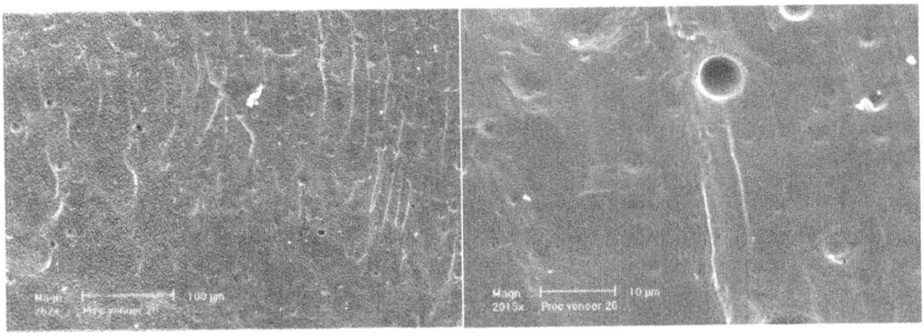

Fig. 15: Higher magnifications of zone 3 showing again evidence of hackle and wake hackle. The crack moved from top to bottom.

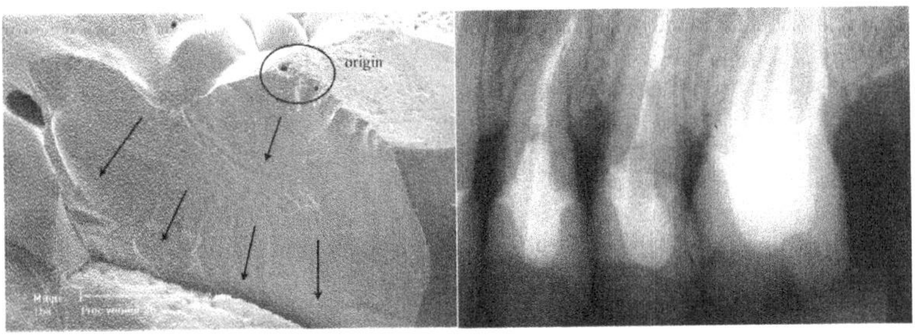

Fig. 16: Synopsis map of the crack path (black arrows) and the origin (circle) at the occluso-palatal edge of the crown. The X ray image shows a large porcelain veneer thickness insufficiently supported by the alumina core material.

In summary, this crown fractured from contact stress damage on the occlusal surface. Damage cracks propagated down and through the veneering porcelain causing a big chip. There was evidence of bruxism. Additional interpretation is in the next section.

CONCLUSIONS

1) The stereo and SEM analysis of three all-ceramic crown failure cases using descriptive fractography provided key information as to the crack propagation direction and location of failure origins. On the recovered none damaged in-service failed parts (Case One and 2) as well as on the replica resin-poured epoxy model of an in-situ taken impression of a partially fractured ceramic crown (case three) identification of fracture markings such as wake hackle, hackle, arrest lines and compression curl allowed mapping of the crack propagation and determination of the sequence of the crack travel history. In all three cases, the more glass-containing veneering ceramic had a fractographic favorable microstructure on which the markings were easily distinguishable. Key features were wake hackle trails left behind pores and inclusions existing in the veneering porcelains. Very fine fracture surface details could be reproduced with the replica technique thus encouraging its easy use in all clinical failures for which the sole existing ceramic fractured part is still cemented on the tooth as it is nearly impossible to retrieve the remainder of

the failed restoration without damage to the fracture surface. Drawbacks from the replica technique such as artifacts (air bubbles) or epoxy pouring imperfections were noted but not critical for the overall fracture surface analysis.

2) The origin of fracture was different for the two Procera crowns (Case One and 3) that failed each after 4 years of intra-oral use. In Case One (recovered broken crown half) the origin was located at the gingival margin in form of a veneering ceramic chip which must have occurred during the cementation procedure when finger pressure was exercised vertically to seat the crown on the tooth shoulder preparation. A predominant local contact stress must have been generated on a slightly overhanging veneering ceramic causing it to chip off. The veneer ceramic is a rather weak material (bending strength = ~100 MPa, Young's modulus = ~70 GPa, K_{Ic} = ~ 1 MPa√m). Hence, any overhanging build-up of this ceramic over the high-strength, high-density alumina core margin will be at risk to early failure during cementation or initial chewing cycles and should be avoided during fabrication of the crown. The margin chip acted as a predamaged site on which hoop stresses (generated during mastication) concentrated and from which a critical crack started and propagated until cleaving the crown in two parts.

The second Procera crown (case three) had a major veneer fracture without exposing the alumina core. It could be inferred from the worn occlusal (chewing) surface seen on the SEM (Fig.12) that this patient had strong bruxing habits (that means a highly stressed crown). It appears that the fairly thick layer of unsupported veneering ceramic (2-3mm) (Fig.16, right image) combined to the repetitive loadings generated while the patient was bruxing promoted the initiation of a surface crack which propagated downwards to the gingival margin. The two indents at the origin of the fracture on the occlusal surface may be explained by the patient inadvertently biting on a hard particle. With respect to the unsupported veneering ceramic, newer core design procedures such as double scanning should prevent excessive thicknesses of the mechanically weak veneering ceramic.

3) The choice of the restoration material was the main reason for the early failure in case two. Indeed, the use of a feldspathic porcelain as a complete crown coverage on a posterior tooth was a wrong decision. The mechanical properties of porcelain are too weak to withstand the masticatory stresses generated in the posterior region (premolar, molar). They do not exceed a bending strength of ~100 MPa, a Young's modulus of ~70 GPa and a K_{Ic} of ~ 1 MPa√m. Also, the internal resin composite core build up may not have provided adequate support for the porcelain crown under high masticatory loads. In addition, the rather steep cusp design added to build-up maximal contact pressure during lateral chewing (Fig.11). Although the origin could not precisely be located, the crack propagation direction was clearly defined as moving from the occlusal side downwards to the root. The catastrophic fracture splitting the crown with its root in two parts is due to the excellent adhesive bond between all structures (i.e. ceramic crown , internal resin core build-up and dentin). Knowing that an alumina core of an all-ceramic crown (i.e. Procera) breaks having properties of S = ~600 MPa, E = ~ 360 GPa, and K_{Ic} = ~ 4.5 MPa√m, one has to know the risk in using weaker but more esthetic glassy ceramic materials in posterior teeth.

[a] Certain commercial materials or equipment are identified in this paper to specify adequately the experimental procedure. Such identification does not imply endorsement by the National Institute of Standards and Technology nor does it imply that these materials or equipment are necessarily the best for the purpose.

ACKNOWLDEMENTS
This work is supported in part by NIH Grant DE 14534.

REFERENCES

[1]J.R. Kelly, "Dental Ceramics: Current Thinking and Trends" *Dent. Clin. N. Am.*, 48, 513-30 (2004).

[2]J.R. Kelly, S.D. Campbell, H.K. Bowen, "Fracture-Surface Analysis of Dental Ceramics" *J. Prosthet. Dent.*, 62, 536-41 (1989).

[3]J.R. Kelly, S.D. Campbell, H.K. Bowen, "Fracture Surface Analysis of Dental Ceramics: Clinically Failed Restorations" *Int .J. Prosthodont.*, 3, 430-40 (1990).

[4]J.Y. Thompson, K.J. Anusavice, A. Naman, H.F. Morris, "Fracture Surface Characterization of Clinically-Failed All-Ceramic Crowns" *J. Dent Res.*, 73, 1824-32 (1994).

[5]J.R. Kelly, "Fractography of Dental Ceramics. In: Varner JR, Quinn GD, editors. *Fractography of Glasses and Ceramics IV*. Westerville, OH, American Ceramic Society; Ceramic Transaction, 122, 241-51 (2001).

[6]J.B. Quinn, G.D. Quinn, J.R. Kelly, S.S. Scherrer, "Fractographic Analyses of Three Ceramic Whole Crown Restoration Failures" *Dent. Mater.*, 21, 920-29 (2005).

[7]B. Taskonak, J.J. Mecholsky, Jr., K.J. Anusavice, "Fracture Surface Analysis of Clinically Failed Fixed Partial Dentures" *J. Dent. Res.*, 85, 277-81 (2006).

[8]S.S. Scherrer, J.B. Quinn, G.D. Quinn, J.R. Kelly, "Failure Analysis of Ceramic Clinical Cases Using Qualitative Fractography" *Int. J. Prosthodont.*, 19, 151-58 (2006).

[9]S.S. Scherrer, J.B. Quinn, G.D. Quinn, H.W.A. Wiskott, "Fractographic Ceramic Failure Analysis Using the Replica Technique" (submitted to *Dent. Mater.* June 2006).

[10]R. W. Rice, "Ceramic Fracture Features, Observations, Mechanisms and Uses. In: Mecholsky JJ Jr, Powel SR Jr, (eds). Fractography of Ceramics and Metal Failures, ASTM STP 827, Philadelphia: American Society for Testing and Materials, 5-103 (1984).

[11]V.D. Fréchette, Failure Analysis of Brittle Materials. vol. 28, Advances in Ceramics, Westerville, OH: *American Ceramic Society*, (1990).

[12] J.R. Varner, Descriptive Fractography. In: Schneider SJ (ed). Ceramics and Glasses, Vol. 4, Engineered Materials Handbook. Metals Park, Ohio: *ASM International*, 635-44 (1991).

[13] G.D. Quinn, Recommended Practice Guide for Fractography of Ceramics and Glasses. Special Publication 960-XX, National *Institute of Standards and Technology*, Gaithersburg, MD (2006).

[14] ASTM C1322-05a, "Standard Practice for Fractography and Characterization of Fracture Origins in Advanced Ceramics." *ASTM International*, West Conshohocken, PA, (2005).

SECTION VI

FRACTOGRAPHY OF COMPONENTS

DEALING WITH COMPONENT FAILURES

Roger Morrell
Division of Engineering and Process Control
National Physical Laboratory
Teddington, Middlesex, UK, TW11 0LW

ABSTRACT

Component evaluation is frequently more challenging than that of small test-pieces, principally because of the huge variety of reasons for the failure, the additional damage that often arises, and, not infrequently, the appearance of fractographic features which are not immediately recognised or interpretable. This paper provides examples in three areas in some of which the author found fracture features not previously recognised by him during many years of evaluations, for which the explanations remain tentative. The first concerns CaF_2 windows suffering from thermal stress failure in a laser flash thermal diffusivity apparatus; the second concerns a series of proof-test failures in large SiC gas seal rings; and the third concerns novel fracture surface features in alumina ceramic femoral failures. These examples demonstrate that the fractographer has to keep an open mind at all times, must not jump to conclusions, but must seek to ensure all the evidence fits with his fracture hypothesis, and must admit uncertainty rather than side with any pre-conceived view.

INTRODUCTION

The majority of evaluations published and the majority of the guidelines issued for fractography of ceramics[1-4] and glasses have focused on detailed origin identification and fractographic assessments of fractured test-pieces, usually prepared and broken under known conditions. To the inexperienced, many of the lessons learned from such work become confusing when confronted with service fractures, especially when flaw types are involved which do not appear in test-bars, when the stress distributions are unknown, when there are multiple origins arising because of the geometry of the component, and when there are factors such as subcritical crack growth, handling damage, abnormal environment, or missing fragments. Thus, although the same principles apply to larger components, they must be applied with an open mind.[5]

This paper describes a number of fractographic investigations on ceramic components which have posed interesting problems for the author over the years, either in terms of understanding the crack pattern or understanding how and why fracture occurred in the first place. The key to solving more complex problems encountered with components is to think more laterally about the circumstances of fracture and the path that fracture has taken.

EXAMPLE 1 – CaF_2 WINDOWS

A series of fractures in CaF_2 transparent windows used in a laser flash thermal diffusivity apparatus occurred unexpectedly. The windows, about 48 mm diameter and 6 mm thick, allow the radiation from the rear face of the test-piece to pass from the surrounding furnace to the infra-red detector during high-temperature experiments. The window itself is clamped between O-rings supported on both sides by water-cooled parts of the system. In principle there should be no hard contact between the window and its surroundings.

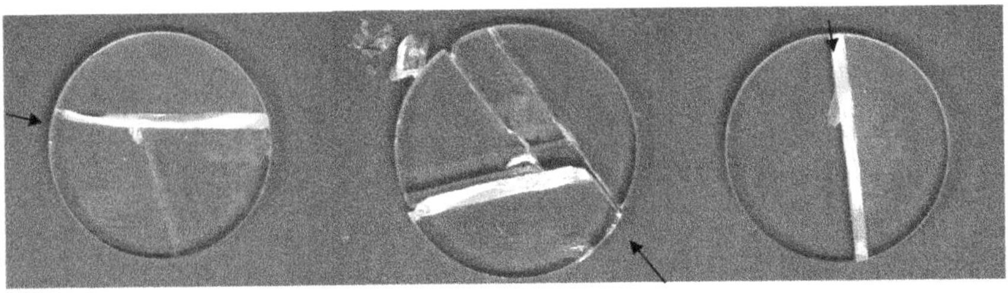

Figure 1. Three failed CaF_2 windows, (left to right) A, B and C. Note angled cleavage planes. Principal fracture planes are arrowed.

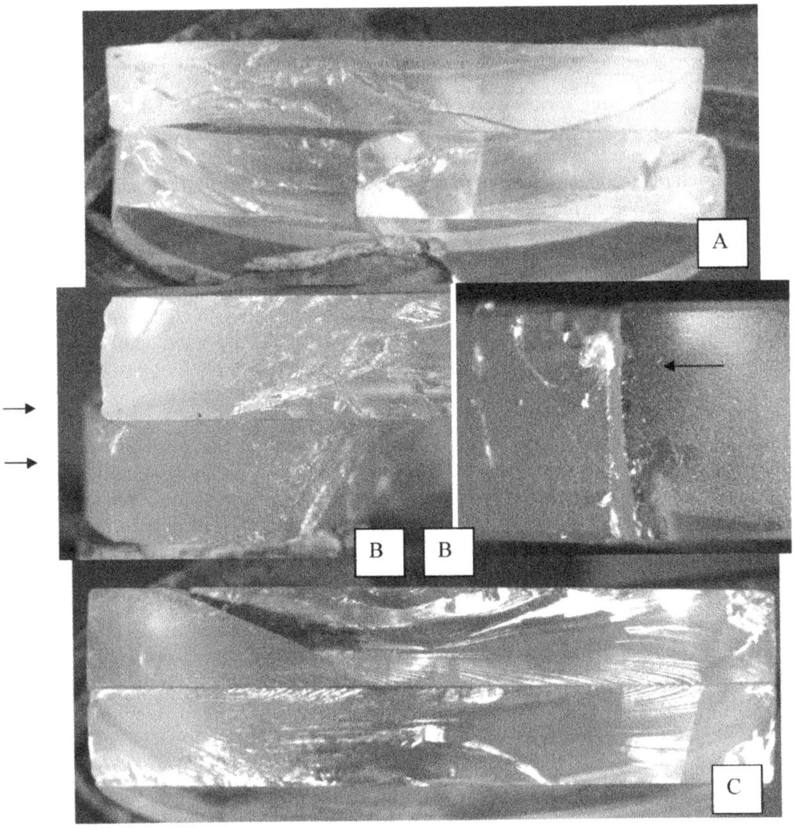

Figure 2. Principal fracture planes of (top) disc A with corner chamfer origin, (centre) disc B showing crush-type damage zone on periphery, and (bottom) disc C with periphery origin.

The failures (Figure 1) took the form primarily of diametral or near diametral primary cracks, in two cases with subsidiary cracking of varying degrees. The planes of the principal diametral fractures were canted at angles of approximately 12°, 27° and 32° to the disc normal in the three examples, correlating with the reported preferred (111) cleavage plane of the fluorite cubic structure,[5] but indicating no controlled orientation for the window. Transmission optical microscopy revealed no evidence for twinning or residual stress related birefringence.

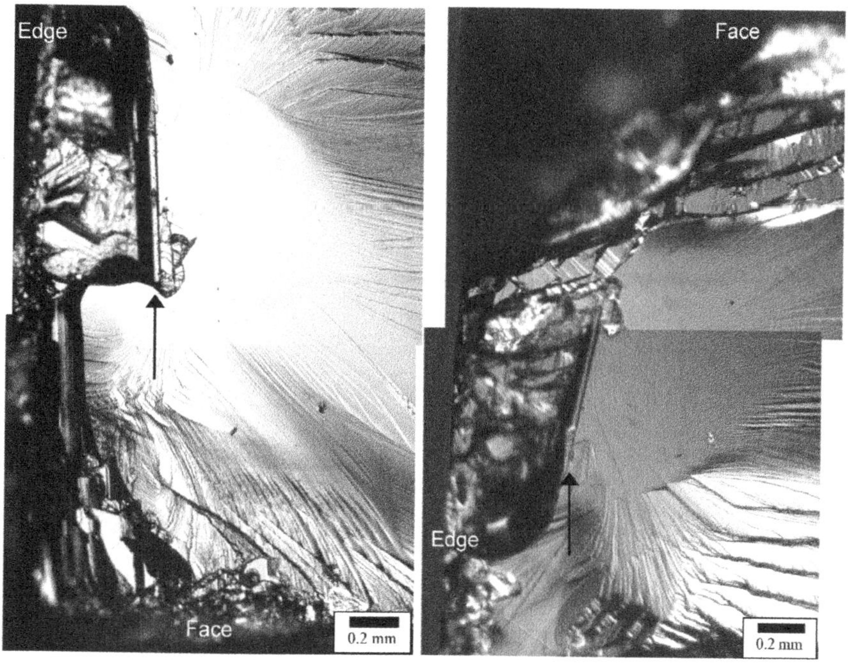

Figure 3. Matching halves of the fracture origin of disc C showing equivalent features. The fracture seems to initiate from a linear crystal defect (arrowed) close to the external wall, which could be a pre-existing cleavage crack from machining.

Visual examination using a macroscope suggested edge failures, producing the strongest and most distinct marking. Where the crack had moved immediately into the cleavage plane, the initiation zone was clear (Figure 2) and produced a distinct asymmetric mirror with radiating hackle, but in one example, disc A, this zone was rough and stepped, suggesting that the orientation of the stress and the initiating flaw were aligned steeply to the nearest cleavage plane. However, photomicrography was difficult owing to the material's transparency. The best detail could be obtained only by use of a conventional optical microscope with through-lens illumination to take best advantage of angular reflectivity off the surface, but only if the fragment could be carefully and slowly orientated under the lens to obtain the most appropriate appearance.

Examination of the origin locations on the exterior surface showed that each seemed to have a different cause. The item that split into two pieces (C) seemed to have some form of crystallographic damage, possibly a pre-existing crack, near, but not at, the edge chamfer (Figure 3). In the item that split into three fragments (A), the stepped fracture origin was at the chamfer, while the multi-fragment item (B) appeared to fracture from an indentation site on the circumference away from the chamfer, with missing flakes on both sides (Figure 4). It was also noted that the quality of surface finish, the direction of peripheral grinding, and the size of the chamfer varied markedly between the items.

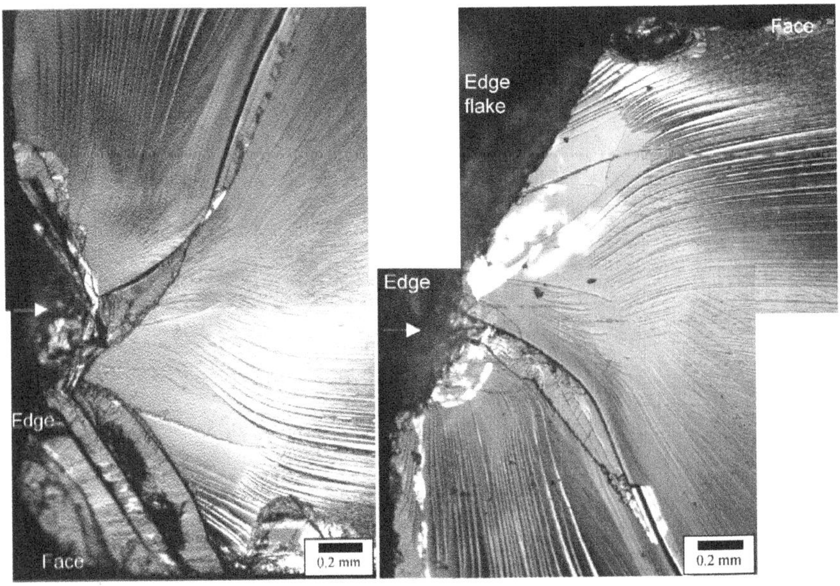

Figure 4. Matching halves of the fracture origin of disc B showing equivalent features. The site of the suspected periphery contact damage is arrowed.

Since CaF_2 is a good infra-red transmitter (typically >90% for 6 mm thickness over the wavelength range 0.3 μm to 7 μm [6]), for the most part the windows survive in this application. However, its thermal expansion coefficient (isotropic) is high, about 19 x 10^{-6} $°C^{-1}$ at room temperature, so mechanically significant thermal strains readily develop should temperature inhomogeneities arise. To estimate this, a simple model for the thermoelastic stress arising from a hot-spot of radius a in the centre of a disc of radius r shows that the edge hoop stress σ_t is tensile and of equal and opposite magnitude to the radial stress σ_r:

$$\sigma_t = -\sigma_r = \alpha \Delta T E \left(a^2 / 2r^2\right) \quad (1)$$

For a typical brittle material failure strain (σ_f/E) of 0.05%, the required temperature rise is only 300 °C. This appears to be eminently achievable in the apparatus should the window become slightly fogged with evaporation products from the furnace region and hence become infra-red absorbent, or even when the apparatus is heated quickly to very high temperatures. By comparison, mechanically induced stresses from O-ring clamping are likely to be rather smaller unless window on metal contact is developed, but these seem to be geometrically unlikely in the particular design.

It can be concluded that where such windows might be subject to significant thermal stress, care needs to be taken concerning the processing and final quality of the periphery and chamfer edge finish. These features cannot be deemed to be unimportant. Consideration needs to be given to edge polishing or to removal of damage by etching.

EXAMPLE 2 – LARGE GAS SEAL RINGS

A series of five failures of 300 mm diameter profiled gas seal rings in sintered silicon carbide were examined.[3] The rings had failed during over-speed proof-testing, and had failed at a variety of rotation speeds. Most of the fragments had been collected, although many suffered from subsequent impact damage. Once the fragments had been sorted and their relative locations identified it was possible to identify the crack patterns (Figure 5). By noting the direction of crack propagation on each fragment surface it was then possible to obtain a unique sequence of failure events. In particular, all rings had two or more possible fracture origins, so it was necessary to identify uniquely the sequence of events and ensure that they were consistent with the crack pattern.

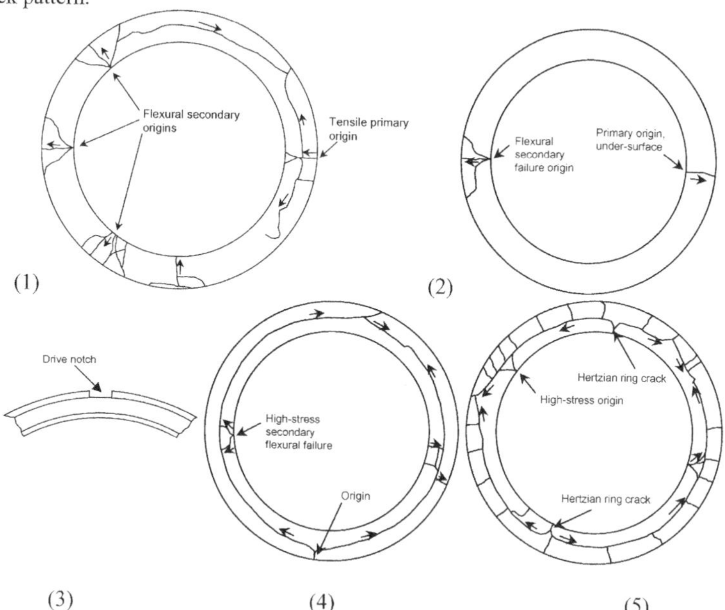

Figure 5. Crack patterns in five large gas seal rings.

A review of the stress distribution in spinning rings[7] shows that hoop tensile stress is a maximum on the inner surface of the ring, while there is a tensile peak in the radial stress near the centre of the ring width. In practice, in a narrow ring of the present geometry, the hoop tensile stress is only about 20% lower on the ring outside surface than on the inner surface. This stress distribution explains the crack pattern observed in the more highly fragmented rings where there appears to be a 'tyre' effect, with the ring fracturing along its centre of radius. While the hoop tensile stress will dominate at crack initiation, as soon as the hoop constraint is lost with propagation, the much smaller radial tensile component, which still exists because of the persistence of angular momentum, takes over and deflects the bifurcating elements of the crack to take a circumferential path.

Next, the nature of the fracture origins could be determined, and the initiating origin identified. In all the items examined, the crack patterns demonstrated multiple origins.

Case 1: At least three similar high-energy origins with crack bifurcation opposite a single planar origin suggests the latter to be the tensile initiating failure which then allowed the ring to open, and the remaining three developed as flexural fractures. In this case, all the flexural failures occurred from the inner ring radius, while the initiating defect was on the outer radius. The fracture origin (Figure 6) was missing as a result of chipping damage, but the radiating fracture markings indicated that it was located on one of the external chamfers. The large mirror radius indicated a low stress fracture compared with the much smaller mirrors from associated with the flexural mode failures opposite.

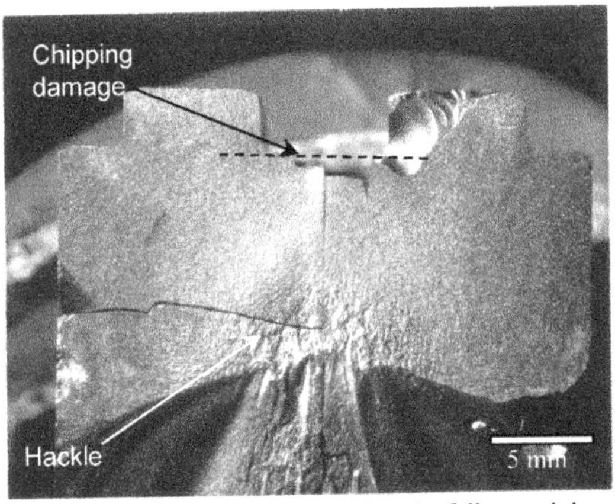

Figure 6. Gas seal ring Case 1, exterior chamfer failure, origin missing.

Case 2: This example had a single high-energy origin with crack bifurcation opposite a planar crack origin, the latter being the initiating site. A localised, strongly kinked origin (Figure 7) indicated a pre-existing crack, possibly from mechanical handling. There is some evidence of a contact region on the planar surface.

Case 3: Similar to Case 2, with failure from the outside radius. In this case the origin was clearly an axial machining groove located to one side of the drive notch, the radiating fracture marks with a focus beyond the immediate surface indicating the extended nature of the initiating flaw (Figure 8).

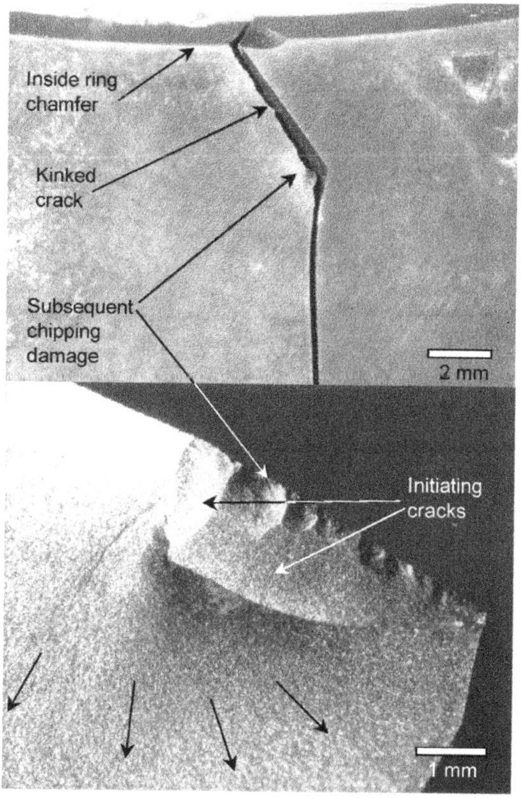

Figure 7. Gas seal ring Case 2. A pre-existing crack near the inner chamfer.

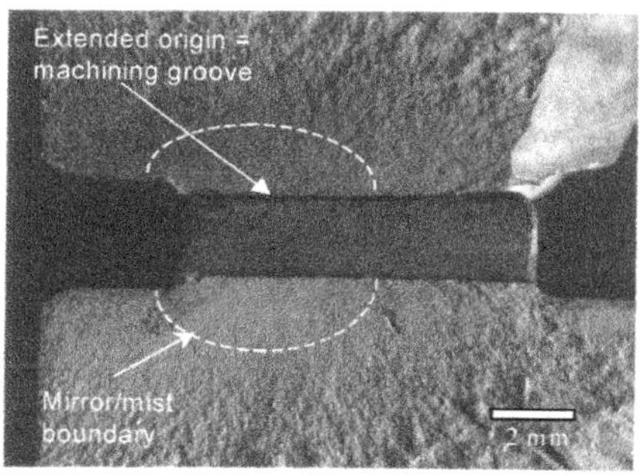

Figure 8. Gas seal ring Case 3. Drive notch failure from extended machining flaw.

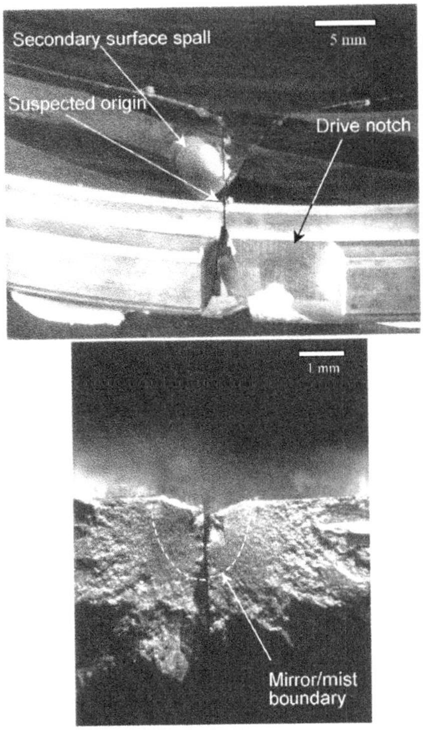

Figure 9. Gas seal ring Case 4. Failure from the external chamfer near a drive notch.

Case 4: An initial bifurcating crack from the outside radius running inwards and then circumferentially in both directions to initiate flexural failures (Figure 9). Despite a small amount of chipping damage, the origin was clearly located on an external chamfer producing a medium sized mirror.

Case 5: The highly fragmented crack pattern indicated at least four potential origins on the inner radius requiring individual identification to establish which initiated the fracture process. Careful examination showed that this particular ring had two regions of handling damage on the same face opposite each other across the ring diameter. One of these had produced a clear cone crack of fracture initiation (Figure 10). It appeared that there had been high-load blunt contact at these two points during handling. The other origins were much smaller and were more discrete machining flaws.

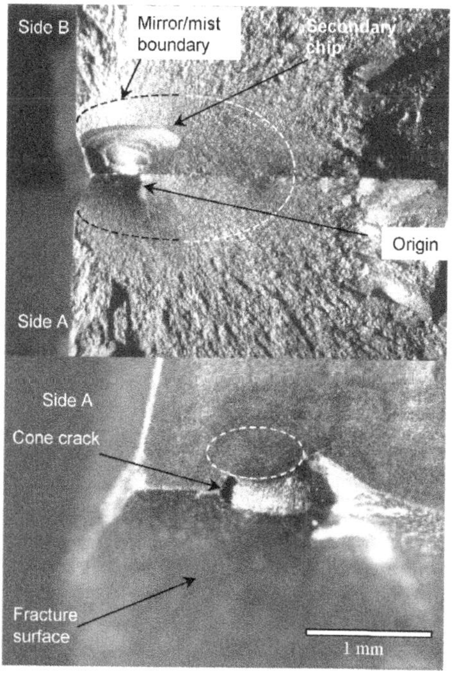

Figure 10. Gas seal ring Case 5. Failure from a ring crack from mechanical contact.

The variety of fracture origins found in these rings indicated a clear need for very careful handling of such large monolithic components that were to be subject to high tensile stress. Of special concern were the contact damage type cracks which need to be avoided. The clear necessity of over-speed proof-testing for this type of product had been demonstrated, although this would only provide some assurance immediately at the end of the test and not, of course, protection against further mishandling in transport and installation.

EXAMPLE 3 – FAILURES IN CERAMIC FEMORAL HEADS

Although ceramic femoral heads provide an excellent joint-replacement option for the surgeon, the minimal wear debris produced reducing the risk of fixation loosening and other joint problems, there is a continuing low percentage of premature failures in vivo. This is occurring despite manufacturers' attempts to provide assurance through pressurisation proof-testing and through testing programmes for head type approval. One of the continuing issues is the fixation of the ceramic head onto a metallic stem via a taper geometry cone fit, the slight deliberate taper mismatch employed being designed to ensure that the first contact is made at the innermost end of the taper to ensure locking, and the contact then spreads further along towards the bore mouth as the axial force is increased and the metallic stem deforms to conform to the ceramic taper profile. While this geometry generally gives satisfactory ultimate fracture force and fatigue behaviour during axial in vitro testing, the loading in vivo is not axial, but at a variable angle of physiological inclination. It also does not normally achieve such high force levels as required to cause failure in testing, and the metallic contact is thus less homogeneous and less widely spread. Understanding performance in such conditions is not trivial, and there remains much work to be done to reveal the true in vivo stress distribution and the consequent failure risk.

The fractography of axially loaded laboratory test heads is usually straightforward. The maximum tensile stress developed in axial loading is usually associated with the bore region of the head. In geometries in which the stem contact region is in the geometrical centre of the head (so-called "short-neck" designs), the maximum stress is generally distributed in the centre of the head, and most failures occur from bore machining flaws, or internal defects very close to or intersected by the machined bore surface. The principal fracture plane is axial, but in high-energy failures, there is typically strongly hackled multiple crack bifurcation. Fractures opposite this initial split tend to show signs of a non-axial flexural nature due to the wedging effect of the test stem down the head bore, with compression curl close to the outside surface usually being evident. In geometries in which the stem contact region is towards the bore mouth (so-called "extended neck" designs), the hoop tensile stresses are a maximum around the bore mouth. Fracture origins tend to be of a more variable type, and can be associated with the bore machining, the bore mouth chamfer (machined green or fired), the flat region around the bore (machined green or fired) or even the external taper (usually green machined) on which head coding details are engraved mechanically or by laser etching. Fracture initiation at high energy is again usually axial, but bifurcation tends to lead to confusing crack patterns unless one seeks the geometrical symmetry of fracture surfaces. Overall, from the fractographic point of view it is uncommon to lose completely information on the flaw type except perhaps in the most explosive of cases.

In contrast, fractography of in vivo failures is more problematical. Typically these are low energy failures with only a small number of initial fragments, often only two or three. They tend to remain located in the femoral cup, pinned by the stem which has moved up into the head crown. The patient often does not appreciate that failure has occurred, and continues to conduct normal life until an unusual movement or further fragmentation releases one or more fragments, at which point the abnormality is painfully noticed. By this point, the fragments retrieved from the revision operation have suffered extensive post fracture damage in the form of extensive bore edge chipping and mutual rubbing of fracture surfaces. Depending on the post-fracture duration, it may or may not be possible to see any useful original fracture features on the fragment surfaces. A further problem with low-energy failures is that the mirror region tends to be very large, especially in alumina heads, and typically of at least the dimension of the head. There is almost no initial radiating hackle to identify even the location of the origin. Much greater reliance than usual has to be placed on the homogeneity of the metallic bore marking, although even this can be difficult because of post-fracture contacts with the metallic stem, and with the revision surgeon's efforts to remove fragments. Also, typically, the stem is not available for examination; it often remains in the patient, with a replacement metallic head being applied.

For commercial and legal reasons it is usually not possible to discuss in vivo failures publicly. Furthermore, a fractographic examination usually cannot find definitive reasons for the premature failure to assist a legal case. The following example, however, can be discussed. It is unusual for the reason that explanation occurred very soon after the fracture with minimal fragment damage, and thus provides an excellent case study.

In this particular case, the head underwent a medium energy fracture with a clear initial axial split to the head. The hackle markings on both side of the origin remain clear, as do the apparent origin locations, which were initially thought to be on the bore surface near the bore mouth. However, it was noted that there was a distinct, near-annular, 'shiny' region of fracture surface surrounding a near-circular duller region inside the body of the head adjacent to the suspected fracture origin (Figure 11(a)). In addition, the focal origin of the radiating hackle comes from this circular region (Figure (11b)), and not from a location by the bore surface.

(a)

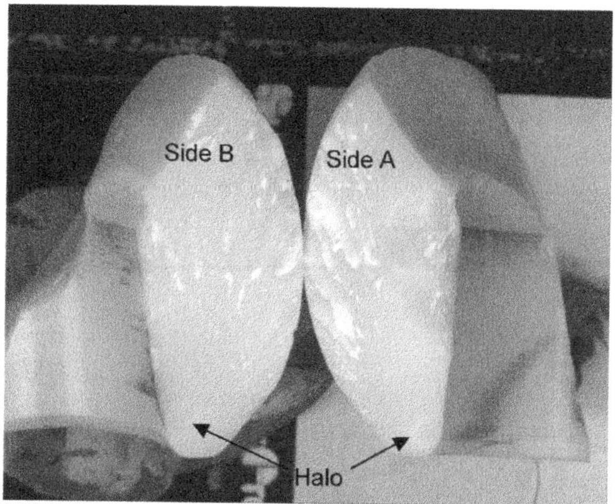

(b)

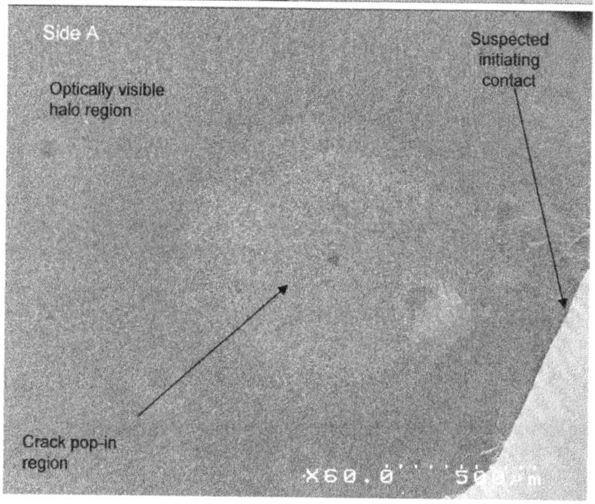

Figure 11. Explanted in- vivo femoral head failure (a) showing an unusual shiny halo of subcritical crack growth, and (b) this region seen by scanning electron microscopy.

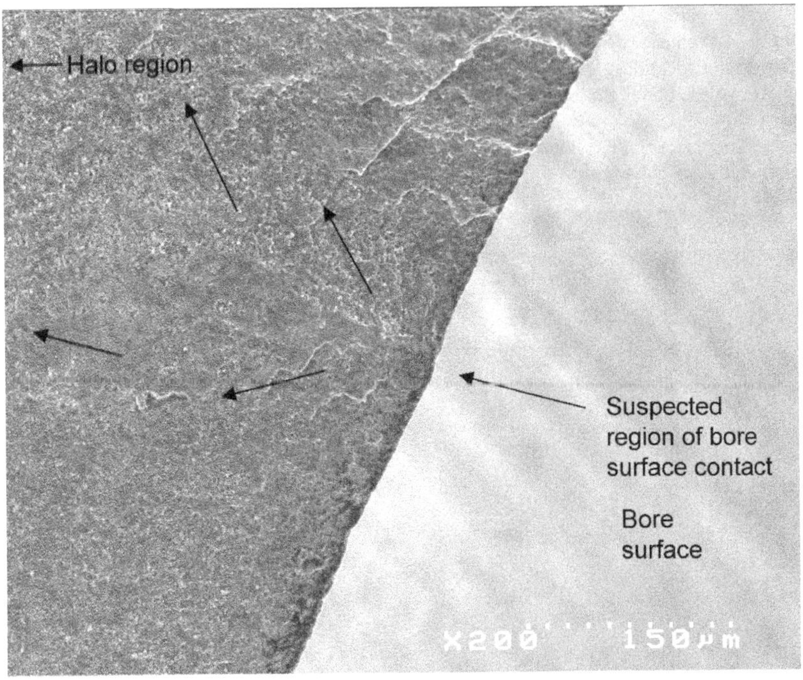

Figure 12. Suspected fracture-initiating site on the bore surface giving a 'flow' pattern towards the halo region.

Such shiny halos are common around fracture origins in fine-grained high-purity alumina, and have been seen frequently on laboratory test heads. They can be used to locate clearly the precise origin, even with the naked eye. They are usually associated with a region of subcritical crack growth from the initial defect in which the fracture mode tends to be mixed intergranular and transgranular, compared with fast fracture or crack pop-in regions which are almost overwhelmingly intergranular. The flat transgranular regions give the additional directional reflectivity when illuminated by an off-axis light source.

It is recognised that delayed failures of implanted heads must occur through some form of subcritical crack growth, but the question is: how do the necessary in vivo stresses arise that allow it to occur from defects which have survived internal pressurisation proof-testing? In the present case, how does this curious circular internal region of intergranular fracture develop, and from what? One possibility is that the growth occurred from a surface-initiated defect. The localised 'flow' fracture markings (Figure 12), which appear to have propagated from a small kinked region of fracture in the bore surface, are reminiscent of a localised contact-initiated crack, 'popped in' to the bulk of the head. Had this occurred before or during surgery, it is conceivable that under patient loading a halo region of crack growth could then have occurred from the popped-in crack periphery. The next question concerns how such damage could have been obtained in the head bore. Possibly, it could have occurred during the proof-test procedure by accidental handling. Equally it might have happened during surgery as the head was located onto the stem, especially if it was struck at an inappropriate angle before being properly seated on the taper. From the fractographic point of view, we have clear evidence of a popped-in event followed by a growth phenomenon occurring in vivo, although the exact shape of this popped-in region remains difficult to explain in detail.

As mentioned above, similar growth regions have been seen in laboratory test heads. In the next example of an alumina head, the fracture origin is clearly from the inner bore region, but the appearance of the subcritical growth region is unusual. It appears to be a series of small regions that develop between individual locations on the head bore, correlating with the local

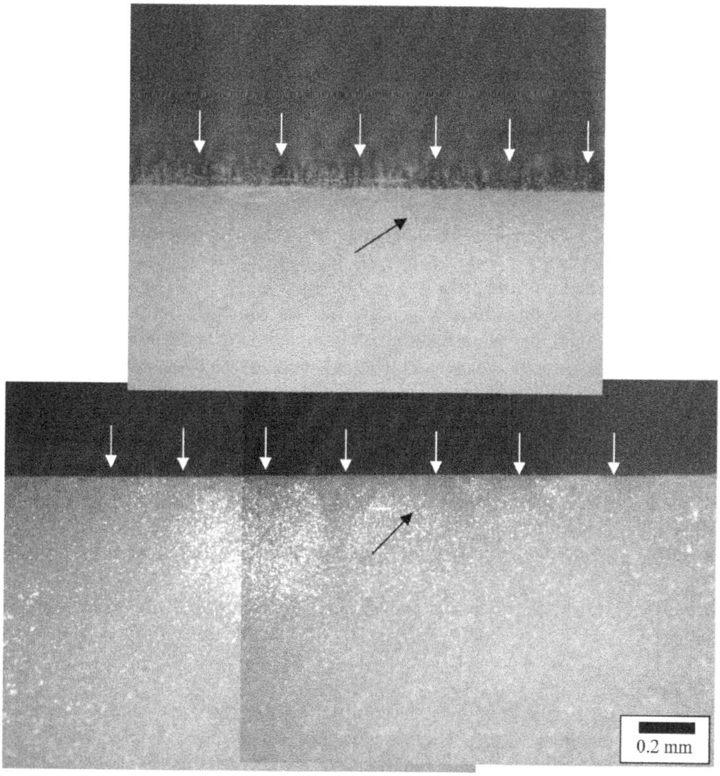

Figure 13. (Lower) small regions of transgranular fracture in an alumina test head loaded with a cylindrically ridged Co-Cr alloy stem, correlating (upper) with the location of the stem contact lines (white arrows). The black arrows identify the same faint twist hackle line in both images.

ridged-stem rings of contact (Figure 13). Note that these actual locations produce dull less-reflective areas, suggesting intergranular 'pop-in' type features, with the shiny subcritical crack growth regions between. The commercial use of deliberately ridged machining of stem tapers is intended to provide locations of high stress in the stem which yield more readily, and thus allow the stem to conform more readily to the ceramic surface than if the contact was spread more widely. However, in this case using a hard Co-Cr alloy stem, the localised contact pressure associated with the series of ridges appears to have induced a series of crack pop-in regions, the largest of which subsequently determined the fracture force in the test. This particular item clearly demonstrates the limitations of head/stem contact mechanics in determining the mechanical stress resistance of femoral head systems. Some reconsideration of the contact philosophy could be needed.

CONCLUSIONS

The examples of fractographic studies undertaken over the past few years and described above demonstrate a number of issues for the fractographer that take the skill requirements beyond those needed for simple test-bar fractures.

- There is a need to keep a completely open mind on what happened, and not to jump to premature conclusions.
- Use the eyes first, an optical macroscope second, and an optical microscope third – don't bother with the SEM unless you need a really local explanation. Evidence may be destroyed preparing a test sample for the SEM.
- Crack patterns must be interpreted correctly, such as which fracture directions are possible and which are not, and the sequence of crack branching and pattern development. One must also decide what is original damage and what is secondary damage, and ignore the latter.
- As much circumstantial evidence as possible should be obtained to help rule out inappropriate yet conceivable scenarios.
- Understanding of possible stress distributions in the component at the point of failure and immediately thereafter is important for explaining and correlating with crack patterns.
- Use Occam's razor (*i.e.* investigate and eliminate the simple before looking at the complicated).
- The resulting hypothesis must match all the facts. If there are few facts, the hypothesis will be inconclusive.
- Don't get fixed on existing personal knowledge and experience – there's always something new to learn!

ACKNOWLEDGEMENTS

The author would like to acknowledge permissions to publish the above examples from a variety of sources including: Steve Hughes, Morgan Advanced Ceramics Ltd., Rugby, UK; Ingrid Milosev, Jozef Stefan Institute, Ljubljana, Slovenia; Ray Wallis, John Crane UK Ltd., Slough, UK. In addition, the author would like to acknowledge the collaboration and corroboration of conclusions with George Quinn of NIST, USA.

REFERENCES

[1]J. Swab and G.D. Quinn, "Fractography of Advanced Structural Ceramics: Results from the VAMAS Round Robin Exercise," US Army Technical Report, ARL TR-656, (Dec 1994). Also published as VAMAS Technical Report No 19.

[2]ASTM Standard C1322-02, "Standard Practice for Fractography and Characterisation of Fracture Origins in Advanced Ceramics", Annual Book of ASTM standards Vol. 15.01 (2006).

[3]R. Morrell, "Fractography of Brittle Materials," NPL Good Practice Guide no. 15, National Physical Laboratory, Teddington, UK, (August 1999).

[4]CEN Standard ENV 843-6, "Advanced Technical Ceramics- Monolithic Ceramics – Mechanical Properties at Room Temperature – Part 6: Guidance for Fractographic Investigation."

[5]G.D. Quinn, NIST Recommended Practice Guide, "Fractography of Ceramics and Glasses," Special Publication 960-16, NIST, Gaithersburg, MD, 20899 (2006).

[6]Saint-Gobain Crystals website, Photonic Materials at http://www.photonic.saint-gobain.com/media/documents/S0000000000000001020/CaF2_Excimer.pdf

[7]S. Timoshenko, "Strength of Materials, Part II Advanced Theory and Problems", Third Edition, R E Krieger Publ. Co., New York, (Reprinted 1976).

ADHESIVE INDUCED FRACTURE OF AUTOMOTIVE GLASS

Thomas M. Cleary
Automotive Components Holdings, LLC
17333 Federal Drive
Allen Park, MI 8101

R. Terry Nichols
RTN Services
19850 Merriman Rd.
Romulus, MI 48174

ABSTRACT

A fracture mechanism that was prevalent in several different types of automotive glass has been identified. Fractography played an essential role in determining the cause. The root cause of the fractures was traced to a rigid adhesive bond between a polymeric molding and the edge of the glass, and subsequent stress induced by thermal expansion mismatch of the two materials during temperature fluctuations in service. The tensile stress induced on the glass (by the high expansion molding via the adhesive) has been found to be sufficiently high to overcome the surface compressive stresses of tempering and cause component fracture. The failure mode was replicated in the laboratory by thermal cycling of test parts. Furthermore, it was found that the severity and type of the fracture formed was a function of the location of the adhesive.

INTRODUCTION

In a typical modern vehicle, glass comprises approximately 15-20% of its exterior surface[1] and is becoming more and more prominent in vehicle design and functionality. Looking forward, there is an apparent trend towards increased glass usage[2,3] with greater complexity as glazing suppliers are challenged to provide products with more intricate shapes, tighter tolerances, more specifications, and increased content.[4] Performance features such as solar control, acoustic performance, bonded on brackets for sensors and self-defrosting windshields are becoming more prevalent, making supplied glazing more systems rather than simple components of the vehicle. As glazing becomes more complex and the value-added content increases, it becomes increasingly important for manufacturers to understand and anticipate potential failure modes of these systems.

Glass breakage is, by far, the most common cause for automotive glazing replacement and understanding the exact nature of the fracture is imperative in its reduction. While the vast majority of breakage can be attributed to road hazard damage (stone impacts, etc.),[5] in order to meet customer's expectations for improved quality, it is important to understand as many causes of breakage as possible. Analysis of fractured field parts is an invaluable tool to gain insight as to the various causes of breakage events. This paper documents several unique fracture cases with similar root causes and the methods used to replicate the fractures in a controlled laboratory environment.

ANALYSIS OF FIELD PARTS

Several different cases are presented below. In each case, the under-lying cause of fracture was determined to be the same; however, the fractures manifested themselves in unique and different ways. Prior to discussion of the cases, it is beneficial to understand the vehicle glazing systems and some related terminology.

A typical automotive windshield is comprised of two pieces of annealed glass joined together by an intermediate layer of poly-vinyl-butyral. For nomenclature purposes, a simple numbering system is employed to denote the four surfaces of the discrete glass panes. The surfaces are increasingly numbered from the outside of the vehicle to the inside. As such, the outer-most exterior surface is referred to as "surface 1" and the inner-most surface is denoted as "surface 4". In many applications, the windshield has a "U-shaped" lip-molding encasing the peripheral edge of the part. The purpose of this molding is to conceal the edge of the glass while providing an aesthetic transition from glass to vehicle body. To install a windshield into a vehicle body opening, typically an isocyanate-based primer is applied to the number four surface of the windshield and a subsequent polyurethane bead is placed on top of the primed surface. The windshield is then decked into the body opening. The primer and polyurethane systems act to bond the windshield in place and provide sealant from wind and water. Figure 1 shows a cross-sectional schematic of a windshield with attached molding and bonding components.

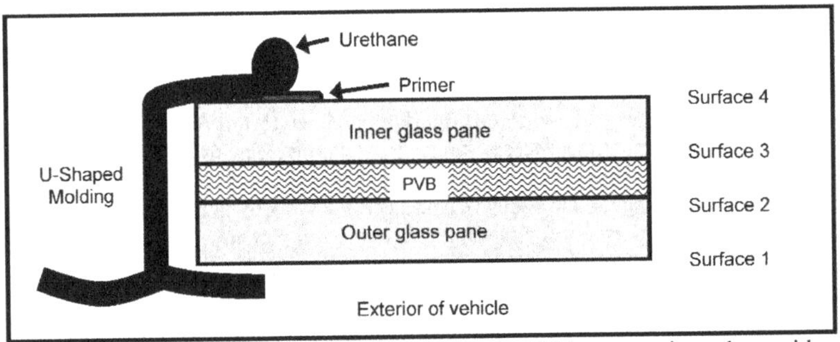

Figure 1: Schematic denoting nomenclature used for windshield constructions along with a "U-shaped" peripheral molding and various bonding components.

In the case of traditional single ply tempered safety glass (commonly used for body glass applications), the surfaces of the single glass ply are referred to simply as surface 1 (exterior) and surface 2 (interior). In this application, the primer and polyurethane are applied to surface number two (as opposed to surface four in the case of windshields).

CASE 1: Analysis of Single Lead Fractures in Windshields

Multiple occurrences of single-lead fractures were found on windshields of several different types of vehicles. The fractures were discovered at relatively low time in service (i.e., less than 6 months of usage) and in some cases occurred prior to the vehicles being shipped to dealerships. These fractures were at first, somewhat mysterious because they were not accompanied by evidence of mechanical damage due to impact or scratches on the glass surface. One common characteristic observed was the presence of a hardened black material on the number one (exterior) surface of the windshields in the vicinity of the fracture. Chemical analysis confirmed that the black substance was windshield primer that had contaminated the number one glass surface. To more closely examine the fracture, the polyurethane was trimmed away and the U-shaped lip molding was removed. In each case it was observed that the molding was very difficult to remove and that a glass chip was adhered inside the molding. Figure 2 shows a photograph of a representative fracture with the associated chip in the molding. Upon examination of the cross-section of the fracture, Wallner lines and other fracture markings indicated that the break had initiated at the edge of the windshield where a V-shaped chip was missing from the outer glass pane. This missing chip corresponded to the chip found adhered to the inside of the molding. Figure 3 shows a photograph of the windshield edge where the fracture originated.

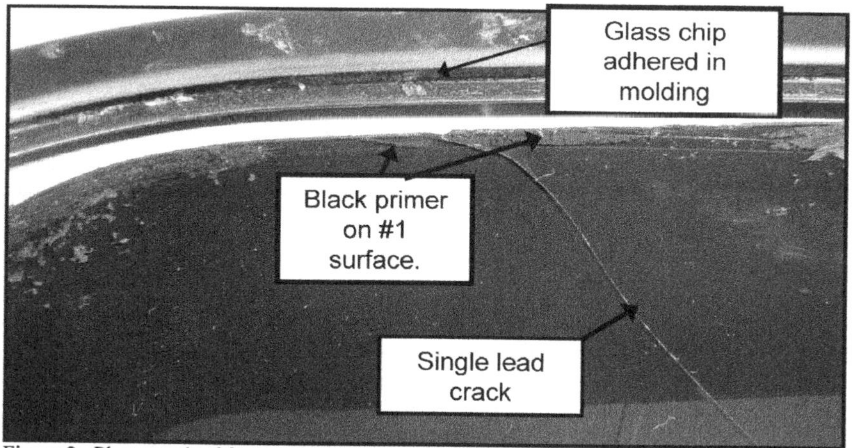

Figure 2: Photograph of fractured windshield. Note the presence of the black primer on the outer (#1) surface and the glass chip found in the molding.

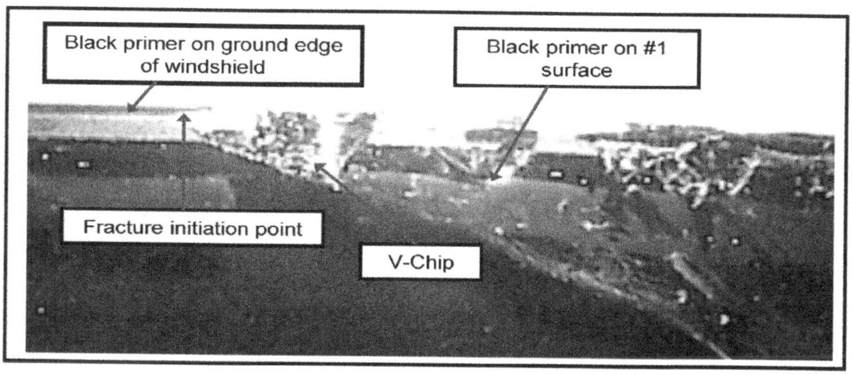

Figure 3: Photomicrograph of the edge V-chip that initiated fracture. Note the presence of black primer on both the #1 surface and on the ground edge of glass at the crack origin.

The presence of the chip inside of the molding was evidence that the damage occurred during or after installation of the molding (i.e., the windshield was not damaged during its manufacture or in handling prior to assembly of the molding). Given the found evidence, a fairly complex hypothesis of fracture was developed. It was believed that the cause of the fracture was that the liberally applied glass primer had carried over onto the edge of the windshield and had rigidly bonded the U-shaped lip molding to the edge of the windshield. Subsequent thermal expansion mismatch between the glass and molding caused the V-chip to be pulled free from the edge of glass. This V-chip substantially weakened the glass and subsequent stresses (mechanical, vibrational and/or thermal) applied to the windshield caused the single lead to propagate. While at first it may be difficult to conceive that a plastic molding could be responsible for tearing a chip free from glass, it becomes a plausible explanation when the difference in thermal expansion between the glass and plastic molding (approximately 10X) is considered. Subsequent laboratory tests validated the hypothesis (see section below on lab testing).

In a similar case, another windshield product line was found to have single lead cracks induced not by windshield primer, but rather a rigid adhesive system that was purposely applied by the customer to bond the molding to the windshield edge. In this case, the formation of the chip by the adhesive was even more apparent. Figure 4 shows a close-up of the fracture induced on the edge. In Figure 4, repetitive arrest lines are evident, perhaps as the result of numerous thermal cycles needed to grow the crack.

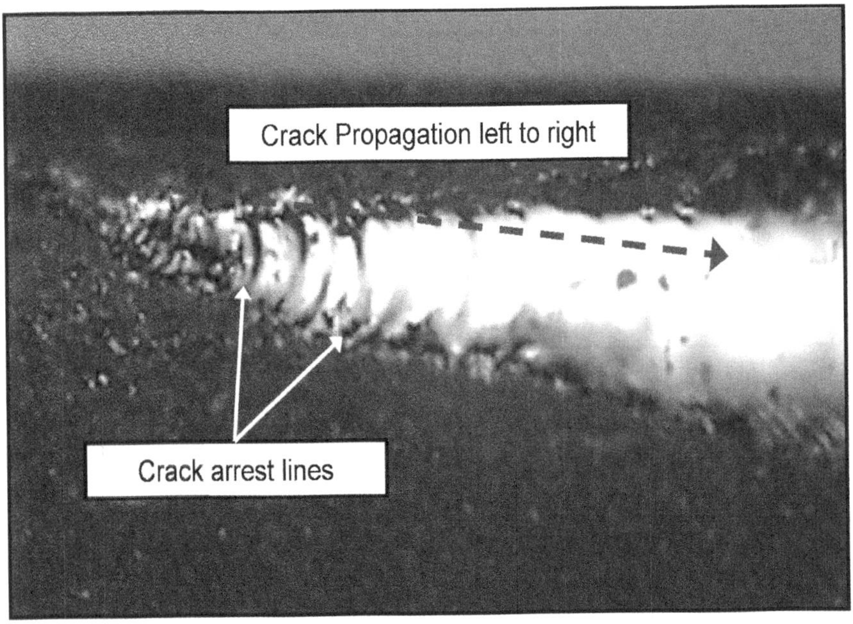

Figure 4: Photomicrograph of an adhesive induced chip on the ground edge of a windshield.

CASE 2: Analysis of Spalling Chips on Enameled Surfaces of Windshields

In this case, numerous windshields were found with a cosmetic anomaly in the enameled region of the parts. In each case, the anomaly was found to be near the edge of the windshield adjacent to the edge of the U-shaped lip molding. Initially, the anomaly was perceived to be bubbles present in the poly-vinyl butyral (PVB) interlayer of the laminate windshields. In most cases, the anomalies were found to occur at the top or bottom corners of the windshields. Figure 5 shows a photograph of one of these cosmetic anomalies.

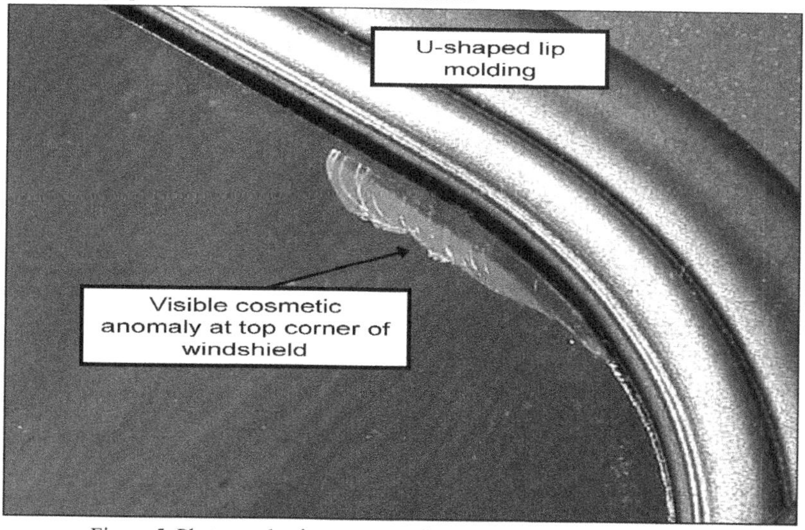

Figure 5: Photograph of a representative cosmetic anomaly found.

Upon dissection of the parts it became evident that the anomalies were not bubbles in the vinyl, but rather shallow chips spalled from the number four enameled surface of the windshield. The spalled chips were found to be bonded to the U-shaped lip moldings by windshield primer. The mechanism was virtually the same as detailed in case one above, with the exception that the windshield primer was on the number four, enameled surface of the windshield (as opposed to being on the edge of the windshield as in case one above). Figure 6 shows a schematic representation of a spalled chip bonded to the molding. Numerous crack arrest lines were evident in the spalled chips which again may be evidence of the repetitive thermal cycling causing the chip to propagate.

A review of the assembly process gave further insight as to how the chips were formed. In the process, the U-shaped molding was first applied to the windshield edge and then used as a locating guide for the primer application. This practice virtually assured intimate contact between primer and molding thus allowing the primer to wick under the molding and form a bond between the enameled number four surface of the windshield and the molding. The prevalence of the chips to form at the corners of the windshield was likely due to the fact that during priming, the operators would have to reduce the application rate at these transition points, thus resulting in a larger puddle of primer to be formed at these regions. Additionally, it is known that the enameled surface of the glass is weaker than non-enameled areas[6] thus allowing for easier fracture of the glass when subjected to stress. This failure mode was also duplicated in the laboratory (see section below on lab testing).

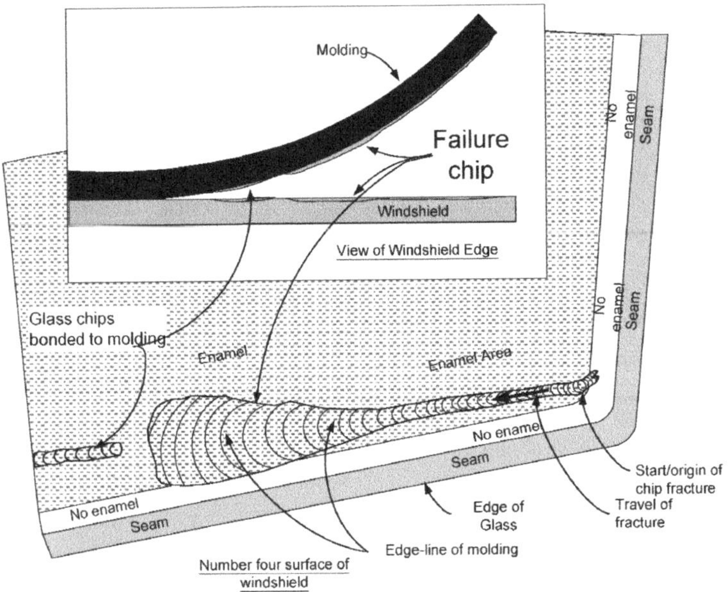

Figure 6: Schematic of a spalled chip created on interior surface of windshield (figure courtesy of Roger Johnson).

CASE 3: Occurrence of Shattered Tempered Safety Glass

In the third case, several tempered rear-windows, or backlites were found to have shattered. In each case, the origin of fracture appeared to be near the top corners of the parts. This was evidenced by the divergence of the fracture lines from this point. Figure 7 shows a photograph of one of the fractured backlites.

Analysis indicated that the fracture initiated underneath the U-shaped molding. The glass fragments near the molding showed that the black ceramic enamel on the inside (#2) surface of the glass had been cleaved free of the glass surface and were still bonded to the molding by the urethane primer. It appeared that the primer over-lapped the molding and had migrated underneath the molding to form a bond. This was evident in all samples analyzed. Figure 8 shows a photomicrograph of the number two surface of an origin fragment with the ceramic enamel missing. In this figure, crack arrest lines are evident which indicates the crack was moving from left to right in the photo and that the crack stopped and restarted several times before reaching critical size. Initially the crack ran parallel to the surface of the glass resulting in glass "spalling" or chipping. Eventually the crack turned in towards the body of the glass at which point it reached the tensile-core of the part and caused the part to shatter. The arrest lines shown in Figure 8 suggest that this is a repetitive and time-dependant process.

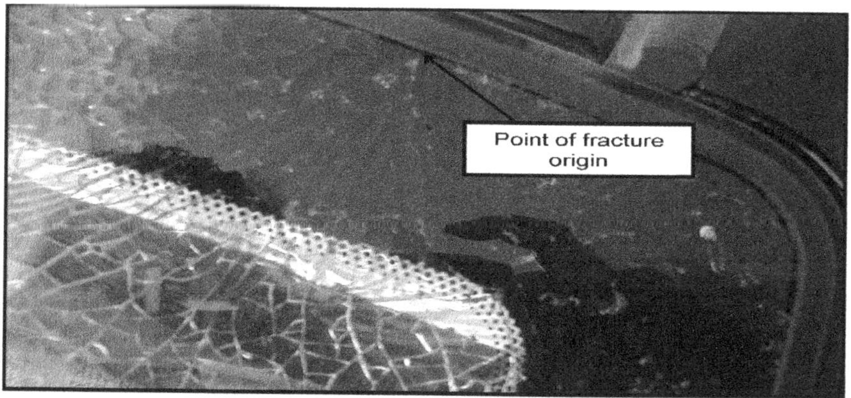

Figure 7: Photograph of shattered backlite. Note how the fractures diverge from a point near the peripheral molding.

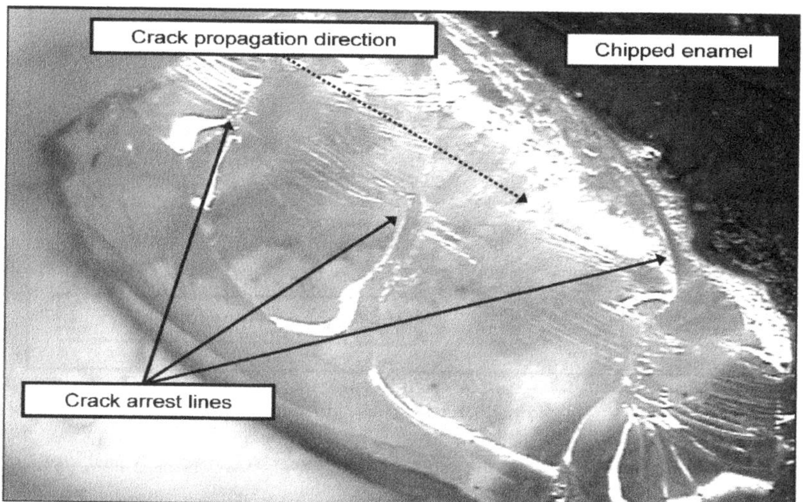

Figure 8: Photomicrograph of the #2 surface of the origin fragment. Note that the black ceramic enamel is torn free of the surface. Crack arrest lines indicate the fracture was moving from left to right and that it stopped and restarted several times.

The actual fracture origin itself was somewhat atypical in its features in that it lacked the classic "mirror-region"; however, features present indicated the crack initiated and diverged from this location. Figure 9 shows a cross-sectional view of the fragment from Figure 8. In Figure 9, it is evident that the fracture initiated as a surface chip that propagated from left to right in the photograph. The crack then turned inward towards the core of the glass at which point it encountered the tensile region of the tempered glass. The fracture then spontaneously bifurcated into the typical diced fracture pattern of tempered glass. Wallner lines at the bottom left of the photo indicate that the crack was moving from right to left. Additionally, mist & velocity hackle on the right indicate that the crack moved from left to right and that the crack had reached terminal velocity. This evidence indicates that the fracture diverged from this location. Similar markings were evident on several of the samples analyzed. The fracture mechanism was once again attributed to the rigid bond formed by the urethane primer and the U-shaped molding and subsequent thermal cycling of the parts. It is truly remarkable, and perhaps incredulous at first, that the tensile stress induced by this system can be substantial enough to overcome the compressive residual stresses created by the tempering process used in manufacturing these parts. Strength degradation due to the enamel likely aided in the facilitation of the fracture.

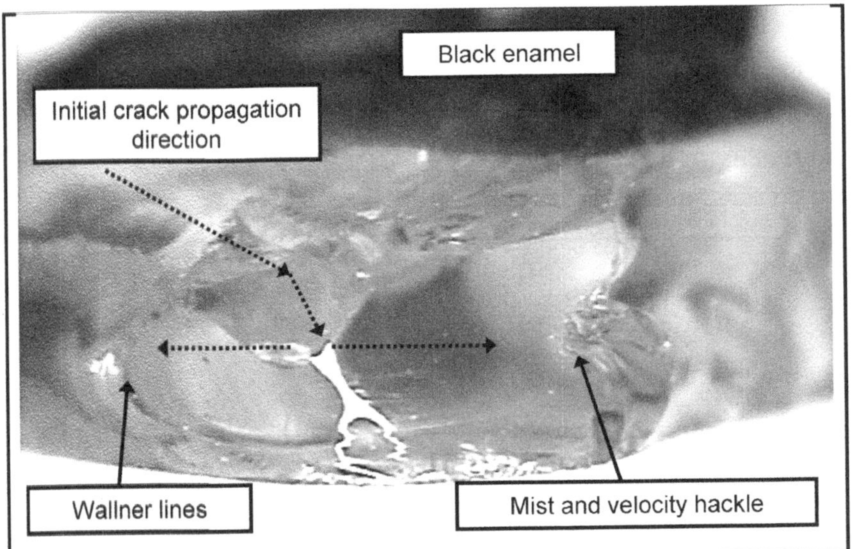

Figure 9: Cross-sectional view of the fracture origin for one of the tempered parts. The dashed arrows indicate the propagation path of the fracture. The fracture started at the enamel (top of the photomicrograph) and then proceeded into the tensile core of the glass. Wallner lines on the left indicate that the fracture was moving from right to left while mist and velocity hackle on the right indicate the fracture was moving from left to right at terminal velocity.

LABORATORY DUPLICATION OF FRACTURES

Assorted laboratory tests were conducted to verify the hypothetical failure mode. Tests utilized glass coupons, full windshields, backlites and several competitor products to verify the failure mode.

Coupon Tests

For the initial tests, 6 inch x 6 inch (non-tempered) glass specimens were used as bonding substrates for the adhesive (urethane primer)-plastic-glass system. The glass samples were not enameled for the tests. The glass and molding materials were cleaned and primed with the designated materials used in production processes. The primer was applied in liberal amounts in approximately 2 inch intervals on the glass surface adjacent to the edge of the substrates. The plastic strip was placed on top of the primer while it was still wet so as to facilitate bonding of the materials and allowed to cure for 30 minutes. The specimens were then placed inside a drying oven operating at approximately 82°C and allowed to soak overnight. The next morning the samples were transferred directly to a chamber operating at -20°C and allowed to soak for thirty minutes. When the specimens were removed it was found that the primer/adhesive system had spalled large chips from the glass surface at each of the bonding locations. The chips were rigidly adhered to the plastic strip. Figure 10 shows a photograph of a substrate with the spalled chips on the polymer strip.

Figure 10: Photograph of test specimen showing the large spalled chips resulting after thermal cycling.

While it is recognized that these conditions of thermal exposure were extreme, the test provided validation that given the proper combination of thermally and mechanically induced stresses, this combination of materials is susceptible to fracture.

A second test using 12 inch x 12inch thermally tempered glass coupons and strips of actual U-shaped molding was conducted using a similar thermal cycle. In this testing, several thermal cycles were required to generate chipping. While the parts did not fully shatter, the test again provided further evidence to support the failure mechanism theory.

Full Windshield and Backlite Tests

For this testing, an assortment of windshield and tempered backlites, including competitor products was used. The samples included several different enamel compositions and were produced on several different manufacturing processes. The samples were prepped with production materials in a manner consistent with production processes. The samples were allowed to cure and placed in a thermal chamber where they were repetitively cycled (10 hot-cold cycles) from 90°C to -20°C. The parts were then observed for chipping. Virtually all of the samples showed some degree of chipping or fracture. While none of the tempered parts fully shattered, it is easy to conceive that if the chips had propagated into the tensile core of the glass, spontaneous bifurcation of fractures could have occurred. Figure 11 shows a photograph of one of the windshield samples with the associated chipping post testing.

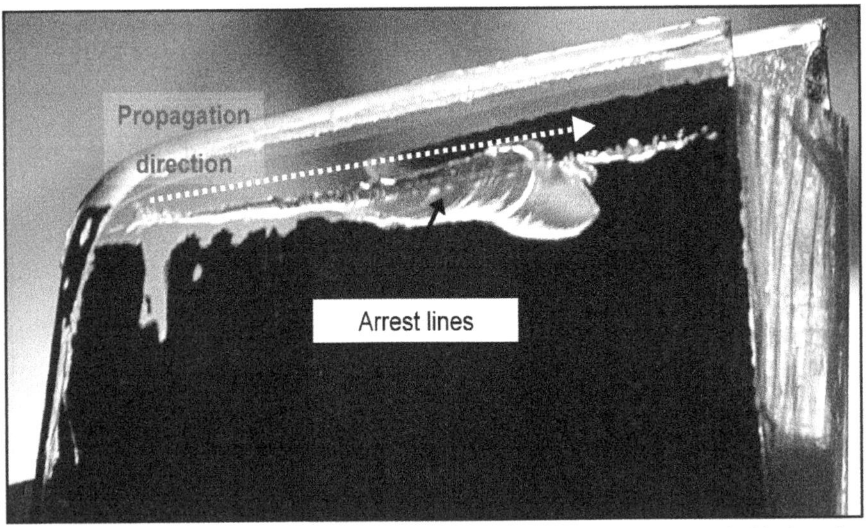

Figure 11: Photograph of a windshield test sample showing the adhesive induced chipping after thermal cycle testing. Crack propagated left to right in the photograph. View is of the #4 (interior), enameled surface. (Photograph courtesy of W. Kern).

DISCUSSION

The evidence gathered through the analysis of field parts as well as through laboratory testing confirms that the combination of materials and subsequent thermally induced stress can cause fracture of the glass. Perhaps contributing to the condition is the fact that the dark molding material absorbs heat at a much more rapid rate than the glass. Additionally, it is believed that the fracture is likely induced during the cold cycle stage as the polymer is much more rigid at lower temperatures and therefore can more effectively transfer stress to the glass. While it may be somewhat surprising that the glass surface gives way before the adhesive bond is severed, an early glass decorating technique know as "glue-chipping"[7, 8] provides further support to explain the fracture mechanism theory.

Given that the fracture was found and recreated on both tempered and annealed glass; and on a variety of different enamel compositions (as well as un-enameled glass), it would appear that the actual manufacturing process of the glass components did not play a critical role in the formation of the fractures.

Moving forward, as automotive glass products become more complex and include more value-added content (which is often bonded to the glazing products), awareness of this failure mode will become more and more important as new materials and bonding methods will be employed. Proper selection of materials and careful assembly processes will be necessary to avoid failure of products via adhesive induced stresses.

SUMMARY & CONCLUSIONS

Fractography of field parts was conducted and revealed a unique failure mode for both annealed and tempered automotive parts. Evidence from the fracture analysis indicated that the underlying root cause of fracture was a rigid bond formed between the glass and U-shaped polymeric molding on the glass edge. Thermal expansion mismatch between the materials resulted in induced tensile stresses that were sufficiently large to cause rupture of the glass. Furthermore, it was found that the location of the bond on the glass played a role in the type of fracture formed. If the bond occurred on the edge of the glass product, the glass rupture resulted in the formation of a V-chip in the edge of glass that subsequently resulted in a crack in annealed products. If the bond occurred on the surface of the glass (inboard from the edge), the rupture resulted in a shallow "spalled" chip in annealed glass. In the case of tempered glass, if this chip turned into the core of the glass, it caused shattering of the part. Laboratory testing confirmed that the fracture could be induced on both raw glass and on enameled products.

REFERENCES

[1]G.D. Liamadis, "Role of Glass in Automotive Design Synthesis: Functional and Aesthetic Aspects," Proceedings from 7th International Glass Conference in Tampere, Finland, June 18-21, 2001.

[2]A. Leupolz, "Multifunctional Automotive Glazing – Examples and Trends," Proceedings from 8th International Glass Conference in Tampere, Finland, 2003.

[3]V. Henry, R.W. Davies, J.L. Woods, and K.K. Koram, "Absolute Automotive Glazing," *International Glass Review – Flat Glass*, Issue 1, 41-42 (2000).

[4]E.A. Hahn, "Automotive Glass – The Road Ahead," Proceedings from 7th International Glass Conference in Tampere, Finland, June 18-21, 2001.

[5]Internal Study by R.T. Nichols, S.Culross & H. Goertz, 1997.

[6] M. H. Krohn, J.R. Hellman, D.L. Shelleman, C.G. Pantano, and G.E. Sakoske, "Strength and Fatigue of Float Glass Before and After Enameling," *The Glass Researcher*, Vol. 11 No.2, 24-27 (2002).

[7] V.D. Fréchette, "Failure Analysis of Brittle Materials," *Advances in Ceramics*, Vol. 28, Westerville, OH: *AM. Ceram. Soc.* (1990).

[8] V.D. Fréchette and M. Donovan, "Some Effects of the "Glue Chipping" Process on Strength," pp.407 – 411 in *Fractography of Glasses and Ceramics II*, Eds. V.D. Fréchette and J.R. Varner, Ceramic Transactions Vol. 17, American Ceramic Society, Westerville, OH, 1991.

ACKNOWLEDGMENTS

Numerous individuals contributed to the work included in this paper. While it is not feasible to acknowledge all of them, certainly many of them deserve specific mention. The authors would like to thank Dominic Allam for supplying primer materials for testing; Yvonne Bertrand, Bob Corder, Mark Best, Stan Culross, Roman Surowiec, Richard King, Marcus Tanksley and Bob Hatcher for assistance in laboratory testing; Kris Cockerel, Tom Boettger, Mukta Mehandru, Roger Johnson, Mike Brennan and Rick May for discussions on the issue. Dr. James Varner collaborated on the fracture analysis of the tempered field samples and Rod Baker collaborated on some of the windshield fracture analysis.

AUTOMOTIVE SIDE GLAZING RELATED INJURIES IN ROLLOVER COLLISIONS

Stephen A. Batzer
Engineering Institute
13045 W. Highway 62
Farmington, AR 72730-0610

ABSTRACT
The injury mechanisms associated with tempered and laminated side glazing in automobile rollover collisions are enumerated and analyzed. The effects of containment loss, impact and laceration are considered. It is shown that the greatest statistical threat to the occupant is not from the glazing directly, but from exposure to the exterior coupled with partial or full ejection. The history and relative merits of tempered and laminated side glazing are discussed. The study is supplemented with examples from accident statistics, actual rollovers, and the results of current research.

FRACTURE PATTERNS OF IMPACT RESISTANT GLASS PANEL LAMINATES WITH ANNEALED AND HEAT STRENGTHENED GLASS PLATES

David L. Ahearn, III[*]
[*]M. M. & E
Suwanee, GA., 30024

Justin L. Ladner[×], Stanley E. Jones[×]
Richard E. Wright[×*], Richard C. Bradt[×]
[×]The University of Alabama
Tuscaloosa, AL, 35087

[×*]Oldcastle Glass
Delran, NJ, 08075

ABSTRACT

The fracture patterns which develop during the ballistic impact of two series of three layer glass/polycarbonate/glass impact resistant glass panels laminates were studied for both annealed glass and heat strengthened glass plates in the glass panels. The cracking patterns are qualitatively described. The numbers of radial cracks were related to the projectile kinetic energy and the stored elastic strain energy in the heat strengthened glass. An energy balance revealed that the elastic strain energy and the residual stresses in the heat strengthened glass are only partially relieved by the impact cracking. Results suggest that for higher velocity projectile impacts a greater amount of residual stress relief occurs from a greater fragmentation of the heat strengthened glass plates in the panels.

INTRODUCTION

The utilization of impact resistant glass is continually increasing in applications that range from wind damage prevention during storms, such as the recent hurricanes, to that of bullet proofing applications in automobile windows. Homeland security procedures have been one of the driving forces for the latter increase. Impact resistant glass is a glass/polymer composite of multi-layer laminate design. The exterior layers are usually soda-lime-silicate float glass to resist scratching and maintain transparency. A polymer, often polycarbonate, is sandwiched between the outer glass layers.[1] More than three layers are common for severe impact situations. The polycarbonate layer(s) serve to absorb energy during impact and also to maintain the integrity of the glass panel during and after an impact event that is severe enough to cause multiple fractures within the glass plates of the composite.

Depending on the anticipated severity of the impact event, the thicknesses of the individual glass plates and the polycarbonate layers, and also the numbers of these layers may vary. Thicknesses may be several centimeters. Panel characteristics are specified by Underwriters Laboratories.[2] In addition to the obvious need to maintain transparency, these glass panel laminates are also expected to impede or defeat the impacting object by preventing complete penetration. This is true for an impact from a tree limb in a storm, stones thrown by the demonstrators in a protesting crowd or a projectile that is propelled from a firearm.

One approach to increasing the impact resistance of these glass panels is to simply increase their thicknesses. This is achieved by either increasing the thickness of the individual glass and polycarbonate layers or by increasing the total number of individual layers. That approach results in an increase in the weight of the panel. Because of this resulting weight increase, this panel design modification is not always a viable approach to their impact enhancement. There exist other impact improving alternatives than to just make thicker and heavier glass panels.

The impact damage resistance of any glass panel laminate is determined by a combination of factors. These include not only the individual layer thicknesses and the overall panel thickness, but also the areal size of the panel and the details of the supporting framework structure that secures the panel in place. From the perspective of the penetration of the glass panel by a projectile from a firearm, the geometric design of the projectile and its physical properties are also important. Considering this multitude of factors creates a complex and challenging glass engineering problem for which to specify the different design criteria. It is equally complex for any comprehensive experimental testing program to evaluate the roles of each of the above on the panel impact resistance.

One potential improvement for the design of impact resistant glass panel laminates is to increase the strengths of the glass plates within the panel laminates. The strength of glass is always a concern for all of its many applications, ranging from soft drink containers to storefront windows. Gardon[3] has discussed the glass thermal tempering process and the strengths of both heat strengthened glass and thermally tempered glass relative to that of annealed glass. Gardon confirms that significant glass strength improvements are possible by thermal processing.

The fracture patterns of thermally strengthened glass plate are also significantly altered.[4] The presence of stored elastic strain energy in the compressive surface layers and the tensile central section leads to extensive crack pattern development if failure is initiated. The fracture pattern transforms from one of long sharp lathes or shards for the annealed glass to the morphology of smaller equiaxed dicings for thermally strengthened glass plates. Warren[5] has addressed the theory of this fragmentation process to create equiaxed particles in a quantitative manner. He demonstrates the importance of the stored elastic strain energy from the residual stresses in producing new fracture surfaces and less dangerous fragmentation.

In principle, there may be advantages to the use of strengthened glass plates in impact resistant laminated glass panels. The most obvious is that of improved impact resistance. Stronger glass may elevate the threshold damage level. This could result either by extending the load level for damage initiation beneath the crushed contact area or for fracture of the panel backside glass plate during flexure. Both mechanisms would be most effective within the elastic loading regime prior to damage and fracture initiation.

Another possible benefit is that the fracture pattern of the glass will be modified, perhaps in a fashion similar to the dramatic changes for stand-alone strengthened glass plate. This would consume more energy in the form of fracture surface energy. It is also possible that increased glass plate strengths will impart greater structural damage to the impacting object, whether it be a stone thrown by protestors or the projectile from a firearm. For the latter, this could be a major contribution to the defeat of the projectile. These advantages will have to be balanced with the economics of using strengthened glass plates in the impact resistant laminates.

This paper describes the fracture patterns of two series of glass panel laminates during ballistic impact by 30 caliber flat cylinders of OFHC copper. Only single projectile impacts above the damage initiation threshold, but below the total penetration threshold are reported. One series of the laminates had annealed float glass plates sandwiching a central polycarbonate layer, while the second series incorporated heat strengthened float glass plates within the laminates. Both series of these panels were subjected to a series of increasingly higher projectile velocity impacts to ~300 m/s.

All experimental tests were completed for sub-penetration threshold impacts, thus emphasizing the initiation of the projectile impact damage and allowing for the observation and analysis of the crack patterns during their early stages of development following the impact event. The crack patterns are described qualitatively and also analyzed quantitatively through an energy balance approach. The role of the stored elastic strain energy within the heat strengthened glass plates in the crack development process is revealed.

EXPERIMENTAL PROCEDURES

The ballistic test stand for the experiments is the one described in detail by Stevenson, et al[6,7] and shown in Figure 1. It consists of a smooth bore gun tube, a laser velocimeter to determine projectile velocities, a target mounting facility and a projectile catch system to recover any projectiles that penetrate and defeat the target. Since the impacts in this study were of the sub-penetration threshold variety, the projectiles experienced post impact elastic recoil after impacting the test panel. They were recovered within the armor plated test room facility. Velocities of the 30 caliber OFHC copper ~10g flat cylinder projectiles were controlled by hand loading the appropriate explosive charge level into commercial cartridge cases.

The glass test panels were one square foot laminates that consisted of two external 3/16" thick soda-lime-silica float glass plates sandwiching a central 3/16" polycarbonate sheet. They were bonded with 0.05" urethane. One test set of the laminated panels had both glass plates heat strengthened to compressive surface stress levels of ~5,000–7,000 psi (~30–50 MPa). The other set had annealed glass plates with zero residual stress. The panels were prepared for the impact testing by taping the edges with composite Scotch tape containing fiberglass reinforcement. This assured the integrity of the panels and the glass fracture patterns after impact. Mounting of the test panels consisted of pressure holding the flat panels within a wooden frame in the steel framework setting with sponge supports along the edges to prevent any edge damage to the panels in flexure during the impact.

For the energy balance analysis of the crack patterns, the projectile velocities were directly converted to kinetic energy through $KE = 1/2\, mv^2$. The stored elastic strain energy in the

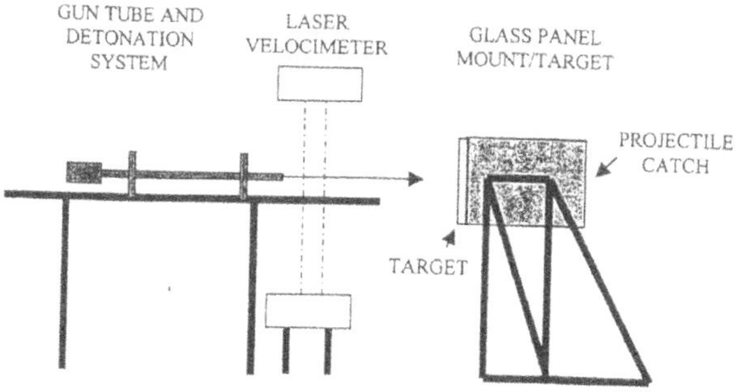

Figure 1. The ballistic impact test stand for glass panel testing.

heat strengthened glass plates was estimated assuming an average surface compressive stress of ~6,000 psi (~40 MPa) applying the modified equations presented by Warren[5] which are just a reduction of the formal equation of Timoshenko and Gere[8] with a number of simplifying assumptions. The result was ~1,000 Joules per heat strengthened glass plate.

RESULTS AND DISCUSSION

Qualitative Observations

Figure 2 is a schematic of the cracking pattern of the projectile impacted glass panel laminate. The primary crack features are identified. Of particular interest are the regions of the central impact zone, the spalled region about the impact area, the radial cracks emanating from the central impact and the circumferential cracks that lead to a spider web-like appearance. Both the front impact side and the backside glass plates are shown although they appear similar at first glance.

Figures 3 through 6 compare the direct impact fracture patterns of the glass panel laminates for the two extreme projectile velocities, ~100 m/s and ~300 m/s, respectively. The top photographs (3 and 5) are of the impacted panels with the annealed glass plates on both front and back. This type of panel is designated throughout as "A-A". The lower photographs (4 and 6) are for the heat or thermally strengthened glass plates and are designated as "T-T" throughout. This A-A and T-T nomenclature is continued throughout the paper.

Several of the important impact created crack pattern features are evident and will be first discussed in a qualitative sense. The most obvious difference between the two sets of impacted panels is that the higher velocity impacts create more cracking damage within the glass panels. It was consistent, even though the panels did not experience full penetration by the projectiles.

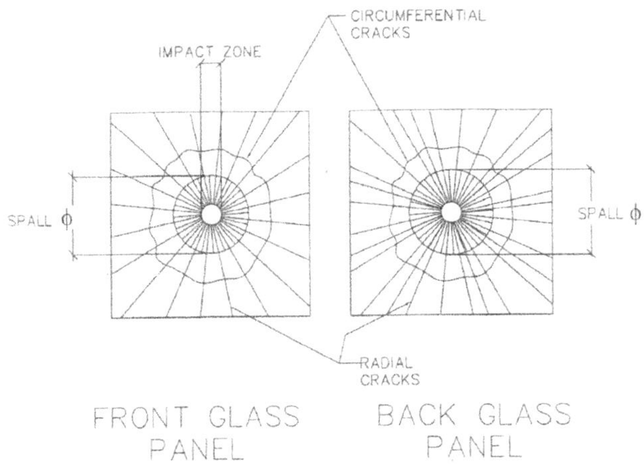

Figure 2. Schematic of the cracking pattern of the impacted glass panels.

This is not surprising, for the kinetic energy of the higher velocity projectiles for the photographs was nine times that of the slower, almost an order of magnitude greater.

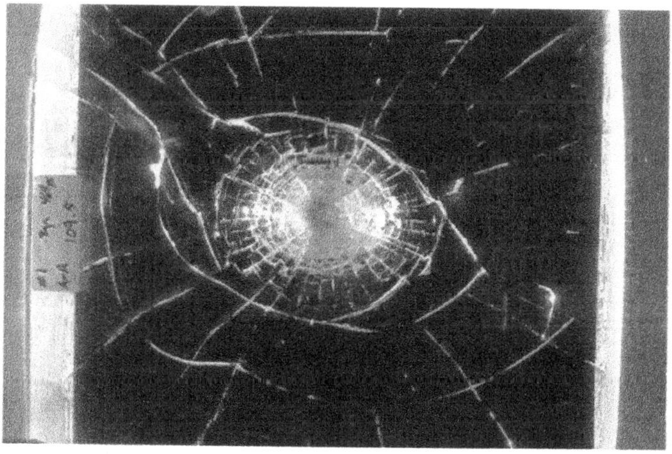

Figure 3. The macrofracture pattern of the glass panel laminate with annealed glass plates, impacted at ~100m/s. Note the small number of radial cracks from the impact point.

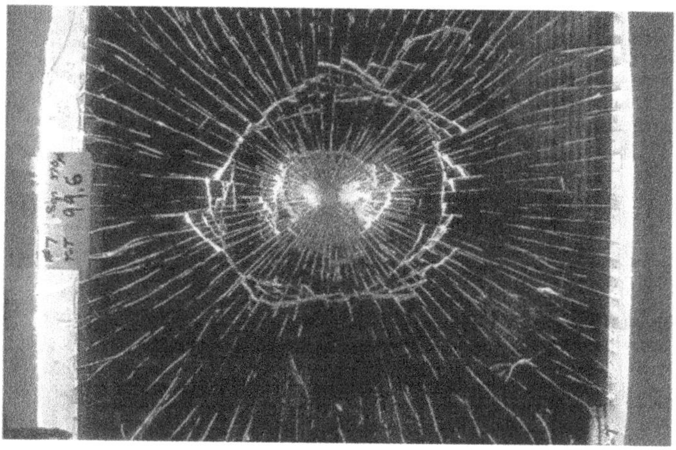

Figure 4. The macrofracture pattern of the glass panel laminate with heat strengthened glass plates, impacts at ~100m/s/ Note the increased number of radial cracks.

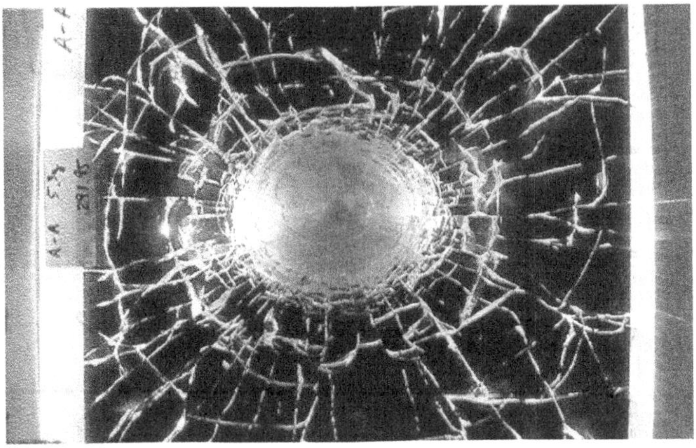

Figure 5. The macrofracture pattern of the glass panel laminate with annealed glass plates, impacted at ~300m/s. Note the small number of radial cracks from the impact point.

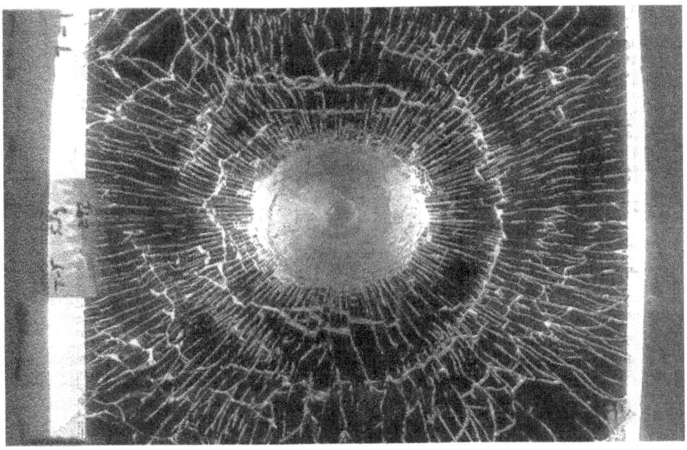

Figure 6. The macrofracture pattern of the glass panel laminate with heat strengthened glass plates, impacted at ~300m/s. Note the increased number of radial cracks.

Although reference is made here to the difference in projectile kinetic energies, it must be noted that not all of the initial kinetic energy of the projectile is actually transformed to the creation of the fracture pattern within the glass plates. Some of the initial projectile kinetic energy is consumed in the compressive deformation of the projectile and some of it is also transferred to the rebound kinetic energy of the projectile. The rebound kinetic energy of the deformed projectile originates from the elastic springback of the panel after the impact causes its flexure. During the experimental testing, it was not unusual to hear the rebounding projectile bounce around the confined test facility. Neither of these two forms of dissipating energies could be quantitatively estimated for the subsequent energy balance calculations. They are mentioned for describing the situation.

It also appears that the heat strengthened glass panels have an ever slightly smaller spalled region about the central impact point. Although this observation cannot be confirmed though actual measurements because the velocities are not always identical and the outer boundary of the spall is rather irregular, it nonetheless appears to be the case. This suggests an advantage for the use of strengthened glass plates in the laminates and that it may extend to greater damage resistance for multiple impact events. Further testing is required to confirm this speculative point.

In the extensive cracking away from the central crushed and spalled region, it is evident that the higher projectile velocities create a greater number of individual radial cracks and also more circumferential cracks. Relative to the panels with annealed glass, the panels with heat strengthened glass plates experience a much greater number of radial cracks during fracture which is later addressed quantitatively. There is also more circumferential cracking to create the spider web-like pattern, but this is not treated quantitatively for the number of circumferential cracks is small. It appears that the stored elastic strain energy in the heat strengthened glass panels contributes to more extensive fracturing of those panels with strengthened glass.

Crack Development Sequence

Stevenson, et al[6] have described the sequence of crack formation events for another type of impact resistant glass panel laminates. The impact/cracking sequence for the laminates is first one of an initial purely elastic flexure of the panels away from the impacting object. At very low projectile velocities, <50 m/s, there is no cracking, or even any evidence of the impact as the glass plate on the impact side does not experience any damage from the projectile. The projectile also experiences purely elastic compression and is undamaged as well, although it experiences a rebound event. This is followed by the Hertzian crushing of the impact side glass beneath the projectile and then the radial fracture of the backside glass panel by flexure of the flat panel. The latter occurs through the generation of a state of equibiaxial tensile stress during the flexure of the panel during the impact event. From the high-speed photography of a glass impact from a projectile reported on the internet,[9] it is evident that the radial cracks form first and that the circumferential cracks created by the reflected stress waves follow. Espinosa, et al[10] attribute the spalling to a stress wave propagation mechanism.

Quantitative Analysis of the Crack Patterns

The photographs in Figures 3 through 6 are quite revealing, but they are not convenient to observe as to whether the individual radial cracks are on the impact side glass plate or on the backside glass plate. Their actual locations are easily discerned on the actual panel laminates. Their numbers are recorded as the radial cracks occur just outside of the impact spall damage zone. Figure 7 presents the numbers of radial cracks on the impact side glass plate versus the number on the backside glass plate. This illustration reveals several important features of the radial cracks, their patterns and their numbers.

The first characteristic of the radial cracks is that more of them are formed on the backside glass plates of the panels than form on the front side. Once the projectile velocity is exceeded where the radial cracks begin to form, there are consistently about 10% more cracks on the backside glass plate. The backside radial cracks form when the composite panel flexes during the impact event such that there develops a stress state of equibiaxial tension on the backside glass plate, and it is sufficiently high to cause the radial cracking and the attendant backside spalling that is also observed.

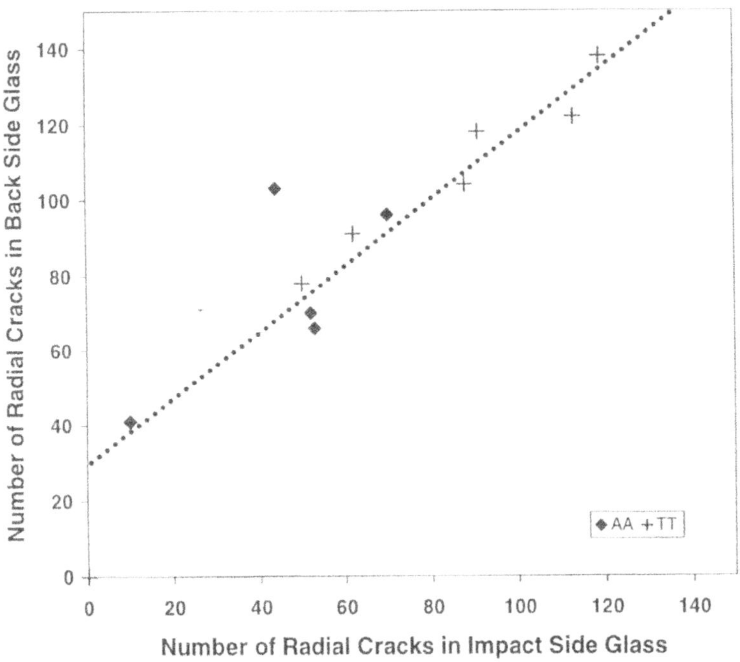

Figure 7. Numbers of radial cracks on the back side glass plate versus those on the impact side glass plate. Note the overlap of the annealed and heat strengthened glass panels.

When the actual numbers of impact side and backside radial cracks for both the annealed and the heat strengthened glass plates are presented, as they are in Figure 7, it is observed that all data points fall on the same straightline relationship. This occurs regardless of the individual projectile velocity over the range that was studied and it is independent of whether the plates are annealed or heat strengthened. There is a linear relationship between the numbers of radial cracks on the two glass plates, with the number on the backside glass plate always exceeding the number on the front impact side, but maintaining a constant ratio. It appears that the backside glass plate radial cracking occurs first. At the same time, the glass plate on the impact side is experiencing crushing beneath the projectile and compression during the initial flexure. The impact side radial cracks may even form during the springback of the flexed panel into reverse tension following the initial flexure in the direction of the projectile. In any event, the backside glass plate always contains more radial cracks than the front impact side after the impact event.

The single straight line practically exhibits two distinct regions for the different strength glass plates in the laminates. The heat strengthened glass plates consistently have higher numbers of radial cracks emanating from the impact region than the annealed glass plates for the same projectile velocities. This is true for both the backside and the impact side glass plates. However, there is a regime of overlap for the two types of glass plates in the impact resistant panels. Those panels with annealed glass that are impacted at high projectile velocities and those panels with heat strengthened glass that are impacted at low projectile velocities have similar radial crack numbers where this overlap region occurs.

Application of an Energy Balance

The heat strengthened glass panels obtain much of their energy for crack formation from the stored elastic strain energy of the residual stresses. The annealed glass panels obtain their energy for crack formation only from the kinetic energy of the projectile, although it is first transformed into stored elastic strain energy during the flexure of the panel. This suggests that the original source of the energy for crack formation does not by itself affect the radial cracking relationship between the front impact side and backside glass panels, for the energy must all be transformed into stored elastic strain energy during the flexure of the panel. This is confirmed by the overlap of the two types of panels with regard to their radial cracking. This suggests that examining an energy balance for the cracking may be revealing.

An energy balance for the impact process requires the knowledge of the two sources of energy in the impact event and the energy dissipation sinks as well. The first of the energy sources is the kinetic energy of the projectile, which is easily estimated from: $E = 1/2\ mv^2$. This calculation yields energies of a few hundred Joules, the highest velocities having ~600 Joules. Although this is the original kinetic energy of the projectile as it passes through the laser velocimeter, all of this energy is not converted just to the energy of crack formation in the glass plates. Some of the energy is lost in the deformation energy in the mushrooming of the projectile and additional energy is lost in the elastic recoil of the projectile when it does not penetrate the glass. However, in spite of these two losses of energy, the original kinetic energy of the projectile must be the starting point for any energy balance.

The other source of energy is the stored elastic strain energy of the compressive and tensile stresses in the heat strengthened glass plates. This energy term is zero for the annealed glass plates. However, following the approach of Warren,[5] it is possible to estimate an upper limiting value to this energy. It does, however, suffer from a similar problem to the kinetic energy of the projectile with regard to a total accountability of the energy. Even though the stored elastic strain energy in the heat strengthened glass plates no doubt contributes significantly to the crack formation in those glass plates, the formation of the radial and circumferential cracks will not release all of the stored elastic strain energy. Crossed polar examination of fractured glass shards that had been thermally strengthened reveals that some of the residual stress remains even after extensive fracture. Thus it is only reasonable to apply the stored elastic strain energy as an upper limit to the availability of energy for crack formation.

It seems almost intuitive that the kinetic energy of the projectile and the stored elastic strain energy of the residual stress in the heat strengthened glass should be additive. Supporting this concept is the plot in Figure 7. It has a significant region of overlap where the energy is

predominantly either one source or the other, yet the impact side to backside radial crack ratio, that of the numbers of these cracks remains constant, plotting on the same straight line. This suggests that the mechanisms of the radial crack formation on the backside glass plate and the front side glass plate have some form of an interdependent relationship and that it does not depend on the specific source of the energy to form the cracks.

The stored elastic strain energy in a heat strengthened glass panel laminate can be estimated as previously stated. For an average heat strengthened surface compressive stress of ~6,000 psi, there are ~1,000 Joules contained within a single heat strengthened glass plate. An impact resistant glass panel laminate with two glass plates, one on each side would therefore contain about 2,000 Joules of stored elastic strain energy. Relative to the kinetic energy of the highest velocity projectiles tested, ~ 600 Joules, this is more than three times as much energy for crack formation, once the glass panel laminates begin to fail. Neither the kinetic energy of the projectile, nor the stored elastic strain energy is converted 100% to crack formation. It nevertheless clearly reveals that there is considerably more energy available from the stored elastic strain energy of the residual stress in the heat strengthened glass panel than there is in the kinetic energy of the projectile. It is not surprising that the heat strengthened glass should generate considerably more cracking upon failure as is visually apparent in Figures 3 through 6, and the plot in Figure 7 as well.

Knowledge of the magnitudes of these energy sources allows one to analyze the crack patterns on the basis of energy, even though one cannot account for 100% of the energies. Figure 8 presents the total numbers of radial cracks formed at the edge of the impact zone versus the kinetic energies of the impacting projectiles. Both the annealed and the heat strengthened glass plates in the two series of glass panel laminates are presented. When presented in this manner, the annealed and the heat strengthened glass panels do not fall on the same line as they did when the back and front glass plate radial cracks were compared in Figure 7. Rather the heat strengthened glass plates have a much higher number of radial cracks. This is not surprising, for the heat strengthened glass plates also contain significant stored elastic strain energy that is not accounted for in the calculation of the kinetic energy of the projectile. It is many times more than the kinetic energy of the projectile. This form of presentation clearly illustrates the importance of the stored elastic strain energy of the residual stresses in the crack formation process. It also contains information about the recovery of the stored elastic strain energy for crack formation processes.

Figure 8 allows for an estimate of the upper limiting value to the amount of stored elastic strain energy that still remains in the heat strengthened glass plates after fracture. This is the stored elastic strain energy that remains as the unrelieved residual stresses after fracture. Some, but not all of the residual stress is relieved by the fracture process. The accompanying strain energy is converted to fracture surface energy by the crack formation process. The amount of energy from the stored elastic strain energy in the heat strengthened glass that contributes to the fracture process can be estimated. Simply shifting the line for the heat strengthened glass panel line to the right along the energy axis until it appears as an extension of the annealed glass line allows this. Although the slopes are not identical, that shift illustrates the approximate amount of energy expended for the additional cracking in the heat strengthened glass plates. The extent of the shift required for the two lines to appear almost as one is ~600 Joules. This suggests that the quantity of elastic strain energy that contributes to the crack formation is a little less than one

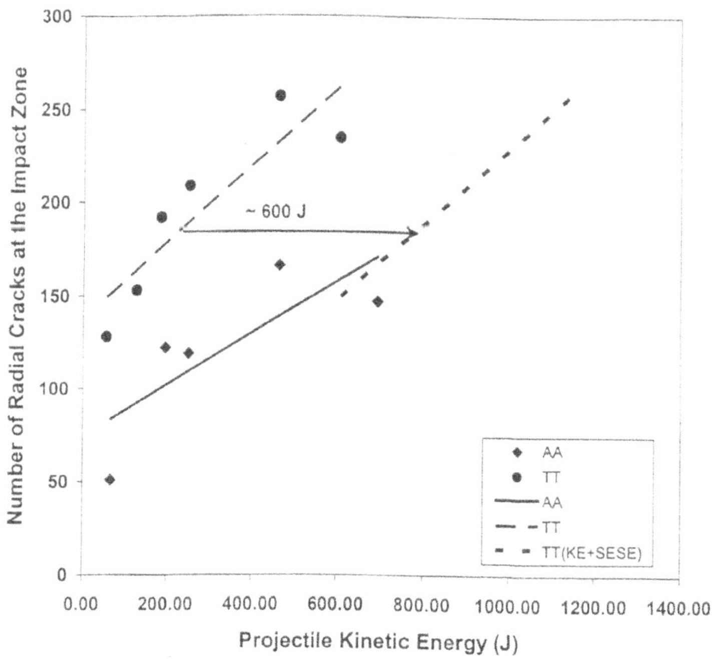

Figure 8. Number of radial cracks versus the kinetic energy of the projectile. Note the projected extension of the heat strengthened line after a shift of ~600 J.

fourth of the stored elastic strain energy in the heat strengthened glass plates prior to impact fracture.

It was noted that the slope of the heat strengthened glass containing panels in the Figure 8 presentation was not the same as that of the annealed glass containing panels. The slope is actually slightly larger. This indicates that as the projectile velocities are increased, which increases the numbers of radial cracks that are formed, then the amount of energy recovered from the stored elastic strain energy for fracture surface energy increases. This appears logical for as the numbers of radial cracks increase, the ability to support higher residual stresses and thus maintain the stored elastic strain energy of the residual stress in the strengthened glass must decrease, for a greater volume of fragments is nearer to a free surface, and free surfaces do not support stress.

CONCLUSIONS

It has been previously mentioned that there are several unquantifiable energy contributions to the energy balance. These include the deformation energy for the mushrooming of the projectile, the rebound kinetic energy of the projectile, the kinetic energy of any glass fragments and the remaining unrelieved residual stress strain energy in the cracked glass plates. In spite of the point that several energy terms cannot be fully accounted for, a couple of conclusions can be reached from energy balance estimates.

The kinetic energy of the projectile and the stored elastic strain energy of the residual stress field of the heat strengthened glass are additive in their contributions to the fracture crack patterns of these impacted glass panels. The point that both the strengthened and the annealed glass containing panels fall on the same line of the plot for the impact side to backside radial crack numbers confirms this point. However, it appears that neither 100% of either energy source is converted completely to fracture surface energy during the impact fracture event.

It is evident that the stored elastic strain energy in the residual stress field of the heat strengthened glass plates contributes significantly to the development of the crack patterns of the heat strengthened glass plates in the laminated panels. For the study reported in this paper, that contribution appears to equal or exceed the energy contribution from the kinetic energy of the projectiles that impact the glass panels. The glass panels with stored elastic strain energy consistently exhibit more extensive radial cracking and circumferential cracking once failure is initiated. This is because of the additional stored elastic strain energy for fracture processes.

The numbers of radial cracks as a function of the kinetic energies of the projectiles illustrates several important features of the energy condition. Two distinct lines result on the plot for the annealed glass plates and the heat strengthened glass plates, respectively. The numbers of radial cracks in the heat strengthened glass containing panels are consistently much larger for the same projectile velocities. This confirms the importance and contributions of the stored elastic strain energy of the residual stresses in the heat strengthened glass plates to the overall fracture of the impact resistant panels.

The slopes of the lines for the two glass plates are different. However, a translational shift of ~600 Joules to extend the annealed glass panel line suggests that only ~25% of the total stored elastic strain energy in the two heat strengthened glass plates is actually converted to fracture surface energy during the cracking process. This suggests that ~75% of the original total stored elastic strain energy in the two plates remains in the fractured glass plate fragments of the original heat strengthened glass plates. The plates must either then still contain considerable residual stress, or the energy is dissipated in some other manner. Perhaps the heating of the panel during the impact fracture is another form of dissipation.

For the plot of radial crack numbers versus the projectile kinetic energy, the slope of the line for the heat strengthened glass is greater than that for the annealed glass. This suggests that at higher projectile velocities, there is more of the stored elastic strain energy in the residual stress of the heat strengthened glass transferred to fracture surface energy during fracture. This probably occurs because the fragmentation of the heat strengthened glass is greater during the higher projectile velocity impacts. Not only are more radial cracks formed, but there also appear to be more intermediate circumferential cracks. This results in smaller fragments with more free surfaces per unit volume and a reduced ability to support any remanent residual stresses.

ACKNOWLEDGEMENTS

The authors acknowledge the manufacture and the supply of the impact resistant glass panels by Oldcastle Glass Co. Partial support of this research was provided by the Alton N. Scott endowment fund of the College of Engineering of the University of Alabama.

REFERENCES

[1]N. Shinkai, "Fracture and Fractography of Flat Glass," 253-297 in *Fractography of Glass"*, edit by R.C. Bradt and R.E. Tressler, Plenum Pub. Co., NY, NY (1994).

[2]Underwriters Laboratories, Inc., ULC 752, "Bullet Resistant Equipment," can be found at the website: http://www.database.ul.com.

[3]R. Gardon, "Evolution of Theories of Annealing and Tempering: Historical Perspective," *Amer. Cer. Soc. Bulletin* 66 (11) 1594 –1601 (1987).

[4]R.A. McMaster, D.M. Shetterly and A.G. Bueno, 453-459, "Annealed and Tempered Glass" in the *Engineered Materials Handbook, Vol. 4*, "Ceramics and Glasses", edit by S. J. Schneider, Jr., *ASM*, Materials Park, OH.(1991).

[5]P.D. Warren, "Fragmentation of Thermally Strengthened Glass," 389-400 in *Fractography of Ceramics and Glasses IV*, Ceramic Transactions 122, edit by J.R. Varner and G.D. Quinn, *Am. Ceram. Soc.*, Westerville, OH.(2001).

[6]M. E. Stevenson, S.E. Jones and R.C. Bradt, "Fracture Patterns of a Composite Safety Glass Panel during High Velocity Projectile Impacts," 473-488 in *Fractography of Ceramics and Glasses IV*, Ceramic Transactions 122, edit by J.R. Varner and G.D. Quinn, *Am. Ceram. Soc.*, Westerville, OH (2001).

[7]R.C. Bradt, M.E. Barkey, S.E. Jones and M.E. Stevenson, "Projectile Impact – A Major Cause for the Fracture of Flat Glass," *The Glass Researcher* 11(2), 20-23 (2002).

[8]S.P. Timoshenko and J.M. Gere, *Theory of Elastic Instability,* McGraw-Hill Pub., NY, NY. (1961).

[9]Projectile Impact on Glass Panel, Slow Motion Photography, can be viewed at the website: http://www.corin.com/impink.html

[10]H.D. Espinosa, Yueping Xu and N.S. Brar. "Micromechanics of Failure Waves in Glass." J. *Am. Ceram. Soc.* 80(8) 2061-2073 (1997).

ANALYSIS OF PROJECTILE IMPACT DAMAGE IN BULLETPROOF GLASSES

Humberto Naoyuki Yoshimura
Institute for Technological Research of the State of São Paulo
Av. Prof. Almeida Prado 532, Butantã
São Paulo, SP, Brazil, 05508-901

Ricardo Guimarães Morrone
Gepco Ind. e Com. Ltda.
R. Prof. João Cavalheiro Salem 186
Guarulhos, São Paulo, Brazil, 07243-580

ABSTRACT

The damage evolution in bulletproof glasses subjected to ballistic tests was investigated. Laminates were prepared with soda-lime glass and polycarbonate (PC) laminae using PVB and PU films as adhesives. Two laminates were subjected to single hit ballistic tests varying the velocity of the projectile. A panel composed by three 6mm glass laminae and a 3mm PC laminae was hit by .44 Magnum projectile and a second panel composed of four glass laminae (one 5mm and three 8mm) and two PC laminae (6 and 3mm) was hit by a 7.62mm FMJ projectile. The crack patterns of each of the glass lamina were recorded and analyzed. Both laminates presented similar and complementary damage evolution. Main radial cracks formed first, before the conical cracks. The increase of kinetic energy of the projectile (E_p) caused an almost linear increase of the number of radial cracks and an exponential increase of the quantity of crack bifurcation. Concentrical cone cracks formed the intensely damaged envelope under the impact point. The sequence of failure events observed with the increase of E_p was classified in three steps: first, the cone crack envelope advanced gradually in the glass laminae up to the PC lamina; subsequently, the material in the cone crack envelope was crushed and eroded; then, a highly localized plastic deformation of PC lamina formed a protuberance. The protuberance depth increased linearly with the increase of E_p. The crater area formed in the frontal glass lamina increased with the increase of E_p during the two first steps and then decreased in the third step.

INTRODUCTION

The increase of violence and criminality in Brazil has increased the demand for armored structures mainly for the protection of human lives. The bulletproof glasses are ballistic resistant materials with high light transmission capacity, so they are applied where the visual observation through the panel is important, like vehicle and building windows. The use of bulletproof glass or bullet-resistant glazing, however, increases the weight of the vehicle, which causes, among other factors increases in the fuel consumption and the emission of pollutants into the atmosphere. Efforts have been made in order to diminish the weight of the armor by means of the development of materials and assembly systems.[1-3]

The bulletproof glasses are basically comprised of one or more laminae of glass bonded together with one or more laminae of a transparent polymer, usually polycarbonate (PC) at the backface. These laminate composites combine a hard external layer (glass) with a more flexible and tough internal layer (PC), conjugating synergistically their properties and resulting in a superior ballistic performance to the isolated components. In general, the strike face (first glass lamina) has the function of flattening the projectile, eliminating its conical shape, and slowing down its rotation movement, in both cases to diminish its capacity of perforation. The intermediate glass laminae act to absorb the kinetic energy of the projectile and to diminish its velocity. The PC laminae performs to absorb the residual kinetic energy of the projectile and to avoid the trespassing of the projectile and glass fragments that can cause injury to the persons behind the transparent armor. The main mechanisms of the absorption of kinetic energy of the projectile are the generation of fracture surfaces, in the case of the glass laminae, and plastic deformation, in the case of the PC laminae. Another possible mechanism of the composite is the delamination between the laminae, which depends on the characteristics of the polymeric adhesive films used for the bonding of the PC and the glass laminae.

The understanding of the damage mechanisms in ballistic armors by the impact of firearm projectiles can assist with the development of lighter armor structures. The aim of this work was to analyze the damage evolution in bulletproof glasses subjected to ballistic tests.

EXPERIMENTAL

The bulletproof glass samples were prepared with soda-lime-silica float glass laminae of 5, 6 and 8 mm thickness (Cebrace), PC laminae of 3 and 6 mm thickness (GE Lexan), and polyvinyl butyryl (PVB) and polyurethane (PU) films of 0.38 mm thickness. The laminates were prepared with area of 50 x 50 cm^2 using a film of PVB between two glass laminae and a PU film between the glass and PC laminae and between the PC laminae. The assembly was mounted manually, placed inside a vacuum bag and submitted to a cycle of internal vacuum, external pressure, and heating (~120°C) in an autoclave. For level III-A (NIJ), the laminates were prepared with three 6 mm glass laminae and one 3 mm PC lamina in the backface. The sample named 666-3 presented 21.8 mm thickness and 49.1 kg/m^2 of areal density. For level III (NIJ), the laminates were prepared with one 5 mm glass lamina at the strike face followed by three 8 mm glass laminae, one 6 mm PC lamina and one 3 mm PC lamina in the backface. The sample named 5888-63 presented 39.2 mm thickness and 82.9 kg/m^2 of areal density.

The samples were submitted to the ballistic test based on the NIJ 0108.01 (1985) standard. The laminate was fixed in a metallic support covered with rubber with an open area of ~40 x 40 cm^2 located 5 or 15 m (levels III-A or III, respectively) ahead of the muzzle of the test weapon. The velocity of the projectile was determined with a chronograph (Crony F-1), with distance between sensors of 0.305 m, located 2.0 m ahead of the muzzle of the test weapon. The test ammunition for level III-A was a .44 Magnum with the mass of the projectile of 15.55 g and for level III was a 7.62mm caliper (308 Winchester) with the mass of the projectile of 9.6 g. The velocity of the projectile, v, was varied by using different amounts of gunpowder (CBC ref. 102), in order to vary the kinetic energy of projectile ($E_p = mv^2/2$). One shot was fired in the center of each laminate.

After the shot, the frontal and back images of the sample were registered with a digital camera. The images were analyzed with an image analyzer (Leica QWin) for quantification of the damage. The depth of penetration in the PC lamina was measured with a caliper. The crack pattern of each glass lamina was manually copied in acetate films. In each crack pattern, the number of radial cracks crossing the circles with 10, 15, and 20 cm radii and the amount of crack bifurcation in these sectors were counted and recorded. The samples were transversally cut with a diamond blade through the center of the impact point in order to analyze the cross-section of the damage zone.

RESULTS AND DISCUSSION

Fig. 1 presents images of the region close to the impact where intense damage occurred. In the frontal view a crater caused by the impact of the projectile can be observed (Fig. 1a), and in the back view a whitened area (Fig. 1b) and a PC protuberance (Fig. 1c) can be observed. Conical cracks were formed from the impact region and were approximately concentrical to the impact axis (Fig. 2), where the largest base of these cracks defined the whitened area (Fig. 1b). This area is partially opaque and partially translucent because of the light scattering caused by the intensely damaged region of the glass. In this region many radial, lateral, circumferential, and conical cracks were observed. The intersection of these cracks formed loosed glass pieces. These easily detached in the first glass lamina in the region near the impact location.

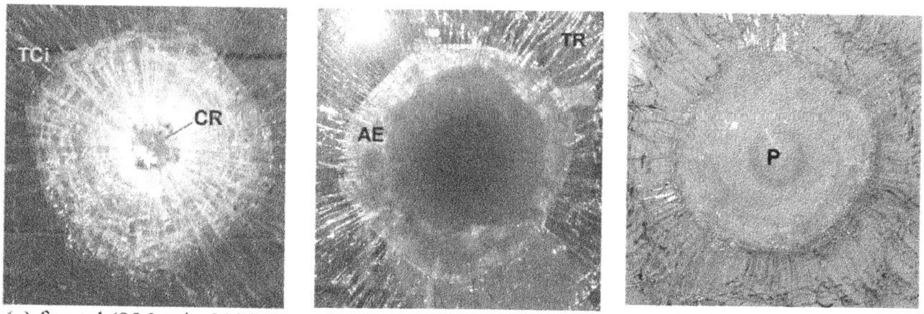

(a) frontal (906 m/s, 3903 J) (b) back (906 m/s, 3903 J) (c) back (513 m/s, 2054 J)

Fig. 1 – Images of the region close to the impact location (highly damaged region) of the sample: (a,b) 5888-63 (level III); and (c) 666-3 (level III-A). CR - crater, AE - whitened area, TR - radial crack, TCi - circumferential crack, and P - PC protuberance.

(a) 463 m/s (1053 J) (b) 657 m/s (2118 J) (c) 906 m/s (3903 J)

Fig. 2 – Cross-sectional views of the highly damaged region of the sample 5888-63 subjected to different projectile velocity. SP - semi-spherical plug, LC - lateral crack, and CC - conical crack.

With sufficiently high kinetic energy, the impact of the projectile caused the formation of a plug in each damaged glass lamina. Its center was aligned to the impact axis (Fig. 2c). The plug presented the geometry of a dome, with a spherical surface, and consisted of glass particles with columnar shapes adhered to the PVB film (Fig. 3). At high projectile velocities, the plug was comminuted during the impact. In Fig. 2c, the plugs of the two first glass laminae were comminuted, remaining plugs of the two last glass laminae. The semi-spherical plug seems to be similar to the conical plug or conoid observed in ceramic materials. It is a consequence of the tensile stress similar to that causing Hertzian contact damage.[4,5]

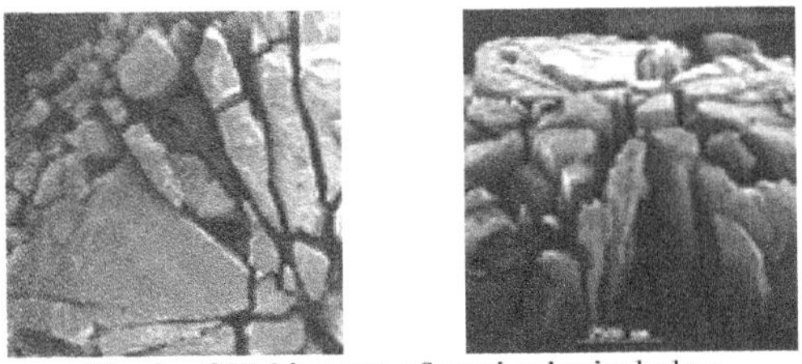

Fig. 3 – SEM images of semi-spherical plug.

In the sample 666-3 (level III-A), at the lowest impact velocity (225 m/s, 396 J), there was no perforation of the first glass lamina (strike face), but the impact caused intense damage in the three glass laminae with the formation of an envelope containing many concentrical cone cracks. The increase of the projectile velocity caused the increase of the penetration of the projectile into the glass laminae causing mass loss of the envelope (comminuted zone) and defined a crater in the first glass lamina. At the velocity of 439 m/s (1502 J) and above, it was observed that there occurred a penetration of PC lamina with the formation of a protuberance. This penetration corresponds to absorbed energy by plastic deformation of the PC lamina. Up to the highest velocity of this series (513 m/s, 2054 J), perforation of the PC lamina was not observed.

In the sample 5888-63 (level III), the damage from the projectile impact in the lower velocity (348 m/s, 568 J) was superficial, with the formation of two radial cracks in the first glass lamina (5 mm thickness). At the velocity of 463 m/s (1052 J), the damage was restricted to the two first glass laminae, where radial, lateral, circumferential, and conical cracks were observed. Partial comminution and spalling of the first glass lamina in the region close to the impact were also observed. With the increase of the velocity of the projectile, the area and the depth of the damage region intensified and the numbers of cracks increased. At the velocity of 529 m/s (1305 J) and above, the damage region was observed in all four glass laminae. Up to the velocity of 657 m/s (2117 J) the perforation of the first glass lamina was not observed (Fig. 4a), which initiated from 740 m/s (2544 J) (Fig. 4b), with the formation of the crater. Except for the lower velocity, the impact of the projectile caused the formation of a semi-spherical plug in each damaged glass lamina. It was observed that the comminuted region around the plugs formed a cone profile with the smaller base being the crater and the biggest base the whitened area (Fig. 2c). In all samples tested, no plastic deformation of the PC laminae was observed (Fig. 2c). The maximum velocity reached in this series (906 m/s, 3903 J) corresponded to the maximum mass of gunpowder that was possible to load into the cartridge.

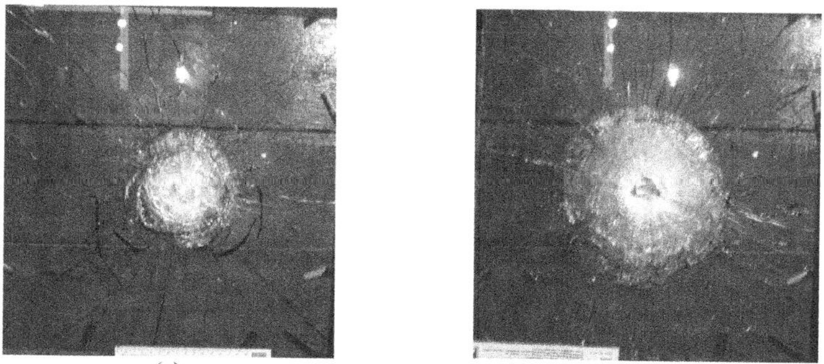

(a) (b)

Fig. 4 – Images of the sample 5888-63 subjected to the projectile velocities of (a) 586 m/s (1681 J) and (b) 740 m/s (2544 J).

Circumferential cracks, which were restricted to the region delimited to the whitened area, and radial cracks were perpendicular to the surface of the glass laminae. The formation of radial cracks out of the region of the whitened area indicates that they propagated more rapidly (and probably formed before) than the circumferential cracks (it was also observed that circumferential cracks were discontinuous in the intersection with radial cracks). Some cracks with circular shape, concentric to the position of impact, were observed out of the region of the whitened area (Fig. 4a). They were deflected radial cracks or branched from them, and looked like to be "wide", since they were inclined in relation to the surface of the glass lamina. These cracks have been called wing cracks and are inclined 45° in relation to the surface of the glass. It seems to be a form of shear crack resultant of the reaction with the tension waves reflected from the edges of the glass laminae.[6] Wing cracks were mainly observed in the second glass lamina of the samples tested with the velocity between 529 and 657 m/s (1305 to 2117 J), which correspond to the samples with damage in the four glass laminae without the perforation of the first lamina.

Fig. 5 and 6 show, respectively, the crack patterns of the 666-3 and 5888-63 samples hit by projectiles with different velocities (kinetic energies). The central region of the crack patterns could not be reproduced because the highly damaged region made visualization difficult. Most of the samples presented patterns with radial cracks starting from the impact point and also circumferential cracks. Only the first glass lamina of the 666-3 sample hit by the lowest velocity (225 m/s) presented a different pattern (Fig. 5a). It was observed that, above a certain projectile velocity (368 m/s for 666-3 sample and 841 m/s for 5888-63 sample), some of the radial cracks deflected and circumferential cracks formed among the radial cracks, out of the whitened area (Fig. 5c and 6c). Some deflected cracks were slightly inclined, but most of them were perpendicular to the surface of the glass. Circumferential cracks were also perpendicular to the glass surface. These cracks seem not to be of the same nature as the wing cracks, and seem to have been formed as a result of the deflection of the glass lamina, when the laminate was impacted with high kinetic energy of the projectile. The presence of both deflected and non-deflected radial cracks indicated that these cracks had different velocities of propagation or that they passed in different times during the moment of the deflection.

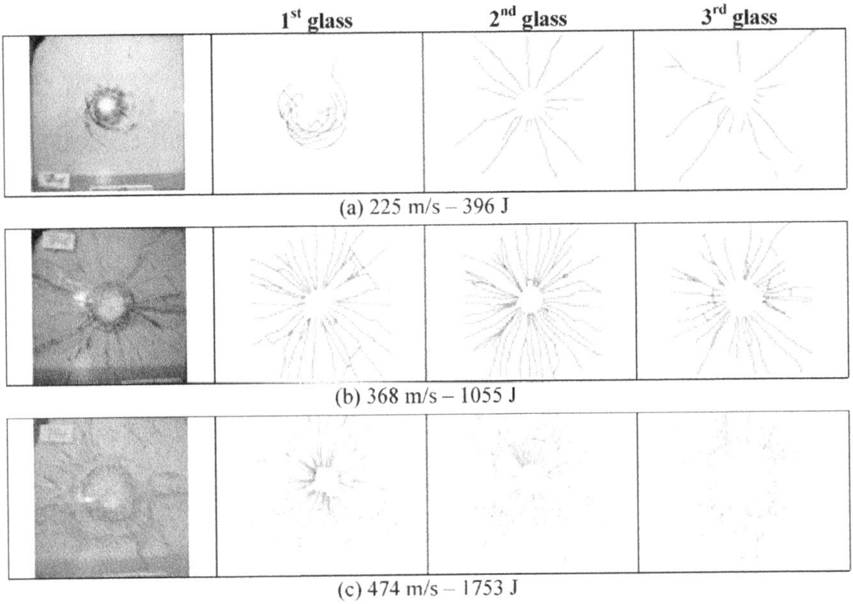

Fig. 5 - Crack patterns of 666-3 sample subjected to level III-A ballistic test with varying projectile velocity (kinetic energy).

Basically, two types of crack patterns were observed in the 5888-63 samples with damage in all four glass laminae. In the projectile velocity range of 529 to 657 m/s (1305 the 2117 J), which corresponded to the samples that presented no perforation of the first glass lamina, the amount of radial cracks gradually increased from the first glass (strike face) to the fourth glass (Fig. 6a). Above the velocity of 740 m/s (2544 J), which corresponded to the beginning of the formation of the crater, there were practically no differences among the quantities of radial cracks of the four glass laminae (Fig. 6b).

With increasing projectile velocity many radial cracks bifurcated. The numbers of radial cracks crossing the circumferences of 10, 15, and 20 mm radii was a balance between deflected cracks and bifurcated cracks. Fig. 7a shows that the total number of radial cracks, considering the three glass laminae in the sample 666-3, that crossed the circumference of 10 mm radius increased linearly with the kinetic energy of the projectile. At the circumferences of 15 and 20 mm, the numbers of radial cracks were lower than at the 10 mm radius in the high kinetic energy range, because of the deflection of the cracks (Fig. 5c), although the bifurcated radial cracks increased exponentially with the kinetic energy of the projectile (Fig. 7b). The total number of radial cracks, considering the four glass laminae in the sample 5888-63, also increased linearly with the kinetic energy of the projectile (Fig. 8a) and the number of bifurcations/ramifications of the radial cracks increased exponentially with the increase of the kinetic energy of the projectile (Fig. 8b).

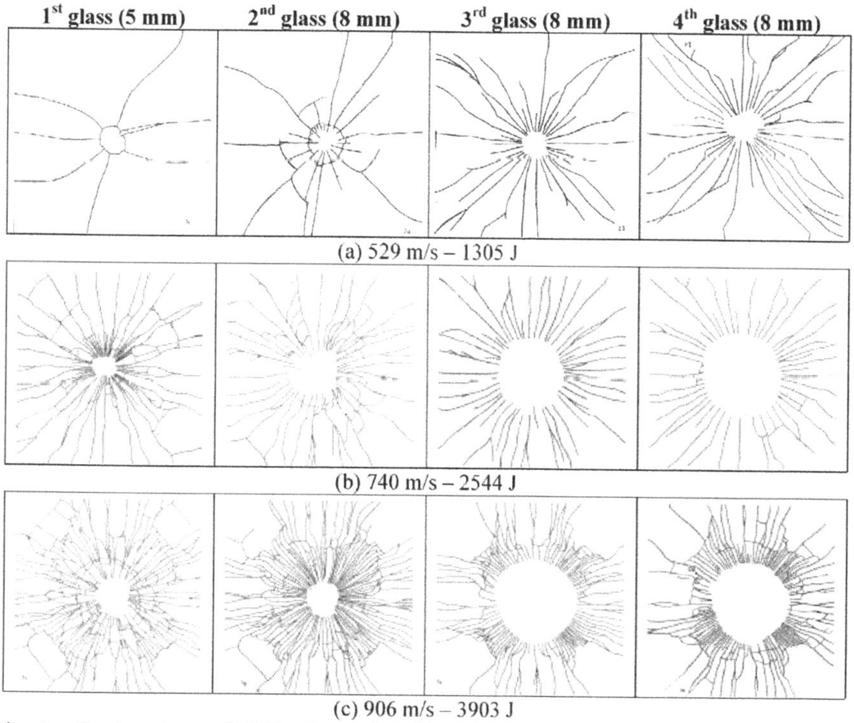

Fig. 6 - Crack patterns of 5888-63 sample subjected to level III-A ballistic test with varying projectile velocity (kinetic energy).

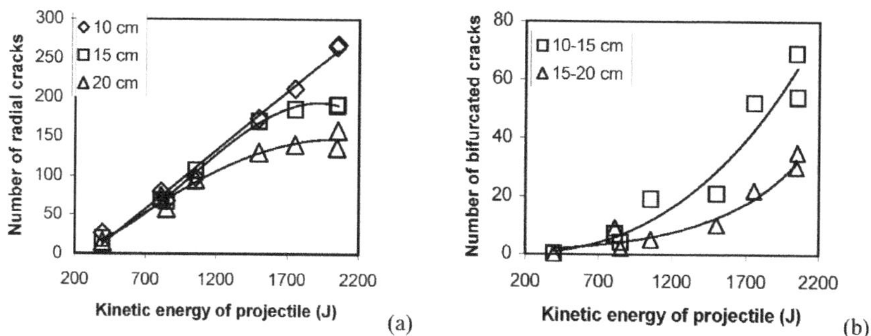

Fig. 7 – Number of radial cracks, which crossed the circumference with indicated radius, and number of bifurcations of radial cracks, in the circular section indicated, of the sample 666-3 as a function of kinetic energy of the projectile.

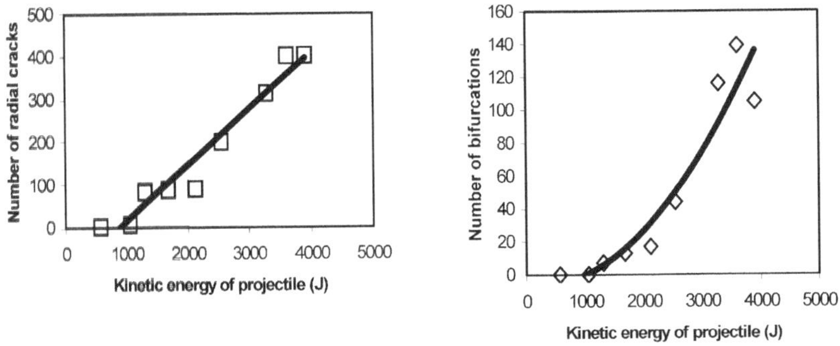

(a) (b)

Fig. 8 – Number of radial cracks, which crossed the circumference of 15 cm radius, and the number of bifurcations of the radial cracks, in the circular section between the 10 and 20 cm radii, of the sample 5888-63 as a function of kinetic energy of the projectile.

Fig. 9 shows the results of the whitened area and the crater area of the sample 5888-63 as a function of the energy of the projectile. Before the formation of the crater, the area of the semi-spherical plug slowly increased with the increase of the kinetic energy of the projectile between 1305 and 2117 J (529 and 657 m/s). Above 2544 J (740 m/s) the area of the crater increased quickly with the increase of the kinetic energy. The continuity of the results in the ranges of the area of the semi-spherical plug and the crater indicated that the semi-spherical plug is the precursor of the crater. The whitened area initially increased rapidly with the increase of the kinetic energy and then tended to stabilize with the formation of the crater. This result showed that initially the cone cracks propagated into the glass laminates defining the highly damaged envelope. Subsequently, the volume of the envelope is comminuted and eroded, increasing the crater area. In this regime, the mass of the cone crack envelope absorbs some of the kinetic energy of the projectile.

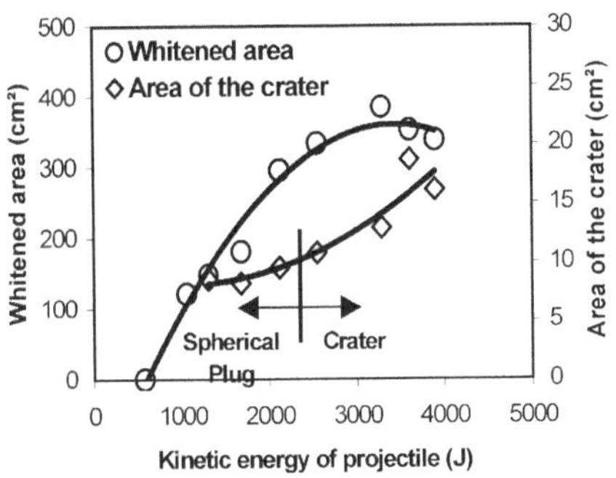

Fig. 9 - Results of whitened area and the area of the crater of the sample 5888-63 as a function of the kinetic energy of the projectile.

Fig. 10 shows the results of the whitened area, the crater area, and the depth of PC penetration of the 666-3 sample as a function of the energy of the projectile. Also in this sample, the whitened area tended to stabilize with the formation of the crater, and it tended to increase again when the penetration of PC started (Fig. 10a). The crater area initially increased with the increase of the kinetic energy of the projectile up to 1055 J (368 m/s) and then it lowered (Fig. 10b). This change of tendency was coincident to the initiation of penetration of the PC lamina at the kinetic energy of 1502 J (439 m/s) (Fig. 10c). This result showed that the mass of the glass in the cone crack envelope has limited capacity to stop the projectile. Above this limit, the mass of the envelope has less effectiveness to absorb the kinetic energy of the projectile. The depth of the PC penetration tended to increase linearly with the kinetic energy of the projectile (Fig. 10c).

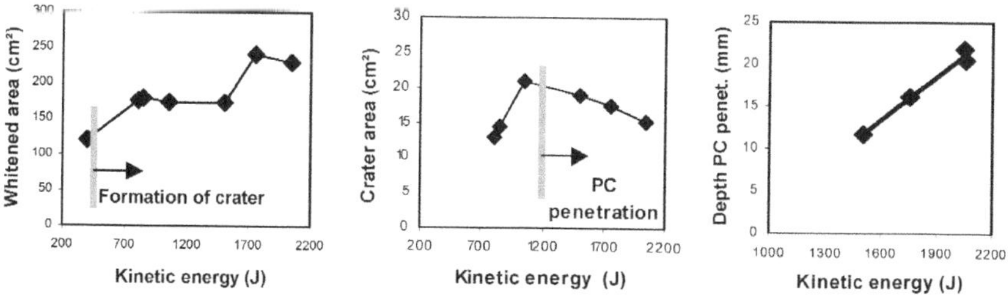

(a) (b) (c)

Fig. 10 - Results of whitened area, crater area, and depth of PC penetration of the sample 666-3 as a function of the kinetic energy of the projectile.

Based on the results of Fig. 9 and 10, a schematic diagram showing the sequence of failure events in the highly damaged region was constructed (Fig. 11). In stage I, the main event is the advancing of the cone crack envelope into the glass laminae. In stage II, the crushing and erosion of the glass occur in the cone crack envelope; and in stage III, the formation of the PC protuberance occurs.

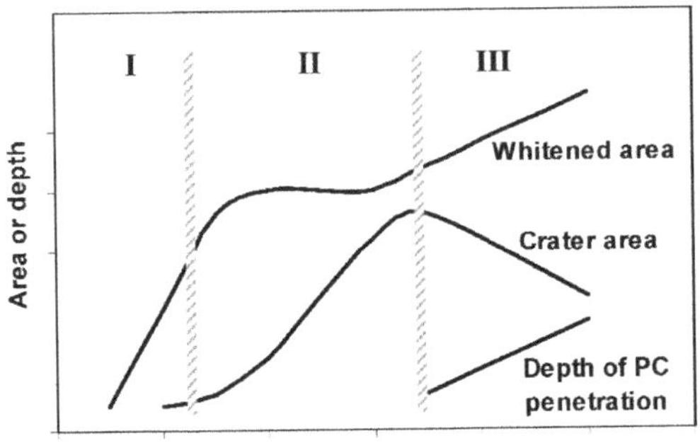

Fig. 11 - Sequence of failure events in the highly damaged region.

CONCLUSIONS

The following damage evolution in bulletproof glasses subjected to ballistic tests was observed. Main radial cracks form first, before the conical cracks. An increase of the kinetic energy of the projectile (E_p) causes an almost linear increase of the number of radial cracks and an exponential increase of the quantity of crack bifurcation. Concentric cone cracks formed the intense damaged envelope under the impact point. The sequence of failure events observed with the increase of E_p was classified in three steps: first, the cone crack envelope advanced gradually in the glass laminae up to the PC lamina. Subsequently, the material in the cone crack envelope was crushed and eroded. Finally, the highly localized plastic deformation of the PC lamina formed a protuberance. The protuberance depth increased linearly with the increase of E_p. The crater area formed in the frontal glass lamina increased with the increase of E_p during the two first steps and then decreased in the third step.

REFERENCES

[1]C.A. Smith, "Multilayered glass laminate having enhanced resistance to penetration by high velocity projectiles", USP 5,567,529, 1996.

[2]H. Charrue, F. Rifqi, R. Crepet, and C. Guillement, "Chemically toughened pane", USP 5,773,148, 1998.

[3]J.R. Manheim, "Method to manufacture a blindaged glass", USP 4,879,183, 1989.

[4]V.D. Fréchette and C.F. Cline, "Fractography of ballistically tested ceramics", Ceramic Bulletin 49, 994-97 (1970).

[5]D. Sherman and D.G. Brandon, "The ballistic failure mechanisms and sequence in semi-infinite supported alumina tiles", J. Mater. Res. 12, 1335-43 (1997).

[6]R.C. Bradt, M.E. Barkey, S.E. Jones, and M.E. Stevenson, "Projectile impact – a major cause for fracture of flat glass", The Glass Researcher, 11, 20-23 (2002).

CHARACTERIZATION OF CRACK BRANCHING AND FRAGMENTATION PATTERNS FOR ION-EXCHANGED GLASS

S. Jill Glass and Rajan Tandon
Materials Reliability Dept.
Sandia National Laboratories
Albuquerque, NM 87185

Daimon Heller
University of Missouri, Rolla
Rolla, MO 65409

ABSTRACT

The crack branching behavior of residually stressed glass disks was analyzed as a function of the residual central tension. Sets of glass disks were ion-exchanged in potassium nitrate for a range of times, resulting in central tension values from 44.4 to 63.4 MPa. Fragmentation of samples was initiated with a Vickers diamond indenter. The methodology used to characterize the fragmentation patterns is described. Over the stress range probed, the crack branching angle was not a function of the central stress, averaging around 40° for 1st order alpha cracks and 30° for second order alpha cracks. At each branching event trifurcation was common, with one new crack forming on each side of the initial crack.

INTRODUCTION

Fragmentation behavior of brittle materials is of interest in applications such as powder comminution, personal and vehicle armor, fracture of architectural and automobile glazing, and the fracture and dispersal of mining and construction materials. Stressed glass is a model material for studying fragmentation of brittle materials because its failure is linear elastic with no plastic deformation, it is transparent and there are no complications introduced by a microstructure, the strength and residual stress profile can be controlled by thermal tempering or ion exchange, and multiple fragmentation events can be observed in a single sample. The fragmentation behavior of stressed glass depends on both its residual stress profile and the manner in which it is loaded to induce fracture. With sufficient central tension a crack in stressed glass will branch repeatedly; the result is often relatively uniformly-sized dicing (cube-like) fragments. The greater the amount of energy available from the residual stress profile and the stored elastic energy due to loading (determined by the glass strength and loading conditions), the smaller the fragments. In contrast annealed glass typically breaks into large elongated shards. There is little or no crack branching as a result of the limited amount of energy stored in the glass prior to fracture.

The process of pre-stressing glass by ion exchange involves placing the glass into a molten salt where smaller ions (e.g., Na^+) in the glass are exchanged with larger ions (e.g., K^+) from the salt. The exchange produces an ion stuffing effect on the glass surface, which results in a compressive stress at the surface and a tensile stress in the interior. The stress profile for ion-exchanged glass differs considerably from that of thermally tempered glass (Fig. 1). For ion-exchanged glass diffusion depths are typically 200 μm or less, the magnitude of the surface compression can be very high (up to 1000 MPa), the magnitude of the surface compression decreases rapidly as a function of distance into the surface, and the tensile stress profile is flat across a significant region of the glass interior. A thermally tempered glass sample has a parabolic stress profile, with the magnitude of the surface compression value being twice the central tension value. Much higher levels of compression and tension can be achieved with the ion-exchange process than for the thermal tempering process.

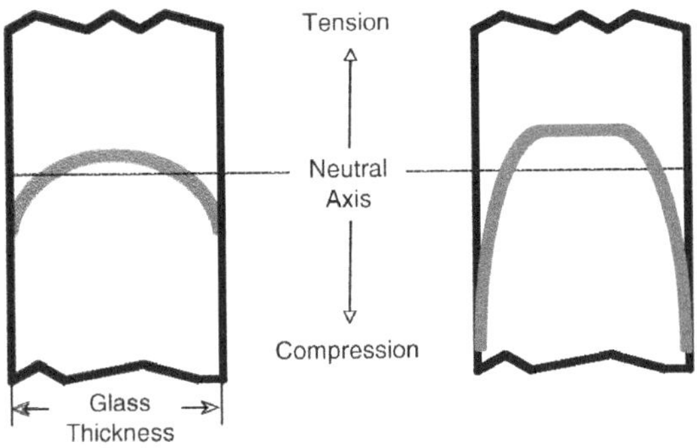

Figure 1: Comparison of residual stress profiles for thermally tempered (left) and ion-exchanged glass (right).

In dynamic crack propagation, a crack grows until it reaches a critical stress intensity, K_{IB}, at which point it branches (bifurcates) and generates two or more new cracks. In the cracking that occurs during the failure of stressed glass, there is enough stored energy from the stress profile to produce multiple branching events, leading to complete fragmentation of the glass. The higher the central tension, the shorter the distance each crack will propagate before branching, and the smaller fragment size.

Characterizing and understanding the fragmentation behavior of stressed glass are essential ingredients for predicting and simulating the manner in which it fails. There has been considerable interest in selecting compositions and/or manipulating the processing of ion-exchanged glasses to enhance their strength and to control their fragmentation behavior. This includes controlling the size and/or shape of the fragments produced upon glass failure. This control is of interest for architectural glazing because large shards of glass, sent airborne when an explosion or severe weather occurs, can cause numerous injuries and casualties.[1] In response to this problem, Glass et al.[2] evaluated using engineered stress profile (ESP) glasses, which fail over a narrow range of stresses,[3] and break into small fragments.[4]

Several studies have characterized and simulated fragmentation for brittle materials, including those by Sakai et al. on thermally tempered glass,[5] Quinn et al. for annealed glass,[6] and Mecholsky et al. for MgF_2.[7] The Sakai study measured crack branching distances and angles and used these in conjunction with fractal dimensions to simulate fragmentation patterns.

Experimental data and models are available for crack branching distances and fragment sizes for ion-exchanged glass;[8] however, there is little or no information regarding the number of branches and branching angles, how they change over the sample area, and how they are affected by the residual central tension. Our observations of fragmented ion-exchanged glass samples have shown that although cracks occasionally bifurcate (one crack forks to become two), 'trifurcation' (two cracks initiate on each side of the original crack, while the original crack continues with little or no change in direction) seems to occur most frequently. In some instances three or four branches from the main crack are observed.

In a parallel study we have been using simulations to evaluate the effects of various fragmentation parameters (number of initial cracks, branching distance, number of branches, branching angle, and fracture initiation location) on the fragment size, size distribution, and shape.[9] The fragmentation simulations were designed to analyze the effects of varying one parameter of the fragmentation pattern, while holding the others constant. Data for branching angles and number of branches were needed to produce realistic fragmentation simulations for ion-exchanged glass. By determining whether changes in central tension affect the crack branching angle and number of branches, more realistic simulations can be produced.

The objective of the present study was to characterize the crack branching behavior in ion-exchanged glass samples with a range of central tension values, to determine the branching angles and number of branches, and to determine if these parameters are affected by the residual central tension. Samples were supported on a foundation with no bending loads and fracture was initiated using a Vickers diamond indenter. Therefore the energy to drive crack propagation was due only to the residual stress profile. Because there are no formal characterization rules governing how fragmentation events should be described and categorizing them is a subjective process, nomenclature and a methodology for characterizing the patterns were developed.

METHODOLOGY

Sample Preparation

The glass used in this study was Corning 0317, a sodium aluminosilicate composition engineered to facilitate ion exchange. Disks that were 2.54 cm in diameter were machined from 1.8 mm thick glass plates. The disks' surface was left in the as-received condition; no additional polishing or grinding was performed.

From our previous studies samples exchanged beyond twenty-two hours at 450°C had fragments so small and numerous that the recorded image did not possess sufficient detail to be able to identify fragmentation events or measure branching angles. Samples exchanged for less than ten hours did not exhibit sufficient branching. Hence seven sets of five samples each were exchanged at 450°C for times ranging from ten to twenty-two hours, in two-hour increments.

A commercial ion-exchange bath[&] with laboratory grade KNO_3 was used for the ion exchange. The glass disks were suspended above the molten salt bath for 10-15 min, allowing them to heat slowly before immersion. This minimized the chance of thermal shock. At the end of the exchange period the samples were removed from the bath and suspended above it for another 10-15 min before being removed to cool to room temperature. They were then rinsed with distilled water and dried.

Stress Measurement

Prior to fragmenting the samples, their central tensile stresses were measured using an optical birefringence technique.[10] Stress in glass causes birefringence, a quantifiable optical property. The technique for determining the stress involves measuring the optical retardation, R_{Total}, in the central region of the disk cross section, and relating that value to the tensile stress, $\sigma_{Tensile}$, using Eqn. 1. The optical path length, d (the disk diameter), and the stress optic coefficient, K (27.5 nm/cm·MPa), for the Corning 0317 glass disks are needed to calculate the stress.

[&] CH-60 Chemical Lens Tempering Unit, Kirk Optical Co., Inc., Freeport, NY

$$\sigma_{Tensile} = \frac{R_{Total}}{(d)(\mathrm{K})} \qquad (1)$$

Fragmentation Initiation

Successful fragmentation pattern characterization depends on being able to fragment a sample using the same loading conditions, keeping the pieces intact, and then obtaining a clear image of the pattern. A Vickers diamond indenter* at a load of 294 N was used to initiate surface cracking and fragmentation of the glass disks. Fragmentation normally occurred when the indenter was being unloaded from the glass surface. To keep the samples intact for analysis, they were taped with clear plastic tape on both sides prior to indentation. The tape on the side of the sample to be indented had a small hole to allow uninhibited access to the glass surface.

Optical Imaging

After fragmentation, digital images of the samples were captured with an optical microscope# using digital capture software.@ Color images were taken using transmitted light and then converted to black and white. Figure 2 shows representative images.

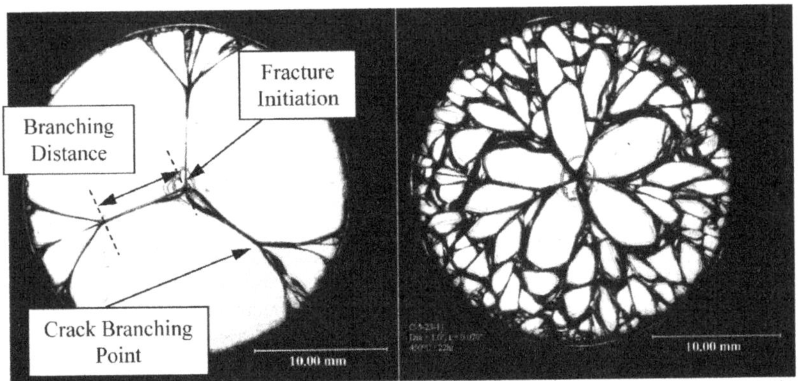

Figure 2: Left: a 450°C/10hr exchange sample with fracture initiation, crack branching distance, and one crack branching point labeled. Right: a 450°C/22hr exchange sample

Nomenclature and Characterization Rules

Fracture patterns are shown in Figures 2 and 3. The fracture initiation location in the center of the disk is labeled in Fig. 2. There are usually four cracks that propagate from the Vickers diamond (4-sided) indentation site; however, there are also examples of three and five cracks as shown in Fig. 2. The cracks propagate in nominally straight lines until they reach the sample edge or they intersect another crack and stop. The distance a crack travels between

* Zwick Model 3212 Indenter, Kennesaw, GA

M7S Stereomicroscope, Wild Heerbrugg Ltd., Heerbrugg, Switzerland

@ Spot Advanced 4.1 Software, Diagnostic Instruments, Inc., Sterling Heights, MI fracture initiation and the next branching point is the 'Branching Distance' and varies with the central tensile stress and sample thickness.[11-13]

To distinguish where specific angles were recorded and to organize and analyze the data, a naming system that included three kinds of cracks was adopted (Fig. 3). Initial cracks (green) are those that originate from the indentation. Alpha cracks (blue) are branches off of initial cracks. Beta cracks (red) branch off of alpha cracks. The terms 1st, 2nd, and 3rd order are used to describe the sequence of branching events when there is more than one level of alpha or beta cracks. For example, cracks that form at the first branching point of an initial crack are called 1st order alpha cracks, whereas cracks forming at the second branching point are called 2nd order alpha cracks. Only 1st and 2nd order alpha branching angles were measured for this study. Beta crack branching angles were not measured because there are fewer of them and they are generally located close to the edges of the sample where other phenomenon (e.g., changes in the stress profile or the reflection of stress pulses) may affect them. Because of occasional difficulties in differentiating between a branching event and the intersection of cracks, and in identifying where a branching point is located, it was important to clearly identify, mark and record all events so that the values and their locations on the disk could be easily identified if a discrepancy in measured values or observations should arise.

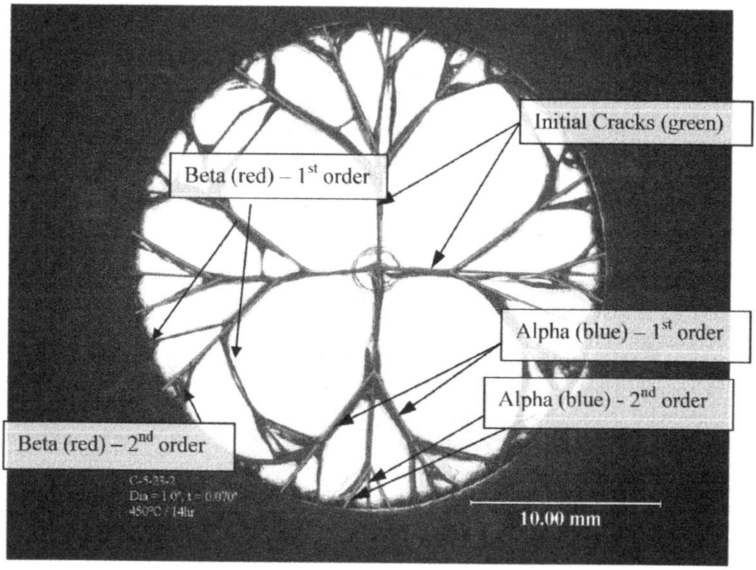

Figure 3: A 450°C/14 hour sample with examples of the nomenclature used in this study. Initial originate at the fracture origin, alpha cracks are those stemming from initial cracks, and beta cracks stem from alpha cracks.

Image Processing and Pattern Analysis

The black and white images used for pattern analysis were enhanced by removing extraneous features such as tape marks and smudges, and by adjusting brightness, contrast, and sharpness. Branching angle measurements were made using skeletonized images of the fragmentation patterns rather than the enhanced images. Two approaches were attempted to create the 'skeletonized' images. Both started with the cleaned up black and white images. The first approach used partially automated image processing and analysis routines available in ImageJ, freeware available from the National Institute of Health (Approach 1), while the second relied on manual line drawing of the cracks (Approach 2).

Both surface chipping and cracks that are not perpendicular to the surface cause the microscope's light path to reflect light away from the microscope column. This produces features that are similar to the cracks themselves or wide-looking black cracks. Efforts to manipulate the images into a form more conducive for graphical filters were never satisfactory. Additionally expert knowledge is required to accurately identify where the crack is located and where it branches, especially when surface chipping occurs. Therefore automated image analysis and reconstruction algorithms (Approach 1) could not be used to produce skeletonized images that were accurate representations of the fragmentation patterns.

The final methodology used to skeletonize and analyze the fragmentation patterns used a mixture of image processing and manual construction of crack traces (hybrid of Approach 1 and 2). The black and white images were imported into a software package[#] with layering capabilities. The software was used in conjunction with manual reproduction of the crack patterns with a series of straight lines. Line segments were color-coded according to whether they were initial (green), alpha (blue), or beta cracks (red). Anything beyond beta cracks was also colored red and included to obtain a sense of the complete fragmentation pattern. A representative sample of one of the skeletonized images is shown in Fig. 4. Angles were designated as 1st, 2nd, 3rd, or n^{th} order alpha crack branching angles on the images and then measured using the linear crack representations. The branching angles, ϕ, for 1st and 2nd order alpha cracks, were measured from the initial crack (green) to each alpha crack (blue), as shown in Fig. 4. The surfaces of several disks were analyzed with scanning electron microscopy (SEM) to help identify more precisely the location of trifurcation' events and to examine them in greater detail.

RESULTS AND DISCUSSION

The average central tension as a function of the square root of the exchange time for samples (five in each set) exchanged at 450°C is shown in Fig. 5. For each exchange time the standard deviation of all stress measurements was < 0.33 MPa. The central tension increases monotonically with the square root of the exchange time; the values range from 44 to 62 MPa. Images of the sets of fragmented samples are shown in Fig. 6. The samples within sets are similar in appearance, but each is unique. Sample sets for increasing ion-exchange times and residual tensile stresses can be distinguished by their increasing crack density. The branching angles, ϕ, for 1st and 2nd order alpha cracks, measured from the initial crack (green) to each alpha crack (blue), as shown in Fig. 4, are plotted in Fig. 7 as a function of central tensile stress. Image analysis and crack branching angle measurements were not possible for the highest stress (61.5 MPa) sample set because of its high crack density and complexity.

[#] Adobe Photoshop, Adobe Systems Inc., San Jose, CA

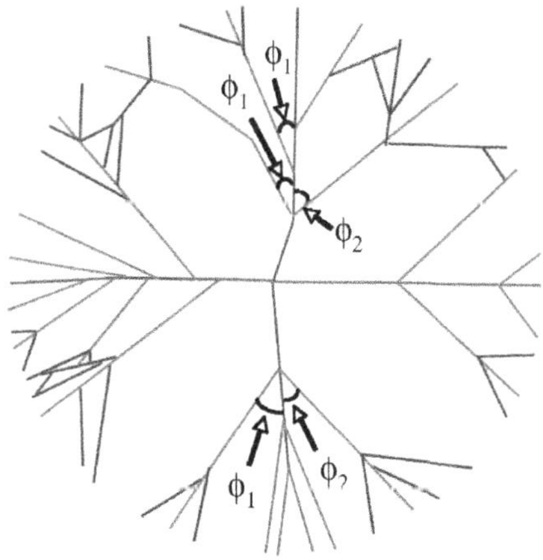

Figure 4: Example of 'skeletal' image used for measurement and analysis. The image is for a sample that was exchanged at 450°C for 14 hr.

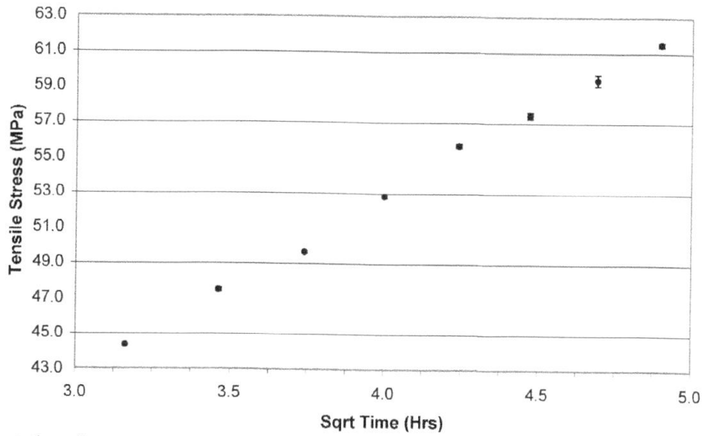

Figure 5: Central tensile stress vs. square root of the exchange time, including error bars (one standard deviation). Exchange temperature=450°C, glass disk thickness=1.8 mm.

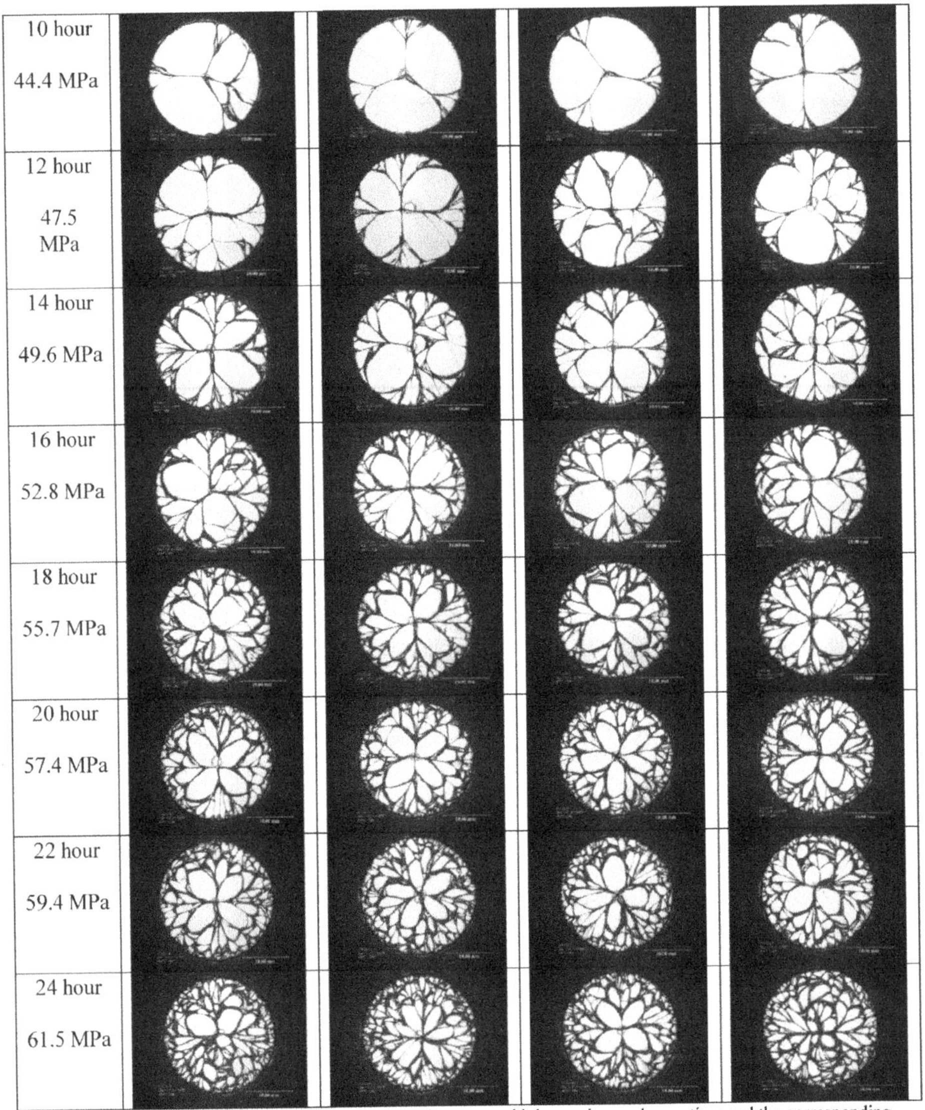

Figure 6: Crack density increases and fragment size decreases with increasing exchange time and the corresponding increase in central tension (shown to left of each sample set).

For the ~15 MPa range of tensile stress for the samples in this study there was no change in branching angle; however, the relatively large standard deviations may mask subtle changes in the branching angle over this range. The origin of this variability is likely due to variations in the orientation of the local stress field. The stress field is affected by nearby cracks and the proximity to the sample edge. The smaller the number of adjacent cracks, the larger the angle tends to be. The average first order alpha angle is 40°; the average second order alpha angle is 30°. The lower value for the second order alpha crack branching angle may be related to both the larger number of cracks that already exist nearby when these cracks are forming, and to the proximity of these to the edges. Both of these crack branching angles are larger than the branching angle of 22.26° determined by Sakai et al. for thermally tempered glass.[5] Their crack branching angle measurements were made for a large sample of thermally tempered glass with a well-developed fragmentation pattern. As tensile stresses from a few MPa to in excess of 125 MPa can be produced in Corning 0317 glass, the stress range in this study represents a very small subset of the range of possible stresses for ion-exchanged glass.

The relatively small diameter of the samples used in the present study only allowed us to view and analyze a small part of what would be a much larger fragmentation pattern. Larger diameter samples would allow many more branching events to be observed and would minimize the edge effects that may contribute to the lower angles measured for 2nd order alpha cracks. An alternate approach for small samples would be to initiate fracture in a region of the sample other than the center. Because the initial crack will have a considerably longer path to travel before reaching the edge, the creation of a greater number of higher order cracks would be possible.

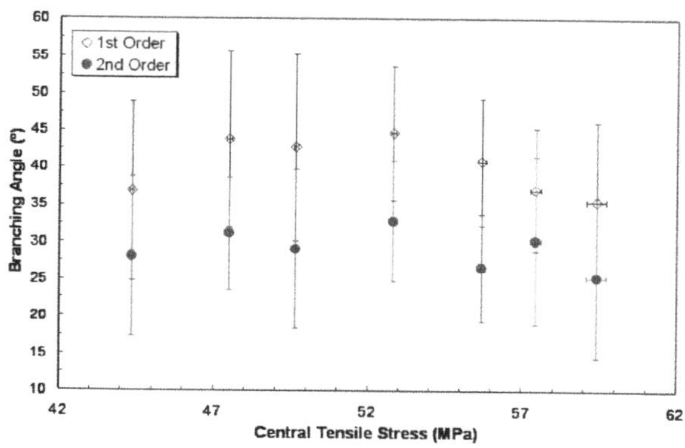

Figure 7: First and second order alpha crack branching angles, (average of all ϕ_1 and ϕ_2), vs. central tensile stress for ion-exchanged Corning 0317 glass disks (diameter=2.54 cm, thickness=1.8 mm).

As this was primarily a study to estimate the crack branching angles for use in fragmentation simulations, only a limited examination of the physical branching phenomenon was performed using the SEM. The extensive surface chipping that occurs when the crack intersects the sample surface makes it difficult, if not impossible, to observe the crack behavior before it reached the surface. By removing these chips, a better representation of the underlying crack and its behavior could be obtained (Fig. 8 and 9). What sometimes appear to be multiple branches at the surface is an artifact of the surface chipping. Figure 8 shows a low magnification picture of the initial cracks that emanate from the indentation. After the chips are removed from the right-hand crack it is clearer that this is only one crack. Figure 9 shows a close-up of a crack branch from the top left hand corner of Fig. 8. Once the surface chips are removed it is clearer that the initial crack forms two branches, one on each side of the initial crack as highlighted with the dotted lines.

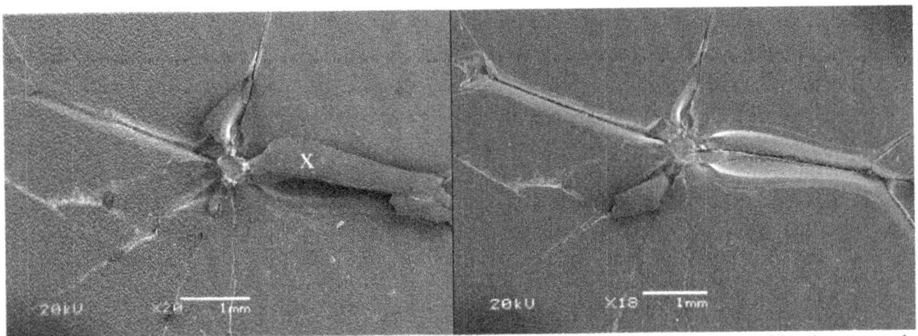

Figure 8: Before (left) and after (right) SEM images depicting the change in appearance caused by the removal of surface chips (marked with x) around the fragmentation origin (indentation).

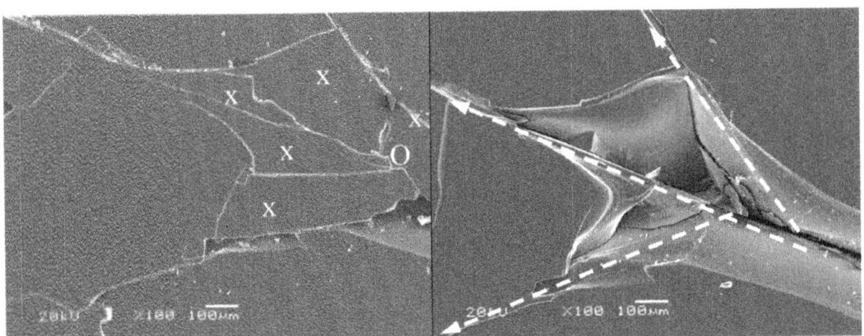

Figure 9: Before (left) and after (right) SEM images depicting the change in appearance highlighted by the removal of surface chips (shown with x's) at the first crack branching event shown in the top left corner of Fig. 8. The O in the left image was originally thought to be a 'trifurcation. After the chips are removed the occurrence of two branches is clearer at the intersection of the dotted lines.

The crack branching angle as normally understood, and the crack branching angle measured in this study, is the angle of intersection of the crack branch with the initial crack on the disk surface as shown schematically in Fig. 10. In stressed glass, crack propagation and crack branching occur throughout the tensile zone before intersecting the compressive surface. Beauchamp[14] showed the presence of fracture 'tongues' or 'ribbons'; these are precursors of macroscopic crack branching. The branching angle measured on the disk surface results from the orientation of the tensile stress field after the fracture tongue reaches criticality and becomes a macroscopic crack branch. Several fracture tongues may develop off the initial crack near the critical crack branching distance, but some do not reach criticality, where they would then turn back to intersect with the initial crack as shown in Fig. 10. Although Beauchamp's observations and study were on thermally tempered glass (Fig. 11), similar behavior is observed in ion-exchanged glass.

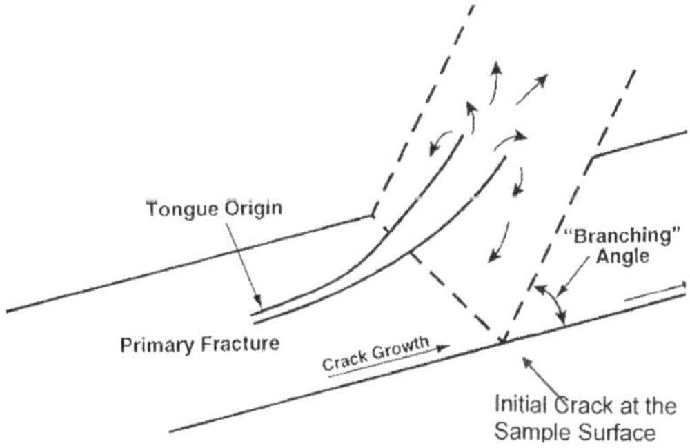

Figure 10: Diagram showing the development of a fracture tongue that becomes critical and produces a macroscopic crack branch that intersects the initial crack as depicted by Beauchamp.[14]

Fig. 11: Crack branching in thermally tempered glass. The failure origin is at the position marked with O. Fracture tongues that reach criticality can be seen at positions marked with arrows.

A determination of the number of branches at each branching event as a function of the tensile stress could not be made because of several difficulties. Frequently it was impossible to determine where a branch was located relative to other branches, primarily because of surface chipping and the thickness of the crack trace. Also, because crack branching starts below the surface in the tensile region, and we were observing a crack branch only after it has propagated forward and laterally toward the surface, what we observe on the surface is not representative of what occurs during the original branching.

At this time we can qualitatively only say that there are usually two branches that form at each branching event, and that the initial crack nominally continues in the same direction it was propagating prior to branching. However, there are instances where either one or three branches are formed. If there was only one branch at a surface branching location, there would often be another branch soon after it. This may be related to what happens to two branches that initiate at the same or nearly same time, but propagate in a different manner (due to differences in the local stress field), and then intersect the surface at a different time and place. Some branches even stop propagating before they intersect the surface as the initial crack moves forward and the magnitude of the local tensile stress drops.

The occurrence of two or more branches originating at each branching point is contrary to what is usually reported for crack branching during uniaxial or biaxial flexure loading of glass and ceramics samples. In these samples the initial crack usually generates one branch, giving the appearance of the crack forking or splitting into two. In these tests, in which the tensile stress is a maximum on the surface, there may be a strong bias towards only one branch developing because of the presence of substantial surface flaws that locally raise the stress intensity on one side of the initial crack. Once one branch has formed the stress intensity drops significantly precluding the formation of another. In stressed glass samples in which the crack branching occurs inside the glass (away from the surface), the size of the intrinsic flaws is usually very small and much more uniform. Therefore once the initial crack reaches K_{IB}, it is as likely to branch on one side of the crack as the other; the result is usually two "simultaneous" branches forming before the stress intensity drops.

Rice[15] suggested that cracks will branch to fragment a sample in a manner that most equally divides the sample surface. In light of the results of this study, this behavior appears to be accomplished by a combination of number of cracks and crack branching angle. If the number of branches is smaller, the branching angle is larger and vice versa. If the number of branches does increase with increasing tensile stress, this suggests that the branching angle will be smaller with increasing stress. The results of this study do not show that, but may have been measured over too small a stress range.

By analyzing 2.54 cm diameter samples, only a few higher order branching events could be examined. Our limited observations indicated that branching angle decreases with proximity to the disk edge, from 40° to 30°. To determine if this is a real effect of increasing branching order as the initial crack grows longer, or an edge effect, larger diameter samples need to be analyzed. Two 7.6 cm diameter disks exchanged for two different times were qualitatively examined to check for consistency with the previous observation. The limited results indicate that higher order alpha branching angles are not smaller than first order alpha branching angles. Additional analysis of larger samples with more branching events would also allow for the collection of statistically significant results.

CONCLUSIONS

Fragmentation was initiated in ion-exchanged glass disks using a Vickers indentation. Automated image processing and analysis routines could not be used to replicate or analyze the fragmentation patterns for these samples because of surface chipping and image artifacts, and because of the expert knowledge required to interpret the sequence of fragmentation events. The combined image processing/manual method used for fragmentation pattern replication and analysis would not be efficient for large sample sets, even for the moderately stressed samples used in this study. For highly stressed samples, i.e., those above ~60 MPa, this method cannot be used because of the high crack density and pattern complexity and the difficulty of obtaining images of sufficient clarity.

The alpha crack (crack branches off the initial indentation cracks) branching angles measured for a limited range of central tension values (44 to 63 MPa) showed no dependence on the residual tensile stress of the ion-exchanged samples. The branching angle, ϕ, was 40° for 1st order alpha cracks, and 30° for 2nd order alpha cracks. The standard deviation of the measured angles was around 30% of the actual values, primarily due to variability in the number of branches. The presence of nearby branches affects the stress field and orientation for subsequent branches. If there were fewer branches the branching angle tended to be larger. The branching angle for second order cracks was approximately 25% less than first order cracks. This lower value for the higher order alpha cracks may be related to their proximity to the disk edges where other effects may influence the stress field and branching behavior. Our examination of crack branching led us to the conclusion that there are usually two crack branches occurring within a very small distance of each other on opposite sides of a crack. There is a tendency for a higher number of branches at higher stresses.

REFERENCES

[1]S. J. Glass, "Assessment, Development, and Testing of Glass for Blast Environments," Sandia National Labs Report, SAND2003-2149, June 2003.

[2]S. J. Glass, M. B. Abrams, and R. V. Matalucci, "New Glass Technologies for Advanced Architectural Surety®: Engineered Stress Profiles in Soda-Lime-Silica Glass," Sandia National Labs Report, SAND2000-3001, Dec. (2000).

[3]M. B. Abrams, D. J. Green, and S. J. Glass, "Fracture Behavior of Engineered Stress Profile Soda Lime Silicate Glass," *J. Non-Cryst. Solids*, 321, pp. 10-19, (2003).

[4]D. J. Green, R. Tandon and V. M. Sglavo, "Crack Arrest and Multiple Cracking in Glass Using Designed Residual Stress Profiles," *Science*, 283, pp, 1295-97 (1999).

[5]T. Sakai, M. Ramulu, A. Ghosh, and R. C. Bradt, "A Fractal Approach to Crack Branching (Bifurcation) in Glass," pp. 131-146 in Fractography of Glasses and Ceramics II, *Ceram. Trans.,* Vol. 17. Edited by V. D. Fréchette and J. R. Varner, The American Ceramic Society, (1991).

[6]G. W. Quinn, J.B. Quinn, J.J. Mecholsky, and G.D. Quinn, "Computer Modeling of Crack Propagation Using Fractal Geometry," *Ceram. Eng. and Sci. Proc.*, 26 [2] 77-84 (2005).

[7]J. J. Mecholsky Jr., R. Linhart, B.D. Kwitkin, and R.W. Rice, "On the Fractal Nature of Crack Branching in MgF_2," *J. Mater. Res.,* 13, 11 (1998).

[8]R. Tandon and S. J. Glass, "Controlling the Fragmentation Behavior of Stressed Glass," pp. 77-91 in Fracture Mechanics of Ceramics, Vol. 14, Active Materials, Nanoscale Materials, Composites, Glass, and Fundamentals. Edited by R. C. Bradt, D. Munz, M. Sakai, and K. W. White, Springer, (2005).

[9]J. L. Rygel, R. Tandon, M. Braginsky, and S. J. Glass, "Simulation of Fracture Patterns Obtained During Fragmentation," presentation at the 107th ACerS Annual Meeting, Baltimore, MD, Apr. 11-13, (2005).
[10]J. W. Dally and W. F. Riley, *Experimental Stress Analysis*, McGraw-Hill, Inc., USA (1978).
[11]R. A. McMaster, D. M. Shetterly, and A. G. Bueno, Annealed and Tempered Glass, pp. 453-459 in *Engineered Materials Handbook*, Vol. 4, ASM International, (1991).
[12]P. D. Warren, "Fragmentation of Thermally Tempered Glass," pp. 389-400 in Fractography of Glasses and Ceramics IV, *Ceram. Trans.* Vol. 122. Edited by J. R. Varner and G. D. Quinn, The American Ceramic Society, Ohio, (2001).
[13]E. Bouyne and O. Gaume, "Fragmentation of Thin Chemically Tempered Glass Plates," *Glass Technol.*, 43C 2002, Proc. XIX Int. Congr. Glass, Edinburgh, 1-6 July, 2001.
[14]E. K. Beauchamp, "Mechanisms for Hackle Formation and Crack Branching," pp. 409-445 in Fractography of Glasses and Ceramics III, *Ceram. Trans.*, Vol. 64. Edited by J. R. Varner, V. D. Fréchette, and G. D. Quinn, The American Ceramic Society, (1996).
[15]R. W. Rice, "Ceramic Fracture Features, Observations, Mechanisms, and Uses," Fractography of Ceramic and Metal Failures, ASTM STP 827, J.J. Mecholsky Jr., and S.R. Powell Jr. eds., ASTM 5-103 (1984).

CHARACTERIZATION OF DYNAMIC FAILURE PROCESS OF Si_3N_4 CERAMICS, PART I: TEST PROCEDURES, FRACTURE ENERGIES, AND FRACTOGRAPHIC ANALYSIS

Maria B. Maros
Associate Professor
University of Miskolc, Department of Mechanical Engineering,
Mechanical Technological Research Group of HAS UM, 3515 Miskolc – Egyetemváros, Hungary

Nikoletta Kaulics
Ph.D. candidate
University of Miskolc, Department of Mechanical Engineering, 3515 Miskolc – Egyetemváros, Hungary

Peter Arató
Scientific Adviser
Research Institute for Technical Physics and Materials Science of Hungarian Academy of Sciences, RITPMS, HAS, H-1525 Budapest, Pf. 49.
Hungary

ABSTRACT

Brittleness of engineering ceramics is one of the most critical properties limiting their application as structural materials. Therefore mechanical tests providing toughness characteristics of these materials have special importance. The instrumented impact test represents a quick and simple method for characterizing the brittle/ductile behavior of different materials and provides valuable information about the damage process that take place over time. While this method is widely applied for metallic and polymeric materials, it is still rarely used for examining ceramics. This paper introduces new results on instrumented impact testing of Si_3N_4 based ceramics. Different aspects of the dynamic failure process are considered, providing various mechanical characteristics of the tested material. These results are completed with analysis of the fracture surfaces and failure mechanism using fractography.

The main objectives of the first part of the experimental and theoretical work were as follows:

- Determining various dynamic mechanical characteristics of Si_3N_4 ceramics and investigating the effect of certain test parameters on the measured data;
- Introducing the problem of determination of the force belonging to the unstable crack propagation at high loading rates for notched and unnotched specimens;
- Analyzing the failure origin and fracture mechanisms of the investigated Si_3N_4 ceramics by fractography.

A companion paper[1] presents results for determination of dynamic fracture toughness, K_{Id}.

INTRODUCTION

Technical ceramics in structural applications face the problem of their brittleness. Their ability to sustain mechanical loading is limited especially under dynamic loading conditions.

Testing methods providing characterization of brittle/ductile behavior may contribute to more reliable estimation of their performance in such conditions. However, reproducibility of the impact tests of brittle materials is poor due to the statistical nature of their mechanical behavior. Testing statistically reliable sample sizes is rather expensive. Today there is no commonly used, accepted, and standardized method for mechanical characterization of ceramics under dynamic loading. Therefore, it is of special importance to develop reliable testing procedures. Due to the above mentioned gap in appropriate standards related to engineering ceramics, we have used an existing standard procedure recommended for metallic materials.[2]

During our experimental work we applied instrumented impact tests. Various dynamic strength and toughness features such as three-point bending strength, total fracture energy, and impact work have been determined. The effects of certain test parameters such as the pendulum mass, loading rate, and notches on the dynamic mechanical characteristics were also investigated.

This paper presents results on the mechanical characterization of the tested silicon nitride ceramic in dynamic loading conditions. Results of morphological analysis of the fracture surface by scanning electron microscopy that reveal the characteristic regions of the fracture surface as well as mechanisms of crack propagation are also presented. Part II[1] presents results for dynamic fracture toughness including a novel application of an electro emission technique to ceramics.

EXPERIMENTAL WORK

The basic principle of the instrumented impact test is similar to the traditional Charpy impact test. Using special instrumentation from complementary devices the amount and quality of the information that could be derived from the classical test, may be multiplied.[3]

In case of ceramics the following characteristics may be determined from the measurement:

- Load-time (F-t) and load-deflection (F-s) diagrams derived from the F-t diagram. These curves provide valuable information about the nature of the damage process, as well as serve as a basis for determining further important material data;
- Loading characteristics: F_{max}, the maximum force of the fitted curve through the oscillations, $F_{i,u}$, force, corresponding to the initiation of unstable crack propagation;
- Toughness features: KV: impact energy based on the potential energy of the pendulum, shown by the machine dial; E_{max}: total fracture energy, calculated from the F-s diagram; K_{Id}: dynamic fracture toughness (provided that $F_{i,u}$ or t_F is measurable);
- Strength feature: $M_{TPB,d}$: dynamic three-point bending strength;
- Other characteristics: t_F: fracture time; s_{max}: maximum deflection of specimen.
- Dynamic Young modulus: E_{dyn} by supplementary strain measurement on the specimen.

Instrumented impact test machines characteristically consist of four main parts (see Fig. 1.): The impact pendulum (1) or specimen-striker system with the instrumented tup, the load cell (2) for measuring the force indirectly by two strain gauges on two sides of the pendulum, the amplifier (3) to magnify the detected signals to a reliably measurable value, (e.g., N=1000x) and appliances for data-storage, i.e., data-acquisition card and/or digital storage oscilloscope (4).

Our experiments have been performed on a CEAST 25/15/2 type machine, that was supplemented with an electric-emission measuring system, to expand information on the dynamic failure process, that can be derived from the instrumented impact test.[3]

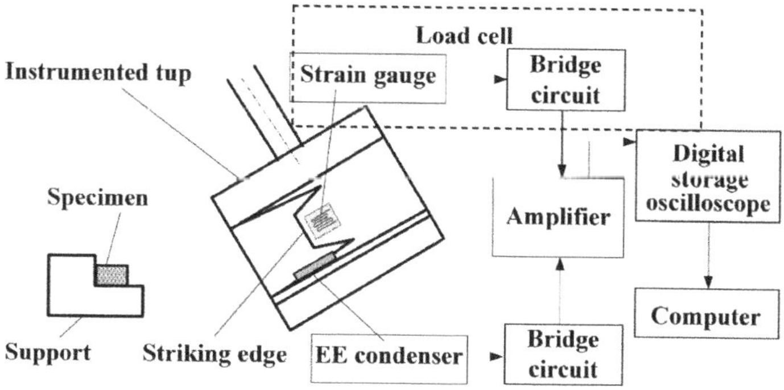

Fig. 1. The instrumented impact test machine and the measuring circuit

Si_3N_4 based ceramics, having composition of 90 % Si_3N_4, 4 % Al_2O_3 and 6 % Y_2O_3, were investigated. Specimens of the same nominal composition have been produced by the Research Institute for Technical Physics and Materials Science of Hungarian Academy of Sciences (RITPMS, HAS). All the specimens were produced by uniaxial pressing using with polyethylene-glycol additive, which were extracted on air at 600 °C, after pressing. Firing of the pieces occurred in two-stage gas-pressure sintering in a hot isostatic press (HIP).

The specimens were rectangular prisms with dimensions of 4.9 mm x 3.2 mm x 49 mm. We performed measurements with impact energy of 15 J and 2J. The pendulum masses were 2.219 kg and 0.473 kg, respectively. In both cases, two loading rates were applied. They were 3.68 m/s and 0.99 m/s for the bigger pendulum and 2.88 and 0.78 m/s for the smaller pendulums. In case of the 2J pendulum, unnotched and notched specimens (with a notch depth to specimen depth ratio a/W = 0.5) were tested. The distance between the supports was l = 40 mm. For each testing regime, two sets of specimens were investigated.

TEST RESULTS AND EVALUATION

Load-Time (F-t) Diagrams

Load-time (F-t) diagrams were recorded. Characteristic force-time diagrams for unnotched specimens taken by 2J max. impact energy pendulum for v = 0.78 m/s and 2.88 m/s loading rates are illustrated in Fig. 2. Various information on the material behavior and the failure process could be derived from F-t diagrams. F-t curves illustrated in Fig. 2 relate to a brittle, elastic material, as the initial stage of the diagrams has no diminishing section. At the higher loading rates, strong force oscillation caused by inertia effects at the initial stage of the diagrams could be observed as seen in Fig. 2b. These F-t diagrams were investigated according to the following aspects:

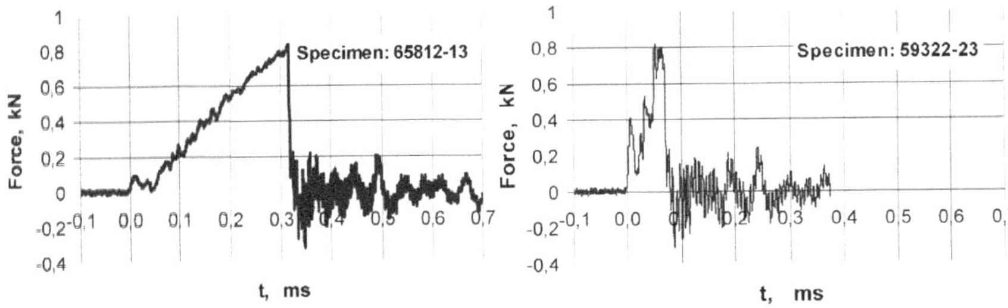

a) v = 0.78 m/s b) v = 2.88 m/s

Fig. 2. Typical force-time diagrams of unnotched Si_3N_4 specimens during impact loading with the 2 J pendulum

1. Effect of the Pendulum Mass. In the case of pendulum with 15 J impact energy, the F_{max} values fall into the range of 0.53 to 1.01 kN for unnotched specimens. For the 2 J pendulum, these forces ranged from 0.55 to 1.05 kN. Thus, it can be concluded that F_{max} is independent of the pendulum mass. The shapes of the F-t curves for the 15 J pendulum were similar to those from the 2 J pendulum maximum impact energy. The fracture time had the same order of magnitude for the smaller and bigger pendulum. Hence, similar to the case with the maximum loading force, the fracture time was independent of the mass of pendulum.
2. Effect of Notch. Compared to the values of the E_{max} total fracture energy and t_F fracture time measured on unnotched samples, these characteristics were reduced by one order of magnitude when we used notched samples with an a/W = 0.5 ratio. The related fracture time values (t_F = 0.04 and 0.006 ms for v = 0.78 and 2.88 m/s loading rates, respectively) are presented by F-t diagrams in Fig. 3. In addition, the $F_{i,u}$ force which corresponds to the initiation of unstable crack propagation, cannot be clearly recognized in the force-time diagrams since force oscillations due to inertia effects remain undamped. Consequently, the K_{Id} dynamic fracture toughness unfortunately cannot be determined using static equations,[4-5] since that would require the use of $F_{max} = F_{i,u}$ value for calculations as discussed in Part II.[1]

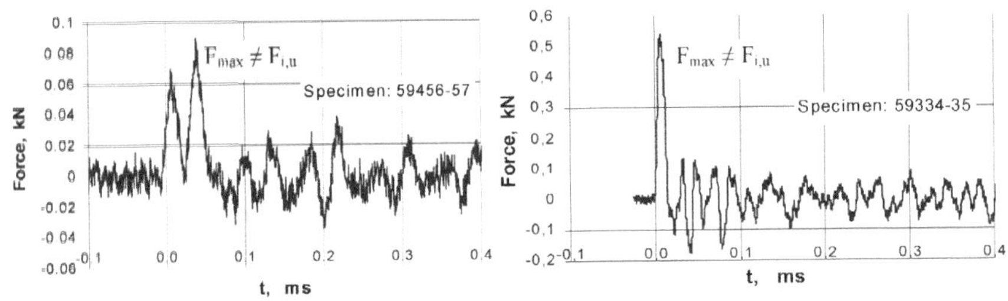

a) v = 0.78 m/s b) v = 2.88 m/s

Fig. 3. Force-time diagrams measured on notched Si_3N_4 specimens during impact loading (2 J). Force oscillations interfere with determination of the force at the onset of unstable fracture.

3. Effect of the Loading Rate. At higher loading rates the load oscillation and inertia effects were more pronounced for both unnotched and notched specimens as seen in Fig. 2. and Fig. 3. These diagrams relate to tests executed by pendulum with the 2 J maximum impact energy. The fracture time was characteristically t_F = 0.3 and 0.08 ms in case of the v = 0.78 and 2.88 m/s loading rates for unnotched specimens using the smaller pendulum, while it was t_F = 0.2 and 0.07 ms for the case of the v = 0.99 and 3.68 m/s loading rates using the bigger pendulum. The reduction of the fracture time with increasing loading rate was less pronounced in case of the bigger pendulum.

The characteristic fracture time values for the different experimental circumstances are summarized in Table I.

Table I. The Characteristic Fracture times for the Different Test Procedures

Experimental circumstances			
Loading rate, m/s	Specimen type	Max. impact energy of pendulum 2 J	Max. impact energy of pendulum 15 J
		Fracture time, ms	Fracture time, ms
0.78	unnotched	0.30	—
	notched	0.04	—
0.99	unnotched	—	0.20
	notched	—	—
2.88	unnotched	0.08	—
	notched	0.006	—
3.68	unnotched	—	0.07
	notched	—	—

Based on these values, a good correlation (see Fig. 4) can be found between fracture time and loading rate for unnotched specimens, in the form of relationship of $t_F = 0.2187v^{-0.912}$, with correlation coefficient of R = 0.993.

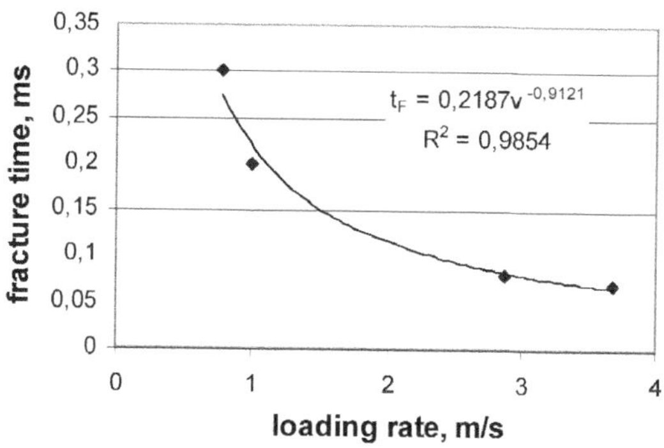

Fig. 4. Correlation between fracture time and loading rate during impact test of Si_3N_4 ceramics

The Problem of Determination of the $F_{i,u}$

For unnotched specimens the F_{max} force in all cases can be considered as identical with the force corresponding to the unstable crack propagation. However, in the case of notched specimens it was possible in only one of ten trials (five experiments at each of two loading rates) to associate F_{max} with the onset of unstable crack propagation. The maximum force values in Fig. 3a and 3b do not correspond to the unstable crack propagation. This fact causes a basic problem if we would like to determine the dynamic fracture toughness using quasi-static relationships.

Load-Deflection (F-s) Diagrams

From load-time (F-t) diagrams load-deflection (F-s) curves could be derived using kinetic equations taken from standard[2]:

$$v(t) = v_0 - \frac{1}{m} \int_{t_0}^{t} F(t)\, dt \tag{1}$$

$$s(t) = \int_{t=0}^{t} v(t)dt \tag{2}$$

where m is the mass of pendulum in kg, v_0 is the velocity of the pendulum in m/s at the moment of contact, and F(t) is the load-time function in Newtons and seconds recorded during testing.

The load-deflection F-s diagrams show a similar character as the load-time F-t diagrams as seen in Fig. 5. An important feature is the maximum fracture energy, E_{max}, which is the area below the F-s curve.

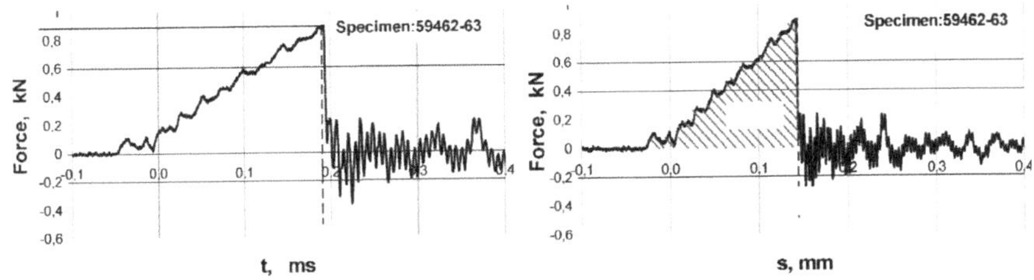

Fig. 5. Load-time and load-deflection diagrams for an unnotched Si_3N_4 specimen

Fracture Energy Data

The total fracture energy, E_t absorbed during the failure process is consumed partly by the deformation and fracture of the specimen (E_f), but also into subsidiary energies like kinetic energy consumed by the moving parts of the broken specimen (E_k) or the work used for the elastic deformation of the testing equipment (E_m).[6,7] The total fracture energy can be expressed as:

$$E_t = E_f + E_k + E_m. \tag{3}$$

The sum of the $E_f + E_k$ is the energy stored by the specimen, E_S:

$$E_f + E_k = E_S. \quad (4)$$

The stored energy of the measuring system and the specimen can be determined by calculations given in the references cited above.[6,7]

While the initiation part of load force-deflection diagrams for plastic materials consist of linear and non-linear sections, in the case of brittle materials such as ceramics load increases linearly to fracture (if force oscillations are screened out). Since in ceramic materials the macroscopic deformation is linear, after the point of maximum force the curve steeply decreases to zero. The initial part of the F-t curve represents the crack initiation period, while the second part, when force falls down to zero, is the period of crack propagation. Thus, total fracture energy absorbed during the failure of the specimen is the sum of the energy consumed by crack initiation, E_i, (up to F_{max}) and energy consumed by propagation, E_p, (after F_{max}):

$$E_t = E_i + E_p \quad (5)$$

The second part can be taken as zero for ceramics. Thus, E_{max} can be approximated by the E_i energy, belonging to the crack initiation, and $E_{max} = E_t \cong E_i$.

For the comparison of different energy data that can be derived from the test, the relative difference between the KV impact energy and E_{max} was characterized by the ΔE_S difference:

$$\Delta E_S = |E_{max} - KV| / KV, \ [\%] \quad (6)$$

The E_{max} total fracture energy calculated from the F-s diagrams showed no variation with the pendulum mass (or maximum impact energy) at a given loading rate, while the average KV impact energy in case of 15 J pendulum and v= 3.68 m/s was 40 % lower than the E_{max} and it was strongly influenced by the pendulum mass, as shown in Table II. (Here it should be noted that KV energy can be measured only at the maximum loading rate for a certain pendulum).

Table II. Values of the Measured (KV) and Calculated (E_{max}) Energy, and their Differences (ΔE_S)

Pendulum	Number of specimens	Loading rate [m/s]	$KV_{average}$ [J]	$E_{max,ave}$ [J]	ΔE_S [%]
15J	5	v = 3.68	0.078	0.113	40.3
	5	v = 0.99	–	0.086	–
2J	5	v = 2.88	0.115	0.104	11.8
	5	v = 0.78	–	0.085	–

This significant difference is caused mainly by the lower (i.e., 10^{-2} J) reading accuracy of KV values in case of the bigger pendulum. Since the measured values fell into the range of 0.01 to 0.1 J, the accuracy and measured values had the same order. In case of the smaller 2 J pendulum that provides evidently more accurate data for such brittle materials (reading accuracy is 10^{-3} J), KV and E_{max} were comparable in magnitude. Furthermore, at lower loading rates the total absorbed energy up to failure (E_{max}) did not depend on the pendulum mass. Thus, E_{max} seems to be a more reliable material feature.

The E_{max} values in Table II are mean values of the population tested at the certain pendulum and loading rate combination. The variation coefficient of E_{max} values for unnotched specimens was 11.3 % in case of v = 0.78 m/s, and 7.1 % in case of v = 2.88 m/s, thus considering dynamic loading conditions and the statistical nature of the failure process in ceramics, the reproducibility of this feature is acceptably good. In case of notched specimens, the total fracture energy could be derived from the F-s diagrams in only one case of the ten measurements when the $F_{max} = F_{i,u}$ force could be taken reliably from the diagram. In that case, the E_{max} energy was one order of magnitude less than that for unnotched bars (0,009J).

Three-Point Bending Strength

The dynamic three point bending strength ($R_{TPB,d}$) data were calculated from test results gained by 2J pendulum, using the general strength formula suggested by a relevant standard.[8] The strengths are compared to the quasi-static data for the same material (provided by the producer RITPMS, HAS) in Table III.

Table III. Quasi-static and dynamic three-point bending strength values

Type of specimen	Number of specimens	Loading rate [m/s]	R_{TPB} and $R_{TPB,d}$		
			Mean [MPa]	St. dev. [MPa]	Coef. of Var. [%]
Unnotched	5	quasi-static	916.2	59.4	6.4
	5	v = 0.78	631.7	48.0	7.5
	5	v = 2.88	626.2	39.5	6.3

*only one datum could be evaluated

From Table III it can be seen, that increasing loading rate resulted in a decrease of the mean value of the dynamic bending strength. The coefficients of variation were small (6.3 - 7.6 %) and independent of loading rate.

FRACTOGRAPHY

Fractographic measurements were carried out on a TESLA BS340 type scanning electron microscope at the Department of Ceramics at the Institute of Materials Sciences of Slovak Academy of Sciences.

Fractography of the fracture surface that contains key features of the damage process provides a useful aid in the failure analysis.[9] During macro fractographic analysis we had two main objectives: identification of the fracture cause and determination of the initiation point. So the primary goal was to locate and identify the fracture origin at higher magnifications, as well as to find the characteristic regions of the fracture surface.

The characteristic pattern of crack branching often marked out the vicinity of the fracture origin, as illustrated in Fig. 6.

The total fracture surfaces of unnotched and notched specimens showed significant differences in failure mode and surface morphology as shown in Fig 7.

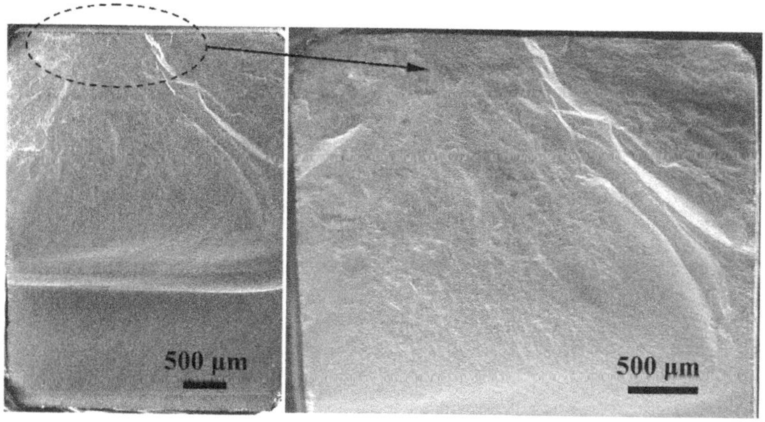

Fig. 6. Macro fractographs of a Si_3N_4 ceramic revealing the region of the fracture origin. (original magnifications: N= 35x and 45x)

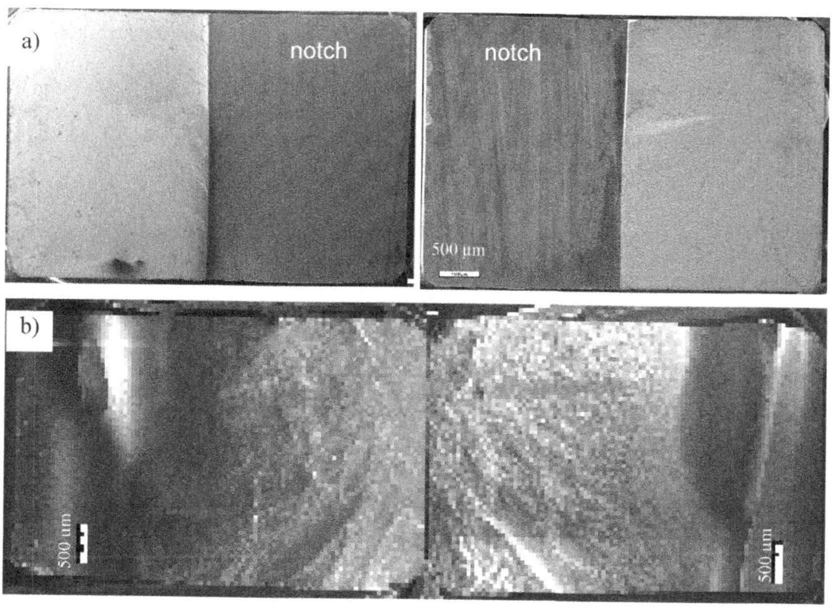

Fig. 7. Matching fracture halves of notched (a) and unnotched (b) specimens at 2.88 m/s

The high stress concentration of the notch results in much faster crack propagation in a smaller cross-section, so the fracture mirror region covers the total surface and ridged morphology didn't appear. The lower energy in the notched specimens also causes fewer markings to form.

On the fracture surface of unnotched specimen the crack branching points out the fracture origin and other morphological features (discussed further on) can also be recognised. At a magnification level of N = 150x, the characteristic regions of the fracture surface can be identified and the fracture mirror size can be determined, as illustrated in Fig. 8. Here a near-surface pore as the fracture origin can also be identified.

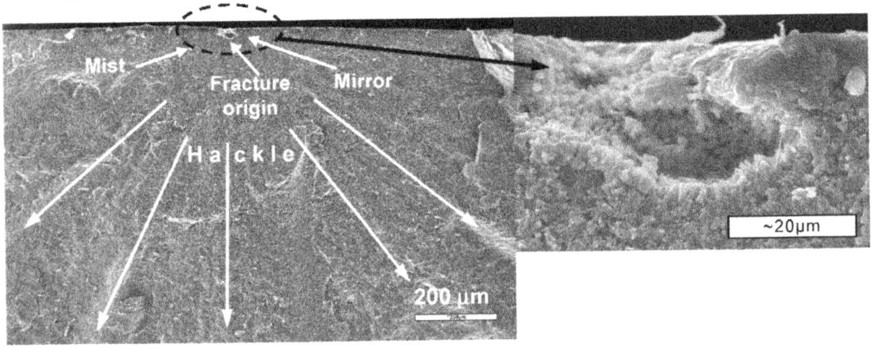

Fig. 8. The characteristic regions of the fracture surface of the impacted unnotched specimen

The characteristic features of the different crack propagation zones and their connection with the loading force-deflection curve can be seen in Fig. 9. The compression curl[10-11-12] appears when the material is loaded in bending. The fracture initiates on the tensile side of the surface, and exits on the compressive side not in a plane perpendicular to the outer surface.

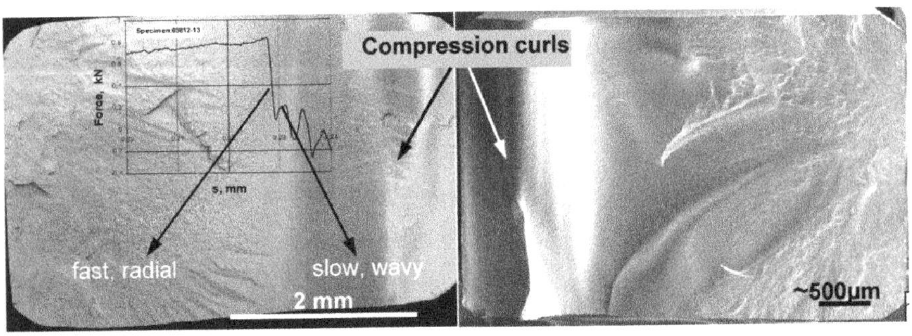

Fig. 9. Characteristic zones of crack advancement with the related load - deflection F-s diagram (left) and the rippled effect from compression curls at the compressive side of two broken unnotched impact specimens.

In order to reveal micromechanisms of the crack advancement, the fracture surface was studied at higher magnifications (N=8000x) in the vicinity of the crack initiation in the mirror zone and also somewhat further away in the hackle region. Fig. 10. shows mixed type mechanisms in both region, but close to the initiation zone it was mainly intergranular (Fig. 10a), while crack propagation in the hackle zone was dominantly transgranular (Fig. 10b).

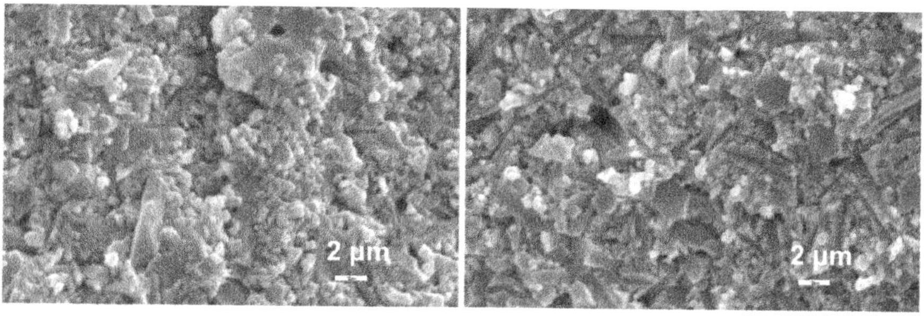

a) b)

Fig. 10. Crack propagation in the mirror zone is dominantly intergranular (a), while in the hackle region it is mainly transgranular (b).

Fractography can provide additional quantitative and qualitative information about the nature and peculiarity of the damage process as presented in the Part II.[1]

SUMMARY

Instrumented impact testing can be successfully used to determine load-time (F-t) and load deflection (F-s) diagrams, the total fracture energy (E_{max}), the dynamic three-point bending strength ($R_{TPB,d}$) and fracture time (t_F) of silicon nitride ceramics with suitable accuracy. The most important features characterizing the investigated material in case of unnotched specimens were as follows: E_{max} = 0.09 - 0.10 J, $R_{TPB,d}$ = 632 - 626 MPa, t_F = 0.3 - 0.08 ms for loading rates between 0.78 - 2.88 (3.68) m/s.

The influence of different test parameters such as the pendulum mass, the loading rate and a notch on the measured characteristics has been investigated.

- Pendulum mass had no influence on the measured F_{max} forces and fracture time. The KV impact energy calculated from alteration of potential energy of the pendulum is strongly dependent of the pendulum mass. The total absorbed energy to failure, E_{max}, derived from the area of the F-s diagram is independent of pendulum mass.
- Loading rate in the applied range strongly affected the character of the F-t diagrams, the magnitude of the fracture time and the dynamic three-point bending strength, but only slightly influenced the magnitude of the total fracture energy. With an increase of the loading rate, the measured E_{max} energies increased, whereas the $R_{TPB,d}$ dynamic bending strength was unambiguously reduced.
- For unnotched specimens, a good correlation between the fracture time and loading rate could be observed in the form of $t_F = 0.2187\ v^{-0.912}$.

- The E_{max} value determined for unnotched samples reduced by one order, when we introduced a notch with ratio of a/W =0.5 into the test pieces.
- With increased loading rates, the load oscillation and inertia effects were more pronounced especially during testing of notched specimens. Thus, the $F_{i,u}$ loading force corresponding to the onset of unstable crack propagation and the related fracture time could not be taken from the F-t diagrams. This precludes the possibility of determining the dynamic fracture toughness using the quasi-static relationships.
- Fractography revealed the cause and origin of the dynamic fracture, as well as it was applied to identify the characteristic regions of the fracture surface of the tested silicon nitride.
- The morphology of the fracture surfaces of notched and unnotched specimens were significantly different.
- The fracture surfaces showed mixed transgranular and intergranular crack propagation modes. A possible change in the crack propagation mode from mostly intergranular near the initiation sites to mostly transgranular in the hackle zones was observed.

ACKNOWLEDGEMENTS

Authors express their thanks for the professional and technical aid given by Gyöngyvér B. Lenkey (BAY-LOGI, Miskolc, Hungary), Árpád Kovács (UM Miskolc, Hungary), Jan Dusza, Emoke Rudnayova and Maria Beliecayova (IMR, Slovak Academy of Sciences). The research work was supported by the OTKA T046467, OTKA NI 61724 and OTKA K63609 projects, the Bolyai János Research Foundation, and the MÖB 2-13-2-193158/2005 Research Grant.

REFERENCES

[1]M. Maros, N. Kaulics, and J. Dusza, „Characterization of the Dynamic Failure Process of Si_3N_4 Ceramics, Part II," this volume.

[2]Proposed Standard Method for Instrumented Impact Testing of Sub-size Charpy V-Notch Specimens of Steels, Draft 10a, European Structural Integrity Society, ESIS, Technical Committee 5, May 23, 2002.

[3]Z. Radakovic, Gy. B. Lenkey, V. Grabulov and A. Sedmak, "Application of Two Independent Measurement Techniques for Determination of Ductile Crack Growth Initiation," *Int. J. of Fract.*, Vol. 96., [2] L23-L28 (1999).

[4]Test Method for Fracture Toughness of Monolithic Ceramics at Room Temperature by Single Edge Precracked Beam (SEPB) Method, ISO 15732, Int. Org. for Stds., Geneva, (1999).

[5]Standard Test Methods for Determination of Fracture Toughness of Advanced Ceramics at Ambient Temperature, ASTM C1421-99, American Society for Testing and Materials Philadelphia, (1999).

[6]T. Kobayashi and H. Ikawa, "Evaluation of Dynamic Fracture Toughness of Brittle Materials," MRS *Int. I. Mtg. On Adv. Mats.*, Vol. 5, pp.343-348 (1989).

[7]T. Kobayashi, M. Niinomi, Y. Koide and K. Matsunuma, "Instrumented Impact Testing of Ceramics," *Transactions of the Japan Institute of Metals*, Vol. 27 [10] pp. 775-783 (1986).

[8]Standard Test Method for Flexural Strength of Advanced Ceramics at Ambient Temperature, ASTM C1161-02ce1, *American Society for Testing and Materials*, West Conshohocken, PA, (2002).

[9]J. Dusza and M. Steen, "Fractography and Fracture Mechanics Property Assessment of Advanced Structural Ceramics,"*International Materials Reviews*, Vol. 44 [5] pp.165-216 (1999).

[10]V. D. Fréchette, *Failure Analysis of Brittle Materials,* Advances in Ceramics, Vol. 28., American Ceramic Society, Westerville, Ohio, (1990).

[11]Advanced Technical Ceramics–Monolithic Ceramics–Mechanical Properties at Room Temperature – Part 6: Guidelines for Fractographic Investigation, ENV 843-6: 200x, Draft 10, N425.

[12]Standard Practice for Fractography and Characterization of Fracture Origins in Advanced Ceramics, ASTM C1322-05b, *American Society for Testing and Materials*, West Conshohocken, PA, (2005).

CHARACTERIZATION OF DYNAMIC FAILURE PROCESS OF Si_3N_4 CERAMICS, PART II: DYNAMIC FRACTURE TOUGHNESS

Maria B. Maros
Associate Professor
University of Miskolc
Department of Mechanical Engineering
Mechanical Technological Research Group of HAS UM
3515 Miskolc – Egyetemváros, Hungary

Nikoletta Kaulics
PhD Candidate
University of Miskolc
Department of Mechanical Engineering
3515 Miskolc – Egyetemváros, Hungary

Jan Dusza
Associate Professor, Head of Department
Department of Structural Ceramics
IMR, Slovak Academy of Sciences, Kosice

ABSTRACT

A comprehensive experimental and theoretical work aiming at studying the dynamic failure process of silicon nitride ceramics has recently been started. The main goal is characterizing the material behavior by mechanical and microstructural features. The basic force,- strength- and energy type characteristics completed with some fractographic analysis of the damage process were discussed and presented in a companion paper.[1] This paper deals with the problem of determining the dynamic fracture toughness of the investigated ceramic, especially at high loading rates. A solution is provided by applying the dynamic key curve (DKC) method that utilizes a recently observed important peculiarity of silicon nitride: it shows electric-emission activity during dynamic failure. This finding makes it possible to determine the fracture time even in cases of impact tests at high loading rates, as well as during investigating notched specimens. The analysis of the mechanical behavior is completed by qualitative and quantitative fractography. Combining the measuring results gained from the instrumented impact tests with the fractographic findings, the dynamic fracture toughness K_{Id} is estimated based on key features of the fracture surface of the dynamically broken specimens. The mirror size constant A_0 is also determined.

INTRODUCTION

The wide choice of testing methods relating to the determination of quasi-static K_{Ic} fracture toughness of engineering ceramics indicates the efforts and difficulties to satisfy simultaneously the demand of determining reliable material data as well as applying economical and technically simple procedures. Advanced, accurate, and sensitive testing equipment are required for the reliable determination of the force and energy type characteristics for small size specimens and for tracing and detecting extremely short time failure processes during impact loading. In case of instrumented impact tests, a special problem during determination of K_{Id} or J_{Id} fracture toughness values is that the force $F_{i,u}$, corresponding to the unstable crack propagation that is necessary for the calculations based on quasistatic relationships, cannot be taken from the load-time diagrams of the notched specimens, due to the undamped force oscillations.[2] This problem was discussed in detail in Part I[1] and related F-t diagrams were shown. In this paper, determination of the dynamic fracture toughness will be discussed. Alternatives for solution of the above problem are shown. Three different methods, namely calculations based on the quasi-static relationships, the procedure of the impact response curve, as well as the dynamic-key-curve method are reviewed and compared from the aspect of applicability for the given material and testing circumstances. An important observation contributing to finding the appropriate solution is that Si_3N_4 ceramics show electric emission activity which makes it possible to determine the fracture time for this material in an alternative way.

Failure analysis of the investigated material has been successfully completed with macro- and micro-fractographic investigations. Determining various qualitative and quantitative characteristics by fractography, the nature and peculiarities of the damage process can be analysed in a more complex way. Furthermore, mechanical properties like the critical stress intensity factor and the fracture stress are also compared. They may be determined both by mechanical testing and by fractographic analysis.

EXPERIMENTAL WORK

Investigated Material and Test Series

The testing equipment, similar to that introduced in the companion paper,[1] was a CEAST 25/15/2 type impact testing machine with the same instrumentation. Tested material was similar to that described in the preceding paper, so Si_3N_4 based ceramics, having composition of 90% Si_3N_4, 4% Al_2O_3 and 6% Y_2O_3 were investigated, produced by the Research Institute for Technical Physics and Materials Science of Hungarian Academy of Sciences (RITPMS, HAS). All the specimens were produced by uniaxial pressing using polyethylene-glycol additive. The additive was extracted by heating in air at 600°C after pressing. The firing of the pieces occurred in two-stage gas-pressure sintering in HIP equipment.

The specimens were rectangular prisms with dimensions of 4.9 mm x 3.2 mm x 49 mm. The investigations were executed in two test series, with different specimen positions and geometry (see Figure 1~~Figure 1Figure 1~~.) according to the different aim of the measurement.

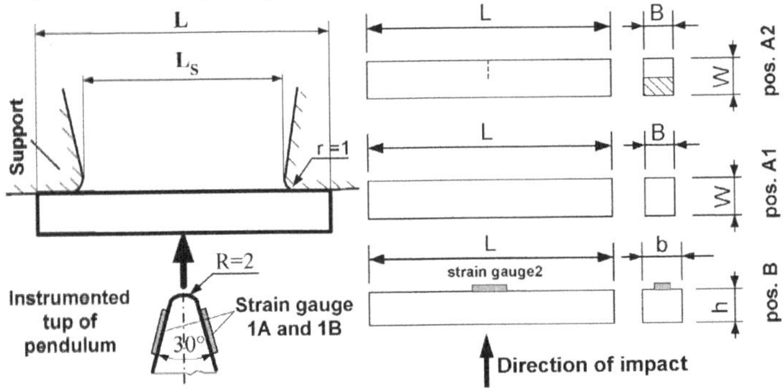

Figure 1. Specimen geometry and different test arrangements for the instrumented impact tests.

Position "A1 and A2" were applied aimed at determining the basic force, strength and energy type characteristics on unnotched specimens (position "A1"), as well as for fracture mechanical test using notched specimens (position "A2"). In both cases A1 and A2, the specimens were impacted on the narrower face. This position differs from that used during standardized three-point bending strength tests when samples are loaded on the wider face. Nevertheless, for comparative measurements on the failure behavior of unnotched and notched specimens, this "A1" set was applied in dynamic bending tests, as well.

Specimens were arranged according to position "B" during tests aimed at determining the dynamic Young's modulus. This required direct strain measurements with an extra strain gauge on the tensile surface of the specimens. Evaluation of the modulus measurement does not constitute an objective of the current work, but some samples and test results from these experiments were used during fractographic analyses described in this paper. Therefore this position is also presented in Figure 1. (Here should be noted that even in these cases, measuring data that have been processed during calculations, were derived only from the instrumented tup.)

Fracture mechanics specimens were created by cutting a notch with a nominal size of a/W ratio ≈ 0.5 and $t=0.2$ mm width into the bars with a diamond grinding wheel. The notch size was chosen according to the usual practice during dynamic fracture mechanical tests.[3,4]

Due to the limited amount of specimens, only two loading rates were applied in tests with position "A", namely $v=0.78$; and 2.88 m/s. The support's span was $L_S = 40$ mm in position "A", while it was $L_S = 38$ mm in position "B", when samples were struck on the wider face. For each test regime 5-5 specimens were tested. The maximum impact energy of the applied pendulum was 2J, and its mass was 0.4734 kg.

Electric-Emission Activity of Si_3N_4 Ceramics

During testing the impact pendulum machine was equipped by a so called electric-emission specimen (EE condenser) and the related measuring system (see Figure 2). This provides the possibility for gaining additional information about the failure process, e.g. determining the moment of crack initiation.

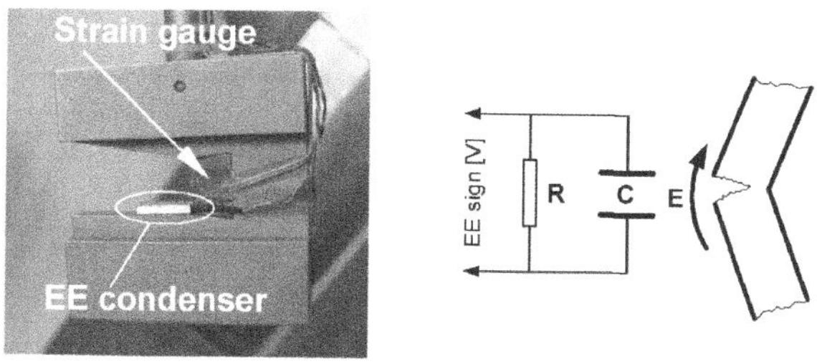

Figure 2. The instrumented tup (a); and the electric measuring circuit with the EE-condenser (b)

The physical background of the electric-emission measurement and its application for assigning the moment of unstable crack propagation is described in other work.[5] In certain materials microstructural changes occur in the moment of fracture, that modifies the properties of the surrounding electric space. Metallic materials always give electric-emission (EE) sign during breaking, while this type of activity can be observed in a limited case during testing polymers and ceramics. An important, novel observation during our experiments has been that the investigated Si_3N_4 ceramics show this type of behaviour,[6] consequently the moment of failure, denoted by t_F is possible to be determined by this technique.

The principle of the method is that at the initiation of the unstable crack propagation, i.e. at the moment of t_F, the electric-emission signal will be changed. The modification can be represented by either changing the character, the magnitude and/or the direction of the EE-sign. In addition, periodicity of the electric noise (comparing to the one registered in the time period preceding the crack initiation) also varies.

Characteristic force-time diagrams and their connection with the EE sign vs. time curves for a low loading rate test of unnotched specimen and a high loading rate test of notched specimens are illustrated in Figure 3. The coincidence of the fracture times read on the one hand from the F-t curve at F_{max} (when $F_{max} = F_{i,u}$ is available) and taken on the other hand from the EE-t curves, (as described above) can clearly be seen in Figure 3a, i.e. in the case of the unnotched specimen. In the case of notched samples, t_F can not be taken from F-t diagrams, since $F_{max} \neq F_{i,u}$, can be read from EE-t curve, as illustrated in Figure 3b.

The exact reason for why the direction of the EE-sign is changing differently has not been cleared up at the moment. Relating our measuring experiences on silicon nitride up to this time (as much as 50 data) no systematic variation has been observed. It is assumed, that during failure a predominance of different electrical charges may develop on the broken surfaces, the sign of which depends on the type of ions being in majority on the free surface that can vary according to the layered crystalline structure of the silicon nitride.

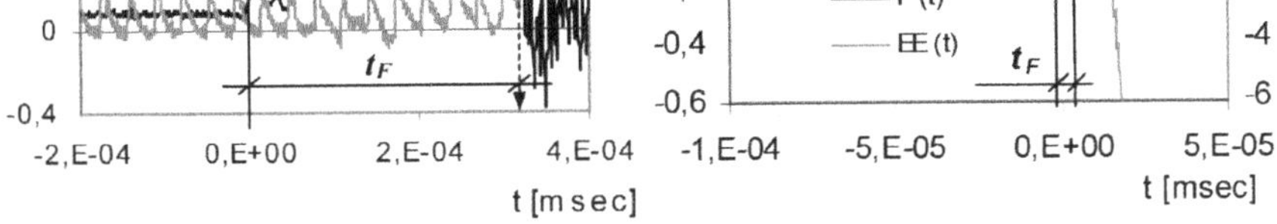

a) Unnotched sample, v = 0.78 m/s

b) Notched sample, v = 2.88 m/s

Figure 3. Electric-emission sign vs. time diagrams and the related F-t diagrams

Determination of the K_{Id} Dynamic Fracture Toughness

Calculation of the dynamic fracture toughness can be done in different ways. If failure process has a quasi-static nature, the dynamic fracture toughness can be calculated based on quasi-static equations. This applies only up to a certain loading rate. The precondition of a failure process to be quasi-static is fulfilled in case of acceptable damping of the force oscillation at the beginning of the fracture, i.e. appearance of at least three peaks of oscillation up to the maximum loading force, that means the fracture time is greater than the treble of the normal oscillation period of the compression wave (see Part I[1]). In these cases F_{max} force represents the initiation of the unstable crack propagation, i.e., $F_{i,u}$ can be taken from the recorded F-t diagram and may be applied during calculations based on a quasi-static relationship,[7,8] as given by Eq. (1).

$$K_{Id} = g\left(\frac{a}{W}\right)\cdot\left[\frac{F_{max}\cdot L_S}{B\cdot W^{3/2}}\right]\left[\frac{3\cdot(a/W)^{1/2}}{2\cdot(1-a/W)^{3/2}}\right] \tag{1}$$

where F_{max} is the maximum loading force,
a is the initial crack length,
L_S, B and W geometrical parameters, as given in Figure 1,
g(a/W) is a geometrical function.

The g(a/W) function is calculated according to Eq. (2). for the usual $L_S/W = 4$ ratio:

$$g\left(\frac{a}{W}\right)=\left(1+2\frac{a}{W}\right)^{-1}\cdot\left\{1.99-\frac{a}{W}\left(1-\frac{a}{W}\right)\left(2.15-3.93\frac{a}{W}+2.7\left(\frac{a}{W}\right)^2\right)\right\} \tag{2}$$

Equations (1) and (2) may also be used for $L_S = 5$ ratio. In case of $5 \leq L_S/W \leq 10$ and $0.35 \leq a/W \leq 0.7$ conditions the $g = g\left(\frac{a}{W}\right)$ function is given in the form of Eq. (3):

$$g\left(\frac{a}{W}\right)=A_0+A_1\left(\frac{a}{W}\right)+A_2\left(\frac{a}{W}\right)^2+A_3\left(\frac{a}{W}\right)^3+A_4\left(\frac{a}{W}\right)^4+A_5\left(\frac{a}{W}\right)^5 \tag{3}$$

where coefficients A_i can be taken from standard.[8] For our case $L_S/W = 8.16$, thus $g = g\left(\frac{a}{W}\right)$ for $L_S/W = 8$ were taken, that gives:

$$g\left(\frac{a}{W}\right)=1.9381-5.0947\left(\frac{a}{W}\right)+12.3861\left(\frac{a}{W}\right)^2-19.2142\left(\frac{a}{W}\right)^3+15.7747\left(\frac{a}{W}\right)^4-5.127\cdot\left(\frac{a}{W}\right)^5 \tag{4}$$

During dynamic loading at high loading rates of very brittle materials like ceramics, determination of the $F_{max} = F_{i,u}$ force is often impossible for notched specimens. This problem was also discussed in Part I. In these cases the impact response curve (IRC) [9,10,1112] or dynamic key curve (DKC)[13,14] methods may be applied. Both procedures require determination of the t_F fracture time.

The IRC method is based on the principle that dynamic fracture toughness (that is changing with the time) takes the same value up-to-the moment of fracture if test circumstances are similar, i.e. a relationship between the dynamic fracture toughness and fracture time can be assumed in the form of Eq. (5).

$$K_{Id} = R \cdot v_0 \cdot t'' \tag{5}$$

where R is constant for a given machine compliance, v_0 the impact velocity, and t''=f(t') is a time parameter given by references.[9,10] t' can be calculated as follows:

$$t' = t_F \left\{1 - 0.62\left(a_0/W - 0.5\right) + 4.8\left(a_0/W - 0.5\right)^2\right\} \tag{6}$$

If the relationship between K_{Id} and t_F time is pre-determined for different loading rates, i.e. the so called impact response curves are known for the given material,[15] the dynamic fracture toughness can be determined simply by measuring the fracture time.

In the dynamic key curve (DKC) method[16,17] K_{Id} is considered as a product of the quasi-static fracture toughness, K_I^{qs} and the dynamic correction function, k^{dyn}:

$$K_{Id} = K_I^{qs} \cdot k^{dyn} \tag{7}$$

The K_I^{qs} is analytically derived from a mass-spring model, while k^{dyn} represents the dynamic effects. For a given test arrangement and machine set-up the K_{Id} can be expressed in the form of Eq. 8.:[13, 14, 18]

$$K_{Id} = \frac{3 \cdot E \cdot L_S \cdot Y\left(a_0/W\right)}{2 \cdot W^{3/2} \cdot C_S^*\left[1 + C_m/C_S\right]} v_0 \cdot t_F \cdot k^{dyn}\left(t = t_F\right) \tag{8}$$

where E: is the elastic modulus of the specimen,
a_0: the initial crack length,
C_S*: the dimensionless specimen compliance,
C_m: the machine compliance,
C_S: the compliance of the notched specimen,
v_0: the impact loading rate,
t_F: the time to fracture,
k^{dyn}: the dynamic key function depending on the t_F,
W: thickness of the specimen,
Y(a_0/W): the geometric function of the static relationship, that can be calculated for L_S/W = 4 ratio and a/W<1 according to Eq. (9):[13, 18]

$$Y\left(\frac{a}{W}\right) = \left(\frac{a}{W}\right)^{1/2}\left(1 + 2\frac{a}{W}\right)^{-1}\left(1 - \frac{a}{W}\right)^{-3/2}\left\{1.99 - \frac{a}{W}\left(1 - \frac{a}{W}\right)\left(2.15 - 3.93\frac{a}{W} + 2.7\left(\frac{a}{W}\right)^2\right)\right\} \tag{9}$$

and for $L_S/W=8$ ratio, for $a/W<0{,}6$ according to Eq. (10):[13, 19]

$$Y\left(\frac{a}{W}\right)=\sqrt{\pi}\left(\frac{a}{W}\right)^{1/2}\left[1.107\quad 2.12\left(\frac{a}{W}\right)+7.71\left(\frac{a}{W}\right)^{2}-13.55\left(\frac{a}{W}\right)^{3}+14.25\left(\frac{a}{W}\right)^{4}\right] \tag{10}$$

The C_S^* dimensionless specimen compliance is related to the C_S specimen compliance as follows:

$$C_S^* = E\cdot B\cdot C_S \tag{11}$$

where B is the thickness of the specimen.

For calculating the specimen compliance, various relationships are suggested by the literatures.[13,14,15,20,21] For notched specimens, the compliance is theoretically given by the following equation:[21]

$$C_S=\frac{2.0475\cdot L_S^2}{E\cdot B\cdot W^4}\cdot a\cdot(W-a)\cdot Y \tag{12}$$

where Y is expressed as a function of the notch depth ratio:

$$Y\left(\frac{a}{W}\right)=1.93-3.07\left(\frac{a}{W}\right)+14.53\left(\frac{a}{W}\right)^{2}-25.11\left(\frac{a}{W}\right)^{3}+25.8\left(\frac{a}{W}\right)^{4} \tag{13}$$

The machine compliance should be pre-determined on an unnotched specimen from a low rate impact test,[14] according to Eq. (14):

$$C_m=\frac{v_0}{dP/dt}-C_{S,0} \tag{14}$$

where v_0: is the low impact loading rate ($v_0 \approx 0{,}5$m/s),
dP/dt: is the slope of the mean load line in the pre-test and,
$C_{S,0}=20.1/(EB)$: is the compliance of the unnotched specimen.

Depending on the fracture time, the dynamic effects are taken into consideration in the following manner:[14]

If $0<t_F<9{,}2W/c_1$, then:

$$k^{dyn}=-0.9096+0.8176\left(c_1t/W\right)-0.1005\left(c_1t/W\right)^2+0.003765\left(c_1t/W\right)^3 \tag{15}$$

If $t_F\geq 9{,}2W/c_1$, than $k^{dyn}=1$, where c_l is the propagation rate of longitudinal wave in the specimen, given by Eq. (16):

$$c_1 = \sqrt{\frac{E}{\rho \cdot (1-\nu^2)}} \quad (16)$$

where ρ is the density, and ν is Poisson's ratio.

Comparing the impact response curve procedure with the dynamic key curve method, the latter has the advantage of avoiding preliminary measurements and empirical correlations. Furthermore, it provides calculating the K_{Id} for specimen of the actual geometry.

EXPERIMENTAL RESULTS

Our previous experiences during instrumented impact testing on Si_3N_4 ceramics proved that F_{max} force could not be taken from the load-time diagrams for notched specimens, due to the increased dynamic effects[1], thus quasi-static theory is not applicable for determination of the dynamic fracture toughness. Therefore K_{Id} was calculated based on the dynamic key curve (DKC) method. According to Eq. (8), only t_F and C_m should be derived from the impact test. The fracture time was measured by electric-emission technique on notched specimens arranged in position A2, while C_m was calculated according to Eq. (14) from an impact test at loading rate v = 0.5 m/s, when the unnotched specimen was in position A1.

The precondition of $t_F \geq 9{,}2W/c_1$ was satisfied for almost all cases. Thus the dynamic key function resulted in $k_{dyn} = 1$ with two exceptions, when k_{dyn} values were 1.17 and 1.16. For Eq. (16) the Poisson's factor was taken as 0.24 for Si_3N_4, and density was 3.24 g/mm^3, representing the average density of the investigated ceramic material.

The dynamic fracture toughness was calculated by applying Eqs. (8), (10), (11), (12), (13) and (14). The resulted K_{Id} values for the applied two loading rates, as well as the related fracture time and the statistical characteristics are shown in Table I.

For comparison, K_{Ic} quasi-static fracture toughness values (see $v \approx 0$ m/s) are also indicated in Table I. These data have been measured by single edge notched beam test done by the producer on specimens of similar geometry, preparation and test position as used for the impact tests. The only difference was the $L_S/W = 4$ ratio and therefore, K_{Ic} values were derived using Eqs. (1) and (2).

Table I. Fracture time and fracture toughness values at different loading rates for notched specimens

Feature	Loading rate, v, m/s			Feature	Loading rate, v, m/s	
	0	0.78	2.88		0.78	2.88
Fracture toughness, K_{Ic} or K_{Id}, MPa $m^{1/2}$	6.81	4.51	3.00	Time to fracture, t_F, ms	0.036	0.007
	7.87	4.75	3.81		0.039	0.008
	7.51	3.60	2.05		0.029	0.004
(sample position: A2)	7.55	4.73	2.07		0.038	0.004
	—	—	3.03		—	0.007
Average, MPa $m^{1/2}$	7.43	4.4	2.79	Mean, ms	0.036	0.006
St. dev., MPa $m^{1/2}$	0.45	0.54	0.74	St. dev., ms	0.005	0.002
Var. coeff., %	6.0	12.3	26.6	Var. coeff., %	12.7	26.1

On the basis of these results the following observations are made:

- The average static fracture toughness of the material was K_{Ic}, = 7.43 MPa $m^{1/2}$, while the mean K_{Id} values at the lower and higher loading rates were 4.4 MPa $m^{1/2}$, and 2.79 MPa $m^{1/2}$, (for v = 0.78 m/s and 2.88 m/s respectively). K_{Id} significantly decreased with increasing loading rate that refers to a rate dependent fracture toughness behavior. This nature of silicon nitride has also been reported by others.[2,22] The obtained K_{Id} results, based on the dynamic key curve method are also in good agreement with those published for silicon nitride for similar impact velocities, determined by the IRC method.[15]
- The variation coefficient for the static case was 6.01 %, while increasing the loading rate it increased considerably, namely up to 12.29 % and 26.61 % for the lower and higher loading rates, that can be explained first of all by the dynamic effects, resulting in a significant variation of the fracture time.
- Fracture time of the investigated material fall into the range of 0.029 to 0.039 ms at v=0.78m/s, the related mean value was 0.036 ms, while at v = 2.88 m/s the fracture time decreased by one order and varied between 0.0039 to 0.0082 ms, with mean value of 0.006 ms.

Fractography

In the case of ceramics, the primary goal of applying fractography during failure analysis is to locate and identify the fracture origin and find the characteristic regions of the fracture surface. These qualitative features for the investigated silicon nitride were presented in Part I[1]. Fractography also ensures several quantitative characteristics be determined. The critical stress intensity factor can be estimated from geometrical parameters of the fracture origin. In addition, based on the mirror size of the stress at the failure, or a material characteristic, called mirror size constant can be determined, deriving numerical information on the quality and mechanical performance of material simply from morphological features of the fracture surface.

Assessment of the K_{Id} Dynamic Fracture Toughness using Fractography

Various structural defects may be encountered as potential origin of the crack propagation, leading to brittle failure of ceramics. Among the others single pore, set of pores, individual inclusion or its agglomerated form, different inhomogeneities, etc. can be the initiation point of the failure. Based on their geometry these defects are characterized by a so called flaw shape factor, Y.[23,24] Considering that each of these material flaws act as a stress concentrator during loading, the fracture strength of the material and the critical stress intensity factor can be correlated based on a fracture mechanics approach:

$$K_{Ic} = \sigma_f \cdot Y \cdot a^{1/2}, \quad MPa\ m^{1/2} \tag{17}$$

where a is a measure of the size of fracture origin. Supposing that Eq. (17) applies for dynamic loading conditions as well, the K_{Id} dynamic fracture toughness of silicon nitride samples was estimated by this relationship, applying σ_f values determined for unnotched specimens by instrumented impact test.

The characteristic defect type from which the fracture was originated was predominantly a single, near-surface pore, as shown in Figure 4.

Flaws representing the fracture origin are commonly described by their geometrical features. Beside measuring the 2a and 2c axis, representing the direction parallel and perpendicular to the fracture direction, the flaw shape factor should be determined, by which the ellipticity of the origin is taken into account. Furthermore, Y depends on the location of flaw (whether intersected by the surface or being beneath it in the bulk). In case of surface related origins Y varies by the nature of fracture initiation, i.e., whether it is a so called central initiation, remote from the surface, or it is a surface-edge initiation.

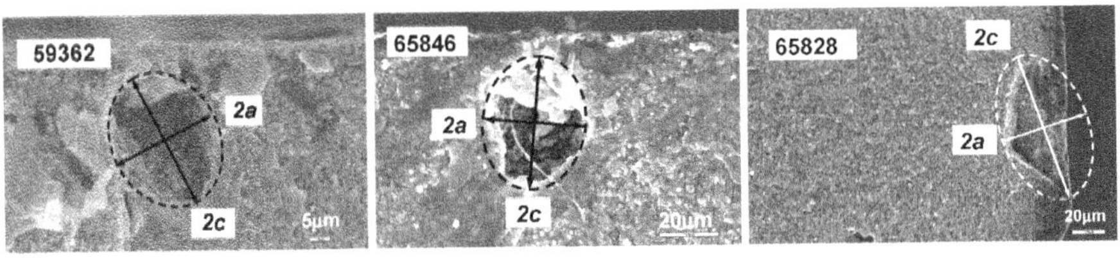

Figure 4. Near surface elliptical pores at tensile surface, as the cause of the brittle failure.

The approximate nature of the procedure is increased by the uncertainty during measuring the 2a and 2c axis, the value of which depends on the section of the three dimensional flaw, detected in planes of the two neighboring fracture surfaces. Due to loss or chipping of small particles during failure the measured values can be different on the two matching broken parts, as illustrated in Fig. 5.

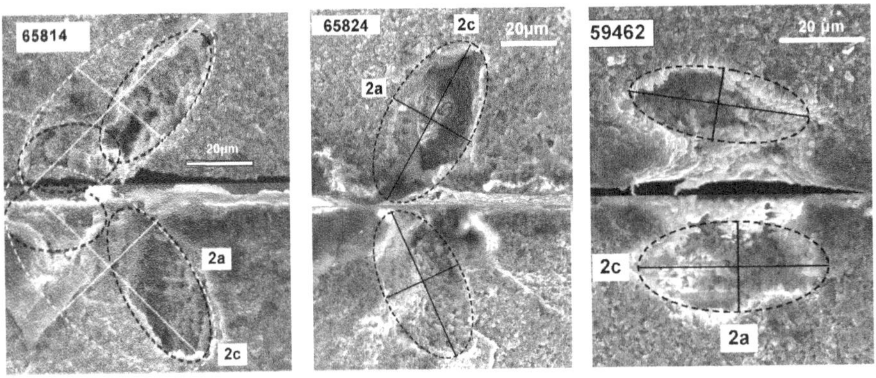

Figure 5. Determination of the flaw size on the surface of pairs of broken halves

Using fracture stress values that were determined by instrumented impact test in different test series (unnotched specimens, positions A1 and B), the dynamic fracture toughness could be estimated from the flaw characteristics. The obtained results are presented in Table II, based on which some establishments can be made.

- The K_{Id} dynamic fracture toughness values determined by fractography fell into the range of 2.24 to 3.94 MPa $m^{1/2}$ for impact loading rates of v = 0.78 and 2.88 m/s. The average values were 3.56 MPa $m^{1/2}$ for v = 0.78 m/s and 2.66 MPa $m^{1/2}$ for 2.88 m/s.
- A clear tendency of decreasing dynamic fracture toughness with increasing loading rate could be observed for fractographic evaluation.
- Relating both the magnitude, and the loading rate dependence, the material behavior corresponds to results determined using notched specimens and evaluated by the dynamic key curve method. In that case the average K_{Id} values were 4.4 MPa $m^{1/2}$ and 2.79 MPa $m^{1/2}$ for v = 0.78 and 2.88 m/s respectively.

Table II. Determination of K_{Id} fracture toughness by fractography on unnotched specimens

Loading rate [m/s]	Position	Specimen N°	Max. stress, σ_{max}, MPa	Failure stress σ_f, MPa	Fracture origin	2c, μm	2a, μm	Y	c/a	$K_{Id,\ fracto}$, MPa $m^{1/2}$	Average of $K_{Id,fracto}$ MPa $m^{1/2}$
0.78	A1	59462	694	672	bulk	44.7	21.0	1.47	2.13	3.20	
0.78	A1	65814	569	547	surface center	82.0	41.0	1.59	2.00	3.94	
0.78	A1	65824	622	599	surface center	65.0	31.0	1.59	2.10	3.75	3.56
0.78	B	65828	557	410	surface center	94.0	53.0	1.59	1.77	3.36	
2.10	B	59524	540	461	surface center	72.0	31.0	1.59	2.32	2.89	
2.88	A1	65846	657	641	bulk	46.2	33.7	1.26	1.37	3.32	
2.88	A1	59322	596	547	bulk	39.3	10.7	1.77	3.67	2.24	2.66
2.88	A1	59362	634	610	bulk	32.5	25.0	1.13	1.30	2.44	

The averages and some statistics of the dynamic fracture toughness results determined by the different methods, completed with K_{Ic} quasi-static fracture toughness values are summarized in Table III.

Table III. Comparison of dynamic fracture toughness determined by different procedures

Test method	Evaluation method	Feature	Statistics	Loading rate, v, m/s 0	0.78	2.10	2.88
Single edge notched beam	Quasi-static relationship Eqs. (1), (2)	K_{Ic}	Average, MPa $m^{1/2}$	7.43			
			St. dev., MPa $m^{1/2}$	0.45	–	–	–
			Var. coeff., %	6.0			
			Number of data	4			
Instrumented impact test	Dynamic key curve method Eqs. (8) (10) (11) (12) (13)	K_{Id},	Average, MPa $m^{1/2}$		4.40		2.79
			St. dev., MPa $m^{1/2}$	–	0.54	–	0.74
			Var. coeff., %		12.3		26.6
			Number of data		4		5
Instrumented impact test	Fractography Eq. (17)	K_{Id},	Average, MPa $m^{1/2}$		3.56	2.89	2.66
			St. dev., MPa $m^{1/2}$	–	0.34	–	–
			Var. coeff., %		9.6	–	–
			Number of data		4	1	3

It is seen that the average data for a given loading rate are close for the two different methods, and a general tendency of decreasing toughness with increasing loading rate still holds for the combined mass of data. Visualizing K_{Ic} and K_{Id} values reported by Table III., a comparison is possible based on Fig. 6, deriving the following conclusions:

The proximity of solid and dotted lines connecting the averages for the two measuring methods (initiating obviously from average K_{Ic} value at $v \approx 0$ m/s), alludes to acceptably good coincidence of the DKC and fractographic procedure. A significance of this finding consists in not only having a simple, quick alternative method (i.e., by fractography) for assessing the dynamic fracture toughness, but to have the possibility to estimate this important material characteristic without executing expensive fracture mechanical procedures and without the time consuming preparation of notched specimens.

It should also be noted that the assessment of fracture toughness of ceramics based on fractographic method of the presented procedure requires a reliable and accurate determination (by instrumented impact test) of the failure stress acting at the fracture origin. Further experimental procedures based on a greater number of test samples is suggested to improve the reliability of the measured material characteristics for both the fractographic and DKC methods.

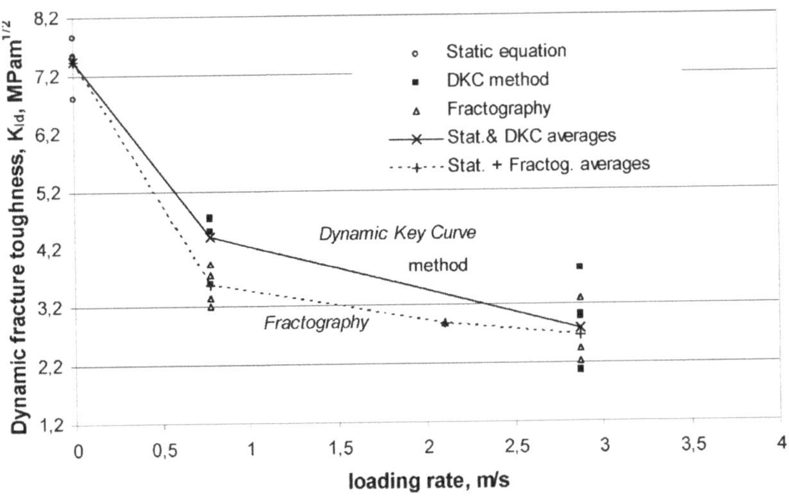

Figure 6. Loading rate dependence of K_{Id} fracture toughness of the Si_3N_4 based ceramic

Determination of the Mirror Size Constant, A_0.

Fractography provides a simple and easy way for estimating the stress at failure, σ_f relating it to the radius of fracture mirror, r through an empirical relationship according to Equation (18). This formula, suggested by the related standards[23,24] incorporates a material parameter, the so called A_0 fracture mirror constant that is found experimentally.

$$\sigma_f = \frac{A_0}{\sqrt{r}}, \tag{18}$$

Although the cited standards recommend the relationship for quasi-static loading conditions, it is assumed that such correlation applies for the dynamic loading circumstances as well. During instrumented impact test procedure the σ_f fracture stress values are known from the F-t diagrams, so we have the possibility to estimate the A_0 material constant for our test material using Eq. (18).

A significant subjectivity of the measurement consists in determination of the fracture mirror size. Due to the very rough fracture surface of the investigated silicon nitride, correct measurement of it is eventually rather difficult, especially if a distinct mirror zone does not appear on the broken surface. Many times we face also the problem that increasing the magnification the characteristic pattern of the fracture surface appears in another scale. As a general rule, mirror size is suggested to be determined by macro-fractography.

Locating and measuring the fracture mirror zone for a low and a high loading rate case is introduced in Figs. 7 and 8. The area of the fracture mirror is indicated by broken line circles on the pictures. In Fig. 7(a) hackle lines are radiating from a focus located at the surface. There is no clear evidence of the mirror region. In Fig. 7 (b) at a higher magnification, a near surface pore is visible as the fracture origin. The mirror radius is rather estimated than clearly

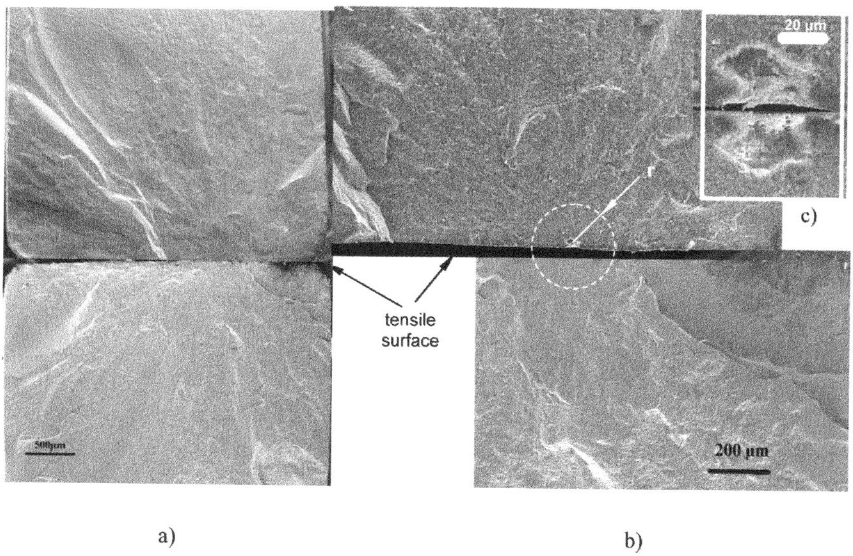

Figure 7. Macro-fractography of Si_3N_4 ceramic. a) Overall fracture surface morphology, (original magnification, 45x). b) Close up for mirror size measurement, (original magnification, 150x; Specimen N°: 59462; loading rate: v = 0.78 m/s; Fracture mirror size: r = 128μm; Stress at failure: σ_f = 672 MPa; A_0 = 7.6 MPa $m^{1/2}$

recognized, since the mist/mirror boundary cannot be clearly seen due to a progressive transition of roughness with increasing distance from the pore. Fig. 7 (c) illustrates the matching halves of pore at the tensile surface of the test piece. This difficulty should be encountered as an uncertainty factor influencing the A_0 constant to be determined.

Strong hackle marks in Fig. 8 (a) are commonly observed at higher loading rates and clearly point out the location of the fracture origin. Fig. 8b which is at a slightly higher magnification shows the initiation region with a triangular shape that appears on both halves of the matching pairs.

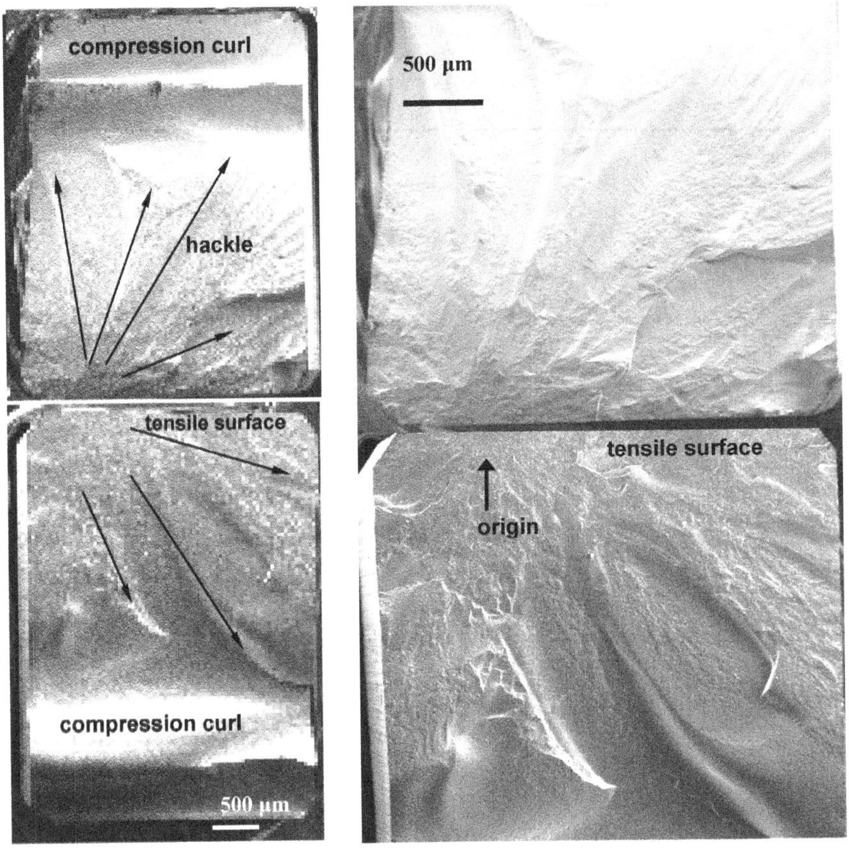

(a) (b)

Figure 8. Characteristic regions of fracture surface of an unnotched Si_3N_4 impact specimen: (b) is a close-up of (a.) Specimen N°: 59322; Loading rate: v =2.88 m/s.

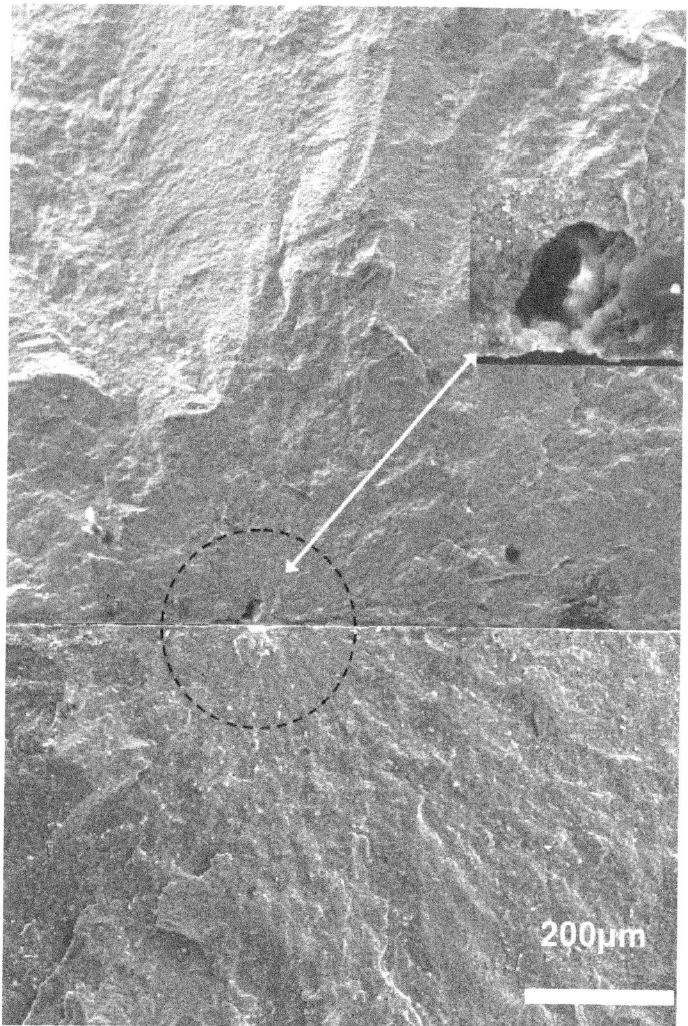

Figure 9. Close up views of the origin of the specimen in Figure 8. The mirror size was r = 127 μm. The insert shows the elliptical pore with a glassy inclusion.

A large surface connected pore is identified in Fig. 9 as the fracture origin in the higher magnification photo. The close-up view in the insert shows the origin is a blurred inclusion suggests a presence of glassy phase, probably from sintering additives that had weakened the subsurface region in tension.

The slightly non-symmetric crack advancement indicates the varying stress field through the sample. The fracture stress calculated from the F-t diagram (recorded by previous instrumented impact test) was σ_f = 547 MPa. The calculated mirror size constant has a value of A_0 = 6.16 MPa $m^{1/2}$.

Another aspect of determining the mirror size is shown in Figure ~~Figure Figure~~ 10. In this case the fracture origin is a semi elliptical pore close to the chamfer of specimen. Here the stress concentrating effects of both the material flaw and the sample geometry (i.e., chamfer) are combined. The mirror shape is more like an ellipse than a circle and therefore its size was estimated by the mean of the longer and shorter axes. This results in r=205 µm giving a mirror size constant value A_0 = 5.87 MPa $m^{1/2}$ at the fracture stress of 410 MPa.

Table IV~~Table IVTable IV~~. gives some examples for the measured mirror sizes and the related A_0 constants for the applied two loading rates.

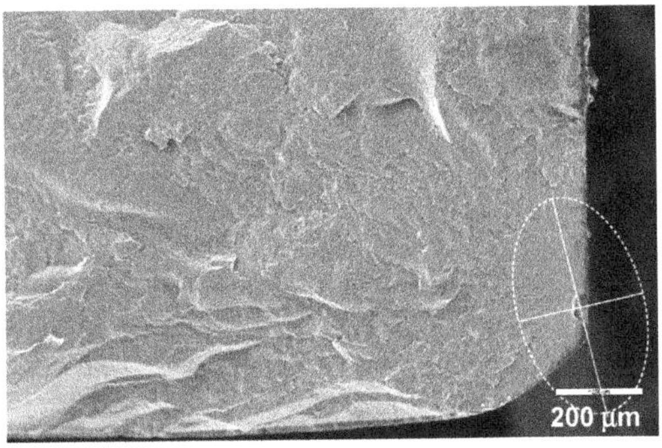

Figure 10. Determination of the mirror size on specimen N° 65828, (v = 0.78 m/s)

Table IV. The A_0 mirror size constant for dynamically impacted silicon nitride ceramic

Loading rate v, [m/s]	Sample	Position	Failure stress, σ_f, [MPa]	Mirror size, r, [µm]	Mirror constant, A_0, [MPa $m^{1/2}$]	A_0, mean [MPa $m^{1/2}$]
0.78	59462	A1	672	128	7.60	
0.78	65814	A1	547	142	6.52	6.63
0.78	65824	A1	599	119	6.53	
0.78	65828	B	410	205	5.87	
2.10	59524	B	461	165	5.92	
2.88	65846	A1	641	133	7.39	
2.88	59322	A1	547	127	6.16	6.60
2.88	59362	A1	610	105	6.25	

The fracture mirror constant A_0 values ranged from 5.9 to 7.6 MPa $m^{1/2}$. This range falls into the interval reported by the referenced ASTM and CEN standards (5.9 to 18.1) for different dense silicon nitride ceramics. Based on the available data, fracture mirror constants do not show a loading rate dependence. It should also be noted that additional testing is underway in order that more reliable conclusions can be drawn from a wider database.

SUMMARY

1. Dynamic failure behaviour of a two-stage gas-pressure sintered Si_3N_4 ceramic was investigated by instrumented impact test completed with electric-emission technique using different (unnotched and notched) samples in varying loading positions.
2. A novel observation of the experimental procedure is that Si_3N_4 ceramics show electric-emission (EE) activity during brittle failure. This provides a possibility to determine the fracture time, t_F even in such problematic cases of the impact test as the high loading rates, or application of notched specimens, when the usual test procedure does not supply this information.
3. Characteristic fracture time of the investigated material was 0.04 ms, at v = 0.78 m/s loading rate, while it decreased by one order, up to the mean value of 0.006 ms at v = 2.88 m/s.
4. Utilizing the EE technique for measuring the fracture time, the K_{Id} dynamic fracture toughness for different loading rates has been determined based on the dynamic key curve method. Comparing these data with the K_{Ic} quasistatic fracture toughness an unambiguous tendency of decreasing toughness with increasing loading rate has been demonstrated by test results, characterized by mean fracture toughness values of 7.43 MPa $m^{1/2}$, 4.4 MPa $m^{1/2}$ and 2.79 MPa $m^{1/2}$ for $v \approx 0$ m/s, v = 0.78 m/s and v = 2.88 m/s loading rates, respectively.
5. Macro- and micro-fractography were applied for qualitative and quantitative characterization of the damage process. Dynamic fracture toughness values estimated from flaw characteristics of unnotched impact specimens fell into the range of 2.24 to 3.94 MPa $m^{1/2}$, for loading rates of v = 2.88 and 0.78 m/s. Relating both the magnitude, as well as the rate dependence of this feature, a good agreement is seen with K_{Id} values, determined using notched impact specimens evaluated by the DKC method.
6. Important benefit of combination of fractography and instrumented impact test relating to estimating fracture mechanical feature, as K_{Id}, consists in providing a quick and simple method of acceptable reliability without executing expensive fracture mechanical procedures, avoiding preparation of notched or precracked specimens.
7. Applying a standardized empirical relationship to correlate the fracture stress, derived from impact tests with the radius of fracture mirror, the A_0 fracture mirror constant could have been determined for the investigated ceramic. The calculated A_0 values for both loading rates varied between 5.9 to 7.6 MPa $m^{1/2}$ with a total average of 6.53 MPa $m^{1/2}$. This falls to the lower boundary of the interval, reported by standards for different dense silicon nitride ceramics in quasi-static flexural loading condition.
8. The observed A_0 constants showed no variation with the loading rate. Reliability of the reported A_0 values should be substantiated through evaluating fracture surface of similar material of a considerably higher population of investigation.

ACKNOWLEDGEMENTS

The authors express their thanks for the professional and technical aid given by Péter Arató (RIMSF, HAS), Gyöngyvér B. Lenkey (BAY-LOGI, Miskolc), Géza Csukás, Emoke Rudnayova and Maria Beliecayova (IMR, SAS). The research work was supported by the OTKA T046467, OTKA NI 61724 OTKA K63609 projects, the Bolyai János Research Foundation and MÖB2-13-2-19-3158/2005 Research Grant.

REFERENCES

[1]M. B. Maros, N. Kaulics, and P. Arató, "Characterization of Dynamic Failure Process of Si_3N_4 Ceramics, Part I: Test Procedures, Fracture Energies, and Fractographic Analysis," this volume.

[2]T. Kobayashi, K. Matsunuma, H. Ikawa and K. Motoyoshi, "Evaluation of Static and Dynamic Fracture Toughness in Ceramics," *Eng. Fract. Mech.*, Vol. 31, No. 5, 873-885 (1988).

[3]J. Fengchun, L. Ruitang, Z. Xiaoxin, K. S. Vecchio and A. Rohatgi, "Evaluation of Dynamic Fracture Toughness K_{Id} by Hopkinson Pressure Bar Loaded Instrumented Charpy Impact Test," *Eng. Fract. Mech.*, 71, Issue 3, February 2004, 279-287 (2004).

[4]G. A. Gogotsi, "Fracture Toughness of Ceramics and Ceramic Composites," *Ceramics International*, Vol. 29, 777-784 (2003).

[5]Gy. B. Lenkey, S. Winkler, Z. Major and I. Lévay, "Applicability of Magnetic and Electric-Emission Techniques for Detecting Crack Initiation in Impact Tests," *Proceedings of the 11th European Conference on Fracture*, Vol. III. Poitiers, 3-6 September 1996, 1989-1994 (1996).

[6]M. B. Maros, N. Kaulics, P. Arató and Gy. B. Lenkey, "Determination of Dynamic Fracture Toughness of Si_3N_4 Ceramics By Instrumented Impact Test," *Key Eng. Mat.*, Vol. 290, 304-307, ISBN: 0-87849-973-3 (2005).

[7]"Fine ceramics — Test Method for Fracture Toughness of Monolithic Ceramics at Room Temperature by Single Edge Precracked Beam (SEPB) Method," ISO/DIN 15732, Int. Org. For Standard (1999).

[8]"Standard Test Methods for Determination of Fracture Toughness of Advanced Ceramics at Ambient Temperature," ASTM C1421-99, Am. Soc. for Test. and Matls., Philadelphia (1999).

[9]J. F. Kalthoff, S. Winkler and W. Böhme, "A Novel Procedure for Measuring the Impact Fracture Toughness K_{Id} with Precracked Charpy Specimens," *Journal de Physique*, Vol. 8, 179-186 (1985).

[10]J. F. Kalthoff, "Concept of impact response curves," *ASM Handbook*, Vol. 8., ASM, 269-271 (1995).

[11]T. Kobayashi, M. Niinomi, Y. Koide and K. Matsunuma, "Instrumented Impact Testing of Ceramics" *Transactions of the Japan Institute of Metals*, Vol.27, No.10, 775-783 (1986).

[12]T. Kobayashi, Y. Koide, Y. Daicho and R. Ikeda, "Dynamic Fracture Toughness Testing of Epoxy Resin Filled with SiO_2 Particles" *Eng. Fract. Mech.*, Vol. 28, No. 1., 21-29 (1987).

[13]W. Böhme, "Determination of the Impact Fracture Toughness K_{Id} of Plastics at High Rates of Loading >1m/s," *Fract. Mech. Testing Methods for Polymers, Adhesives & Composites*, ESIS 28, 73-89 (2001).

[14]W. Böhme, "Application of Dynamic Key Curves for the Determination of the Impact Fracture Toughness of Polymers at High Rates of Loading," *Impact and Dynamic Fracture of Polymers and Composites*, (Edited by J.G. Williams and A. Pavan), ESIS 19, Mechanical Engineering Publications, London, 59-71 (1995).

[15]T. Kobayashi, M. Niinomi and I. Yamamoto, "What Does it Tell Us? - Computer Aided Instrumented Charpy Impact Testing System," *Proc. Oji Int. Seminar on Dynamic Fracture*, 203-214 (1989).

[16]W. Böhme, "Dynamic Key-Curves for Brittle Fracture Impact Tests and Establishment of a Transition Time," *ASTM STP 1074*, American Society for Testing and Materials, Philadelphia, 144-156 (1990).

[17]S. W. Böhme, "Application of the Method of Dynamic Key Curves to Determination of the Impact Fracture Toughness, K_{Id}," *Fraunhofer-IWM Report Z 18/92*, Freiburg, February(1992).

[18]"Standard Test Method for Plain-Strain Fracture Toughness of Metallic Materials," ASTM E399-90, *Am. Soc. for Test. and Mats.*, Philadelphia (1997).

[19]"Stress Intensity Factors Handbook," (in two volumes) Vol.1., Editor-in-Chief: Y. Murakami, Pergamon Press, ISBN 0-08-034809-2, page13, (1987).

[20]T. Kobayashi, T. Kazino, M. Kamimura, H. Ikawa, "Basic Principle of Dynamic Fracture Toughness Evaluation by Computer Aided Instrumented Impact Testing (CAI) System," *Advances in Fracture Research ICF7*, 651-658, (1989).

[21]T. Kobayashi, Y. Koide, Y. Daicho, R. Ikeda, "Dynamic Fracture Toughness Testing of Epoxy Resin Filled with SiO_2 Particles", *Engineering Fracture Mechanics*, Vol. 28, No.1, 21-29 (1987).

[22]T. Kobayashi, H.Ikawa, "Evaluation of Dynamic Fracture Toughness on Brittle Materials, *MRS Int'l. Mtg. on Adv. Mats.* Vol.5, Materials Research Society 343-348, (1989).

[23]"Standard Practice for Fractography and Characterization of Fracture Origins in Advanced Ceramics," ASTM C 1322-05b, *Am. Soc. for Test. and Mats.*, Philadelphia (2005).

[24]"Advanced Technical Ceramics – Monolithic Ceramics – Mechanical Properties at Room Temperature – Part 6: Guidelines for Fractographic Investigation," ENV 843-6: 200x, Draft 10, N425.

SECTION VII

FRACTURE PHENOMENA IN GEOLOGY

PROPAGATION VELOCITY OF JOINTS: A DEBATE OVER STABLE VS. UNSTABLE GROWTH OF CRACKS IN THE EARTH

Terry Engelder
Department of Geosciences
334A Deike Building
The Pennsylvania State University
University Park, PA 16802

ABSTRACT

One of the hotly debated questions in the geoscience literature as of 2006 concerns the velocity of joint (i.e., a mesoscopic crack in rock) propagation in the crust of the Earth. Earthquake rupture loads rocks at stress rates of many MPa/sec, whereas plate tectonic deformation yields long-term loading rates as much as ten orders magnitude slower. This large range in loading rates leaves open the possibility of joint propagation in rocks anywhere from subcritical (i.e., stable growth) to critical and post-critical (i.e., unstable growth). Several facts are relevant during adjudication of the propagation-velocity debate: 1.) joints are inherently planar but may propagate as gradually curving single surfaces on scales > 10 m, 2.) hackle fringes are relatively rare, and thus, the exception to the planar surface (i.e., << 1% of all joint surface area in the crust of the earth consists of hackle fringe), 3.) fixed-grips loading is the normal configuration for propagation of joints in the brittle crust of the Earth if there is a fatigue limit, 4.) plume morphology on planar surfaces of joints is consistent with a velocity = $f(K_I)$ relationship, 5.) fracture under fixed-grips loading is inherently stable for several reasons, mainly, 5A.) developing elastic properties of a bulk rock with fractures favors stable propagation, 5B.) fluid-drive mechanisms in the earth favor stable propagation. The fractography of rock can demonstrate facts 1, 2, and 4 directly, and facts 3, 5A, and 5B indirectly. In summary, these facts point to a large role for subcritical propagation (i.e., stable growth) in the crust of the Earth whereas critical and post-critical propagation (i.e., unstable growth) in natural rock is exceptionally rare.

INTRODUCTION

Geoscientists use the term "joint" when referring to a macroscopic crack cutting rock as viewed on foot, within a mine, within core, or in a borehole image. A joint is a single discontinuity in rock that propagated from some smaller crack or flaw when its tip was subject to opening mode displacement.[1,2] Rupture to generate a brittle discontinuity is complex and leads to the rich textures found on the surfaces of both joints and cracks in ceramics and other materials. Larger-scale discontinuities are common in the Earth, but these consist of faults or joint zones growing from a more complex rupture sequence.[3,4] One of the earliest attempts to describe the texture of brittle rupture focused on the morphology of joint surfaces in the Cambridge slate quarries of Somerville, Massachusetts.[5] Observations elsewhere more or less reaffirmed that the surface morphology of joints in the Cambridge slate is representative of most joints.[6,7,8,9,10,11,12]

Joints are the product of prehistoric brittle fracture. This means that the rate of propagation of these ancient cracks cannot be measured by direct observation. Therefore, in situ propagation velocity must be inferred using models and experiments. Small samples are tested in the laboratory to measure fracture toughness, but joints are so large that 1:1 scale testing is not possible. Because joints and cracks in glasses, ceramics and metals have similar surface morphology, it is reasonable to draw upon cracks in the latter materials as analogs when it comes to inferring the velocity of ancient joints in rock.

The use of surface morphology on cracks in the engineering and ceramics realm as analogs for the surface morphology on rock joints is not as straightforward as it may seem for two reasons. First, in the case of joints, it is difficult to distinguish between surface morphology that is a product of single rupture moving through a change in orientation of the Earth's stress field in space, and morphology that is a product of a rupture that arrests and reinitiates after a change in the orientation of Earth stress with time. In the ceramics and engineering realm, rupture morphology may reflect a change in stress orientation in space as a consequence of odd boundary conditions, but rarely, if ever, is the rupture a product of arrest and reinitiation after boundary conditions change to cause a reorientation of stress. Second, it is difficult to gauge the velocity of joint propagation in the Earth, whereas in the ceramics and engineering realm, the velocity of the brittle rupture is often the product of unstable crack propagation.

BACKGROUND

One general conclusion arising from the analyses of joints in the Cambridge slate is that after more or less planar growth, the rupture often departed from the original plane of growth to form a fringe of multiple cracks.[5] These fringe cracks commonly consist of a set of parallel, en echelon planes that are systematically misoriented from the plane of the parent joint. When viewed normal to the parent joint, the en echelon cracks appear as river lines.[13,14] Engineering mechanics offers a theory for the departure from planar growth based on a stress-induced displacement of the crack tip.[15,16,17] As long as the crack tip is subject to pure opening mode displacement, propagation remains in its plane. Tearing or sliding mode displacements induce out-of-plane propagation and the generation of fringe cracks. In early studies of rock joints, there was no sense of whether the tearing and sliding mode displacements were the manifestation of a continuous rupture passing through a non-rectilinear stress field with curving stress trajectories or whether these latter displacements were the manifestation of a rupture that arrested and reinitiated after shifting in the Earth changed the orientation of the stress field.[18] In summary, fringe cracks on joints reflect either a spatial or temporal change in orientation of the stress field responsible for joint growth. Drawing a distinction between these two cases is difficult when using surface morphology alone.

Difficulty in the interpretation of joint surface morphology goes beyond the debate concerning a spatial vs. temporal stress reorientation. Without any direct observation of growth rate, connection between joint surface morphology and the velocity of rupture must be inferred. The first success in making a correlation between fracture surface markings and rupture velocity came in engineering fractography where rupture velocity could be controlled experimentally.[19,20] Glass was particularly suitable for producing a diagnostic set of surface features during high-speed propagation.[21] Later, a distinction was drawn between high-speed rupture at critical stress intensity, K_{Ic}, and chemically-aided low-speed rupture.[22] A complication arose because the glass literature adapted terms first used to describe the surface morphology of joints in the Cambridge slate where rupture velocity was unknown. For example, the original use of the term, 'hackle', was purely descriptive. Only later did hackle become linked to high-speed fracture in glass at or above K_{Ic} where it was used to describe the gradual roughening of fractures from a mirror to mist and beyond.

The similarity between failure surfaces in engineering applications and those on natural rocks was not lost on 20th Century fractographers working with rock.[23] One of the earliest attempts to match joint surfaces and propagation velocity was a study of blast-induced fractures that display a much different morphology than found on the surface of natural joints.[24] Other attempts to understand the velocity of natural joints were based on a possible link between the mirror-mist-hackle morphology of high-speed rupture in glass and analogs on the surfaces of joints assumed to be growing during the high-speed rupture of rocks.[25,26] This analog is valid only if joints are the product of high-speed or critical rupture.

In comparing joint propagation in the Earth and crack propagation in engineering materials, a number of general statements apply. Joints propagate in rocks, a natural polycrystalline ceramic, that consists of relatively large grain size with some notable exceptions.[27] In the case of sedimentary rocks, the 'ceramic' is a layered composite.[28] Failure is commonly a low stress event and generally not transgranular.[29] Extrapolating the morphological features on the surface of glass to polycrystalline aggregates including rock is not easy because rupture involves a combination of intergranular and transgranular fracture that inevitably roughened the fracture surface regardless of rupture velocity. From these general statements it seems prudent to use caution before drawing an analogy between the morphology induced by high-speed rupture in glass and the morphology seen on rock joints, particularly when data on the velocity of joint propagation are not available.

OBJECTIVES

Despite the link between hackle and high-speed rupture in engineering materials, mainly glass, it is not clear that there is an immediate connection between surface roughness and high-speed fracture at or above K_{Ic} in rocks.[30] There are aspects to rupture and joint growth in the Earth aside from roughening that guide an interpretation of rupture velocity. The purpose of this paper is to briefly describe several critical aspects of rock fracture and its concomitant surface morphology that make joint propagation different from the high-speed rupture of glass and other ceramics. In brief, current theory points to a subcritical growth rate for propagation of an overwhelming majority of rock joints. With this current understanding, the fractography of glass is not a particularly useful analog for joint propagation in rocks.

Comparatively little attention has focused on the morphology of surfaces that are produced during low-speed, subcritical rupture in engineering materials. In this regard, it may be appropriate for the engineer to refer to geological experience in searching for analogs to slow rupture, particularly in polycrystalline materials. This paper attempts to summarize the geological literature regarding the fractography of slow crack growth in the Earth as a means of introducing that literature to the engineering community. Some of the best evidence supporting inferences about rupture velocity during joint propagation comes from the sandstone, siltstone and shale of the Devonian Catskill Delta Complex in a swath of New York State that includes the towns of Alfred, Corning, Watkins Glen, and Ithaca and this review focuses on that swath. Finally, this paper draws on an interpretation of joints in the Borsov Granite of the Czech Republic to illustrate pitfalls associated with making inferences about joint rupture velocity based on glass as an analog.

LOAD SYSTEM FOR FAILURE IN ROCKS VS. ENGINEERING MATERIALS

Placement of the fractography of joints in the proper context requires an understanding of Earth stress and how this stress ultimately leads to crack propagation. This understanding is best achieved with a model that expresses the nature of the Earth's loading system.

The Earth is a self-gravitating body which means that both horizontal principal stresses (i.e., S_H – the maximum horizontal principal stress and S_h – the minimum horizontal principal stress) are tied to the vertical principal stress, S_v.[31] Gravitational stresses are compressive and become even more compressive with depth, z, in the Earth. The coupling between the principal stresses [i.e, $S_h = f(S_v)$] is governed by constitutive equations for rock. On a short time-scale these are the equations of elasticity that can be solved to show that $S_h < S_v$ unless tectonic stresses are present. At first glance it is a paradox that the Earth contains any cracks because its interior is in a highly compressive state. However, thermoelastic contraction and generation of high fluid pressures are two mechanisms by which the Earth can relieve gravitationally induced compressive stresses and fail by crack propagation.

Time scales for Earth processes are orders of magnitude longer than those found in engineering applications. Thermally activated processes generally govern the rates of Earth deformation with mantle convection driving the deformation at strain rates < 10^{-13} sec^{-1}. Some brittle processes such as long-term slip across major lithosphere bounding fault zones like the San Andreas are consistent with this mantle deformation rate.[32] However, little is known about crack propagation at rates less than 10^{-12} m/sec.[33] While experiments on both granite and silica glass leave open the possibility for slower crack growth, other materials appear to have a fatigue limit.[34,35] Some have suggested that under slow growth conditions, crack tips are blunted by dissolution and thus prevent crack velocity less than 10^{-11} m/sec.[36]

Although the reality is unknown, hallway discussion among Earth Scientists at national meetings admits the possibility of a fatigue limit so that crack propagation in the Earth is always faster than the Earth's innate loading rate of 10^{-13} sec^{-1}. Loading systems for crack propagation are distinguished by whether the loading rate is fast enough to maintain a load during crack propagation.[17] The dead-weight loading system can maintain a load during crack propagation whereas the fixed-grips loading system is incapable of maintaining a load (Fig. 1). The distinction between these loading systems is important because dead-weight loading drives many fracture toughness experiments and other types of engineering failures which tend to be unstable. Assuming joint propagation is faster than the Earth's innate loading rate, failures within the Earth are driven by fixed-grips loading which produced stable crack growth. If there is no fatigue limit, crack propagation can slow to the point that the Earth's innate loading rate of 10^{-13} sec^{-1} maintains a load. In this latter case, dead-weight loading drives joint propagation but such propagation is in a subcritical or stable regime.

According to the Griffith energy balance for crack propagation the two loading systems provide energy for crack propagation assuming that the total energy within the loading system and test specimen does not change during crack propagation.[37,38] The energy to drive most engineering failures comes from work by external boundaries in dead-weight loading systems. A stress-strain curve shows that the dead-weight system loses potential energy, $-U_w$, whereas the host rock gains strain energy, U_E (Fig 2a). If there is a fatigue limit to subcritical propagation, the immediate energy for crack propagation comes from the internal strain energy, U_E, within the rocks and not the work at the external boundaries. A stress-strain curve shows that the fixed-grips system does not give up potential energy at its boundary, but that the host rock loses strain energy, $-U_E$ (Fig. 2b). This point is important because fixed-grips loading leads to stable (i.e., low-speed) crack propagation in the Earth, whereas dead-weight loading of glass rods, for

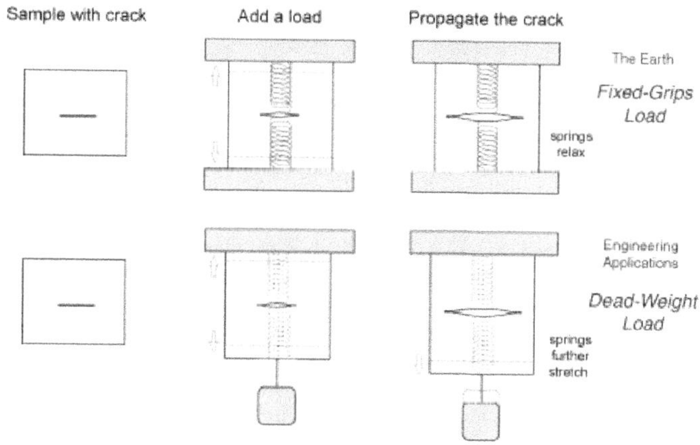

Figure 1. Basic elements of fixed-grips and dead-weight loading systems.[38] The sample material is elastic (represented by springs). Prior to loading a small crack is placed within the elastic medium. Loading the sample has the effect of stretching the sample and opening the crack regardless of the loading configuration. In fixed-grips loading surface energy for crack propagation comes from the elastic relaxation of the rock whereas in dead-weight loading the surface energy for crack propagation comes from the potential energy given up by the falling dead weight.

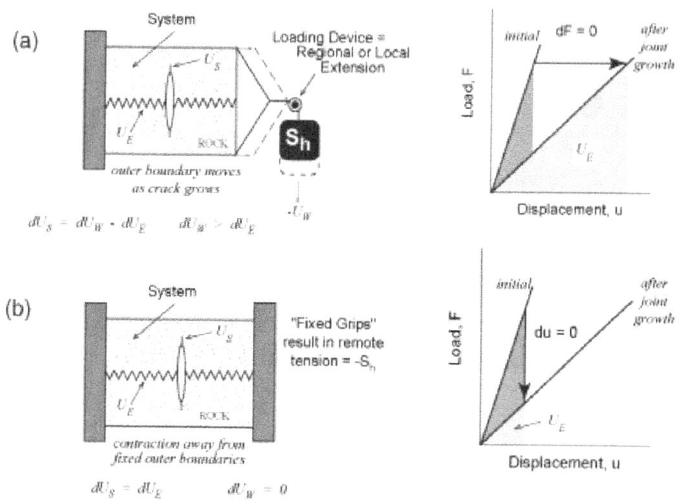

Figure 2. Loading systems and force-displacement graphs illustrates the amount of strain energy before (dark shade) and after (light shade) joint propagation under each loading system.[38] Rock modulus is the slope of the force-displacement curve. (a) Thermodynamic energy balance diagram for dead-weight loading. During joint propagation the strain energy (light shade) increases. (b) Thermodynamic energy balance diagram for the joint-driving mechanism under a fixed-grips loading system. During joint propagation the strain energy (light shade) decreases.

example, leads to unstable (i.e., high-speed) crack propagation.[39] We can anticipate that there are distinct and predictable differences in the fractography of stable and unstable propagation. Rock joints provide a natural laboratory (although not quite controllable) for establishing the differences in fractography between stable and unstable crack propagation.

THE FRACTOGRAPHY OF JOINTS DURING IN-PLANE PROPAGATION

A fine roughness appears on fracture surfaces during initial growth of joints. The fine roughness of the surface of joints during larger-scale in-plane propagation depends on the lithology of the host rock. Ironically, siltstone and sandstone of intermediate grain size carry some of the best-developed examples of the 'feather' or 'plumose' morphology first described on the surfaces of the Cambridge slate.[5] Plumose surfaces on siltstone beds have an rms height of about 10 μm for a profile length of 10 cm.[40] Cooling joints in diabase, an igneous rock, can have an rms height of an order of magnitude higher over the same profile length, whereas joints in shale have a smaller rms height. Joints cutting interlayered shale beds in the Catskill Delta Complex appear smooth to the unaided eye. The plume morphology on siltstones and sandstones allows mapping of the rupture history by tracing 10 μm ridges left by the rupture process back to the initiation point (Fig. 3).

Figure 3. Joint initiation at a stress concentration (worm borrow) inside a siltstone bed, in the Devonian Ithaca Formation exposed along Highway I-81 near Marathon, NY. For scale, the edge of a Swiss army knife is seen below the bed.

Despite the fine roughness, joints in Cambridge slate remain inherently planar during initial growth.[5] Larger-scale planarity is common during the initial stages of joint propagation in the Earth. Modern fracture mechanics explains the tendency for in-plane propagation based on shape of the crack-tip stress field through theories such as the maximum circumferential stress criterion,[41] the maximum strain energy release rate criterion,[42] and minimum strain-energy density criterion.[43]

In mapping rupture back to initiation points, it is seen that in natural layered composites of shale, siltstone, and sandstone beds of the Catskill Delta Complex, joint initiation takes place at a variety of locations both inside and at the bottom of layers a few cm to > m thick. One feature that initiation points in rocks possess is a dimension on the order of one to ten cm.[44] Initiation points internal to bedding include fossils, worm burrows, concretions, and sedimentary structures such as ball-and-pillow structures (Fig. 3). Regarding joint initiation, one important characteristic of a bedded composite is that the bottom of siltstone beds makes a sharp contact with underlying shale whereas the top contact between siltstone and overlying shale is gradual. Flute casts and gutter casts mark initiation points at the bottom of siltstone beds. Joint growth commonly emerges upward from these structures on the underside of siltstone beds (Fig. 4).

Like any layered composite, joints in sedimentary rocks commonly cut the stiffer layer with the interlayered, less-stiff shale remaining intact (Fig. 5). Preferential joint growth is found in the siltstone layers because these are the stiffer member relative to the shale which tends to stretch without early failure.[45,46] The spacing of joints in the siltstone beds is approximately equal to the thickness of the beds.[47,48] This behavior is explained by the shear-lag model for layered composites.[49,50,51]

Not all siltstone beds display initiation points at cm-scale structures or inclusions. Invariably, these beds fracture by joints crossing shale interfaces from beds where initiation points are found. Joint propagation across shale interfaces leads to extensive vertical growth of joints.[28]

In engineering applications, tensile stress may exceed 100 MPa and under these circumstances pores and other flaws on the order of 10 μm are sufficient to initiate crack propagation. The Earth is so completely broken that it has no large-scale tensile strength and, hence, has no gripping mechanism that permits the generation of such large tensile stress. Rupture in the Earth starts from macroscopic stress concentration points that are several cm in size and 3-4 orders of magnitude larger than internal pores.[44] In the Earth macroscopic stress concentration points superimposed on microscopic porosity in rocks enable failure events at low tensile stress (Fig. 6).

THE FRACTOGRAPHY OF JOINTS DURING OUT-OF-PLANE PROPAGATION

The question of what morphological features constitute out-of-plane propagation is scale dependent. Clearly the fine mist on glass rods is a manifestation of out-of-plane propagation at a microscopic scale. The plume structures on joints are also out-of-plane features at a scale large enough to be visible to the unaided eye (i.e., rms height = 10 μm at profile length of 10 cm). Yet, in the case of joints, the out-of-plane rupture to generate plume structures returns to the parent plane within a few grain diameters of the twist or tilt that generated the local out-of-plane propagation. The rupture growing from true out-of-plane propagation never returns to the plane of the parent joint.

Out-of-plane propagation is common in the sandstone, siltstone and shale of the Devonian Catskill Delta Complex in the vicinity of Corning, Watkins Glen, and Ithaca, New York.[18] This is largely a consequence of the shift in tectonic stress during the period of 50 million years or so when, according to plate tectonic theory,[52,53] Africa collided with North America. The vast majority of joints exhibiting out-of-plane propagation are a manifestation of reinitiation of propagation from parent joints after a shift in the tectonic stress field. There are,

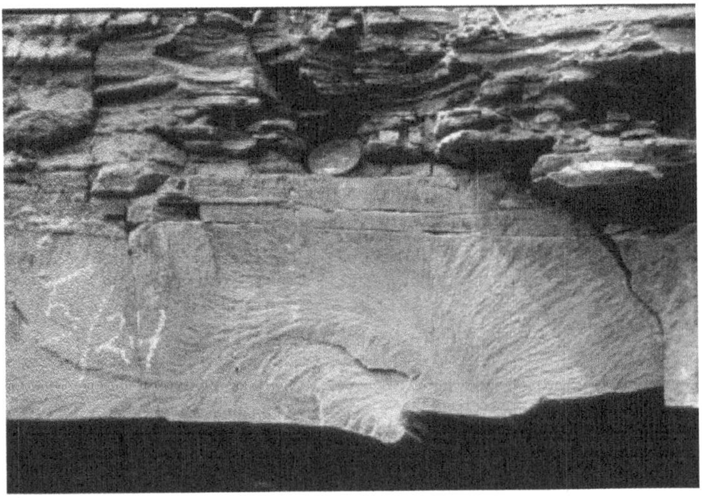

Figure 4. Joint initiation at an asymmetrical stress concentration (flute cast) at the base of a bed. The outcrop is the Devonian Ithaca Formation exposed along Highway 14 between Montour Falls and Watkins Glen, NY. For scale, the coin above the bed is an American quarter dollar.

Figure 5. An interbedded composite of siltstone (the thicker layers) and shale (the thin layers) beds. Systematic joints occur in the siltstone beds with a spacing approximately that of the bed. This outcrop is the Devonian Braillier Formation near Huntington, PA.[54]

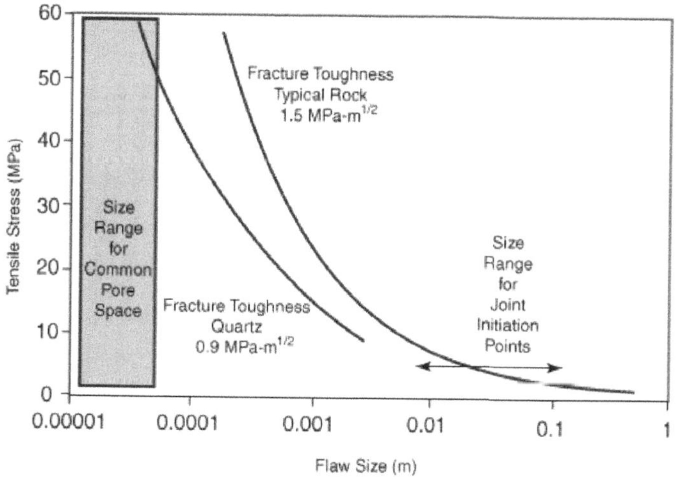

Figure 6. Failure curves for typical rocks and quartz plotted as tensile stress vs. flaw size.

however, joints that exhibit out-of-plane propagation because neighboring joints gave rise to an asymmetric crack-tip stress field.

Twist Hackles

Early descriptions of joint surfaces divided them into a planar portion and a rim of fringe cracks. With the evolution of terminology over time, these two parts of a joint surface are now described as the main joint face, or parent joint, with its characteristic plumose structure and the fringe, with its en echelon fringe cracks. The boundary between the main joint face and its fringe cracks may be either an abrupt transition, known as a shoulder, or a smoothly curving transition. Some descriptions of fringe cracks implicitly recognized the genetic similarity between the abrupt and smooth transitions by referring to both types of en echelon cracks as twist hackles.[10] The fringe is called a gradual twist hackle (Fig. 7) if individual cracks emerge from the tip line of the parent joint face in a smooth, uninterrupted manner, whereas the fringe is known as an abrupt twist hackle (Fig. 8) if a series of planar en echelon cracks abut the joint tip line. The former are known as river lines in the ceramics literature.[13] Other names for these structures include hackle zone, hackle marks, dilatant fringe cracks, fingers, and fracture lances.[55] Unlike its counterpart on glass fractures, hackle in joint fringe zones is usually not diagnostic of fracture velocity.

Gradual twist hackles result from a continuous breakdown of the parent joint, whereas abrupt twist hackles stem from a discontinuous breakdown. The twist angle is a function of change in remote stress orientation, stress magnitude, and elastic properties.[56] Presently, there is

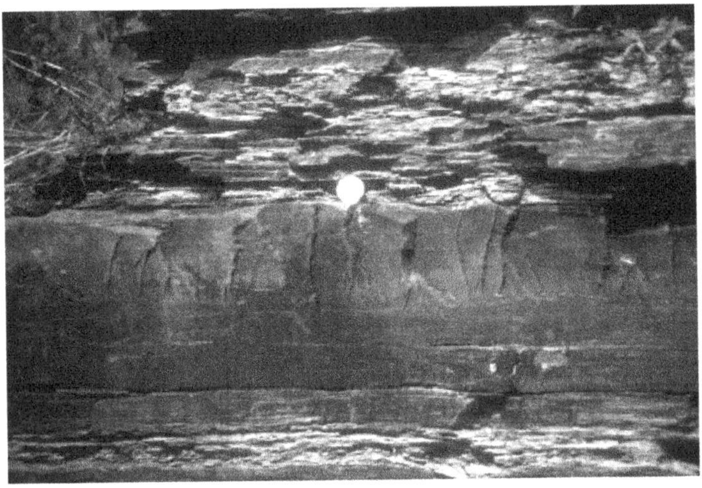

Figure 7. Gradual twist hackles. A parent joint and a twist hackle are carried within a single bed of siltstone of the Ithaca Formation at Taughannock Falls State Park, New York. Propagation direction for the twist hackle is upward. The sense of stress field rotation in this example is clockwise. The scale is an American quarter.

Figure 8. Abrupt twist hackles. This set of fringe cracks propagated downward into a thick shale bed from a thinner siltstone bed hosting the parent joint at Taughannock Falls State Park, New York. These rocks are part of the Ithaca Formation. The sense of stress field rotation in this example is clockwise. The scale is a geologic compass with an 8 cm base.

general agreement that the breakdown of the parent joint into fringe cracks is a consequence of a stress field with principal components that are neither orthogonal nor parallel to the tip line of the parent joint.

Kinks

The tip line of the parent joint can be decorated with a single fringe crack known as a tilt or kink (Fig. 9). A kink is a planar crack that propagates laterally at some angle from the edge of a parent joint.[57] Often, kinks form after parent joints have arrested. If the parent joint had not arrested prior to kink formation, there would have been a smooth curving or hooking of the crack path (Olson and Pollard, 1989).[58]

The Complete Fringe

If a joint grows and then arrests in a homogeneous medium without the anisotropy that bedding introduces, its tipline may approximate a circle. Such macroscopic penny-shaped cracks are found in granite (Fig. 10). In some instances, crack propagation is reinitiated after the remote stress field has rotated. A fringe will develop that completely surrounds the joint but the nature of the fringe depends on the crack-tip displacement caused by this new stress field orientation. The fringe is a twist hackle where the crack tip is subject to mode III displacement and a kink where subject to mode II displacement (Figs. 11, 12).

"SELF-CORRECTING" JOINT GROWTH AND THE VELOCITY = f(K_I) RELATIONSHIP

Both joint growth in nature and crack growth in engineering materials are largely a function of the velocity of the crack tip-line, v_{tl}. If v_{tl} is equal in all directions at the onset of rupture, radial growth takes place with the expansion of a circular crack tip-line to create a macroscopic penny-shaped crack. Classic examples of fracture in glass rods are consistent with the growth of circular tip-lines even during unstable propagation.[59] Geological examples of a circular tip-line include isolated planar joints in granite where c is the radius of the joint (Fig. 10). Joints in the Mrákotín granite, Borsov, the Czech Republic, are remarkable examples of initial tip-line shapes in a homogeneous and relatively isotropic rock. While nature rarely provides a perfect circular tip-line, the best examples of tip-lines have long-to-short axes of no more than 1.37 on vertical joints with diameters in excess of two meters.[26]

The Earth contains fluids (mainly water with various salts) that are highly corrosive. When loads are applied slowly as is the case for the Earth, stress corrosion creep is an active deformation mechanism with static fatigue common.[60] During static fatigue, the magnitude and distribution of the stress intensity K_I along the crack tip-line is the most important parameter in dictating v_{tl}. Such a law was postulated for subcritical crack growth where n is the subcritical

$$v_{tl} = AK_I^n \tag{1}$$

fracture growth index and A is a constant of proportionality.[61,62,63] By Eq. 1, circular ruptures grow only if the instantaneous K_I and dK_I/dc have constant values at all points along the tip-line.

Joint growth in the Earth is often very complicated because both K_I and dK_I/dc are not equal along joint tip-lines. Stress gradients or crack geometry effects can alter dK_I/dc at different points on a joint front periphery. For example, in clastic, interbedded sedimentary sequences, differences in the elastic properties of adjacent siltstone and shale layers, weak bedding planes, or the interaction with nearby crack tip stress concentrations introduce a mechanical

Figure 9. Kink in a siltstone bed near Whitney Point, New York. Parent joint is on the left and the kink tilts away from the viewer on the right.

Figure 10. Penny-shaped joint in the Mrákotín granite of the Borsov Quarry, Czech Republic. This is close-up view of the J5 joint with an initiation point at o (see Fig 9. in Bahat et al.[26])

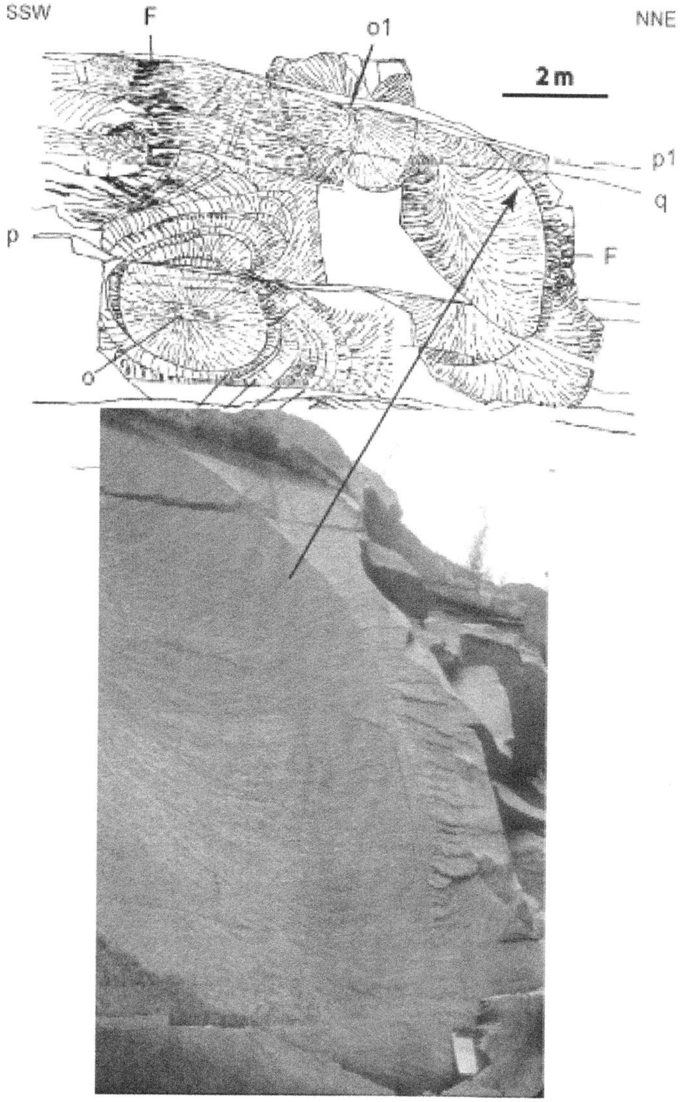

Figure 11. Fringe cracks on a joint in the Mrákotín granite of the Borsov Quarry, Czech Republic with the transition from a kink on the fringe to a mixed mode kink-twist fringe. Kink angle is consistent with the sense of twist. The picture is a close-up view of the fringe on the right side of J5[26] with an initiation point at o1. The line drawing is taken from Fig 9. in Bahat et al.[26]

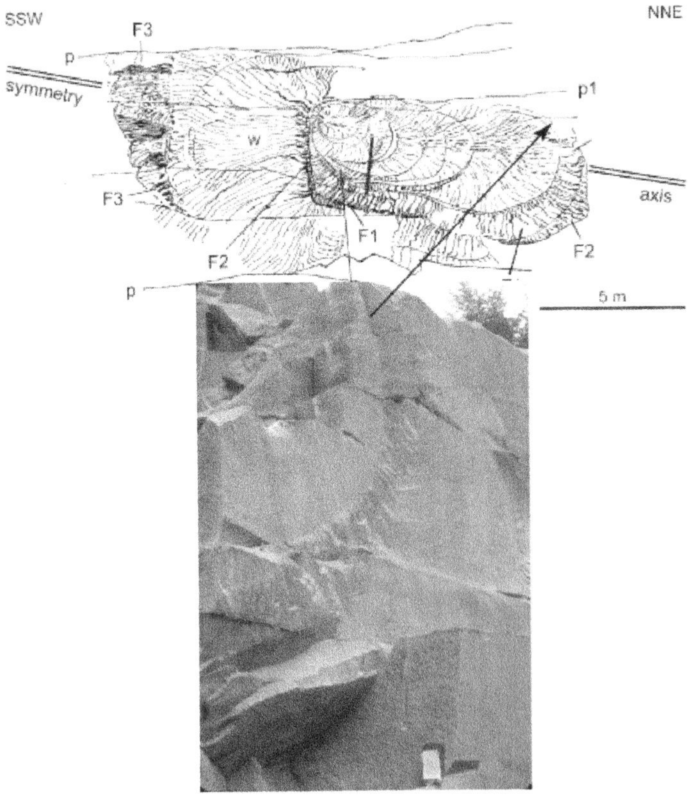

Figure 12. Fringe cracks on a joint in the Mrákotín granite of the Borsov Quarry, Czech Republic with the transition from a kink on the fringe to a mixed mode kink-twist fringe. Kink angle is consistent with the sense of twist. The picture is a close-up view of the fringe on the right side of J6.[26] The line drawing is taken from Fig 10. in Bahat et al.[26]

heterogeneity that modifies K_I along the crack tip-line. During subcritical propagation variation in K_I brings about differential or non-uniform instantaneous v_{II} that causes ruptures to evolve from circular to elliptical and then to irregular shapes. However, there is also a tendency to for an elliptical shape to redistribute K_I in a manner that returns the rupture to its stable, circular shape.[16] This leads to the phenomenon of a self-correcting rupture and the concomitant cycling of rupture shapes.[30] Hence, the circular or penny-shaped rupture is the stable shape from which more complex ruptures evolve providing rupture is subcritical. Eventually, the rupture will become irrevocably complicated at bed boundaries where the moving portion of the tip-line may

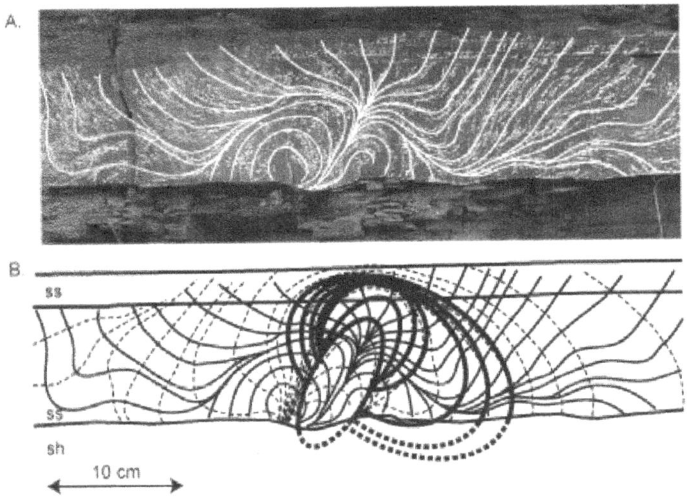

Figure 13. (a) Joint surface exposed along Highway 14 between Watkins Glen and Montour Falls, NY. (b) Interpretation of the joint surface (rupture shapes and tip-lines indicated).

Figure 14. Arrest lines in the Ithaca Formation along Route 414, Watkins Glen, New York. These arrests are part of a joint that propagated in 68 discrete increments.[68]

Figure 15. Arrest lines on cooling joints of the Snake River Basalt flow along the Boise River at Lucky Peak Dam, Idaho.[65]

granite intrusions and these microscopic fractures are also a manifestation of thermoelastic contraction. In both instances, the energy for crack propagation comes from a decrease of the internal strain energy within the body (Fig. 2b).

Thermoelastic contraction promotes stable crack propagation because the elastic properties of the rock change during crack propagation. The stiffness of the host rock is a function of joint length with the stiffness decreasing upon joint growth.[66] A changing modulus causes a load drop and concomitant drop in K_I at the crack tip (Fig 2b.). K_I rapidly drops and will become insufficient to drive joint propagation on the time scale of the fracture process. For cooling basalts, fracture growth stops until further cooling reloads the crack tip and fracture is again initiated.

Fluid Decompression

Fluids within pore space is pressurized by a number of mechanisms. One of interest in the Catskill Delta Complex is the volume-increase reaction during the coversion of organic material to hydrocarbons in a process called maturation.[67] During maturation the microscopic pore space within the rock does not expand so that a volume increase reaction leads to a pressure increase. Pressures may increase to the extent that they become larger than the gravitationally-produced compressive stresses in the Earth, thus producing a net effective tensile. This effective tension will drive crack propagation in a process called natural hydraulic fracturing. The driving mechanism, fluid decompression, is similar to dead-weight loading with the pressure against the inside of the joint (Fig. 16). The question is whether such joint propagation is unstable like engineering failures under dead-weight loading.

The Catskill Delta Complex is characterized by joints with fractographic evidence for cyclic propagation which means they are stable despite that dead-weight-like loading system (Fig 14). Although cyclic propagation could result from periodic external forcing by for-field stress changes, the regularity and rhythmic nature of joints within the Delta Complex suggest that fluid-pressure pulsation is the cause.[68] A joint more than 30m long by 44cm high that propagated within a single bed near Watkins Glen, New York has a plumose morphology with multiple arrest lines indicating the cracking occurred in increments rather than in one smooth rupture. The crack increments increase in overall length in the propagation direction over the final 28m portion of the exposed end of the joint with the largest increments increasing in length from 0.6 m to 1.0 m (Fig. 14). This increase in size of each increment suggests that propagation is tied to the crack length at the beginning of each increment.

At least three conceptual models based on linear elastic fracture mechanics and fluid flow along joints can be imagined to explain incremental crack growth under conditions of constant stress (i.e., fixed-grips loading) and pore pressure: The compressibility-limited propagation model; the flow-limited propagation model; and the infiltration-limited propagation model.[66] This surface morphology of the study joint provides constraints on the propagation process so that the growth of the joint may be analyzed in terms of these three models. Based on quantitative evaluation of the cracking process, the compressibility-limited propagation is favored and the driving fluid is identified as a gas like CH_4 rather than a brine.

Fluid decompression is the joint-driving mechanism for natural hydraulic fracturing. Cyclic propagation takes place for the simple reason that the fluid volume driving joint growth from the inside of the crack increases during each increment of propagation. CH_4 is a highly compressible gas that will decompress as the volume of the crack increases with each cycle. Internal crack pressure as exerted by the CH_4 on the inside of the crack decreases upon

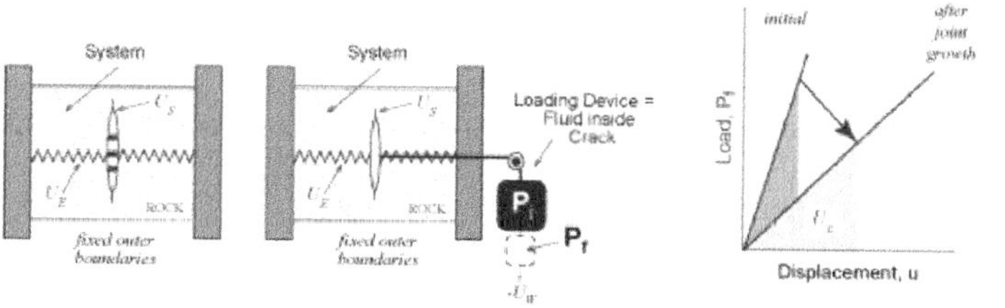

Figure 16. Thermodynamic energy balance diagram for the joint-driving mechanism, fluid decompression. The system is depicted in two ways. The first shows the load of pore pressure in the form of springs on the inside of the joint. As the joint grows, the springs (pore pressure) relax. The second shows the inside wall of the crack loaded by a dead-weight that is connected through the rock mass. The initial pore pressure is P_i and the final pore pressure is P_f. Displacement in this case is the displacement of the internal wall of the joint because the external boundaries are fixed. Schematic force-displacement graphs illustrate the amount of strain energy before (dark shade) and after (light shade) joint propagation. Rock modulus is the slope of the force-displacement curve.

decompression. Hence, the crack driving stress and concomitant K_I at the crack tip drop to stabilize the joint. The joint will recharge from rock porosity and gradually pressurize until K_I initiates propagation for another cycle.

THE MRAKOTIN GRANITE CONTROVERSY

The use of fractographic markings on glass (i.e., mirror, mist, and hackle) as analogs for the surface morphology of joints is controversial for two reasons. First is the question of whether or not it is possible to recognize fracture mirror and mist on joints (i.e., critical and post-critical rupture) when joint surfaces are overprinted by the plumose morphology first described in the Cambridge slates. Second is the question of whether well-organized, systematic fringe cracks are equivalent to the hackle that follows fracture acceleration through the mirror and mist stages of critical and post critical rupture. One interpretation of the fracture patterns in the Mrákotín granite, Borsov, Czech Republic is that joints in the Borsov quarry carry a mirror-hackle surface morphology indicative of unstable, critical and post-critical rupture.[26]

There are a number of facts that point to stable joint propagation in the Mrákotín granite. First, unstable rupture in glass and ceramics is achieved over very short distances within which there is an evolving surface morphology from mirror to mist to hackle. An increase in crack length under dead-weight loading causes unstable crack propagation because excess energy is available for crack propagation. Growth or the rupture under these conditions means that K_I becomes much larger than K_{Ic}. Glass and ceramics literature is replete with examples showing a gradual progression and increase in roughness as the crack radiates outward from the origin under dead-weight loading.[14,58,69] In contrast with glass, the surface morphology of the planar joints of the Mrákotín granite does not evolve until the fringe is reached (Figs. 10–12). The simplest explanation for a non-evolving surface morphology is that the crack-tip K_I does not undergo the radical change necessary for producing the evolving mirror-mist-hackle morphology. A non-evolving surface morphology is most likely to take place at $K_I \leq K_{Ic}$ where $K_I \approx const.$ This latter behavior requires that the load is reduced as the crack grows, a characteristic of local fixed-grips loading.

Second, the fringe cracks around several of the initial joint planes in the Mrákotín granite all have a consistent sense of twist or tilt (Figs. 11, 12). If the tilt on the fringe of one corner of the parent joint is forward, the tilt on the opposite corner is back (Fig. 10). Fringe at corners found at right angles to the tilt boundaries twist with the same sign. Furthermore, the twist angle is consistent with the tilt angle. This behavior is different from the morphology of joints that were generated by high-velocity fracturing associated with blasting at quarries or road cuts. In following around the rupture front of a high-speed fracture there is no consistent twist and tilt angle. The surface markings are formed by random and local deviations of the crack up and down out of the main propagation plane. During unstable rupture out-of-plane propagation along one portion of the fracture is inconsistent with out-of-plane propagation at other positions along the rupture front.

Third, a consistent sense of twist and tilt around the joint fringe, requires that joint propagation arrests so that the remote stress has time to reorganize and reorient itself to control the orientation of a consistent fringe. Such arrest is consistent with point #1 which is that the load comes off the joint during propagation in the Mrákotín granite. Load cycling is consistent with a low-stress, fixed-grips loading system that does not drive crack propagation to critical velocities.

Fourth, the majority of large fringe cracks in the Borsov quarry are systematically oriented clockwise from NNE-striking parent joints.[70] This is the pattern that exists throughout the Southern Bohemian pluton, an intrusive body more than 100 km long. Like the fringe cracks of the Catskill Delta complex, these are the manifestation of a regional rearrangement of Earth's stress.[18] Systematically oriented large fringe cracks are not part of a hackle fringe zone that is the manifestation of unstable joint growth. In a drawing of joint "J9" in the Borsov quarry, abrupt twist hackles of the regional set are mistakenly identified as rotating counter-clockwise from the parent joint.[26]

Fifth, growth of the joints beyond the initial planar joint is asymmetric. Forward progress is more efficient in the direction of the tilt boundaries than in the direction of the twist boundaries, and hence, the joint tip-lines evolve into elliptical shapes (Figs. 11, 12). This creates the impression that the joints are not self-correcting. However, all this means is that once arrested, resumed progress is more efficient when a single joint surface is initiated from a tilt boundary whereas further growth from numerous twist boundaries slows progress and exaggerates the tendency for elliptical growth. This is a pattern that is consistent with slow incremental growth with the slight adjustments in the Earth's stress field between increments.

Sixth, a thoroughly-cracked Earth cannot generate and maintain tensile stresses large enough to drive cracks unstably largely because unstable propagation requires a dead-weight load. A fractured Earth has no means of gripping smaller volumes of rock in tension. While these statements are subjective, they are supported by the fact that not even a small fraction of thousands of stress measurements in the Earth should have detected tensile stress of the type that would have driven joints in the Mrákotín granite to unstable velocities. In fact, tensile stress on the scale of the Borsov quarry has never been detected below the surface of the Earth.[31]

Joints in the Mrákotín granite, Borsov, Czech Republic are well documented in pictures.[26] It is left to the reader to test these six points of fact against the hypothesis that the surface morphology is indicative of unstable, critical and post-critical rupture.

CONCLUSIONS

This paper argues that stable, subcritical crack growth is the dominant mechanism for opening mode crack development in the brittle crust of the Earth.[71,62,72] To date, the majority of work on the subject is based primarily on the development of multi-joint growth patterns with minor input from joint surface morphology. Subcritical growth produces a distinct rupture pattern on joint surfaces and the presence of this rupture pattern is further evidence for stable growth within the subcritical regime. Six major characteristics of rupture in layered rocks support the stable growth hypothesis for propagation of joints in the Earth.

First, unstable, post-critical rupture leads to out-of-plane cracking that produces a surface topography that far exceeds the small-scale, local inhomogeneities seen on the plumose morphology of joints in layered rock. Wholesale out-of-plane cracking includes hackle fringe and branching, both of which result in very irregular joint surfaces.[25] Many joints in the Earth are planar and without a hackle fringe, thus devoid of any direct evidence for unstable rupture.[39]

Second, all evidence from joint surface morphology points to a self-correcting rupture shape in an isotropic, homogeneous rock. A K_I-dependent v_{tl} is a necessary condition for self-correction. K_I-dependent v_{tl} is one of the characteristics for rupture growth within region I of the subcritical regime but not a characteristic of post-critical propagation (Fig. 17).

Third, the whole notion of 'self-correcting' geometry is itself indicative of slow crack growth. Such a geometry is not consistent with unstable growth.

Fourth, the surface roughness of joints often does not increase with growth away from an initiation point. Post-critical propagation, as shown through experiments in glass and ceramics, is characterized by an increase in roughness with growth. A non-evolving surface morphology is indicative of a steady-state behavior consistent with stable growth.

Fifth, the surface roughness on a single joint may vary cyclically (Fig. 13). This behavior is consistent with alternating periods of propagation followed by hesitation or arrest.[66] Hesitation or arrest ($v_{tl} \rightarrow 0$) occurs when the crack-tip K_I drops below a certain threshold in the subcritical regime. For incremental or cyclic growth to take place there must be a clear relationship between crack-tip K_I and v_{tl}.

Sixth, joints with variable roughness are also likely candidates for the transition from region I to region II behavior ($v_{tl} \approx 10^{-3}$ - 10^{-2} m/sec). In region II, K_I can increase a great deal as the joint grows while $\Delta v_{tl} = 0$. Most clastic rocks make a transition to region II behavior at $v_{tl} \approx 10^{-2}$ m/sec.[61] This leaves a broad range of K_I that must be crossed to reach critical behavior at K_{Ic} (Fig 17b). This plateau serves as a large 'barrier' separating stable and unstable behavior. Critical and post-critical behavior in the crust of the Earth is rare as a consequence of this 'barrier'.

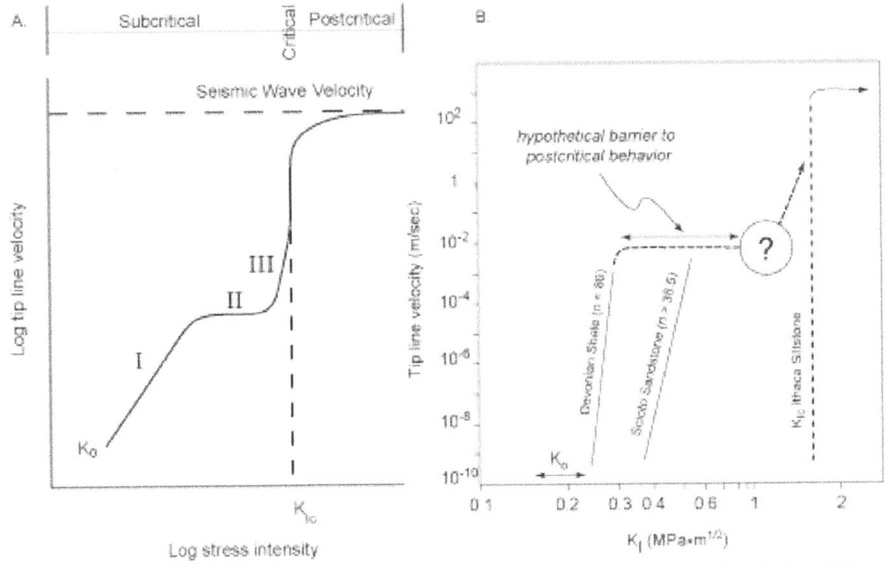

Figure 17. (a) Rate of crack growth as a function of stress intensity at the crack tip for subcritical (Regions I-III), critical, and post critical behavior. (b) Same plot with data from Devonian Shale,[73] Scioto Sandstone,[74] and Ithaca Siltstone.[29].

ACKNOWLEDGMENTS

Support came from National Science Foundation grant EAR-04-40233 and Penn State's Seal Evaluation Consortium (SEC). George Quinn is thanked for a careful review of a draft of this paper.

The author (T.E.) grew up within 15 miles of the Alfred University campus. As a high school scholar he benefited from weekly science seminars at Alfred University and as a high school athlete, he competed against Van Fréchette's son.

REFERENCES

[1] Engelder, T., Joints and Shear Fractures in Rock: in Atkinson, B., ed., *Fracture Mechanics of Rock*, Academic Press, Orlando, p. 27-69 (1987).

[2] Pollard, D.D., and Aydin, A., Progress in Understanding Jointing over the Past Century, *Geological Society of America Bulletin*, v. 100, p. 1181-1204 (1988).

[3] Rogers, C.M., and Engelder, T., Dynamic Analysis of Closely Spaced Joint Propagation in Regularly Spaced Joint Zones of Zion National Park: Joint localized Slot Canyons Masquerading as Blunt-Tipped Edge Cracks in the Navajo Sandstone, Utah, in Cosgrove, J. W., and Engelder, T., eds., The Initiation, Propagation, and Arrest of Joints and other Fractures, *Geological Society of London* Special Publications, v. 231, p. 49-71 (2004).

[4] Rogers, C.M., Myers, D.A., and Engelder, T., Kinematic Implications of Joint Zones and Isolated Joints in the Navajo Sandstone at Zion National Park: Evidence for Cordilleran Relaxation, *Tectonics*, v. 23, TC1007, doi:10.1029/2001TC001329 (2004).

[5] Woodworth, J.B., On the Fracture System of Joints with Remarks on Certain Great Fractures, *Boston Society of Natural Historical Proceedings* 27, p. 163-183 (1896).

[6] Hodgson, R.A., Classification of Structures on Joint Surfaces, American Journal of Science, v. 259, p. 493-502 (1961).

[7] Roberts, J.C., Feather-Fracture and the Mechanics of Rock Jointing: American Journal of Science, v. 259, p. 381-408 (1961).

[8] Bankwitz, P., Uber klufte; I. Beobachtungen im Thuringischen Schiefergebirge: Geologie, v. 14, p. 241-253 (1965).

[9] Gash, P.J.S., A Study of Surface Features Relating to Brittle and Semi-Brittle Fracture: Tectonophysics, v. 12, p. 349-391 (1971).

[10] Kulander, B.R., and Dean, S.L., Hackle Plume Geometry and Joint Propagation Dynamics: in Stephansson, O., ed., *Fundamentals of Rock Joints*, Sweden, Centek Publishers, p. 85-94 (1985).

[11] Bahat, D., Tectonofractography: Amsterdam, Elsevier Press, p. 250 (1991).

[12] Kulander, B.R., and Dean, S.L., Observations on Fractography with Laboratory Experiments for Geologists: In Ameen, M.S., ed., Fractography: Fracture Topography as a Tool in Fracture Mechanics and Stress Analysis, *Geological Society Special Publication* No. 92, p. 59-82 (1995).

[13] Pugh, S.F., The Fracture of Brittle Materials: *British Journal of Applied Physics*, v. 18, p. 234-269 (1967).

[14] Hull, D., Fractography Observing, Measuring and Interpreting Fracture Surface Topography, Cambridge, Cambridge University Press, p. 366 (1999).

[15] Irwin, G.R., Analysis of Stress and Strains near the End of a Crack Traversing a Plate: *Journal of Applied Mechanics*, v. 24, p. 361-364 (1957).

[16] Broek, D., *Elementary Engineering Fracture Mechanics*, Martinus Nijhoff,Boston, p. 516 (1987).

[17] Lawn, B., Fracture of Brittle Solids: Second Ed., Cambridge, Cambridge University Press, p. 378 (1993).

[18] Younes, A., and Engelder, T., Fringe Cracks: Key Structures for the Interpretation of Progressive Alleghanian Deformation of the Appalachian Plateau, *Geological Society of America Bulletin*, v. 111, p. 219-239 (1999).

[19] Fréchette, V.D., The Fractology of Glass: in Pye, L.D., ed., Introduction to Glass Science, p. 432-450 (1972).

[20] Rice, R.W., Ceramic Fracture Features, Observations, Mechanisms and Uses: in Mecholsky, J.J., and Powell, S.R., Fractography of Ceramic and Metal Failures, ASTM Special Technical Publication, v. 827, p. 1-103 (1984).

[21] Preston, D.A., A Study of the Rupture of Glass, *Journal of the Society of Glass Technology*, v. 10, p.234-269 (1926).

[22] Wiederhorn, S.M., Subcritical Crack Growth in Ceramics: in Bradt, R.C., Hasselman, D.P.H., and Lange, F.F., Mechanics of Ceramics: Plenum, New York, v. 2, p. 613 (1972).

[23] Bahat, D., Theoretical Considerations on Mechanical Parameters of Joints Surfaces Based on Studies on Ceramics, *Geological Magazine*, v. 116, 81-92 (1979).

[24]Lutton, R.J., Tensile Fracture Mechanics from Fracture Surface Morphology, in, G.B., Clark, ed., *Dynamic Rock Mechanics*, Proceedings of the 12th U.S. Symposium on Rock, Mechanics, Port City Press Inc. Baltimore, Md., p. 561-571 (1971).

[25]Bahat, D., and Rabinovitch, A., Paleostress Determination in Rock by a Fractographic Method, *Journal of Structural Geology*, v. 10, p. 193-199 (1988).

[26]Bahat, D. Bankwitz, P. and Bankwitz, E., Preuplift Joints in Granites: Evidence for Subcritical and Postcritical Fracture Growth, *Geological Society of America Bulletin*, v. 115, p. 148-165 (2003).

[27]Aydin, A., Chapter 2: Physical Characteristics of Fractures and Fracture Patterns: in Long, J.C.S., ed. Rock Fracture and Fluid Flow, National Research Council, Washington, D.C., p. 29-101 (1996).

[28]Helgeson, D. and Aydin, A., Characteristics of Joint Propagation Across Layer Interfaces in Sedimentary Rocks: *Journal of Structural Geology*, v. 13, p. 897-911 (1991).

[29]Scott, P.A., Engelder, T., and Mecholsky, J.J., The Correlation Between Fracture Toughness Anisotropy and Surface Morphology of the Siltstones in the Ithaca Formation, Appalachian Basin: in Evans, B. and Wong, T-F., eds., Fault Mechanics and Transport Properties of Rocks, Academic Press Ltd., London, p. 341-370 (1992).

[30]Savalli, L., and Engelder, T., Mechanisms Controlling Rupture Shape During Subcritical Growth of Joints in Layered Rock, *Geological Society of America Bulletin*, v. 117, p. 436-449 (2005).

[31]Engelder, T., Stress Regimes in the Lithosphere: Princeton Press, Princeton, New Jersey, p. 451 (1993).

[32]Scholz, C.H., The Mechanics of Earthquakes and Faulting, Cambridge University Press, New York, p. 439 (1990).

[33] Wilkins, B.J.S., Slow Crack Growth and Delayed Failure of Granite, *International Journal of Rock Mechanics and Mining Science*: v. 17, p. 365-369 (1980).

[34]Wiederhorn, S.M., and Bolz, L.H., Stress Corrosion and Static Fatigue of Glass, *Journal of the American Ceramic Society*, v. 53, p. 543-548 (1970).

[35]Wiederhorn, S.M., Guin, J-P., and Wallace, J.S., Surface Roughness in the Mirror Region: in Quinn, G., and Varner, J.R., Fractography of Glasses and Ceramics V Abstracts: Rochester, New York (2006).

[36]Simmons, C.J., and Freiman, S.W., Effect of Corrosion Processes on Subcritical Crack Growth in Glass, *Journal of the American Ceramic Society*, v. 64, p. 683-686 (1981).

[37]Griffith, A.A., Theory of Rupture, *Proc. First International Congress Applied Mechanics*, Delft, p. 55-63 (1924).

[38]Engelder, T., and Fischer, M.P., Loading Configurations and Driving Mechanisms for Joints Based on the Griffith Energy-Balance Concept, Tectonophysics, v. 256 , p. 253-277 (1996).

[39]Segall, P., Rate-Dependent Extensional Deformation Resulting from Crack Growth in Rock, *Journal of Geophysical Research*, v. 89, p. 4185-4195 (1984).

[40]Brown, S.R., and Scholz, C.H., Broad Bandwidth Study of the Topography of Natural Rock Surfaces, *Journal of Geophysical Research*, v. 90., p. 12,575-12,582 (1985).

[41]Erdogan, F., and Sih, G.C., On the Crack Extension in Plates Under Plane Loading and Transverse Shear, *Journal of Basic Engineering*, v. 85, p. 519-527 (1963).

[42]Nuismer, R.J., An Energy Release Rate Criterion for Mixed-Mode Fracture, *International Journal of Fracture*, v. 11, p. 245-250 (1975).

[43]Sih, G.C., and MacDonald, B., Fracture Mechanics Applied to Engineering Problems - Strain Energy Density Fracture Criterion, *Engineering Fracture Mechanics*, v. 6, p. 361-386 (1974).

[44]McConaughy, D.T., and Engelder, T., Joint Initiation in Bedded Clastic Rocks: *Journal of Structural Geology*, v. 23, p. 203-221 (2001).

[45]Warpinski, N.R., Determining the Minimum in situ Stress from Hydraulic Fracturing through Perforations, *International Journal of Rock Mechanics and Mining Science*, v. 26, p. 523-532 (1989).

[46]Evans, K., Engelder, T., and Plumb, R.A., Appalachian Stress Study 1: A Detailed Description of in situ Stress Variations in Devonian Shales of the Appalachian Plateau, *Journal of Geophysical Research*, v. 94, p. 1729-1754 (1989).

[47]Narr, W. and Suppe, J., Joint Spacing in Sedimentary Rocks: *Journal of Structural Geology*, v. 13, p. 1037-1048 (1991).

[48]Gross, M.R., Fischer, M.P., Engelder, T., and Greenfield, R.J., Factors Controlling Joint Spacing in Interbedded Sedimentary Rocks: Integrating Numerical Models with Field Observations from the Monterey Formation, USA: In Ameen, M.S., ed., Fractography: Fracture Topography as a Tool in Fracture Mechanics and Stress Analysis, Geological Society Special Publication No. 92, p. 215-233 (1995).

[49]Cox, H.L., The Elasticity and Strength of Paper and other Fibrous Materials, *British Journal of Applied Physics*, v. 3, p. 72-79 (1952).

[50]Hobbs, D. W., The Formation of Tension Joints in Sedimentary Rocks: An Explanation, *Geological Magazine*, v. 104, 550-556 (1967).

[51]Ji, S. and Saruwatari, K., A Revised Model for the Relationship Between Joint Spacing and Layer Thickness, *Journal of Structural Geology*, v. 20, p. 1495-1508 (1998).

[52]Engelder T., and Whitaker, A., Early Jointing in Coal and Black Shale: Evidence for an Appalachian-Wide Stress Field as a Prelude to the Alleghanian Orogeny, *Geology*, v 34, p. 581-584 (2006).

[53]Whitaker, A.E., and Engelder, T., Plate-Scale Stress Fields Driving the Tectonic Evolution of the Central Ouachita Salient, Oklahoma-Arkansas, *Geological Society of America Bulletin*, v 118, p. 710-723 (2006).

[54]Ruf, J.C., Rust, K.A., and Engelder, T., Investigating the Effect of Mechanical Discontinuities on Joint Spacing, *Tectonophysics*, v. 295, p. 245-257 (1998).

[55]Purslow, D., Matrix Fractography of Fiber-Reinforced Epoxy Composites, Composites, v. 17, p. 289-303 (1986).

[56]Pollard, D. D., Segall, P., and Delaney, P., Formation and Interpretation of Dilatant Echelon Cracks, *Geological Society of America Bulletin*, v. 93, p. 1291-1303 (1982).

[57]Cottrell, B., and Rice, J. R., Slightly Curved or Kinked Cracks, *International Journal of Fracture*, v. 16, p. 155-169 (1980).

[58]Olson, J.E., and Pollard, D.D., Inferring Paleostress from Natural Fracture Patterns: A New Method, *Geology*, v. 17, p. 345-348 (1989).

[59] Johnson, J.W., and Holloway, D.G., On the Shape and Size of the Fracture Zones on Glass Fracture Surfaces, *Phil. Mag.*, v. 14, p. 731-743 (1966).

[60]Scholz, C.H., Static Fatigue of Quartz, *Journal of Geophysical Research*, v. 77, p. 2104-2114 (1972).

[61]Charles, R.J., Dynamic Fatigue of Glass, *Journal of Applied Physics*, v. 29, p 1657-1662 (1958).

[62]Atkinson, B.K., and Meredith, P.G., The Theory of Subcritical Crack Growth with Applications to Minerals and rocks, Atkinson, B.K., ed., Fracture Mechanics of Rock. London, England, Academic Press, p. 111-166 (1987).

[63]Olson, J.E., 1993, Joint Pattern Development: Effects of Subcritical Crack Growth and Mechanical Crack Interaction, *Journal of Geophysical Research*, v. 98, p. 12,252-12,266 (1993).

[64]Engelder, T., and Lacazette, A., Natural Hydraulic Fracturing, in Barton, N., and Stephansson, O., eds., Rock Joints, A.A. Balkema, Rotterdam, p. 35-44 (1990).

[65]DeGraff, J.M., and Aydin, A., Surface Morphology of Columnar Joints and its Significance to Mechanics and Direction of Joint Growth, *Geological Society of America Bulletin*, v, 99, p. 605-617 (1987).

[66] Pollard, D.D., and Segall, P., Theoretical Displacements and Stresses Near Fractures in Rock: With Applications to Faults, Joints, Veins, Dikes, and Solution Surfaces, in Atkinson, B., ed., *Fracture Mechanics of Rock*, Academic Press, Orlando, p.277-350 (1987).

[67]Lash, G.G., Loewy, S., and Engelder, T., Preferential Jointing of Upper Devonian Black Shale, Appalachian Plateau, USA: Evidence Supporting Hydrocarbon Generation as a Joint-Driving Mechanism: in Cosgrove, J. W., and Engelder, T., eds., The Initiation, Propagation, and Arrest of Joints and other Fractures, *Geological Society of London Special Publications*, v. 231, p. 129-151 (2004).

[68]Lacazette, A., and Engelder, T., Fluid-Driven Cyclic Propagation of a Joint in the Ithaca Siltstone, Appalachian Basin, New York, in Evans, B. and Wong, T-F., eds., Fault Mechanics and Transport Properties of Rocks, Academic Press Ltd., London, p. 297-324 (1992).

[69]Quinn, G.D., Guidelines for Measuring Fracture Mirrors: in Quinn, G.D., Varner, J.R., and Wightman, M., eds., Fractography of Glasses and Ceramics V, *American Ceramic Society*, Westerville, OH (this proceedings (2007)).

[70]Bankwitz, P. and Bankwitz, E., The Relationship between Tilt and Twist Fringe Cracks in Granite Plutons: In Cosgrove, J.W., and Engelder, T., eds., The Initiation, Propagation, and Arrest of Joints and other Fractures: Geological Society, London, Special Publications, 231, p. 183-208 (2004).

[71] Segall, P., Formation and Growth of Extensional Fracture Sets, *Geological Society of America Bulletin*, v. 95, p. 454-462 (1984).

[72]Renshaw, C.E., and Pollard, D.D., An Experimentally Verified Criterion for Propagation Across Unbounded Frictional Interfaces in Brittle, Linear Elastic Materials, International Journal of Rock Mechanics and Mining Sciences & Geomechanics Abstracts, v. 32, p. 237-249 (1995).

[73]Swanson, P.L., Subcritical Crack Growth and Other Time- and Environment-Dependent Behavior in Crustal Rocks, *Journal of Geophysical Research*, v. 89, p. 4137-4152 (1984).

[74] Holder, J., Olson, J.E., and Zeno, P., Experimental Determination of Subcritical Crack Growth Parameters in Sedimentary Rock, *Geophysical Research Letters*, v. 28, p. 599-602 (2001).

Author Index

Printed and bound by CPI Group (UK) Ltd, Croydon, CR0 4YY
16/09/2026
14963461-0004